Advances in Neurobiology

Volume 8

Series Editor
Arne Schousboe

For further volumes:
http://www.springer.com/series/8787

Vladimir Berezin • Peter S. Walmod
Editors

Cell Adhesion Molecules

Implications in Neurological Diseases

Springer

Editors
Vladimir Berezin
Neuroscience and Pharmacology
University of Copenhagen
Copenhagen, Denmark

Peter S. Walmod
Neuroscience and Pharmacology
University of Copenhagen
Copenhagen, Denmark

ISSN 2190-5215 ISSN 2190-5223 (electronic)
ISBN 978-1-4614-8089-1 ISBN 978-1-4614-8090-7 (eBook)
DOI 10.1007/978-1-4614-8090-7
Springer New York Heidelberg Dordrecht London

Library of Congress Control Number: 2013953878

Printed on acid-free paper

Springer is part of Springer Science+Business Media (www.springer.com)

Contents

Contributors

Alejandro Barrallo-Gimeno, Ph.D. Physiology Section, Department of Physiological Sciences II, School of Medicine, University of Barcelona, Barcelona, Spain

Philip Beesley, B.Sc., Ph.D. School of Biological Sciences, Royal Holloway University of London, Egham, UK

Life Technologies, Paisley, UK
p.beesley@rhul.ac.uk

Vladimir Berezin, Ph.D. Laboratory of Neural Plasticity, Department of Neuroscience and Pharmacology, Faculty of Health Sciences, University of Copenhagen, Copenhagen, Denmark
berezin@sund.ku.dk

Simone Beuter Department of Molecular Biology, NMI an der Universität Tübingen, Reutlingen, Germany

Samuel Bouyain, D.Phil. Division of Molecular Biology and Biochemistry, School of Biological Sciences, University of Missouri-Kansas City, Kansas City, MO, USA
bouyains@umkc.edu

Ana Bribián, Ph.D. Grupo de Neurobiología del Desarrollo-GNDe, Hospital Nacional de Parapléjicos-SESCAM, Toledo, Spain

Institute of Bioengineering of Catalonia, Parc Cientific de Barcelona and Cell Biology Department, Universidad de Barcelona, Barcelona, Spain
abribian@ibec.pcb.ub.es

Fernando de Castro, M.D., Ph.D. Grupo de Neurobiología del Desarrollo-GNDe, Hospital Nacional de Parapléjicos-SESCAM, Toledo, Spain
fdec@sescam.jccm.es

Alain Chedotal, Ph.D., DR1 INSERM. Institut de la Vision, INSERM UMRS 968, CNRS UMR 7210, Paris, France
alain.chedotal@inserm.fr

Diego Clemente, Ph.D. Grupo de Neurobiología del Desarrollo-GNDe, Hospital Nacional de Parapléjicos-SESCAM, Toledo, Spain
dclemente@sescam.jccm.es

Julia Ebel Department of Molecular Biology, NMI an der Universität Tübingen, Reutlingen, Germany

Pedro F. Esteban, Ph.D. Grupo de Neurobiología del Desarrollo-GNDe, Hospital Nacional de Parapléjicos-SESCAM, Toledo, Spain
pfesteban@externas.sescam.jccm.es

Raúl Estévez, Ph.D. Physiology Section, Department of Physiological Sciences II, School of Medicine, University of Barcelona, Barcelona, Spain
restevez@ub.edu

Carl G. Gahmberg, M.D. Faculty of Bio- and Environmental Sciences, Division of Biochemistry and Biotechnology, University of Helsinki, Helsinki, Finland
carl.gahmberg@helsinki.fi

Diego García-González, Ph.D. Grupo de Neurobiología del Desarrollo-GNDe, Hospital Nacional de Parapléjicos-SESCAM, Toledo, Spain
digago@sescam.jccm.es

James S. Hagood, M.D. Division of Pediatric Respiratory Medicine, Rady Children's Hospital of San Diego, University of California-San Diego, La Jolla, CA, USA
jhagood@ucsd.edu

Sheila Harroch, Ph.D. Department of Neuroscience, School of Biological Sciences, Institut Pasteur de Paris, Paris, France
sharroch@pasteur.fr

Takahiro Hirabayashi, Ph.D. KOKORO-Biology group and JST-CREST, Laboratories for Integrated Biology, Graduate School of Frontier Biosciences, Osaka University, Osaka, Suita, Japan
hira@fbs.osaka-u.ac.jp

Michael Hortsch, Ph.D. Department of Cell and Developmental Biology, University of Michigan, Ann Arbor, MI, USA
hortsch@umich.edu

René Jüttner Max-Delbrück-Centrum für Molekulare Medizin, Berlin, Germany

Michaela Kraus, B.Sc., Ph.D. School of Biological Sciences, Royal Holloway University of London, Egham, UK

Martin Kriebel Department of Molecular Biology, NMI an der Universität Tübingen, Reutlingen, Germany

Hanna Langhorst Max-Delbrück-Centrum für Molekulare Medizin, Berlin, Germany

Lisette Leyton, Ph.D. Programa de Biología Celular y Molecular, Facultad de Medicina, Instituto de Ciencias Biomédicas, Universidad de Chile, Santiago, Región Metropolitana, Chile
lisetteleyton@gmail.com; lleyton@med.uchile.cl

Kenji Mandai, M.D., Ph.D. Division of Pathogenetic Signaling, Department of Biochemistry and Molecular Biology, Kobe University Graduate School of Medicine, Kobe, Japan

CREST, JST, Kobe, Japan
mandai@med.kobe-u.ac.jp

Alma N. Mohebiany, B.Sc., M.Sc. Department of Neuroscience, School of Biological Sciences, Institut Pasteur de Paris, Paris, France

María Luz Montesinos, Ph.D. Departamento de Fisiología Médica y Biofísica, Universidad de Sevilla, Sevilla, Spain
mlmontesinos@us.es

Masahiro Mori, M.D., Ph.D. Division of Neurophysiology, Department of Physiology and Cell Biology, Kobe University Graduate School of Medicine, Kobe, Japan

Faculty of Health Sciences, Kobe University Graduate School of Health Sciences, Kobe, Japan

CREST, JST, Kobe, Japan
morima@med.kobe-u.ac.jp

Rula Mualla, B.Sc. Department of Cell and Developmental Biology, University of Michigan, Ann Arbor, MI, USA

Verónica Murcia-Belmonte, Ph.D. Grupo de Neurobiología del Desarrollo-GNDe, Hospital Nacional de Parapléjicos-SESCAM, Toledo, Spain
vmurcia@sescam.jccm.es

Kakanahalli Nagaraj, Ph.D. Department of Applied Zoology, Kuvempu University, Shimoga, India
knagarajv@gmail.com; kvnag@kuvempu.ac.in

Lin Ning, M.Sc. Faculty of Bio- and Environmental Sciences, Division of Biochemistry and Biotechnology, University of Helsinki, Helsinki, Finland

Sonja Paetau, B.Sc. Faculty of Bio- and Environmental Sciences, Division of Biochemistry and Biotechnology, University of Helsinki, Helsinki, Finland

Nathalie Parolaro, B.Sc., Ph.D. School of Biological Sciences, Royal Holloway University of London, Egham, UK

Fritz G. Rathjen Max-Delbrück-Centrum für Molekulare Medizin, Berlin, Germany
Rathjen@mdc-berlin.de

Yoshiyuki Rikitake, M.D., Ph.D. Division of Signal Transduction, Department of Biochemistry and Molecular Biology, Kobe University Graduate School of Medicine, Kobe, Japan

Division of Cardiovascular Medicine, Department of Internal Medicine, Kobe University Graduate School of Medicine, Kobe, Japan

CREST, JST, Kobe, Japan
rikitake@med.kobe-u.ac.jp

Jadwiga Schreiber Max-Delbrück-Centrum für Molekulare Medizin, Berlin, Germany

Department of Neuroscience, Erasmus University Medical Center, Rotterdam, The Netherlands

Yoshimi Takai, M.D., Ph.D. Division of Pathogenetic Signaling, Department of Biochemistry and Molecular Biology, Kobe University Graduate School of Medicine, Kobe, Japan

CREST, JST, Kobe, Japan
ytakai@med.kobe-u.ac.jp

Hansjürgen Volkmer Department of Molecular Biology, NMI an der Universität Tübingen, Reutlingen, Germany
volkmer@nmi.de

Peter S. Walmod, M.Sc., Ph.D. Laboratory of Neural Plasticity, Department of Neuroscience and Pharmacology, Faculty of Health Sciences, University of Copenhagen, Copenhagen, Denmark

Laboratory of Neural Plasticity, University of Copenhagen, Symbion, Denmark
psw@sund.ku.dk

Malene Winther, M.Sc. Laboratory of Neural Plasticity, Department of Neuroscience and Pharmacology, Faculty of Health Sciences, University of Copenhagen, Copenhagen, Denmark

Jennifer Wuchter Department of Molecular Biology, NMI an der Universität Tübingen, Reutlingen, Germany

Takeshi Yagi, Ph.D. KOKORO-Biology group and JST-CREST, Laboratories for Integrated Biology, Graduate School of Frontier Biosciences, Osaka University, Osaka, Suita, Japan
yagi@fbs.osaka-u.ac.jp

Athéna R. Ypsilanti, Ph.D. Institut de la Vision, INSERM UMRS 968, CNRS UMR 7210, Paris, France
athena.ypsilanti@inserm.fr

List of Abbreviations

3′UTR	3′ untranslated region
aa	Amino acid
ACAM	Adipocyte adhesion molecule
AD	Alzheimer's disease
ADHD	Attention-deficit hyperactivity disorder
AIDP	Acute inflammatory demyelinating polyneuropathy
AIS	Axon initial segment
AJ	Adherens junction
Alivin	Activity-dependent LRR and Ig superfamily survival-related protein
AMIGO	Amphoterin-induced gene and ORF
AMPA	α-amino-3-hydroxy-5-methylisoxazole-propionic acid
APP	Amyloid precursor protein
ASD	Autism spectrum disorder
BACE1	β-site APP cleaving enzyme 1
BDNF	Brain-derived neurotrophic factor
BT-IgSF	Brain- and testis-specific immunoglobulin superfamily
CA	Carbonic anhydrase
CAM	Cell adhesion molecule
CAR	Coxsackievirus-adenovirus receptor
CAR-cKO	Cardiac-specific CAR knockout
CAR-KO	Global CAR knockout
Caspr	Contactin-associated protein
CC	Corpus callosum
CD	Cluster of differentiation
CD90	Cluster of differentiation 90
CDKL5	Cyclin-dependent kinase-like 5
CF	C-flanking
CGN	Cerebellar granule neuron
CHD	Congenital heart defect
CHL1	Close Homolog of L1

CLMP	CAR-like membrane protein
CNS	Central nervous system
CNTN	Contactin
CNTNAP	Contactin-associated-like protein
CNTNAP1	Contactin-associated protein 1
CNV	Copy number variations
CPEB1	Cytoplasmic polyadenylation element-binding protein 1
CR	Cysteine rich
CRASH Syndrome	Corpus callosum hypoplasia, retardation, adducted thumbs, spastic paraplegia, and hydrocephalus syndrome
CREB	cAMP response element-binding protein
CSBS	Congenital short-bowel syndrome
CSF	Cerebrospinal fluid
CVB	Coxsackievirus B
CVB3	Coxsackievirus B3
DCC	Deleted in colorectal carcinomas
DCM	Dilated cardiomyopathy
DG	Dentate gyrus
DIV	Days in vitro
DLPFC	Dorsolateral prefrontal cortex
DMD	Duchenne muscular dystrophy
DR	Diabetic retinopathy
DRG	Dorsal root ganglion
DS	Down's syndrome
DSCAM	Down syndrome cell adhesion molecule
DSCAML1	Down syndrome cell adhesion molecule-like 1
DSCR	Down syndrome critical region
EAE	Experimental allergic/autoimmune encephalomyelitis
EC	Extracellular cadherin
ECG	Electrocardiograms
ECM	Extracellular matrix
EGF	Epidermal growth factor
EGFR	Epidermal growth factor receptor
Erk	Extracellular signal-regulated kinase
ERM	Ezrin–radixin–moesin family of cytoplasmic proteins
ESAM	Endothelial cell-selective adhesion molecule
F-actin	Actin filaments
FA	Focal adhesion
FAK	Focal adhesion kinase
FGF	Fibroblast growth factor
FGFR	FGF receptor
FIGLER	Fibronectin type III, immunoglobulin, and leucine-rich repeat domains
FLRT	Fibronectin leucine-rich transmembrane protein
FNIII	Fibronectin type III

GAC1	Glioblastoma amplification on chromosome 1
GAP	GTPase-activating protein
GEF	Guanine nucleotide exchange factor
GFAP	Glial fibrillar acid protein
GPI	Glycosylphosphatidylinositol
GW	Glial wedge
HBD	Heparin-binding domain
HCC	Hepatocellular carcinoma
HDMEC	Human dermal microvascular endothelial cell
HGPPS	Horizontal gaze palsy and progressive scoliosis
Hh	Hedgehog
HS	Heparan sulfate
HSAS	Hydrocephalus with stenosis of aqueduct of sylvius
HSPG	Heparan sulfate proteoglycans
HSV	Herpes simplex virus
ICAM	Intercellular adhesion molecule
ICAM-5	Intercellular adhesion molecule-5 (telencephalin)
Ig	Immunoglobulin
IGG	Indusium griseum glia
IgSF	Immunoglobulin Superfamily
IgSF-CAM	Immunoglobulin superfamily cell adhesion molecule
INL	Inner nuclear layer
IPL	Inner plexiform layer
JAM	Junctional adhesion molecule
KS	Kallmann syndrome
KO	Knockout
LAR	Leukocyte common antigen related
LAT	Linker for activation of T cells
LFA-1	Leukocyte function associated antigen (CD11a/CD18, $\alpha_L\beta_2$)
LGN	Lateral geniculate nucleus
Lib	LRR protein induced by β-amyloid
LINGO	LRR and Ig domain-containing Nogo Receptor-interacting protein
LNX	Ligand-of-numb protein-X
LNX2	Ligand-of-numb protein-*X*2
LOH	Loss of heterozygosity
LOT	Lateral olfactory tract
LPHN	Latrophilins
LRIT	LRR, immunoglobulin-like and transmembrane domains
LRR	Leucine-rich repeats
LRRC	LRR-containing
LRRK2	LRR kinase-2
LRRN	LRR neuronal
LRRTM	LRR transmembrane neuronal
LTD	Long-term depression

LTP	Long-term potentiation
MAG	Myelin-associated glycoprotein
MAI	Myelin-associated inhibitor
MASA Syndrome	Mental retardation, aphasia, shuffling gait, and adducted thumbs syndrome
MCT	Monocarboxylate transporter
MeCP2	Methyl-CpG-binding protein 2
MEK	Mitogen-activated protein kinase kinase
MGE	Medial ganglion eminence
miRNA	MicroRNA
MLC	Megalencephalic leukoencephalopathy with subcortical cysts
MMP	Matrix metalloprotease
MRI	Magnetic resonance imaging
MS	Multiple sclerosis
Myt1l	Myelin transcription factor 1-like
Necl	Nectin-like molecule
NF	N-flanking
Nfasc	Neurofascin
NGL	Netrin-G ligands
NgR	Nogo-66 receptor
NLRR	Neuronal LRR
NMDA	*N*-methyl-D-aspartic acid
NMDAR	NMDA receptor
np	Neuroplastin
NrCAM	Neuron–glia-related cell adhesion molecule
NRG/Nrg	Neuroglian
NRX IV	Neurexin IV
NSCLC	Non-small-cell lung cancer
NT-3	Neurotrophin-3
NT-4	Neurotrophin 4
nucMLF	Nucleus of the medial longitudinal fascicle
OB	Olfactory bulb
OMgP	Oligodendrocyte myelin glycoprotein
ONL	Outer nuclear layer
OPC	Oligodendrocyte precursor cells
OPL	Outer plexiform layer
OSCC	Oral squamous cell carcinoma
P	Postnatal day
PAK	p21-activated kinase
PAR1	Pseudoautosomal region 1
Pcdh	Protocadherin
PDGF	Platelet-derived growth factor
PI3K	Phosphoinositide-3-kinase
PKA	Protein kinase A
PKC	Protein kinase C

PLC	Phospholipase C
PNS	Peripheral nervous system
PSD	Postsynaptic density
PTPδ	Protein tyrosine phosphatase δ
PTPRA	Receptor protein tyrosine phosphatase alpha
PTPRG	Receptor protein tyrosine phosphatase gamma
PTPRZ	Receptor protein tyrosine phosphatase zeta
PVR	Poliovirus receptor
PYK2	Proline-rich tyrosine kinase 2
RCC	Renal cell carcinoma
RGC	Retinal ganglion cells
RMS	Rostral migratory stream
ROCK	Rho kinase
SF	Stress fibers
SFK	Src-family kinases
SGZ	Subgranular zone
Shh	Sonic hedgehog
sICAM-5	Soluble ICAM-5 generated by cleavage of ICAM-5
SICD	SLITRK1 intracellular domain
SJ	Septate junction
SLITRK	SLIT and NTRK-like
SLRP	Small leucine-rich proteoglycans
SM	Synaptic membrane
SNP	Single nucleotide polymorphism
SVZ	Subventricular zone
TCR	The T-cell receptor
TF	Transcription factor
TGFβ	Transforming growth factor β
Thy-1	Thymocyte differentiation antigen 1
TJ	Tight junction
TS	Gilles de la tourette syndrome
TSA	Trichostatin A
USP33	Ubiquitin-specific protease 33
VEGF	Vascular endothelial growth factor
VGAT	GABA vesicular transporter
VGLUT1	Vesicular glutamate transporter 1
VRAC	Volume-regulated anion channel
WAP	Whey acidic protein-like domain
WAT	White adipose tissue
ZO-1	Zona occludens 1

Introduction

The term "cell adhesion molecule" can be traced back to the mid-1970s where it was used to describe molecules forming bonds between cells (Rutishauser et al. 1976). Some of the first CAMs identified, NCAM1 (Rutishauser et al. 1976) and N-cadherin (Hatta and Takeichi 1986), are molecules that mediate cell adhesion through homophilic *trans*-interactions (one protein attached to one cell surface interacting with an identical protein attached to an opposing cell surface). More recently, some CAMs, for instance, neurexins and neuroligins (Craig and Kang 2007), have been found to facilitate cell–cell interactions through heterophilic *trans*-interactions (one protein attached to one cell surface interacting with a different type of protein attached to an opposing cell surface), and many CAMs [for example, integrins (Myers et al. 2011)] do not mediate cell–cell interactions directly but instead mediate interactions between cells and the extracellular matrix (ECM). Moreover, many CAMs form heterophilic interactions not only with components of the ECM, but also with the extracellular domains of a variety of proteins located in the plasma membrane (Comoglio et al. 2003) as well as with intracellular adapter proteins, cytoskeletal elements, and enzymes (Mège et al. 2006; Takai et al. 2008; Buttner and Horstkorte 2010). These interactions not only serve to modulate the adhesive properties of CAMs but also enable CAMs to modulate intracellular signal transduction, cyto-architecture, and gene transcription (Ingber 2003). Moreover, the ectodomains of many CAMs can still modulate signal transduction after shedding (Cavallaro and Dejana 2011), and consequently, CAMs are involved in a variety of biological processes. In the nervous system, CAMs are for instance involved in the migration of neural crest cells (McKeown et al. 2013), the growth, guidance, and regeneration of axons (Kamiguchi 2007; Zhang et al. 2008), and the maintenance of stem cell niches (Marthiens et al. 2010). Not surprisingly, CAMs are therefore also implicated in numerous diseases of the nervous system. Moreover, the adhesion of most cells is a prerequisite for their growth and survival, and an absence of adhesion generally leads to cell death (anoikis) (Zhong and Rescorla 2012). In contrast, one of the characteristics of transformed cells is adhesion-independent growth (Reddig and Juliano 2005), and consequently, CAMs are also tightly linked to processes like tumor suppression (Moh and Shen 2009) and cancer (Makrilia et al. 2009).

The first CAMs described were proteins with no intrinsic enzymatic activity, which generally could be classified as belonging to one of four families: cadherins, integrins, selectins, and members of the immunoglobulin superfamily (IgSF) (Jordan and Morrow 1990). More recently, several proteins with enzymatic activity have also been demonstrated to mediate cell adhesion, for instance, receptor protein tyrosine phosphatases (Beltran and Bixby 2003), and an increasing number of CAMs have been found not to belong to any of the four protein families mentioned above, for example, neurexins/neuroligins (Craig and Kang 2007) and members of the family of proteins with extracellular leucine-rich repeats (Chen et al. 2006).

The purpose of this book is to provide a snapshot of current knowledge about the function of CAMs in the nervous system with an emphasis on their implications for the development or progression of diseases. In recent years many novel CAMs have been identified, and the book to a large extent focuses on these novel proteins of which many have been characterized only to a minor degree. Indeed several of the described proteins have not yet been demonstrated to serve as CAMs; see e.g., Chap. 14 (Winther and Walmod). In contrast, many of the well-characterized CAMs have deliberately been omitted from the book because they recently have been described extensively elsewhere. The book therefore does not provide a complete or a balanced description of CAMs implicated in neural diseases, but instead aims at providing an impression of the variety of CAMs expressed in the nervous system and how they affect the development, maintenance, and regeneration of the nervous system in health and disease.

Notably, members of the cadherin and integrin families are entirely omitted from the book. For recent information about these proteins, see (Hirano and Takeichi 2012; Redies et al. 2012) and (Becchetti and Arcangeli 2010; McGeachie et al. 2011; Myers et al. 2011; Wojcik-Stanaszek et al. 2011), respectively. Likewise, one of the most prominent CAMs of the IgSF, NCAM1, has been omitted from the book. For recent information about NCAM1 and other members of the NCAM family, see (Owczarek et al. 2009; Berezin 2014; Senkov et al. 2012; Winther et al. 2012).

The individual chapters of this book are devoted to individual CAMs or families of CAMs, and the book has been divided into two sections. The first section describes CAMs belonging to the IgSF; the second section describes CAMs not belonging to the IgSF.

Chapter 1 describes Thy-1 and how this protein is implicated in, e.g., pulmonary fibrosis, cancer, Graves' disease ophthalmopathy, and glomerulonephritis (Leyton and Hagood). Chapter 2 describes the three related proteins CAR, NT-IgSF, and CLMP, of which in particular CAR has received attention in relation to gene therapy (Schreiber et al.). Chapter 3 describes GlialCAM/HepaCAM with an emphasis of its implications in the rate neurological disease megalencephalic leukoencephalopathy with subcortical cysts (MLC) (Barrallo-Gimeno and Estévez). Chapter 4 describes neuroplastins and the effects of these proteins on synaptic plasticity and neurite outgrowth (Beesley et al. 2014). Chapter 5 describes the multiple functions of nectins and nectin-like proteins and their effects on, e.g., cancer and Alzheimer's disease (AD) (Mori et al.). Chapter 6 describes the effects of ICAM-5 on spine maturation, synapse formation, and immunosuppression (Gahmberg et al.). Chapter 7 describes the importance of roundabout receptors on a number of processes ranging from cell migration to organogenesis (Ypsilanti and Chedotal). Chapter 8 describes the effects of members of the contactin family on processes like neural cell migration, axon guidance, and the organization of myelin subdomains (Mohebiany et al.). Chapter 9 describes the L1 family of CAMs, and their implications in, e.g., mental retardation, autism spectrum disorders (ASDs), schizophrenia, and multiple sclerosis (MS) (Nagaraj et al.), whereas Chap. 10 gives a more detailed description of one of the members of the L1 family, neurofascin, and its potential involvement in neural diseases (Ebel et al.). Chapter 11 describes the two proteins DSCAM and DSCAML1 and their involvement in, e.g., Down syndrome (Montesinos). Chapter 12 describes the protein anosmin-1, which is implicated in, e.g., Kallmann syndrome and MS (de Castro et al.). Chapter 13 describes the large family of protocadherins, which include members that are implicated in, e.g., ASDs, schizophrenia, Usher syndrome, mental retardation, Huntington disease, and Retinitis pigmentosa (Hirabayashi and Yagi). Finally, Chap. 14 describes several families of CAMs belonging to the family of leucine-rich repeat-containing proteins. These families contain members that are implicated in, e.g., cancer, hearing impairment, glaucoma, AD, MS, Parkinson's disease, ASDs, schizophrenia, and obsessive–compulsive disorders (Winther and Walmod).

In total, the book describes more than 75 individual CAMs, and although many of these proteins only have been studied to a limited extent, the book will hopefully serve to highlight the importance of CAMs in relation to neural diseases and the development, maintenance, and regeneration of the nervous system.

Copenhagen, Denmark Peter S. Walmod

References

Barrallo-Gimeno A, Estévez R (2014) GLIALCAM, a glial cell adhesion molecule implicated in neurological disease. In: Berezin V, Walmod PS (eds) Cell adhesion molecules: implications in neurological diseases. Advances in neurobiology, vol 8. Springer, New York, NY, pp 47–59

Becchetti A, Arcangeli A (2010) Integrins and ion channels. Molecular complexes and signaling. Advances in experimental medicine and biology, vol 674. Springer, New York, NY, pp 1–123

Beesley P, Kraus M, Parolaro N (2014) The neuroplastins: multifunctional neuronal adhesion molecules; involvement in behaviour and disease. In: Berezin V, Walmod PS (eds) Cell adhesion molecules: implications in neurological diseases. Advances in neurobiology, vol 8. Springer, New York, NY, pp 61–89

Beltran PJ, Bixby JL (2003) Receptor protein tyrosine phosphatases as mediators of cellular adhesion. Front Biosci 8:d87–99

Berezin V (2014) Structure and function of the neural cell adhesion molecule NCAM. Advances in experimental medicine and biology, vol 663. Springer, New York, NY, pp 1–333. doi:10.1007/978-1-4419-1170-4

Buttner B, Horstkorte R (2010) Intracelluar ligands of NCAM. Adv Exp Med Biol 663:55–66. doi:10.1007/978-1-4419-1170-4_3

Cavallaro U, Dejana E (2011) Adhesion molecule signalling: not always a sticky business. Nat Rev Mol Cell Biol 12(3):189–197. doi:10.1038/nrm3068

Chen Y, Aulia S, Li L, Tang BL (2006) AMIGO and friends: an emerging family of brain-enriched, neuronal growth modulating, type I transmembrane proteins with leucine-rich repeats (LRR) and cell adhesion molecule motifs. Brain Res Rev 51(2):265–274

Comoglio PM, Boccaccio C, Trusolino L (2003) Interactions between growth factor receptors and adhesion molecules: breaking the rules. Curr Opin Cell Biol 15(5):565–571. doi:http://dx.doi.org/10.1016/S0955-0674(03)00096-6

Craig AM, Kang Y (2007) Neurexin–neuroligin signaling in synapse development. Curr Opin Neurobiol 17(1):43–52. doi:http://dx.doi.org/10.1016/j.conb.2007.01.011

de Castro F, Esteban PF, Bribián A, Murcia-Belmonte V, García-González D, Clemente D (2014) The adhesion molecule Anosmin-1 in Neurology: Kallmann syndrome and beyond. In: Berezin V, Walmod PS (eds) Cell adhesion molecules: implications in neurological diseases. Advances in neurobiology, vol 8. Springer, New York, NY, pp 273–292

Ebel J, Beuter S, Wuchter J, Kriebel M, Volkmer H (2014) Organisation and control of neuronal connectivity and myelination by cell adhesion molecule neurofascin. In: Berezin V, Walmod PS (eds) Cell adhesion molecules: implications in neurological diseases. Advances in neurobiology, vol 8. Springer, New York, NY, pp 231–247

Gahmberg CG, Ning L, Paetau S (2014) ICAM-5 – a neuronal dendritic adhesion molecule involved in immune and neuronal functions. In: Berezin V, Walmod PS (eds) Cell adhesion molecules: implications in neurological diseases. Advances in neurobiology, vol 8. Springer, New York, NY, pp 117–132

Hatta K, Takeichi M (1986) Expression of N-cadherin adhesion molecules associated with early morphogenetic events in chick development. Nature 320(6061):447–449. doi:10.1038/320447a0

Hirabayashi T, Yagi T (2014) Protocadherins in neurological diseases. In: Berezin V, Walmod PS (eds) Cell adhesion molecules: implications in neurological diseases. Advances in neurobiology, vol 8. Springer, New York, NY, pp 293–314

Hirano S, Takeichi M (2012) Cadherins in brain morphogenesis and wiring. Physiol Rev 92(2):597–634. doi:10.1152/physrev.00014.2011

Ingber DE (2003) Tensegrity II. How structural networks influence cellular information processing networks. J Cell Sci 116(8):1397–1408. doi:10.1242/jcs.00360

Jordan VC, Morrow M (1990) Cell adhesion molecules. Breast Cancer 50:4177–4189

Kamiguchi H (2007) The role of cell adhesion molecules in axon growth and guidance. Adv Exp Med Biol 621:95–103. doi:10.1007/978-0-387-76715-4_7

Leyton L, Hagood JS (2014) Thy-1 modulates neurological cell-cell and cell-matrix interactions through multiple molecular interactions. In: Berezin V, Walmod PS (eds) Cell adhesion molecules: implications in neurological diseases. Advances in neurobiology, vol 8. Springer, New York, NY, pp 3–20

Makrilia N, Kollias A, Manolopoulos L, Syrigos K (2009) Cell adhesion molecules: role and clinical significance in cancer. Cancer Invest 27(10):1023–1037. doi:10.3109/07357900902769749

Marthiens V, Kazanis I, Moss L, Long K, Ffrench-Constant C (2010) Adhesion molecules in the stem cell niche–more than just staying in shape? J Cell Sci 123(Pt 10):1613–1622. doi:10.1242/jcs.054312

McGeachie AB, Cingolani LA, Goda Y (2011) Stabilising influence: integrins in regulation of synaptic plasticity. Neurosci Res 70(1):24–29. doi:10.1016/j.neures.2011.02.006

McKeown SJ, Wallace AS, Anderson RB (2013) Expression and function of cell adhesion molecules during neural crest migration. Dev Biol 373(2):244–257. doi:10.1016/j.ydbio.2012.10.028

Mège R-M, Gavard J, Lambert M (2006) Regulation of cell–cell junctions by the cytoskeleton. Curr Opin Cell Biol 18(5):541–548. doi:http://dx.doi.org/10.1016/j.ceb.2006.08.004

Moh MC, Shen S (2009) The roles of cell adhesion molecules in tumor suppression and cell migration: a new paradox. Cell Adh Migr 3(4):334–336

Mohebiany AN, Harroch S, Bouyain S (2014) New insights into the roles of the contactin cell adhesion molecules in neural development In: Berezin V, Walmod PS (eds) Cell adhesion molecules: implications in neurological diseases. Advances in neurobiology, vol 8. Springer, New York, NY, pp 165–194

Montesinos ML (2014) Roles for DSCAM and DSCAML1 in central nervous system development and disease. In: Berezin V, Walmod PS (eds) Cell adhesion molecules: implications in neurological diseases. Advances in neurobiology, vol 8. Springer, New York, NY, pp 249–270

Mori M, Rikitake Y, Mandai K, Takai Y (2014) Roles of nectins and nectin-like molecules in the nervous system. In: Berezin V, Walmod PS (eds) Cell adhesion molecules: implications in neurological diseases. Advances in neurobiology, vol 8. Springer, New York, NY, pp 91–116

Myers JP, Santiago-Medina M, Gomez TM (2011) Regulation of axonal outgrowth and pathfinding by integrin-ECM interactions. Dev Neurobiol 71(11):901–923. doi:10.1002/dneu.20931

Nagaraj K, Mualla R, Hortsch M (2014) The L1 family of cell adhesion molecules – a sickening number of mutations and protein functions. In: Berezin V, Walmod PS (eds) Cell adhesion molecules: implications in neurological diseases. Advances in neurobiology, vol 8. Springer, New York, NY, pp 195–229

Owczarek SE, Kristiansen LV, Hortsch M, Walmod PS (2009) Cell adhesion molecules of the NCAM family and their role at synapses. In: Hortsch M, Umemori H (eds) The sticky synapse, 1st edn. Springer, Berlin. pp 265–299. doi:978-0-387-92708-4

Reddig PJ, Juliano RL (2005) Clinging to life: cell to matrix adhesion and cell survival. Cancer Metastasis Rev 24(3):425–439. doi:10.1007/s10555-005-5134-3

Redies C, Hertel N, Hubner CA (2012) Cadherins and neuropsychiatric disorders. Brain Res 1470:130–144. doi:10.1016/j.brainres.2012.06.020

Rutishauser U, Thiery JP, Brackenbury R, Sela BA, Edelman GM (1976) Mechanisms of adhesion among cells from neural tissues of the chick embryo. Proc Natl Acad Sci USA 73(2):577–581

Schreiber J, Langhorst H, Jüttner R, Rathjen FG (2014) The IgCAMs CAR, BT-IgSF and CLMP: structure, function and diseases. In: Berezin V, Walmod PS (eds) Cell adhesion molecules: implications in neurological diseases. Advances in neurobiology, vol 8. Springer, New York, NY, pp 21–46

Senkov O, Tikhobrazova O, Dityatev A (2012) PSA-NCAM: synaptic functions mediated by its interactions with proteoglycans and glutamate receptors. Int J Biochem Cell Biol 44(4):591–595. doi:10.1016/j.biocel.2012.01.008

Takai Y, Ikeda W, Ogita H, Rikitake Y (2008) The immunoglobulin-like cell adhesion molecule nectin and its associated protein afadin. Annu Rev Cell Dev Biol 24(1):309–342. doi:10.1146/annurev.cellbio.24.110707.175339

Winther M, Berezin V, Walmod PS (2012) NCAM2/OCAM/RNCAM: cell adhesion molecule with a role in neuronal compartmentalization. Int J Biochem Cell Biol 44(3):441–446. doi:10.1016/j.biocel.2011.11.020

Winther M, Walmod PS (2014) Neural cell adhesion molecules belonging to the family of leucine-rich repeat proteins. In: Berezin V, Walmod PS (eds) Cell adhesion molecules: implications in neurological diseases. Advances in neurobiology, vol 8. Springer, New York, NY, pp 315–395

Wojcik-Stanaszek L, Gregor A, Zalewska T (2011) Regulation of neurogenesis by extracellular matrix and integrins. Acta Neurobiol Exp (Wars) 71(1):103–112

Ypsilanti AR, Chedotal A (2014) ROUNDABOUT receptors. In: Berezin V, Walmod PS (eds) Cell adhesion molecules: implications in neurological diseases. Advances in neurobiology, vol 8. Springer, New York, NY, pp 133–164

Zhang Y, Yeh J, Richardson PM, Bo X (2008) Cell adhesion molecules of the immunoglobulin superfamily in axonal regeneration and neural repair. Restor Neurol Neurosci 26(2–3):81–96

Zhong X, Rescorla FJ (2012) Cell surface adhesion molecules and adhesion-initiated signaling: understanding of anoikis resistance mechanisms and therapeutic opportunities. Cell Signal 24(2):393–401. doi:10.1016/j.cellsig.2011.10.005

Part I
Cell Adhesion Molecules Belonging to the Immunoglobulin Superfamily

Chapter 1
Thy-1 Modulates Neurological Cell–Cell and Cell–Matrix Interactions Through Multiple Molecular Interactions

Lisette Leyton and James S. Hagood

Abstract Thy-1, or CD90, is a glycosylphosphatidylinositol-linked cell surface glycoprotein expressed on multiple cell types, including neurons, thymocytes, fibroblasts, endothelial cells, mesangial cells, and some hematopoietic and stromal stem cells. Thy-1 is developmentally regulated and evolutionarily conserved. Its cellular effects vary between and in some cases within cell types, tissues, and species, indicating that its biological role is context dependent. However, it most often seems to affect cell–cell or cell–matrix interactions and cellular adhesion and migration. In the nervous system, Thy-1 mediates bidirectional cell–cell communication, which modulates cell–matrix adhesion. Neurons express high levels of Thy-1, which interacts with $\alpha_v\beta_3$ integrin present in astrocytes and stimulates increased astrocyte adhesion to the underlying surface (*trans* signaling) and in neurites, the same ligand–receptor association triggers neurite retraction and inhibition of axonal growth (*cis* signaling). Although Thy-1 lacks a cytoplasmic domain, it affects multiple intracellular signaling cascades through interaction with a number of molecules within lipid raft microdomains. Improved understanding of how this enigmatic adhesion molecule modulates signaling and cell phenotype may yield novel insights into neurodevelopment and nerve recovery after injury.

L. Leyton, Ph.D.
Programa de Biología Celular y Molecular, Facultad de Medicina,
Instituto de Ciencias Biomédicas, Universidad de Chile, Independencia 1027,
Independencia, Santiago, Región Metropolitana 8380453, Chile
e-mail: lisetteleyton@gmail.com; lleyton@med.uchile.cl

J.S. Hagood, M.D. (✉)
Division of Pediatric Respiratory Medicine, Rady Children's Hospital
of San Diego, University of California-San Diego, 9500 Gilman Drive,
MC 0731, La Jolla 92019, CA, USA
e-mail: jhagood@ucsd.edu

V. Berezin and P.S. Walmod (eds.), *Cell Adhesion Molecules: Implications in Neurological Diseases*, Advances in Neurobiology 8, DOI 10.1007/978-1-4614-8090-7_1,

1.1 Introduction

Thy-1 (thymocyte differentiation antigen 1), also known as CD90 (cluster of differentiation 90), is a highly conserved cell surface molecule that can exist in membrane-bound and soluble forms. Thy-1 is developmentally regulated and expressed in specific cell types, including neurons, retinal ganglion cells, subsets of fibroblasts, vascular pericytes, activated endothelial cells, mesangial cells, and hematopoietic and mesenchymal stem cells. Previous reviews have focused on its immunologic and non-immunologic roles, mechanisms, and consequences of Thy-1-associated signaling and regulation of its expression (Haeryfar and Hoskin 2004; Rege and Hagood 2006a, b; Barker and Hagood 2009a; Bradley et al. 2009). Here, we will consider the role of Thy-1 as an adhesion molecule, mainly in the context of neurobiology.

Thy-1, originally designated as theta (θ) antigen, was initially defined as a leukemia-specific antigen in mice (Reif and Allen 1964; Schlesinger and Yron 1969). There are important species-specific differences in expression (see Sect. 1.4). Thy-1 regulation and signaling have been implicated in several disease states including neuronal injury (Leyton et al. 2001; Schlamp et al. 2001; Barker and Hagood 2009b), pulmonary fibrosis (Rege and Hagood 2006a; Sanders et al. 2007, 2008), some cancers (Abeysinghe et al. 2003; Lung et al. 2005; Fiegel et al. 2008), Graves' disease ophthalmopathy (Khoo et al. 2008), and glomerulonephritis (Minto et al. 2003).

1.2 Role of Thy-1 in Neurobiology

Thy-1 is either absent from or expressed in a restricted manner in neurons during development (Xue et al. 1990), but accounts for 2.5–7.5 % of total protein on axon membranes of mature rat neurons (Beech et al. 1983). Thy-1 has been associated with the resolution of neuronal injury. Nerve injury in animal models is associated with decreased Thy-1 expression, with recovery of expression associated with a partial or complete return to the normal or proper physiological neuronal function (Chen et al. 2005). Thy-1 expression in the nervous system is predominantly neuronal, but some human glial cells also express Thy-1, especially following differentiation (Kemshead et al. 1982). Thy-1 inhibits neurite outgrowth on astrocytes, in neurons transfected with either human Thy-1 or mouse Thy-1.2 (Tiveron et al. 1992). Neurite outgrowth is restored in these studies by antibodies against Thy-1, or by addition of soluble Thy-1, suggesting that blocking Thy-1's interaction with a ligand on astrocytes removes the inhibitory effect of Thy-1 on neurite extension (Tiveron et al. 1992). The inhibitory effect of Thy-1 on neurite outgrowth requires its correct localization to native membrane microdomains (Tiveron et al. 1994). Recent epitope mapping studies have characterized the antibody-binding sites that affect neurite extension and found that they recognize not only amino acid sequences, but also the three-dimensional immunoglobulin-like domains and integrin-binding regions (Kuroiwa et al. 2012).

Remarkably, the phenotype of Thy-1 null mice seems to lack significant functional abnormalities involving the nervous system. The principal abnormalities described thus far are inhibition of long-term potentiation in the hippocampal dentate gyrus, inability to transmit social cues regarding food selection, and impaired cutaneous immune responses (Mayeux-Portas et al. 2000; Nosten-Bertrand et al. 1996; Beissert et al. 1998).

In cultures of dorsal root ganglion neurons, interfering with Thy-1 molecular interactions causes the neurons to grow complex processes on the culture substrates. Signaling pathways leading to neurite outgrowth in this case include the activation of both protein kinase A (PKA) and Src, which affect the activation of the mitogen-activated protein kinase kinase/extracellular signal-regulated kinase/cAMP response element-binding protein (MEK/Erk/CREB) pathway, although a direct link between Thy-1 engagement by antibodies and these signaling cascades has not been confirmed (Chen et al. 2007; Yang et al. 2008). Improved understanding of the molecular mechanisms involved in Thy-1-mediated neurite outgrowth inhibition may help in designing interventions to block the negative effects of Thy-1 on the repair of neuronal processes.

More recently, the ligand of Thy-1 present in mature astrocytes has been revealed. An $\alpha_v\beta_3$ integrin has been reported to bind to Thy-1 and trigger clustering of Thy-1, inactivation of Src, and neurite outgrowth inhibition (Herrera-Molina et al. 2012). The complete flow of signaling events initiated as a consequence of Thy-1 integrin interactions awaits further investigation.

1.3 Thy-1 in Non-neuronal Contexts

The functions of Thy-1 in immunity and inflammation, as well as in regulation of cell adhesion and migration, have been reviewed previously (Bradley et al. 2009; Haeryfar and Hoskin 2004; Rege and Hagood 2006b). Additionally, Thy-1 appears to function as a tumor suppressor in several malignancies, including nasopharyngeal and ovarian cancer (Lung et al. 2005). Loss of heterozygosity (LOH) at 11q23.3–q24.3, where *THY1* is mapped in humans, is associated with poor prognosis for ovarian cancer (Cao et al. 2001; Williams and Gagnon 1982). Forced Thy-1 expression suppresses tumorigenicity in the ovarian cancer cell line SKOV-3 (Cao et al. 2001; Abeysinghe et al. 2003). In neuroblastoma, Thy-1 expression correlates inversely with patient survival (Fiegel et al. 2008).

In fibroblast cells, Thy-1 has significant effects on the cell phenotype depending on the tissue origin and context. In lung fibroblasts, Thy-1 suppresses myofibroblastic differentiation and cell migration through effects on Src-family kinases (SFK) and phosphatidylinositol 3-kinase (PI3K) signaling (Barker et al. 2004a; Rege et al. 2006; Sanders et al. 2007). Thy-1, via interaction with $\alpha_v\beta_5$ integrin, inhibits activation of latent transforming growth factor-beta1 (TGF-β1) and myofibroblastic differentiation in lung fibroblasts (Zhou et al. 2010). Conversely, in Graves' disease ophthalmopathy, Thy-1 (+) orbital fibroblasts differentiate into myofibroblasts, while Thy-1 (−) are incapable of doing so, but can differentiate into mature adipocytes (Koumas et al. 2003). However, more recent evidence suggests that Thy-1 (+) fibroblasts can differentiate into lipofibroblasts upon treatment with a PPARγ ligand, but that they would secrete an as-yet unidentified soluble factor that inhibits such differentiation (Khoo et al. 2008; Lehmann et al. 2010). Thy-1 increases PPARγ, fatty acid uptake, and lipofibroblastic differentiation in fetal lung fibroblasts (Varisco et al. 2012).

1.4 Thy-1 Species Differences and Structural Evolution

Thy-1 is an evolutionarily conserved member of the immunoglobulin superfamily (IgSF) (Chen et al. 2005), with significant homology among tunicates, birds, fish, amphibians, rodents, and humans. Among different species, tissue and cellular distribution of Thy-1 expression varies. Mice express Thy-1 on thymocytes, T-lymphocytes, bone marrow stem cells, neurons, and some fibroblasts. In humans, Thy-1 is expressed on a subset of CD34+ bone marrow cells, on a subset of CD34+ and CD3+/CD4+ lymphocytes, and on hematopoietic cells derived from umbilical cord blood and fetal liver, but is absent from mature T cells. In humans, the highest expression levels are on thymic stromal cells (especially fetal) and most fibroblasts. Thy-1 is also expressed in endothelial cells, smooth muscle cells, and some leukemic and lymphoblastoid cells (Feng and Wang 1988). Thy-1 is expressed in neural tissue of all mammalian species studied. In the human nervous system, Thy-1 is expressed primarily in gray matter and in some peripheral nerve fibers (McKenzie and Fabre 1981). Thy-1 is both spatially and temporally regulated during nervous system development; brain expression levels rise nearly 100-fold during early postnatal development (Morris 1985).

Because Thy-1 functions in both the immune system and the nervous system, it may represent a primordial domain of the IgSF ancestry (Cao et al. 2001). Most studies of gene regulation and structure of Thy-1 have been done in the mouse. Murine *thy1* has two alleles which map to chromosome 9, coding for proteins designated Thy-1.1 and Thy-1.2, which are characterized by either arginine or glutamine at position 89. Human *THY1* has no described allelic variants. It is expressed as a 161 aa pro form with a 19 aa signal peptide, which is removed after targeting Thy-1 to the cell membrane (Williams and Gagnon 1982). Thy-1 is variably N-glycosylated, with differing glycosylation among different tissues (Seki et al. 1985; Almqvist and Carlsson 1988; Barclay et al. 1976; Hoessli et al. 1980). Carbohydrate content makes up a third or more of the mass of Thy-1, which ranges from 25 to 37 kDa (Almqvist and Carlsson 1988; Haeryfar and Hoskin 2004). Following cleavage of the C-terminal transmembrane domain, a glycosylphosphatidylinositol (GPI) anchor composed of two fatty-acyl groups is added at residue 131, so that mature Thy-1 is tethered to the outer leaflet of the cell membrane and targeted to lipid rafts (Seki et al. 1985).

The carbohydrate composition of Thy-1 is also developmentally regulated and varies between and within tissues. For example, in rats, sialic acid is much more prominent in thymic Thy-1 than in brain Thy-1, and galactosamine is restricted to brain Thy-1 (Haeryfar and Hoskin 2004).

1.5 Thy-1 Regulation

Unusual regulatory elements define the unique expression profile of Thy-1. The Thy-1 promoter is found in an area of high G/C content and lacks a TATA box; it contains two elements traditionally attributed to "housekeeping" genes (Giguere et al. 1985; Spanopoulou et al. 1991). Replacement of the Thy-1 promoter with a heterologous promoter does not abolish the tissue-specific or developmental expression profile (Vidal et al. 1990). Thy-1 expression in the mouse thymus and brain relies on specific sequences in intron 3 and at the 3′ end of intron 1, respectively. Deletion of intron 1 eliminates brain expression while leaving thymic expression intact (Spanopoulou et al. 1988). Interaction of transcription factors with elements within the third intron varies among species (Tokugawa et al. 1997).

A murine *thy1.2* expression cassette has been designed to drive nervous system expression. This cassette is void of all Thy-1.2 coding sequences and the thymic enhancer in intron 3, while retaining the neural enhancer element in the first intron (Campsall et al. 2002).

Thy-1 functions as a tumor suppressor in nasopharyngeal cancer and is downregulated in some tumors by methylation of its promoter (Lung et al. 2005). In human and rat lung fibroblasts, CpG (cytosine-guanine) islands in the Thy-1 promoter are hypermethylated in the Thy-1-negative fibroblast subpopulation, but not in the positive. A DNA methyltransferase inhibitor, 5-aza-2′-deoxycytidine, restores Thy-1 expression in Thy-1 (–) fibroblasts (Sanders et al. 2008). Trichostatin A (TSA, a histone deacetylase inhibitor) also restores Thy-1 expression in Thy-1 (–) cells associated with depletion of trimethylated H3K27, enrichment of trimethylated H3K4 and acetylated H4, and demethylation of previously hypermethylated CpG sites, indicating interaction of the DNA methylation and histone modification systems in cell-specific epigenetic silencing of Thy-1 (Sanders et al. 2011).

Posttranscriptional regulation of Thy-1 mRNA also influences the temporal and spatial expression of Thy-1 protein in developing mouse nervous system, though the exact mechanisms are still not well characterized (Xue and Morris 1992). Heterokaryons generated from fusion of mature Thy-1.1-expressing neurons with immature Thy-1.2-negative neurons become Thy-1 negative within 16 h of fusion. However, Thy-1.2 expression becomes evident within 3–4 days in culture coincident with re-expression of Thy-1.1. The initial inhibition of Thy-1.1 expression was concluded to be the consequence of a developmentally regulated diffusible suppressor molecule (Saleh and Bartlett 1989). This lends support to developmental regulation of Thy-1 in the nervous system being, at least in part, a posttranscriptional event.

Soluble Thy-1 has been detected in serum, cerebrospinal fluid (CSF), wound fluid from skin ulcers, and synovial fluid from rheumatoid arthritis (Almqvist and Carlsson 1988; Saalbach et al. 1999). Possible methods for production of soluble Thy-1 include alternative mRNA splicing, omitting addition of the GPI anchor, or enzymatic cleavage of Thy-1 from the cell surface. Interestingly, the soluble Thy-1 detected in CSF has slightly higher MW than cellular Thy-1 in the cerebral cortex, attributed to unique glycosylation patterns and suggesting that soluble Thy-1 in CSF could originate from a region of the brain other than the cerebral cortex. The significance and origin of soluble Thy-1 in CSF are unclear. The susceptibility of Thy-1 to cleavage by phospholipases varies from one cell type to another (Naquet et al. 1989). Localization of Thy-1 to cholesterol-rich lipid rafts is thought to protect it from GPI-PLD present in serum (Bergman and Carlsson 1994). Release of Thy-1 could also result from proteolysis. The exact mechanism(s) of Thy-1 shedding and possible roles of shedding in normal biology and in disease have yet to be determined. It is important to note that Thy-1 lacking the GPI anchor very often becomes unrecognizable by antibodies against the membranous form (Kukulansky et al. 1999). In the human uterine cervix, vascular pericytes expressing Thy-1 appear to secrete Thy-1 (+) vesicles, which communicate with basal epithelial cells (Bukovsky et al. 2001). The biological significance of this intriguing phenomenon is uncertain.

1.6 Thy-1 and Non-neuronal Cell Adhesion Signaling

Focal adhesion assembly/disassembly and additional cell–cell and cell–matrix interactions are generally regulated by integrin signaling. Thy-1–integrin interactions appear to regulate a number of heterotypic interactions between cells. Thy-1 expressed on endothelial cells interacts with β_2 and β_3 integrins on leukocytes and with melanoma cells (Choi et al. 2005; Wetzel et al. 2004; Avalos et al. 2002; Saalbach et al. 2000, 2002, 2005). Thy-1 on endothelial cells interacts with $\alpha_v\beta_3$ on melanoma cells and with leukocyte $\alpha_X\beta_2$ and $\alpha_M\beta_2$ (Wetzel et al. 2004; Saalbach et al. 2000, 2002, 2005; Choi et al. 2005), regulating melanoma and leukocyte *trans*-endothelial migration in vitro (Saalbach et al. 2005; Wetzel et al. 2004). It is unknown whether Thy-1 null mice have abnormal leukocyte recruitment or resistance to melanoma metastases.

In Thy-1 (−) fibroblasts, SFK and p190 Rho GTPase-activating protein (GAP) activation results in inactive RhoA, promoting focal adhesion (FA) disassembly; Thy-1 expression decreases SFK and p190 RhoGAP activation, which activates Rho and promotes FA formation (Barker et al. 2004a). FA are supramolecular complexes containing structural proteins, signaling molecules, and adapter proteins that include talin, vinculin, α-actinin, focal adhesion kinase (FAK), SFK, p130CAS, paxillin, and tensin, which mediate cell adhesion to the extracellular matrix (ECM) [reviewed in Burridge and Chrzanowska-Wodnicka (1996), Lawson and Schlaepfer (2012), Zamir and Geiger (2001)].

Thy-1 expression inhibits fibroblast migration in vitro (Barker et al. 2004a), likely due to enhanced adhesion of cells to the ECM. Thrombospondin-1 promotes FA disassembly and induces migration only in Thy-1 (+) lung fibroblasts (Rege et al. 2006). This differential response may be important in regulating cell migration in response to factors present during early wound healing following injury; absence of Thy-1 expression, such as in a fibrotic condition or tumor, may promote dysregulated cell migration.

Because the GPI anchor of Thy-1 does not cross the cell membrane, it is unclear how Thy-1 activates intracellular kinases. It is possible that Thy-1 GPI interacts directly with palmitoylated and myristoylated cysteines on these kinases, as has been suggested by Kusumi et al. (2004). Upon aggregation of a critical number of GPI-anchored proteins in rafts of the outer leaflet of the cell membrane, lipid rafts present in the inner leaflet and associated proteins are recruited underneath the outer leaflet of the bilayer (Kusumi et al. 2004). In this manner, communication of two proteins that face opposite sides of the cell membrane can associate functionally. A different possibility is that a third component of the protein complexes exists and that the GPI-anchored protein interacts with intracellular kinases through transducers as has been demonstrated for other GPI-anchored proteins, such as CD55 and CD59 (Shenoy-Scaria et al. 1993). Thus, Thy-1 may interact with cytoplasmic kinases through an associated transmembrane protein. Thy-1-CD3 co-activation engages the lipid raft transmembrane adapter protein linker for activation of T cells (LAT) (Leyton et al. 1999). Thy-1 is known to interact with an 85–90 kDa transmembrane phosphorylated protein possessing binding sites for SH2 domain-containing proteins, including Fyn, Csk, PI3K, Ras GAP, Vav, and Lck (Durrheim et al. 2001). Recently, Thy-1 has been shown to interact with other cell surface and membrane-spanning proteins. Endothelial Thy-1 binds to the adhesion G-protein-coupled receptor CD97 on leukocytes (Wandel et al. 2012). Therefore, the possibility that Thy-1 signals to the cell interior via the formation of a complex containing a transmembrane protein with affinity for both Thy-1 and SFK is a likely model.

1.7 Neuronal Thy-1 and Its Interactions with Astrocytes

Astrocytes constitute the most abundant cell population of the brain; they are not merely associated with neurons serving a supportive function, but also establish contacts with the endothelial cells of capillaries and are interconnected through gap junctions, facilitating the communication of other cells with neurons [reviewed in Benarroch (2005)]. Astrocytes can modulate neuronal excitability and synaptic transmission [reviewed in Perea and Araque (2002) and Fellin and Carmignoto (2004)]. In addition, Leyton et al. have reported a bidirectional signaling between neurons and astrocytes. These findings confirmed that astrocytes are much more dynamic components of the central nervous system than originally thought [reviewed in Hansson and Ronnback (2003) and Volterra and Meldolesi (2005)].

The first identified ligand/receptor for Thy-1 was β_3-containing integrin (Leyton et al. 2001). Stimulation of a rat astrocyte cell line with recombinant Thy-1 in vitro causes morphological changes of astrocytes; their fine-branched processes retract, transforming these cells into a fibroblast-type shape. These events involve a number of cytoskeletal changes, establishing a higher number of adhesive sites at the tips of microfilament bundles that extend to the central region of the cells, generating tension and cell contraction. These events were subsequently demonstrated to be caused by $\alpha_v\beta_3$ integrin in astrocytes (Hermosilla et al. 2008).

On the other hand, the integrin–Thy-1 interactions also induce changes in neurons. HEK293-generated $\alpha_v\beta_3$-Fc integrin has been shown to inhibit the growth of neuronal processes and cause the retraction of existing neurites (Herrera-Molina et al. 2012). Thus, neuron–astrocyte interactions mediated by $\alpha_v\beta_3$ integrin and Thy-1 might represent an important form of bidirectional communication in the nervous system that generates signals in both neurons and astrocytes.

Integrins are surface receptors known to mediate the formation of FA formed at the points of interaction of integrin with the ECM proteins. These FA connect the ECM to the cytoskeleton, which facilitates bundling of actin microfilaments to generate stress fibers (SF) and increase cellular tension (Dubash et al. 2009).

Interaction between Thy-1 and $\alpha_v\beta_3$ integrin present in astrocytes promotes (1) tyrosine phosphorylation of proteins present in the FA such as FAK and p130Cas; (2) recruitment of vinculin, paxillin, and FAK to FA; and (3) formation of FA, adhesion, and spreading of astrocytes over a substratum. The formation of FA and SF in astrocytes stimulated with recombinant Thy-1-Fc occurs through the aggregation of $\alpha_v\beta_3$ integrins and activation of PKCα, the small GTPase RhoA, and its effector Rho kinase (ROCK) (Avalos et al. 2002, 2004, 2009).

Fibroblasts adhering to fibronectin substrate via interactions mediated by the $\alpha_5\beta_1$ integrin receptor decrease the activity of Rho to lower the degree of contractility of cells in suspension, thus allowing spreading on fibronectin within 30 min (Arthur and Burridge 2001). The cooperative interaction of fibronectin with integrins and the proteoglycan Syndecan-4, through its RGD domain and heparin-binding domain (HBD), respectively, leads to the activation of RhoA, increasing the formation of FA and SF, thereby strengthening cell adhesion (Couchman and Woods 1999).

In astrocytes, the interaction of Thy-1 with both integrin and Syndecan-4 is required for adhesion and spreading to produce morphological changes in these cells (Avalos et al. 2009; Leyton et al. 2001). Thy-1 sequence possesses an RGD-like tripeptide shown to be the integrin-binding region. Thy-1 mutated in the third amino acid (aa) of the RLD tripeptide to RLE neither binds the $\alpha_v\beta_3$ integrin nor induces formation of FA (Hermosilla et al. 2008); similarly, mutation of the basic stretch of aa, REKRK, in the Thy-1 molecule, to AEAAA, renders the protein unable to bind heparin, indicating that this region is the HBD of Thy-1 (Avalos et al. 2009; Leyton et al. 2001). Moreover, addition of heparin as a competitive inhibitor of Syndecan-4 binding to Thy-1 causes a decrease in the activation of RhoA in astrocytes and the pretreatment of astrocytes with heparitinase, which digests heparan sulfate proteoglycans and reduces the formation of Thy-1-induced FA and SF. Enhanced cell adhesion and the activation of Rho are inhibited also by overexpression of a dominant negative form of Syndecan-4. Altogether, these results suggest that the formation of FA and SF in astrocytes requires Thy-1 interaction with both integrin and Syndecan-4, through its RLD and REKRK domains, respectively [Fig. 1.1; Avalos et al. (2009)].

Thy-1-induced RhoA-GTP formation induces activation of PKCα (Avalos et al. 2009). PKCα may be activated in response to the aggregation of integrins (Erb et al. 2001; Vossmeyer et al. 2002) or in response to signaling downstream of Syndecan-4 (Dovas et al. 2006). The involvement of PKCα in the activation of RhoA-ROCK in

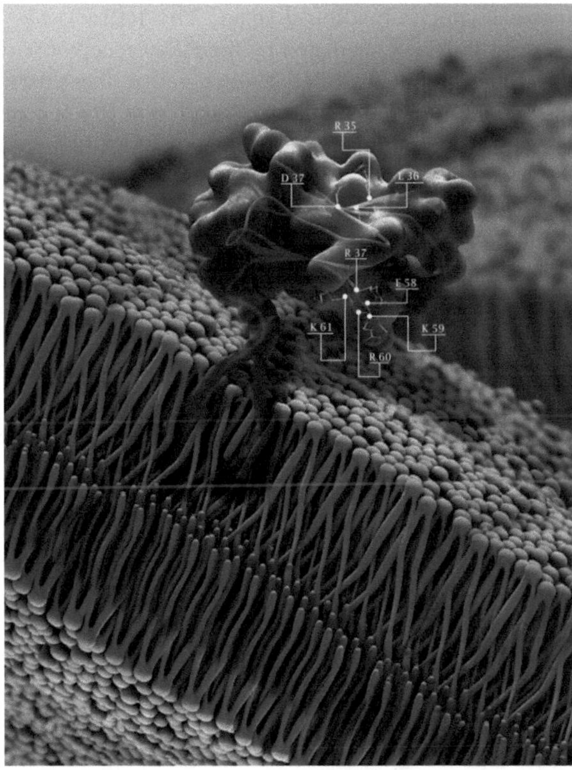

Fig. 1.1 The conserved RLD tripeptide known to interact with $\alpha_v\beta_3$ integrin and the heparin-binding domain (REKRK) of Thy-1, which binds to Syndecan-4, are indicated. Thy-1 is shown inserted in the outer leaflet of the plasma membrane via a lipid anchor. The primary sequence of amino acids 27–161 of human Thy-1 was used to generate a three-dimensional model (Accession number: AAA61180.1). The PBD generated was used to build the Thy-1 molecule using Autodesk Maya mMaya v.1 Molecular Maya toolkit. Graphics and final images were obtained with Adobe Illustrator and Photoshop (Walter Waymann, Designer)

astrocytes has been demonstrated using the inhibitor of classical PKCs (α, β, and γ) Gö6976 and expressing a dominant negative form of PKCα in astrocytes. Activation of RhoA and the formation of FA and SF decrease in both cases (Avalos et al. 2009). These results indicate that the formation of FA and SF in astrocytes stimulated with Thy-1 depends on PKCα. However, whether PKCα is activated downstream of integrin and/or Syndecan-4 remains unresolved.

Due to the fact that PKCα is a calcium/diacylglycerol-activated kinase, these findings opened new questions that are interesting to explore, e.g., Is the release of Ca^{2+} from the endoplasmic reticulum, enough to activate and maintain PKCα at the membrane? Are there other sources of Ca^{2+} involved? How does PKCα activate RhoA? What are the GEFs involved in the activation of RhoA?

In general, increased intracellular Ca^{2+} ($[Ca^{2+}]_i$) implies either the release of calcium from the endoplasmic reticulum or uptake mediated by channels in the membrane. IP3-R inhibition blocks the effect of Thy-1 in astrocytes, and neither RhoA activation nor FA formation is observed in cells treated with this inhibitor (Henriquez et al. 2011). Surprisingly, this cellular response depends also on Thy-1-induced release of ATP from astrocytes, which stimulates influx of extracellular Ca^{2+} through P2X7 purinergic receptors. The increase in $[Ca^{2+}]_i$ is dependent on the interaction of Thy-1 with integrins, because the mutant of the wild-type molecule containing an inactive integrin-binding domain, Thy-1(RLE), has no effect. In addition, the increase in $[Ca^{2+}]_i$ mediated by the P2X7 receptor is required for the formation of FA (Henriquez et al. 2011). Taken together, these data indicate that both Ca^{2+} from the endoplasmic reticulum via activation of IP3-R channels and uptake of extracellular Ca^{2+} are required for the formation of FA induced by Thy-1 in astrocytes. Both the decrease in the expression of P2X7 and inhibition of IP3-R channel block the formation of FA, indicating that both Ca^{2+} currents are necessary for the formation of FA (Henriquez et al. 2011). In retinal ganglion cells, Thy-1 has recently been shown to interact with HCN4, a cation channel subunit (Partida et al. 2012).

It has been suggested that Thy-1 interaction with β_3 integrin may activate bidirectional signaling inducing structural changes in β_3-expressing astrocytes, thus modulating neurite outgrowth of Thy-1-expressing neurons (Avalos et al. 2002). Indeed, recently published studies reported that the $\alpha_v\beta_3$ integrin serves as a ligand for Thy-1 in neurons, which upon binding induces Thy-1 to aggregate in the plasma membrane and leads to neurite outgrowth inhibition and retraction of already formed processes (Herrera-Molina et al. 2012). Astrocytes expressing $\alpha_v\beta_3$, but not those in which the β_3 subunit has been reduced by specific siRNA, inhibit neurite growth of primary cortical neurons maintained for 4–7 days in culture. Recombinant $\alpha_v\beta_3$ expressed as a fusion protein with the Fc portion of IgG1 binds to Thy-1-containing regions of neurons, but remains unbound when these cells have had surface Thy-1 removed by PI-PLC treatment. In addition, $\alpha_v\beta_3$-Fc has no effect on neurons that do not express Thy-1. On the other hand, addition of $\alpha_v\beta_3$-Fc to neurons that have been in culture for more than 13 days, showing a more differentiated phenotype, causes retraction of neuronal processes generating bulb-clubbed endings in these cells (Herrera-Molina et al. 2012).

The mechanisms and consequences of interaction between Thy-1 and $\alpha_v\beta_3$ and modulation of signaling are still unknown and require further investigation, because downregulation of Thy-1 expression or inhibition of Thy-1 signaling on mature neurons may facilitate nerve regeneration.

According to data from antibody cross-linking studies, Thy-1-induced neurite outgrowth requires calcium influx, activation of L- and N-type calcium channels, and G-protein signaling (Doherty et al. 1993). Thy-1 interacts with Fyn and $G\alpha i$ family members in avian neurons and with α- and β-tubulin within lipid rafts (Henke et al. 1997). Thy-1 blocking antibody decreases kinase activity within isolated lipid rafts, and these signaling changes may contribute to Thy-1's effects on neurite outgrowth. In agreement with these data, Thy-1 engaged by its endogenous ligand, $\alpha_v\beta_3$, triggers intracellular signaling that involves the recruitment and inactivation of the non-receptor tyrosine kinase Src (Herrera-Molina et al. 2012); see also Sect. 1.6.

1.8 Thy-1 Adhesive Signaling

Thy-1 can modulate cell signaling "in *trans*" (heterotypically) by Thy-1 on one cell engaging a Thy-1 ligand on another cell. This phenomenon is illustrated by human dermal microvascular endothelial cell (HDMEC) Thy-1 binding to $\alpha_X\beta_2$ (p150, 95,

CD11c/CD18) or $\alpha_M\beta_2$ (Mac-1, CD11b/CD18) on leukocytes, promoting their adhesion and transendothelial migration (Choi et al. 2005; Wetzel et al. 2004; Saalbach et al. 2000), as well as by melanoma cell $\alpha_v\beta_3$ binding to Thy-1 on activated endothelium (Saalbach et al. 2002, 2005), which may promote melanoma metastasis, and by astrocyte–neuron interactions as described above. A number of novel findings are reported in the latter case. The β chain of the $\alpha_v\beta_3$ immunoprecipitated from DI TNC1 rat astrocytes is of smaller-than-expected molecular size, suggesting alternative splicing, posttranslational modification, or increased sensitivity to proteolytic enzymes. However, full-length $\alpha_v\beta_3$ is known to bind Thy-1 in vitro. The RLD sequence in Thy-1 is required for binding to $\alpha_v\beta_3$ integrin and occurs within a highly conserved region (Hermosilla et al. 2008), similar to RGD, the $\alpha_M\beta_2$-binding region in fibrinogen (Altieri et al. 1993). Thy-1 signaling via $\alpha_v\beta_3$ involves FAK and RhoA GTPase, the same pathways activated by "in *cis*" (homotypic) Thy-1 signaling in pulmonary fibroblasts (Barker et al. 2004a, b; Rege et al. 2006). Thy-1 *cis* signaling also occurs in neurons [neurite outgrowth inhibition; Herrera-Molina et al. (2012)]; however, the Thy-1-interacting molecules within neurons, and the downstream signaling pathways activated, have yet to be identified. Others have shown that neurite outgrowth inhibition requires the Thy-1-specific GPI anchor and lipid raft integrity (Tiveron et al. 1992).

Integrity of lipid raft micro- and nano-domains appears important to Thy-1 *cis* signaling. Rafts are microdomains enriched in cholesterol, phosphatidylcholine, and sphingolipids that have been associated with the actin cytoskeleton constituting a platform for signal transduction and communication of the extracellular and the cell interior (Chen et al. 2009a). Thy-1 has been shown to have equal mobility in lipid rafts as it does in the rest of the plasma membrane, which facilitates its trafficking into and out of lipid rafts, whereas transmembrane proteins are less mobile and more spatially constrained (Zhang et al. 1991). Replacement of the GPI anchor with a membrane-spanning domain or disruption of lipid rafts by cholesterol depletion abrogates Thy-1-mediated *cis* signaling (Rege et al. 2006; Tiveron et al. 1992). The residence time of Thy-1 within rafts is controlled by interaction with the cytoskeleton (Chen et al. 2006). To convey messages to the cell interior, Thy-1 clusters in the plasma membrane interacting with itself, adaptors, transducers, or signaling molecules, placing Thy-1 as an important part of a complex that triggers signaling events to the cell interior. In the early 1990s, the presence of multimeric forms of Thy-1 predominantly in differentiated neuron-like cell lines and primary neurons was described. It has been suggested also that these multimers might stabilize the complex formed between Thy-1 and the cytoskeleton (Mahanthappa and Patterson 1992). Thy-1 thus appears to regulate the trafficking and partitioning of signaling molecules into and out of lipid raft domains, thereby modulating signaling networks associated with the cytoskeleton. In this way Thy-1 may act alternately as an inhibitor of signaling molecules such as SFKs, by sequestering them, or as a facilitator of signaling, by regulating their trafficking.

Thy-1 also associates with proteins in the inner leaflet of the plasma membrane via myristoylated and palmitoylated posttranslational modifications, such as the scaffold proteins reggie1 and reggie2 (Neumann-Giesen et al. 2004) and SFKs, as discussed above (Bradshaw 2010). Reggies are key modulators of neuronal process extension and the cytoskeleton, whereas SFKs are non-receptor tyrosine kinases responsible for initiation of cell signaling via tyrosine phosphorylation in response to signals internalized via Thy-1 (Chen et al. 2005, 2006, 2009b; Deininger et al. 2003; Herrera-Molina et al. 2012). Therefore, Thy-1 clusters, lipids, and signaling proteins such as reggies and SFKs are part of these signaling cascades generating signal transduction pathways in *cis*. Other important players in these signaling complexes are Thy-1 transmembrane transducers, which, as indicated earlier, can establish a communication between proteins present in the outer leaflet of the cell membrane with those located in the inner leaflet of the bilayer.

Additional studies are required to better characterize Thy-1 interaction with other molecules in regulation of signaling, particularly in *cis*. Its role in regulating Rho GTPases, focal adhesion turnover, and stress fiber formation is suggestive that Thy-1 interacts in *cis* with integrins, but this has not been shown definitively. Furthermore, the mechanisms regulating Thy-1 localization and trafficking to specialized and distinct lipid nano-domains have not been defined.

1.9 Conclusions

Thy-1 is expressed by a diversity of cell types and has variable effects on cell phenotypes. Hematopoietic and stromal stem cells in an undifferentiated state express Thy-1, whereas in neurons, Thy-1 is developmentally regulated and associated with cessation of neurite outgrowth. In the nervous system, Thy-1 is known to modify the phenotypes of neurons and astrocytes and to mediate their interaction, functioning as an adhesion molecule via interactions with integrins and Syndecan-4. It is known that Thy-1 regulates signaling both in *trans* and in *cis*. Thy-1 is known to interact also with itself, the reggie proteins, and SFK and to modulate intracellular signaling despite lacking a transmembrane domain. Much remains unclear regarding the molecular mechanisms by which Thy-1 modulates intracellular signaling. It is clear, however, that both the expression of Thy-1 and its effects depend a great deal on the cellular and tissue context. Increased understanding of mechanisms of Thy-1 signaling and regulated expression could allow therapeutic manipulation of cell phenotypes in nerve injury, malignancy, and fibrotic disorders.

Acknowledgments LL is supported by FONDECYT 1110149; Fogarty International Center, National Institutes of Health (NIH), Award Number 5R03TW007810; Iniciativas Científicas Milenio: Biomedical Neuroscience Institute P09-015-F; and Proyecto Anillo ACT 1111. JH is supported by NIH Awards HL082818 and HL111169.

Compliance with Ethics Requirements The authors declare that they have no conflicts of interest.

References

Abeysinghe HR, Cao Q, Xu J, Pollock S, Veyberman Y, Guckert NL, Keng P, Wang N (2003) THY1 expression is associated with tumor suppression of human ovarian cancer. Cancer Genet Cytogenet 143(2):125–132

Almqvist P, Carlsson SR (1988) Characterization of a hydrophilic form of Thy-1 purified from human cerebrospinal fluid. J Biol Chem 263(25):12709–12715

Altieri DC, Plescia J, Plow EF (1993) The structural motif glycine 190-valine 202 of the fibrinogen γ chain interacts with CD11b/CD18 integrin ($\alpha_M\beta_2$, Mac-1) and promotes leukocyte adhesion. J Biol Chem 268:1847–1853

Arthur WT, Burridge K (2001) RhoA inactivation by p190RhoGAP regulates cell spreading and migration by promoting membrane protrusion and polarity. Mol Biol Cell 12:2711–2720

Avalos AM, Labra CV, Quest AF, Leyton L (2002) Signaling triggered by Thy-1 interaction with beta 3 integrin on astrocytes is an essential step towards unraveling neuronal Thy-1 function. Biol Res 35(2):231–238

Avalos AM, Arthur WT, Schneider P, Quest AF, Burridge K, Leyton L (2004) Aggregation of integrins and RhoA activation are required for Thy-1-induced morphological changes in astrocytes. J Biol Chem 279(37):39139–39145. doi:10.1074/jbc.M403439200

Avalos AM, Valdivia AD, Munoz N, Herrera-Molina R, Tapia JC, Lavandero S, Chiong M, Burridge K, Schneider P, Quest AF, Leyton L (2009) Neuronal Thy-1 induces astrocyte adhesion by engaging syndecan-4 in a cooperative interaction with alphavbeta3 integrin that activates PKCalpha and RhoA. J Cell Sci 122(Pt 19):3462–3471. doi:10.1242/jcs.034827, jcs.034827 [pii]

Barclay AN, Letarte-Muirhead M, Williams AF, Faulkes RA (1976) Chemical characterisation of the Thy-1 glycoproteins from the membranes of rat thymocytes and brain. Nature 263(5578): 563–567

Barker TH, Hagood JS (2009a) Getting a grip on Thy-1 signaling. Biochim Biophys Acta 1793(5):921–923. doi:10.1016/j.bbamcr.2008.10.004, S0167-4889(08)00349-2 [pii]

Barker TH, Hagood JS (2009b) Getting a grip on Thy-1 signaling. Biochim Biophys Acta 1793(5):921–923

Barker TH, Grenett HE, MacEwen MW, Tilden SG, Fuller GM, Settleman J, Woods A, Murphy-Ullrich J, Hagood JS (2004a) Thy-1 regulates fibroblast focal adhesions, cytoskeletal organization and migration through modulation of p190 RhoGAP and Rho GTPase activity. Exp Cell Res 295(2):488–496

Barker TH, Pallero MA, MacEwen MW, Tilden SG, Woods A, Murphy-Ullrich JE, Hagood JS (2004b) Thrombospondin-1-induced focal adhesion disassembly in fibroblasts requires Thy-1 surface expression, lipid raft integrity, and Src activation. J Biol Chem 279(22):23510–23516

Beech JN, Morris RJ, Raisman G (1983) Density of Thy-1 on axonal membrane of different rat nerves. J Neurochem 41(2):411–417

Beissert S, He HT, Hueber AO, Lellouch AC, Metze D, Mehling A, Luger TA, Schwarz T, Grabbe S (1998) Impaired cutaneous immune responses in Thy-1-deficient mice. J Immunol 161(10): 5296–5302

Benarroch EE (2005) Neuron-astrocyte interactions: partnership for normal function and disease in the central nervous system. Mayo Clin Proc 80(10):1326–1338. doi:10.4065/80.10.1326

Bergman AS, Carlsson SR (1994) Saponin-induced release of cell-surface-anchored Thy-1 by serum glycosylphosphatidylinositol-specific phospholipase D. Biochem J 298(Pt 3):661–668

Bradley JE, Ramirez G, Hagood JS (2009) Roles and regulation of Thy-1, a context-dependent modulator of cell phenotype. Biofactors 35(3):258–265. doi:10.1002/biof.41

Bradshaw JM (2010) The Src, Syk, and Tec family kinases: distinct types of molecular switches. Cell Signal 22(8):1175–1184. doi:10.1016/j.cellsig.2010.03.001

Bukovsky A, Caudle MR, Keenan JA, Upadhyaya NB, Van Meter SE, Wimalasena J, Elder RF (2001) Association of mesenchymal cells and immunoglobulins with differentiating epithelial cells. BMC Dev Biol 1:11

Burridge K, Chrzanowska-Wodnicka M (1996) Focal adhesions, contractility, and signaling. Annu Rev Cell Dev Biol 12:463–518. doi:10.1146/annurev.cellbio.12.1.463

Campsall KD, Mazerolle CJ, De Repentingy Y, Kothary R, Wallace VA (2002) Characterization of transgene expression and Cre recombinase activity in a panel of Thy-1 promoter-Cre transgenic mice. Dev Dyn 224(2):135–143

Cao Q, Abeysinghe H, Chow O, Xu J, Kaung H, Fong C, Keng P, Insel RA, Lee WM, Barrett JC, Wang N (2001) Suppression of tumorigenicity in human ovarian carcinoma cell line SKOV-3 by microcell-mediated transfer of chromosome 11. Cancer Genet Cytogenet 129(2):131–137

Chen Y, Shi-Wen X, van Beek J, Kennedy L, McLeod M, Renzoni EA, Bou-Gharios G, Wilcox-Adelman S, Goetinck PF, Eastwood M, Black CM, Abraham DJ, Leask A (2005) Matrix contraction by dermal fibroblasts requires transforming growth factor-beta/activin-linked kinase 5, heparan sulfate-containing proteoglycans, and MEK/ERK: insights into pathological scarring in chronic fibrotic disease. Am J Pathol 167(6):1699–1711

Chen Y, Thelin WR, Yang B, Milgram SL, Jacobson K (2006) Transient anchorage of cross-linked glycosyl-phosphatidylinositol-anchored proteins depends on cholesterol, Src family kinases, caveolin, and phosphoinositides. J Cell Biol 175(1):169–178

Chen CH, Chen YJ, Jeng CJ, Yang SH, Tung PY, Wang SM (2007) Role of PKA in the anti-Thy-1 antibody-induced neurite outgrowth of dorsal root ganglionic neurons. J Cell Biochem 101:566–575

Chen X, Jen A, Warley A, Lawrence MJ, Quinn PJ, Morris RJ (2009a) Isolation at physiological temperature of detergent-resistant membranes with properties expected of lipid rafts: the influence of buffer composition. Biochem J 417(2):525–533. doi:10.1042/BJ20081385, BJ20081385 [pii]

Chen Y, Veracini L, Benistant C, Jacobson K (2009b) The transmembrane protein CBP plays a role in transiently anchoring small clusters of Thy-1, a GPI-anchored protein, to the cytoskeleton. J Cell Sci 122(21):3966–3972. doi:10.1242/jcs.049346, jcs.049346 [pii]

Choi J, Leyton L, Nham SU (2005) Characterization of alphaX I-domain binding to Thy-1. Biochem Biophys Res Commun 331(2):557–561

Couchman JR, Woods A (1999) Syndecan-4 and integrins: combinatorial signaling in cell adhesion. J Cell Sci 112(Pt 20):3415–3420

Deininger SO, Rajendran L, Lottspeich F, Przybylski M, Illges H, Stuermer CA, Reuter A (2003) Identification of teleost Thy-1 and association with the microdomain/lipid raft reggie proteins in regenerating CNS axons. Mol Cell Neurosci 22(4):544–554

Doherty P, Singh A, Rimon G, Bolsover SR, Walsh FS (1993) Thy-1 antibody-triggered neurite outgrowth requires an influx of calcium into neurons via N- and L-type calcium channels. J Cell Biol 122(1):181–189

Dovas A, Yoneda A, Couchman JR (2006) PKCbeta-dependent activation of RhoA by syndecan-4 during focal adhesion formation. J Cell Sci 119(Pt 13):2837–2846. doi:10.1242/jcs.03020

Dubash AD, Menold MM, Samson T, Boulter E, Garcia-Mata R, Doughman R, Burridge K (2009) Chapter 1. Focal adhesions: new angles on an old structure. Int Rev Cell Mol Biol 277:1–65. doi:10.1016/S1937-6448(09)77001-7

Durrheim GA, Garnett D, Dennehy KM, Beyers AD (2001) Thy-1 associated Pp85-90 is a potential docking site for Sh2 domain-containing signal transduction molecules. Cell Biol Int 25(1):33–42

Erb L, Liu J, Ockerhausen J, Kong Q, Garrad RC, Griffin K, Neal C, Krugh B, Santiago-Perez LI, Gonzalez FA, Gresham HD, Turner JT, Weisman GA (2001) An RGD sequence in the P2Y(2) receptor interacts with alpha(V)beta(3) integrins and is required for G(o)-mediated signal transduction. J Cell Biol 153(3):491–501

Fellin T, Carmignoto G (2004) Neurone-to-astrocyte signalling in the brain represents a distinct multifunctional unit. J Physiol 559(Pt 1):3–15. doi:10.1113/jphysiol.2004.063214

Feng SH, Wang AC (1988) Expression of Thy-1 and effect of phosphatidylinositol-specific phospholipase C on primate and murine cell lines. Cell Immunol 112(2):315–328

Fiegel HC, Kaifi JT, Quaas A, Varol E, Krickhahn A, Metzger R, Sauter G, Till H, Izbicki JR, Erttmann R, Kluth D (2008) Lack of Thy1 (CD90) expression in neuroblastomas is correlated with impaired survival. Pediatr Surg Int 24(1):101–105

Giguere V, Isobe K, Grosveld F (1985) Structure of the murine Thy-1 gene. EMBO J 4(8): 2017–2024

Haeryfar SM, Hoskin DW (2004) Thy-1: more than a mouse pan-T cell marker. J Immunol 173(6):3581–3588

Hansson E, Ronnback L (2003) Glial neuronal signaling in the central nervous system. FASEB J 17(3):341–348. doi:10.1096/fj.02-0429rev

Henke RC, Seeto GS, Jeffrey PL (1997) Thy-1 and AvGp50 signal transduction complex in the avian nervous system: c-Fyn and G alpha i protein association and activation of signalling pathways. J Neurosci Res 49(6):655–670

Henriquez M, Herrera-Molina R, Valdivia A, Alvarez A, Kong M, Munoz N, Eisner V, Jaimovich E, Schneider P, Quest AF, Leyton L (2011) ATP release due to Thy-1-integrin binding induces P2X7-mediated calcium entry required for focal adhesion formation. J Cell Sci 124(Pt 9): 1581–1588. doi:10.1242/jcs.073171

Hermosilla T, Munoz D, Herrera-Molina R, Valdivia A, Munoz N, Nham SU, Schneider P, Burridge K, Quest AF, Leyton L (2008) Direct Thy-1/alphaVbeta3 integrin interaction mediates neuron to astrocyte communication. Biochim Biophys Acta 1783(6):1111–1120. doi:10.1016/j.bbamcr.2008.01.034, S0167-4889(08)00053-0 [pii]

Herrera-Molina R, Frischknecht R, Maldonado H, Seidenbecher CI, Gundelfinger ED, Hetz C, Aylwin Mde L, Schneider P, Quest AF, Leyton L (2012) Astrocytic alphaVbeta3 integrin inhibits neurite outgrowth and promotes retraction of neuronal processes by clustering Thy-1. PLoS One 7(3):e34295. doi:10.1371/journal.pone.0034295

Hoessli D, Bron C, Pink JR (1980) T-lymphocyte differentiation is accompanied by increase in sialic acid content of Thy-1 antigen. Nature 283(5747):576–578

Kemshead JT, Ritter MA, Cotmore SF, Greaves MF (1982) Human Thy-1: expression on the cell surface of neuronal and glial cells. Brain Res 236(2):451–461

Khoo TK, Coenen MJ, Schiefer AR, Kumar S, Bahn RS (2008) Evidence for enhanced Thy-1 (CD90) expression in orbital fibroblasts of patients with Graves' ophthalmopathy. Thyroid 18:1291–1296

Koumas L, Smith TJ, Feldon S, Blumberg N, Phipps RP (2003) Thy-1 expression in human fibroblast subsets defines myofibroblastic or lipofibroblastic phenotypes. Am J Pathol 163(4): 1291–1300

Kukulansky T, Abramovitch S, Hollander N (1999) Cleavage of the glycosylphosphatidylinositol anchor affects the reactivity of thy-1 with antibodies. J Immunol 162(10):5993–5997

Kuroiwa K, Torikai Y, Osawa M, Nakashima T, Nakashima M, Endo H, Arai T (2012) Epitope determination of anti rat thy-1 monoclonal antibody that regulates neurite outgrowth. Hybridoma (Larchmt) 31(4):225–232. doi:10.1089/hyb.2012.0002

Kusumi A, Koyama-Honda I, Suzuki K (2004) Molecular dynamics and interactions for creation of stimulation-induced stabilized rafts from small unstable steady-state rafts. Traffic 5(4): 213–230. doi:10.1111/j.1600-0854.2004.0178.x

Lawson C, Schlaepfer DD (2012) Integrin adhesions: who's on first? What's on second?: connections between FAK and talin. Cell Adh Migr 6(4):302–306. doi:10.4161/cam.20488

Lehmann GM, Woeller CF, Pollock SJ, O'Loughlin CW, Gupta S, Feldon SE, Phipps RP (2010) Novel anti-adipogenic activity produced by human fibroblasts. Am J Physiol Cell Physiol. 299(3):C672–C681. doi:10.1152/ajpcell.00451.2009. Epub 2010 Jun 16

Leyton L, Quest AF, Bron C (1999) Thy-1/CD3 coengagement promotes TCR signaling and enhances particularly tyrosine phosphorylation of the raft molecule LAT. Mol Immunol 36(11–12):755–768

Leyton L, Schneider P, Labra CV, Ruegg C, Hetz CA, Quest AF, Bron C (2001) Thy-1 binds to integrin beta(3) on astrocytes and triggers formation of focal contact sites. Curr Biol 11(13): 1028–1038

Lung HL, Bangarusamy DK, Xie D, Cheung AK, Cheng Y, Kumaran MK, Miller L, Liu ET, Guan XY, Sham JS, Fang Y, Li L, Wang N, Protopopov AI, Zabarovsky ER, Tsao SW, Stanbridge EJ, Lung ML (2005) THY1 is a candidate tumour suppressor gene with decreased expression in metastatic nasopharyngeal carcinoma. Oncogene 24(43):6525–6532

Mahanthappa NK, Patterson PH (1992) Thy-1 multimerization is correlated with neurite outgrowth. Dev Biol 150(1):60–71
Mayeux-Portas V, File SE, Stewart CL, Morris RJ (2000) Mice lacking the cell adhesion molecule Thy-1 fail to use socially transmitted cues to direct their choice of food. Curr Biol 10(2):68–75
McKenzie JL, Fabre JW (1981) Distribution of Thy-1 in human brain: immunofluorescence and absorption analyses with a monoclonal antibody. Brain Res 230(1–2):307–316
Minto AW, Erwig LP, Rees AJ (2003) Heterogeneity of macrophage activation in anti-Thy-1.1 nephritis. Am J Pathol 163(5):2033–2041
Morris R (1985) Thy-1 in developing nervous tissue. Dev Neurosci 7(3):133–160
Naquet P, Barbet J, Pont S, Marchetto S, Barad M, Devaux C, Rougon G, Pierres M (1989) Characterization of Thy-1 with monoclonal antibodies and evidence of Thy-3. Immunol Ser 45:99–117
Neumann-Giesen C, Falkenbach B, Beicht P, Claasen S, Luers G, Stuermer CA, Herzog V, Tikkanen R (2004) Membrane and raft association of reggie-1/flotillin-2: role of myristoylation, palmitoylation and oligomerization and induction of filopodia by overexpression. Biochem J 378(Pt 2):509–518. doi:10.1042/BJ20031100
Nosten-Bertrand M, Errington ML, Murphy KP, Tokugawa Y, Barboni E, Kozlova E, Michalovich D, Morris RG, Silver J, Stewart CL, Bliss TV, Morris RJ (1996) Normal spatial learning despite regional inhibition of LTP in mice lacking Thy-1. Nature 379(6568):826–829
Partida GJ, Stradleigh TW, Ogata G, Godzdanker I, Ishida AT (2012) Thy1 associates with the cation channel subunit HCN4 in adult rat retina. Invest Ophthalmol Vis Sci 53(3):1696–1703. doi:10.1167/iovs.11-9307
Perea G, Araque A (2002) Communication between astrocytes and neurons: a complex language. J Physiol Paris 96(3–4):199–207
Rege TA, Hagood JS (2006a) Thy-1, a versatile modulator of signaling affecting cellular adhesion, proliferation, survival, and cytokine/growth factor responses. Biochim Biophys Acta 1763(10): 991–999
Rege TA, Hagood JS (2006b) Thy-1 as a regulator of cell-cell and cell-matrix interactions in axon regeneration, apoptosis, adhesion, migration, cancer, and fibrosis. FASEB J 20(8):1045–1054
Rege TA, Pallero MA, Gomez C, Grenett HE, Murphy-Ullrich JE, Hagood JS (2006) Thy-1, via its GPI anchor, modulates Src family kinase and focal adhesion kinase phosphorylation and subcellular localization, and fibroblast migration, in response to thrombospondin-1/hep I. Exp Cell Res 312(19):3752–3767
Reif AE, Allen JM (1964) The Akr thymic antigen and its distribution in leukemias and nervous tissues. J Exp Med 120:413–433
Saalbach A, Wetzig T, Haustein UF, Anderegg U (1999) Detection of human soluble Thy-1 in serum by ELISA. Fibroblasts and activated endothelial cells are a possible source of soluble Thy-1 in serum. Cell Tissue Res 298(2):307–315
Saalbach A, Haustein UF, Anderegg U (2000) A ligand of human thy-1 is localized on polymorphonuclear leukocytes and monocytes and mediates the binding to activated thy-1-positive microvascular endothelial cells and fibroblasts. J Invest Dermatol 115(5):882–888
Saalbach A, Hildebrandt G, Haustein UF, Anderegg U (2002) The Thy-1/Thy-1 ligand interaction is involved in binding of melanoma cells to activated Thy-1- positive microvascular endothelial cells. Microvasc Res 64(1):86–93
Saalbach A, Wetzel A, Haustein UF, Sticherling M, Simon JC, Anderegg U (2005) Interaction of human Thy-1 (CD 90) with the integrin alphavbeta3 (CD51/CD61): an important mechanism mediating melanoma cell adhesion to activated endothelium. Oncogene 24(29):4710–4720
Saleh M, Bartlett PF (1989) Evidence from neuronal heterokaryons for a trans-acting factor suppressing Thy-1 expression during neuronal development. J Neurosci Res 23(4):406–415. doi:10.1002/jnr.490230406
Sanders YY, Kumbla P, Hagood JS (2007) Enhanced myofibroblastic differentiation and survival in Thy-1(-) lung fibroblasts. Am J Respir Cell Mol Biol 36(2):226–235

Sanders YY, Pardo A, Selman M, Nuovo GJ, Tollefsbol TO, Siegal GP, Hagood JS (2008) Thy-1 promoter hypermethylation: a novel epigenetic pathogenic mechanism in pulmonary fibrosis. Am J Respir Cell Mol Biol 39(5):610–618
Sanders YY, Tollefsbol TO, Varisco BM, Hagood JS (2011) Epigenetic regulation of thy-1 by histone deacetylase inhibitor in rat lung fibroblasts. Am J Respir Cell Mol Biol 45(1):16–23. doi:10.1165/rcmb.2010-0154OC
Schlamp CL, Johnson EC, Li Y, Morrison JC, Nickells RW (2001) Changes in Thy1 gene expression associated with damaged retinal ganglion cells. Mol Vis 7:192–201
Schlesinger M, Yron I (1969) Antigenic changes in lymph-node cells after administration of antiserum to thymus cells. Science 164(886):1412–1413
Seki T, Spurr N, Obata F, Goyert S, Goodfellow P, Silver J (1985) The human Thy-1 gene: structure and chromosomal location. Proc Natl Acad Sci USA 82(19):6657–6661
Shenoy-Scaria AM, Gauen LK, Kwong J, Shaw AS, Lublin DM (1993) Palmitylation of an amino-terminal cysteine motif of protein tyrosine kinases p56lck and p59fyn mediates interaction with glycosyl-phosphatidylinositol-anchored proteins. Mol Cell Biol 13(10):6385–6392
Spanopoulou E, Giguere V, Grosveld F (1988) Transcriptional unit of the murine Thy-1 gene: different distribution of transcription initiation sites in brain. Mol Cell Biol 8(9):3847–3856
Spanopoulou E, Giguere V, Grosveld F (1991) The functional domains of the murine Thy-1 gene promoter. Mol Cell Biol 11(4):2216–2228
Tiveron M-C, Barboni E, Rivero FBP, Gormley AM, Seeley PJ, Grosveld F, Morris R (1992) Selective inhibition of neurite outgrowth on mature astrocytes by Thy-1 glycoprotein. Nature 355:745–748
Tiveron MC, Nosten-Bertrand M, Jani H, Garnett D, Hirst EM, Grosveld F, Morris RJ (1994) The mode of anchorage to the cell surface determines both the function and the membrane location of Thy-1 glycoprotein. J Cell Sci 107(Pt 7):1783–1796
Tokugawa Y, Koyama M, Silver J (1997) A molecular basis for species differences in Thy-1 expression patterns. Mol Immunol 34(18):1263–1272
Varisco BM, Ambalavanan N, Whitsett JA, Hagood JS (2012) Thy-1 signals through PPARgamma to promote lipofibroblast differentiation in the developing lung. Am J Respir Cell Mol Biol 46(6):765–772. doi:10.1165/rcmb.2011-0316OC
Vidal M, Morris R, Grosveld F, Spanopoulou E (1990) Tissue-specific control elements of the Thy-1 gene. EMBO J 9(3):833–840
Volterra A, Meldolesi J (2005) Astrocytes, from brain glue to communication elements: the revolution continues. Nat Rev Neurosci 6(8):626–640. doi:10.1038/nrn1722
Vossmeyer D, Hofmann W, Loster K, Reutter W, Danker K (2002) Phospholipase Cgamma binds alpha1beta1 integrin and modulates alpha1beta1 integrin-specific adhesion. J Biol Chem 277(7):4636–4643. doi:10.1074/jbc.M105415200
Wandel E, Saalbach A, Sittig D, Gebhardt C, Aust G (2012) Thy-1 (CD90) is an interacting partner for CD97 on activated endothelial cells. J Immunol 188(3):1442–1450. doi:10.4049/jimmunol.1003944
Wetzel A, Chavakis T, Preissner KT, Sticherling M, Haustein UF, Anderegg U, Saalbach A (2004) Human Thy-1 (CD90) on activated endothelial cells is a counterreceptor for the leukocyte integrin Mac-1 (CD11b/CD18). J Immunol 172(6):3850–3859
Williams AF, Gagnon J (1982) Neuronal cell Thy-1 glycoprotein: homology with immunoglobulin. Science 216(4547):696–703
Xue GP, Morris R (1992) Expression of the neuronal surface glycoprotein Thy-1 does not follow appearance of its mRNA in developing mouse Purkinje cells. J Neurochem 58(2):430–440
Xue GP, Calvert RA, Morris RJ (1990) Expression of the neuronal surface glycoprotein Thy-1 is under post-transcriptional control, and is spatially regulated, in the developing olfactory system. Development 109(4):851–864
Yang SH, Chen YJ, Tung PY, Lai WL, Chen Y, Jeng CJ, Wang SM (2008) Anti-Thy-1 antibody-induced neurite outgrowth in cultured dorsal root ganglionic neurons is mediated by the c-Src-MEK signaling pathway. J Cell Biochem 103:67–77

Zamir E, Geiger B (2001) Molecular complexity and dynamics of cell-matrix adhesions. J Cell Sci 114(Pt 20):3583–3590

Zhang F, Crise B, Su B, Hou Y, Rose JK, Bothwell A, Jacobson K (1991) Lateral diffusion of membrane-spanning and glycosylphosphatidylinositol-linked proteins: toward establishing rules governing the lateral mobility of membrane proteins. J Cell Biol 115(1):75–84

Zhou Y, Hagood JS, Lu B, Merryman WD, Murphy-Ullrich JE (2010) Thy-1-integrin alphavbeta5 interactions inhibit lung fibroblast contraction-induced latent TGF-beta1 activation and myofibroblast differentiation. J Biol Chem 285(29):22382–22393. doi:10.1074/jbc.M110.126227, M110.126227 [pii]

Chapter 2
The IgCAMs CAR, BT-IgSF, and CLMP: Structure, Function, and Diseases

Jadwiga Schreiber, Hanna Langhorst, René Jüttner, and Fritz G. Rathjen

Abstract The coxsackie-adenovirus receptor (CAR) is the prototype of a small subfamily of IgCAMs composed of CAR itself, CLMP, BT-IgSF, ESAM, CTX, and A33. These six proteins are composed of one V-set and one C2-set Ig domains and a single transmembrane helix followed by a cytoplasmic stretch. They are localized in several tissues and organs and - except for ESAM, CTX, and A33 - are expressed in the developing brain. CAR becomes downregulated at early postnatal stages and is absent from the adult brain. CAR, CLMP, and BT-IgSF mediate homotypic aggregation. Interestingly, cell adhesion experiments, binding studies, and crystallographic investigations on the extracellular domain reveal a flexible ectodomain for CAR that mediates homophilic and heterophilic binding.

CAR has been extensively investigated in the context of gene therapy and diseases, while research on BT-IgSF and CLMP is at an early stage. Several mouse models as well as studies on patient tissues revealed an essential role for CAR in (1) the development of cardiac, renal, lymphatic, and intestinal tissue; (2) muscle pathology, remodeling, and regeneration; (3) tumor genesis/suppression and metastatic progression; and (4) in virus-mediated infections and gene therapy. Although the in vivo function of CAR in the brain has not been solved its developmentally regulated expression pattern in the brain as well as its function as CAM suggests that CAR might be implicated in neuronal network formation.

J. Schreiber
Max-Delbrück-Centrum für Molekulare Medizin, Robert-Rössle-Strasse 10, 13092 Berlin, Germany

Department of Neuroscience, Erasmus University Medical Center, P.O. Box 2040, 3000 CA Rotterdam, The Netherlands

H. Langhorst • R. Jüttner • F.G. Rathjen (✉)
Max-Delbrück-Centrum für Molekulare Medizin, Robert-Rössle-Strasse 10, 13092 Berlin, Germany
e-mail: Rathjen@mdc-berlin.de

V. Berezin and P.S. Walmod (eds.), *Cell Adhesion Molecules: Implications in Neurological Diseases*, Advances in Neurobiology 8, DOI 10.1007/978-1-4614-8090-7_2,

2.1 The CAR Subgroup of IgCAMs

The coxsackievirus-adenovirus receptor (CAR) was initially identified as a receptor protein for group B coxsackieviruses (CVB) and adenoviruses of the groups A, C, D, E, and F (Bergelson et al. 1997; Tomko et al. 1997; Freimuth et al. 2008; Coyne and Bergelson 2005). It is a prototype of a small structural subgroup of transmembrane proteins of the Ig superfamily in vertebrates which was initially founded by the thymocyte protein CTX (Chretien et al. 1996, 1998) and by the small bowel protein A33 antigen (Heath et al. 1997). Further screens testing endothelial adhesion resulted in the identification of a type I transmembrane protein structurally related to CAR, termed ESAM (endothelial cell-selective adhesion molecule), and whose expression is limited to endothelial cells and platelets but is not found on neural cells (Hirata et al. 2001; Nasdala et al. 2002). Bioinformatics searches in databases then resulted in the cloning of BT-IgSF which is highly expressed in brain and testis (therefore named brain- and testis-specific immunoglobulin superfamily, also termed Igsf11) and CLMP (CAR-like membrane protein) (Suzu et al. 2002; Raschperger et al. 2004; Katoh and Katoh 2003). These six IgSF members are highly related in their overall domain organization and in their primary sequence including the combination of a membrane-distal V-type and a C2-type domain, a short junction between both domains, and an extra pair of cysteines in the C2 domain (Fig. 2.1). The junctional adhesion molecules (JAM) whose extracellular

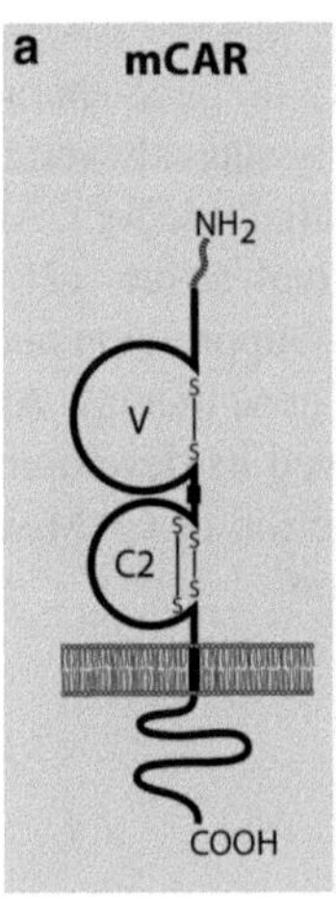

b	mCAR	mCLMP	mBT-IgSF
unprocessed	365 aa	373 aa (30%)	428 aa (29%)
mature	346 aa	356 aa (30%)	406 aa (30%)
SP	19 aa	17 aa (32%)	22 aa (18%)
extra	218 aa	217aa (32%)	218aa (33%)
D1	117 aa	109 aa (35%)	114 aa (33%)
D2	88 aa	90 aa (26%)	91 aa (29%)
TM	21 aa	21 aa (36%)	21 aa (24%)
cyt	107 aa or 94 aa	118 aa (27%)	167 aa (22%)

Fig. 2.1 (**a**) Schematic representation of CAR and (**b**) its relationship to CLMP and BT-IgSF. (**a**) The signal peptide is *shaded*; Ig domains are shown as loops. The junction between the Ig domains is indicated as a *small box*. SP, signal peptide; extra, extracellular domain containing Ig domains D1 and D2; TM, transmembrane segment; cyt, cytoplasmic tail. (**b**) Amino acid identities to CAR are given in percentages. Mouse sequences were used for comparison. ESAM and CTX reveal 29 %, A33 25 %, JAM-A 20 %, JAM-B 21 %, JAM-C 20 %, and JAML 18 % identity to CAR aa, amino acid residues. mCAR, murine Coxsackievirus adenovirus receptor; mCLMP, murine CAR-like membrane protein; mBT-IgSF, murine brain- and testis-specific IgSF member

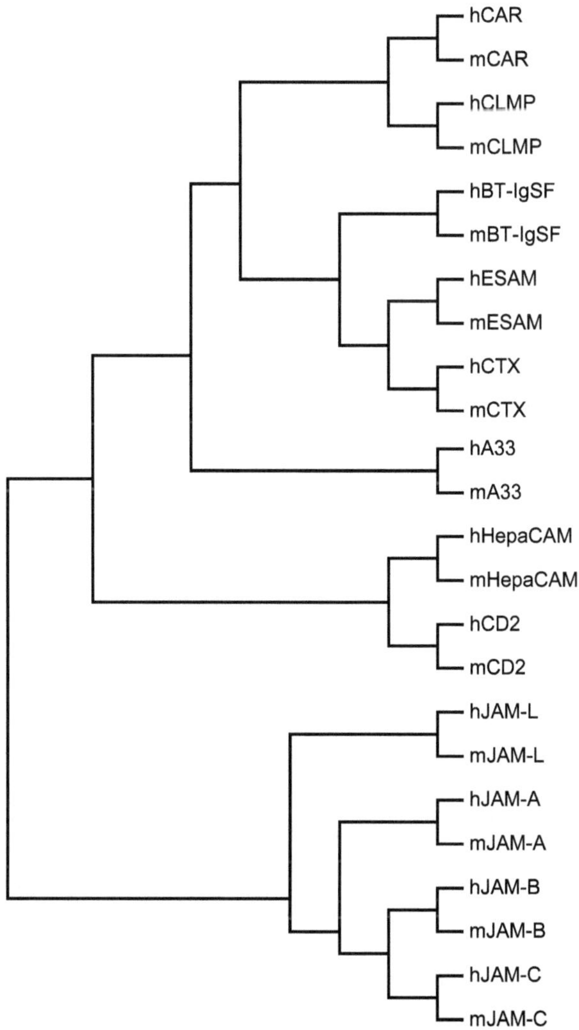

Fig. 2.2 The relationship of the CAR and related proteins is shown. Proteins are grouped according to amino acid sequence similarity. Mouse (m) and human (h) sequences were evaluated. Human and murine sequences of IgSF members were aligned using the Gonnet 250 matrix in ClustalW 2.0.12. Gap-rich columns in the multiple sequence alignment were deleted using Jalview 2.7 program and the edited alignment was analyzed in Protdist 3.67 by the Jones–Taylor–Thornton matrix in order to calculate pairwise distances between the sequences. Bootstrapping was performed to test the reliability of the distance matrix. A consensus tree was computed by Neighbor-Joining 3.67 method by means of bootstrapping and the phylogenetic tree was visualized using TreeView 1.6.6

regions are also composed of two Ig domains, in some cases with two V-type domains instead of the V- and C2-type combination, appear less related to CAR (Weber et al. 2007) (Fig. 2.2).

In this review we discuss primarily CAR which has been extensively investigated in the context of adenoviral gene therapy and heart development. Recent findings on the implication of BT-IgSF and CLMP in diseases are included in our overview of this field while ESAM, CTX, and A33 which are not expressed in the nervous system will not be considered further.

Orthologs of human and mouse CAR have been described in a variety of species including bovine, pig, rat, dog (Fechner et al. 1999; Thoelen et al. 2001a), chick (Patzke et al. 2010), and zebrafish (Petrella et al. 2002). High amino acid sequence identity is observed between mouse and human CAR with 91 % identity in their extracellular domains and 77 % identity within the transmembrane segment and 95 % identity within their cytoplasmic stretch (Wang and Bergelson 1999).

The human *CXADR* gene was initially localized to chromosome 21q11.2 by fluorescence in situ hybridisation analysis (Bowles et al. 1999) and later corrected to 21q21.1 after sequencing of the long arm of chromosome 21 (Hattori et al. 2000). The *CXADR* gene was thought to comprise seven exons, but recent findings revealed the existence of an eighth exon (Excoffon et al. 2010) as found for murine *Cxadr* on chromosome 16 (Chen et al. 2003). Both human and murine genes are transcribed into 6 kb pre-mRNA molecules that are further processed into 2.4 kb and 1.4 kb transcripts, respectively, containing open reading frames of similar size. However, an additional 3′ untranslated region of 1.2 kb is found in the human transcript (Tomko et al. 1997). The human gene encoding BT-IgSF is located on chromosome 3q13.32 while its murine ortholog is found on chromosome 16B4. Two isoforms have been described for BT-IgSF generated by alternative splicing (Katoh and Katoh 2003). The gene of human CLMP was identified on chromosome 11q24.1 and is composed of seven exons. A similar genomic organization was found for the murine gene of CLMP which is located on chromosome 9A5.1 (Raschperger et al. 2004).

2.2 Structural Features of the Extracellular Region of CAR

After cleavage of the signal peptide, the mature CAR exhibits a 218 amino acid extracellular domain, which represents about two-third of the protein and comprises two Ig-like domains: one membrane-distal V-type Ig domain and one membrane-proximal C2-type Ig domain (termed D1 and D2 in the following). These two domains are separated by a short junction and are followed by a linker of five, a transmembrane segment of 21, and a cytoplasmic tail of either 107 or 94 amino acid residues, respectively. Differential splicing of the CAR encoding pre-mRNA leads to two isoforms which are identical except for the extreme C-terminus, which contain class I PDZ-binding motifs ending either in amino acid residues SIV or TVV (also termed form 1 and 2 of CAR, respectively) (Excoffon et al. 2004). Isoform 2 is 13 residues shorter than mCAR isoform 1 (Chen et al. 2003; Excoffon et al. 2010). Further splice variants lacking the transmembrane segment have been described (termed CAR2/7, CAR3/7, and CAR4/7) (Dorner et al. 2004; Thoelen et al. 2001b; Dietel et al. 2011); however, their existence at the protein level has not been confirmed.

In crystallographic investigations the full extracellular region of CAR revealed a U-shaped homodimer through the binding of their N-terminal domains which is reminiscent of JAM-A homodimers (Fig. 2.3a) (Kostrewa et al. 2001; Prota et al. 2003; Patzke et al. 2010). The D1–D1-binding interface of CAR has a size of 684 $Å^2$ and is formed by side chains in β-strands GFCC′ and C″ as well as the FG-connecting loop. Four salt bridges and two hydrogen bonds as well as two hydrophobic interactions are implicated in binding (Fig. 2.3b). Similar observations on the D1 domain were made for the complex of D1 with the fiber knob of the adenovirus and for the homodimer

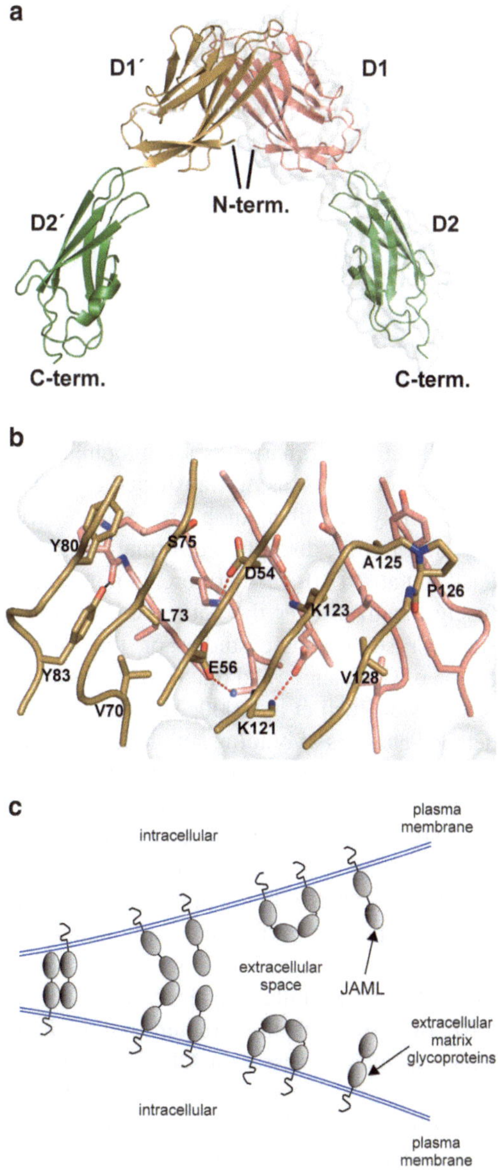

Fig. 2.3 (**a**) Crystal structure of the extracellular region of CAR reveals a U-shaped dimer which is stabilized by binding of their N-terminal D1 domains. D1 domains are colored in either *red* or *brown* and D2 in *green*. *C-term* C-terminus, *N-term* N-terminus. (**b**) Detailed view of interactions inside the dimer interface shown in (**a**). Amino acid residues are given in the *single letter code* and *numbers* indicate their positions. (**c**) Scheme of putative molecular interactions of CAR on the neural plasma membrane. The D1–D1 association observed by the U-shaped crystal structure most likely occurs between two CAR polypeptides associated with the same plasma membrane. Adhesion and binding experiments suggest that homophilic interactions of two CAR polypeptides between neighboring cells might result from an antiparallel D1–D2 binding. Heterophilic bindings of the D2 domain of CAR to ECM glycoproteins or of D1 to JAML are indicated by *arrows* [adopted from (Patzke et al. 2010)]

of the single D1 domain (Bewley et al. 1999; Seiradake et al. 2006; van Raaij et al. 2000). The D2 domain belongs to the C2 type of Ig domains whose two β-sheets are derived from β-strands ABE and CFG, respectively. Two disulfide bonds link the two sheets together, connecting strand A to G and strand B to F. D1 and D2 associate in a head-to-tail manner in the CAR polypeptide and form a rod-like, dumbbell-shaped structure whose protrusions are formed by the globular Ig domains. The two domains are separated by a junction and a linker segment of five residues tethers the extracellular Ig domains of CAR to the cell membrane. Both parts, the junction and the linker, might provide some degree of flexibility to the polypeptide which might influence the interactions of CAR with other proteins (Patzke et al. 2010).

Structural information as well as detailed binding data using extracellular domains is not available for BT-IgSF and CLMP.

2.3 Molecular Interactions of CAR Subgroup Members

2.3.1 The Extracellular Domain of CAR Is Implicated in Homophilic and Heterophilic Binding

CAR functions primarily as a homophilic adhesion molecule as demonstrated by aggregation studies using transfected cells (Honda et al. 2000; Patzke et al. 2010). Consistently, the fiber knob of the adenovirus which binds to the N-terminal Ig domain of CAR disrupts the formation of cell–cell contacts (Walters et al. 2002; Patzke et al. 2010). Furthermore, in non-polarized cells and in neurons CAR is diffusely localized over the entire cell surface but accumulates at cell–cell contact sites. Crystallographic studies on the single D1 and the complete extracellular region of CAR suggested that CAR homophilic binding is mediated by interactions of D1 (van Raaij et al. 2000; Verdino et al. 2010; Patzke et al. 2010). However, a detailed analysis of binding activities of single extracellular domains of CAR, combined with adhesion assays, indicated that homophilic interactions of CAR are also mediated by D1–D2 binding (Patzke et al. 2010).

CAR is also implicated in heterophilic interactions with extracellular matrix proteins fibronectin, tenascin-R, and agrin, which appear to be mediated by D2 of CAR (Patzke et al. 2010). In keratinocytes CAR was found to interact with JAML on γδ T cells in the skin to induce co-stimulation, cytokine production, and activation of the MAP kinase pathway via recruitment of the phosphoinositide-3-kinase (PI3K) to a JAML intracellular sequence motif. The homophilic binding interface of D1 of CAR overlaps with the structure required for the heterophilic binding to JAML which is mainly based on GFCC′C″-sheets packed face to face, and only the A-strand of CAR interacts with the JAML CC′ loop. In contrast to CAR the two V-set domains of JAML associate into a compact assembly, making the extracellular region of JAML more rigid (Verdino et al. 2010). The binding between CAR and JAML is essential for the transmigration of neutrophils across tight junctions (Verdino et al. 2010; Witherden et al. 2010; Zen et al. 2005; Luissint et al. 2008; Guo et al. 2009). In testis CAR interacts with JAM-C (Mirza et al. 2006).

Overall, the binding, structural, and adhesion studies using CAR extracellular domains predict a flexible ectodomain of CAR allowing a conformational shift resulting in either D1–D1 or D1–D2 homophilic binding. A flexible ectodomain has also been observed for CEACAM1 (Klaile et al. 2009) and which might apply to other IgCAMs as well (Volkmer et al. 2013). On the basis of the binding and crystallographic studies a model of molecular interactions of CAR has been proposed in which CAR self-associates on the same cell via D1–D1 as U-shaped dimer and between opposing cells via antiparallel D1–D2 binding (Fig. 2.3c) (Patzke et al. 2010).

BT-IgSF as well as CLMP—similar to CAR—also mediate homotypic aggregation when expressed in heterologous cells, and expression of CLMP increased the transepithelial resistance in MDCK cells, suggesting that CLMP and BT-IgSF might have related functions to CAR (Harada et al. 2005; Eom et al. 2012; Raschperger et al. 2004).

2.3.2 *Molecular Interactions of the Cytoplasmic Segment of CAR*

The two forms of CAR with a cytoplasmic segment differ only in their last 26 or 13 amino acid residues due to alternative splicing. Both isoforms of CAR as well as CLMP and BT-IgSF encode class I PDZ-binding domain sequences. Consequently, several intracellular PDZ domain containing proteins were identified to interact with CAR, suggesting that CAR is part of larger protein complexes. These intracellular binding proteins include ZO-1 (zona occludens 1), MUPP-1 (multi-PDZ domain protein-1), MAGI-1b (membrane-associated guanylate kinase, WW, and PDZ domain containing 1b), and PICK-1 (protein interacting with C kinase 1) and the synaptic scaffolding protein PSD-95 (postsynaptic density protein 95) (Cohen et al. 2001; Raschperger et al. 2006; Coyne et al. 2004; Excoffon et al. 2004; Excoffon et al. 2010). Using a yeast two-hybrid screen, LNX (Ligand-of-Numb protein-X) was found to bind via its second PDZ domain to the cytoplasmic part of CAR. Interestingly, the PDZ-binding segment of CAR was not sufficient for LNX binding, but required a C-terminal upstream sequence (Sollerbrant et al. 2003). Furthermore, LNX2 (Ligand-of-Numb protein-X2) was also discovered as interaction partner by the yeast two-hybrid technique and co-immunoprecipitation. On the basis of truncation and pull-down analyses, the CAR–LNX2 interaction is mediated by the second LNX2 PDZ domain and the CAR PDZ-binding motif, but additional CAR residues 330–339 were also needed for interaction, which is similar to CAR-LNX binding (Mirza et al. 2005). BT-IgSF and CLMP also contain at their C-terminus PDZ-binding motifs and might therefore also interact with several PDZ containing proteins. For example, CLMP was found to associate with ZO-1 on Caco-2, MDCK, and the mouse Sertoli TM4 cells (Raschperger et al. 2004; Sze et al. 2008).

2.4 CAR Is Ubiquitously Expressed but Dominates During Developmental Stages

CAR has been found in numerous organs and tissues during embryonic development including epithelia, myocardium, and the nervous system. Its expression is tightly regulated during development and becomes downregulated at early postnatal stages, in particular in the brain (Hotta et al. 2003; Patzke et al. 2010). In non-polarized cells it is diffusely localized over the entire cell surface but accumulates at cell–cell contact sites. In polarized cells CAR has been found to be associated with tight junctions and co-localizes with the tight junction protein ZO-1 (Cohen et al. 2001; Walters et al. 2002; Nagai et al. 2003; Raschperger et al. 2006). In the developing heart CAR is a component of the intercalated disc (Noutsias et al. 2001; Kashimura et al. 2004), a specialized structure composed of gap junctions, desmosomes, and adherens junctions. CAR is diffusely localized on the plasma membrane of myoblasts and becomes downregulated during skeletal muscle maturation, and in the adult muscle, it is restricted to the sarcolemma of the neuromuscular junction (Shaw et al. 2004; Nalbantoglu et al. 1999). In summary, CAR differs from many IgCAMs in that it is primarily expressed during developmental stages. Its expression is strong on cells with high plasticity. Since CAR is absent from the normal adult nervous system and myocardium it appears to be not a structural protein. It serves as a developmental factor which promotes the formation of cell–cell contacts during development. The strong re-expression of CAR in cardiac diseases associated with heart failure may be considered as an embryonic re-expression (see also below).

Only limited information on the localization of BT-IgSF and CLMP is available. At the transcript level CLMP was detected in several tissues including brain and cell lines (Raschperger et al. 2004; Eguchi et al. 2005; Van Der Werf et al. 2012). BT-IgSF mRNA is highly expressed in the brain, testis, intestinal-type gastric cancers, and melanophores and their precursors of zebrafish (Eom et al. 2012; Suzu et al. 2002; Katoh and Katoh 2003).

2.5 CAR Members and Diseases

Although CAR is strongly expressed in the developing nervous system (Patzke et al. 2010; Hotta et al. 2003), its in vivo role in the brain or in the peripheral nervous system is currently unknown. Due to the very early embryonic lethality of the constitutive CAR knockout (Dorner et al. 2005; Asher et al. 2005; Chen et al. 2006) so far only investigations on immature cultivated CAR-deficient neurons were possible. In vitro experiments revealed that CAR functions in the nervous system as adhesion molecule and might be implicated in the formation of neuronal circuits through homophilic and heterophilic interactions (Patzke et al. 2010). Due to the lack of in vivo studies on CAR, BT-IgSF, and CLMP in the developing brain, we concentrate in the following sections of this article on their functions in nonneural tissues and diseases.

2.5.1 The Interactions of CAR with Adeno- and Coxsackieviruses

The six coxsackie B virus serotypes (CVB1–CVB6) and adenoviruses from subgroups A, C, D, E, and F (but not B) share CAR as a viral receptor; the role of CAR in these viral infections, however, is different. For CVBs, CAR functions for both attachment and viral infection (Martino et al. 2000), while for adenoviruses the major function of CAR is to mediate initial attachment of the virus to the cell surface. Subsequent entry of the adenovirus into cells is mediated by the penton base protein of the virus and by α_V integrin receptors on the cell surface (Wickham et al. 1993; Bai et al. 1993). Because these two virus families are unrelated, their binding specificity to CAR must have evolved independently. The CAR-binding sites on CVB are located in deep crevices or canyons on the capsid surface (Muckelbauer et al. 1995). By contrast, the structures of the adenovirus binding to CAR are surface loops and thus are exposed to immunoselective pressure. Interestingly, the JAML and the homophilic D1–D1-binding site on CAR overlap with the binding site required for the interaction with both viruses (Bewley et al. 1999; Roelvink et al. 1999; van Raaij et al. 2000; Patzke et al. 2010; Verdino et al. 2010). This overlap accounts for the competition of these viruses (and CAR itself) on the cell plasma membrane.

The adenoviruses bind to the N-terminal Ig domain of CAR, D1, by means of trimeric fibers emanating from the vertices of their icosahedral capsid, which terminate in a globular knob domain, termed the fiber knob. The D1 domain of CAR is sufficient for binding of the virus (Bewley et al. 1999; Roelvink et al. 1999). Structural studies indicate that up to three CAR D1 monomers are bound per fiber knob trimer. The AB loop, part of the DE loop, and a short segment of the F-strand of one fiber knob monomer and in addition the FG loop of the adjacent knob interact with strands CC′C″ and the second half of β-strand F in the D1 domain of CAR (Bewley et al. 1999). The AB loop contributes over 50 % of interfacial protein–protein interactions, including the three hydrogen bonds involving conserved atoms in Ad12, Ad2, Ad5, and Ad9 knob and thus may be the key anchor for the complex (Law and Davidson 2005).

The cryo-electron microscopic reconstruction of CVB3 complexes with CAR showed that the D1 domain of CAR binds within the canyon of CVB3. The interface between the virus and CAR consists of the BC and FG loops of D1 binding to the north rim and the floor of the canyon, as well as the A and G β-strands interacting with the south rim of the canyon (He et al. 2001). CVB3 canyon walls are formed by viral structural proteins VP1, VP2, and VP3, and although contact residues are contributed by all three subunits, VP1 dominates the interaction with CAR D1. Several charged residues line the binding interface and provide complementary interactions with CAR. These residues in the virus–receptor interface are moderately well conserved among the six CVB serotypes (He et al. 2001).

2.5.2 CAR as a Viral Receptor

Since CAR functions as receptor for both viruses it has been therefore mainly related to human diseases on the basis of its role as a viral attachment protein. Adenoviruses are non-enveloped DNA viruses, classified into six subgroups (A–F), which are further divided into more than 50 different serotypes based on their immunological properties (Bailey and Mautner 1994; Law and Davidson 2005; Rux and Burnett 2004). Serotypes A, C, D, E, and F are able to attach to CAR (Tomko et al. 1997; Roelvink et al. 1998; Bergelson et al. 1997). The most common, and therefore the most extensively studied, human adenoviruses are those that belong to group C (Rux and Burnett 2004), which predominantly infects the upper respiratory tract. Group C, together with groups B and E, which infect the lower respiratory tract, causes clinical symptoms ranging from mild pharyngitis to acute respiratory disease (common cold syndrome, pneumonia, crop, bronchitis) (Brandt et al. 1969; Figueiredo 2009). Besides respiratory tract infections adenoviruses are recognized as etiologic agents of the gastrointestinal, heart and eye infections (Kaufman 2011; Skevaki et al. 2011), hemorrhagic cystitis, hepatitis, hemorrhagic colitis, pancreatitis, nephritis, or encephalitis (Hayashi and Hogg 2007; Martin et al. 1994; Waldman et al. 2008; Hofland et al. 2004; Lynch et al. 2011). More than 80 % of adenovirus infections occur in young children (<4 years) due to their lack of humoral immunity (Mitchell et al. 2000; Lynch et al. 2011). Epidemics of adenovirus infections may occur in healthy children or adults in closed or crowded settings (e.g., military recruits). Additionally adenovirus infections are more severe, and dissemination is more likely in patients with impaired immunity (e.g., organ transplant recipients, human immunodeficiency virus infection, and congenital immunodeficiency syndromes). Although the vast majority of adenovirus infections are mild and self-limiting, the clinical spectrums are broad, and dissemination or pneumonia can be fatal, in both immunocompetent (particularly infants) and immunocompromised patients (Dudding et al. 1972; Zarraga et al. 1992; Ison 2006; Lynch et al. 2011; Horowitz 2001).

Coxsackie B viruses are non-enveloped RNA viruses and belong to human picornaviruses of the enterovirus group (Melnick 1996). CVB causes a wide range of human and animal diseases such as local myositis, myocarditis (Bergelson et al. 1998; Grist et al. 1975), pancreatitis (Yoon et al. 1986; Imrie et al. 1977), and meningitis (Godman et al. 1952; Melnick 1996). Serotype coxsackievirus B3 (CVB3) is one of the most common pathogens of myocarditis (Bowles et al. 1986; Carthy et al. 1997; Grist and Reid 1993) and its persistent infection may lead to dilated cardiomyopathy (DCM) (Carthy et al. 1997; Wessely et al. 1998; D'Ambrosio et al. 2001; Liu and Mason 2001). Especially in children CVB3 accounts for a significant fraction of cases of terminal heart failure (Shi et al. 2009; Feldman and McNamara 2000). A cardiac-inducible CAR knockout (Shi et al. 2009) provided the first genetic evidence that CAR is the receptor for the coxsackievirus in that the heart was protected from virus entry whereas noncardiac tissues were infected. CVB frequently infects the CNS and, together with other enteroviruses, is the most common cause of viral meningitis in humans. Newborn infants are particularly vulnerable. Moreover, CVB also can infect the fetus, causing death, or neurodevelopmental defects in surviving infants (Feuer et al. 2005).

2.5.3 CAR Is Essential for Embryonic Heart Development

Besides investigations on the role of CAR in viral infections and pathology its function during development as well as its implication in diverse nonviral pathological processes such as cancer or heart arrhythmia has been recently deciphered in loss-of-function and CAR overexpressing mouse mutants. Four constitutive and six conditional CAR knockout mice were generated in different laboratories to gain insight into the physiological function of CAR. These mouse models revealed an essential role of CAR in heart development and in multiple aspects of cardiac function and disease (Fischer et al. 2009). All global CAR knockouts that result in the elimination of all CAR isoforms lead to lethality at midgestation (between embryonic days 11.5 and 13.5) due to cardiac tissue malformation (Asher et al. 2005; Dorner et al. 2005; Chen et al. 2006; Lim et al. 2008; Lisewski et al. 2008; Fischer et al. 2009). Although in all of those mutants the overall outcome was very similar, slightly diverse cardiac phenotypes were reported upon CAR deletion. Dorner et al. (2005) reported that the absence of CAR resulted in enlarged pericards due to edema formation, smaller lumen of the ventricles, and an enlarged cushion. Cardiomyocytes showed reduced density and thickness of myofibrils and their orientation and bundling were disorganized. Neither proliferation nor apoptosis was found to be abnormal in knockout hearts. The authors concluded that the early lethality in these embryos is caused by insufficient heart function and disorganization of Cardiomyocyte structure affecting their contraction capacity. In the study by Asher et al. (2005) CAR-deficient cardiomyocytes exhibited regional apoptosis causing degeneration of the myocardial wall and thoracic hemorrhaging, leading to death at embryonic day 11.5 [see also the reply to (Asher and Finberg 2005)]. Lim et al. (2008) also reported hemorrhage in CAR-deficient embryos and a large pericardial effusion. However, no structural defects, apoptosis, hypertrophy, or ventricular wall thinning in the heart of these CAR-deficient embryos were observed as also described by Dorner et al. (2005). Interestingly, in the study by Chen et al. (2006) CAR deficiency resulted in regional over-proliferation of cardiomyocytes and hyperplasia of the left ventricle. Further analysis revealed that proliferating cardiomyocytes failed to differentiate to form normal trabeculae, which together with poorly organized myofibrils and ill-formed or absent intercellular junctions caused profound heart dysfunction and death at embryonic day 12.5. Surprisingly, when cardiomyocyte-specific deletion occurred not until at E11 [by using an inducible MHC (myosin heavy chain promoter)-Cre recombinase], a significant number (20 %) of conditional mutant animals survived to adulthood and did not reveal cardiac abnormalities. This observation might suggest that CAR is essential during a specific developmental window, and after day 11 when trabeculation is well under way, its loss can be compensated.

In summary, although the phenotype of CAR-deficient embryos appears to be complex, it is clear that CAR is essential for early cardiac development and CAR-mediated signaling is critical for the formation and survival of growing cardiomyocytes (Fechner et al. 2003; Noutsias et al. 2001; Ito et al. 2000).

2.5.4 *CAR Is Implicated in Electrical Conduction in the Developing and Mature Heart*

Conditional ablation of CAR to circumvent embryonic lethality allows the exploration of CAR function at advanced developmental stages and in adult mice. A cardiac-specific CAR knockout (CAR-cKO) showed 70–90 % deletion of CAR at the intercalated discs of cardiomyocytes (Lim et al. 2008). Functional analysis of the heart using electrocardiograms (ECG) of 4- to 5-week-old CAR-cKO mice demonstrated abnormal atrioventricular conduction. While ventricular depolarisations showed no significant difference in the QRS morphology the P-wave was not detectable in the CAR-cKO mice (Lim et al. 2008). A telemetric ECG analysis in awake, conscious mice supported the observation that deletion of CAR leads to an abnormal atrioventricular conduction. In addition, in an embryonic global CAR knockout (CAR-KO) the flow of the blood was assessed by Doppler echocardiography (between E10.5 and 12.5) to study the function of the heart when CAR is absent from the beginning. In CAR-KO embryos the mean PR intervals were significantly enhanced, consistent with a first-degree AV block as shown in CAR-cKO mice (Lim et al. 2008). While the atrial and ventricular action potential generation analyzed by optical mapping studies was not affected, the disruption of CAR leads to a dislocalization of the gap junction protein connexin 45 at the atrioventricular-node cell–cell junctions and a reduced localization of β-catenin and ZO-1 at the ventricular intercalated discs of CAR-cKO at 8 weeks before they developed cardiomyopathy at 21 weeks of age. Similar deficiencies in heart function were uncovered by Lisewski et al. (2008) in a tamoxifen-inducible, cardiac-specific CAR knockout mouse (CAR-cKO). Although the depolarization and repolarization of the CAR-cKO ventricle were normal, the conduction of the electrical activity from the atrium to ventricle was impaired. This is reflected by a prolonged PR interval in CAR-cKO which corresponds with impairments at the level of the atrioventricular node. The changes in electrical conduction are related to a reduced expression and disturbed localization of the gap junction protein connexin 43 which might result in a disturbed communication between cardiomyocytes. Recently, EGC recordings in CAR-cKO mice supported the above findings and revealed a complete atrioventricular block with a temporal dissociation between atrial depolarization (P waves) and ventricular depolarization (QRS complexes) (Pazirandeh et al. 2011). These deficiencies in electrical conductance are associated with a heterogeneous morphology of the intercalated discs in CAR-cKO.

Overall, CAR may be required for the formation of a complex with connexin 45 and 43 at the intercalated discs and might be required for the correct localization of β-catenin and ZO-1 and therefore essential for normal atrioventricular-node conduction. This novel role of CAR in arrhythmia establishes CAR as a potential diagnostic marker for familial cases of atrioventricular block and ventricular dysfunction in genetic and acquired diseases and might also help to explain how, e.g., CVB can cause arrhythmia. Furthermore, arrhythmia could be a potential side effect of therapeutic approaches that target CAR to prevent CVB3-induced myocarditis, pancreatitis, or tumor growth and metastases (see also below) (Fischer et al. 2009).

2.5.5 *CAR Re-expression in Diseased Cardiac and Skeletal Muscle*

In contrast to developing mouse tissue expression of CAR in adult tissue is strongly reduced or even absent in some organs (Tomko et al. 2000); in particular in adult rodents and healthy human hearts CAR expression is very low and restricted to the intercalated discs of cardiomyocytes (Kashimura et al. 2004; Sasse et al. 2003; Ito et al. 2000). Interestingly, a strong CAR re-expression was observed in the intercalated discs and sarcolemma in human DCM, in ischemic cardiomyopathy (ICM) (Noutsias et al. 2001; Poller et al. 2002; Mirza et al. 2006; Tatrai et al. 2011), in mitral/aortic valve diseases (Sasse et al. 2003), and in animal models of cardiac inflammation and myocardial infarction (Ito et al. 2000; Fechner et al. 2003). Ito et al. proposed that re-expression of CAR in experimental autoimmune myocarditis in a rat model is induced by inflammatory cytokines such as interferon-γ, tumor necrosis factor-α, and interleukin-1β and that CAR upregulation might be required for the regeneration of damaged myocardium (Ito et al. 2000; Noutsias et al. 2001). However, induction of CAR re-expression only by humoral factors seems to be unlikely, since CAR upregulation does not occur in all types of heart failures and is often locally confined to, e.g., infarct zone after myocardial infarction (Fechner et al. 2003). Surprisingly, postnatal overexpression of murine CAR in the cardiomyocyte under the control of a tetracycline-responsive α-myosin heavy chain (αMtTA) promoter resulted in inflammatory cardiomyopathy associated with MAPK activation and increased proinflammatory cytokine expression (Yuen et al. 2011). Thus, in contrast to Ito et al. (2000), this study suggests that CAR itself induces inflammation in the heart unrelated to viral infection, rather than just responding to inflammation and injury with upregulation. Therefore, CAR might have a dual function in the pathogenesis of myocarditis: as viral receptor and in addition induction of signals that activate components characteristic of innate immunity. Interestingly, increased CAR expression in the adult heart caused a cardiac phenotype distinct from that observed following embryonic overexpression of CAR which resulted in disrupted cardiomyocyte junctions (Caruso et al. 2010).

An active role of CAR re-expression in diseased cardiac tissue is supported by a similar observation made in regenerating skeletal muscle. CAR is highly expressed in myoblasts and is diffusely distributed on the plasma membrane of immature myofibers. In adult skeletal muscle fibers CAR is confined to the sarcolemma at the neuromuscular junction (Nalbantoglu et al. 1999; Shaw et al. 2004) whereas in diseased muscle with necrosis and regeneration (polymyositis and Duchenne muscular dystrophy, DMD) extrasynaptic sarcolemmal and cytoplasmic CAR is found to be co-expressed with regeneration markers such as desmin and utrophin (Sinnreich et al. 2005). Additionally, similar to cardiac overexpression of CAR, homozygous transgenic mice, in which CAR is regulated by the muscle creatine kinase (MCK) promoter, showed a severe myopathy with a large numbers of necrotic and regenerating fibers and premature death that was associated with an upregulation of caveolin-3 levels and deficiencies in dystrophin and dysferlin (Shaw et al. 2006).

In summary, the general view that emerges from these studies on cardiac and skeletal muscles is that CAR serves as a factor that is transiently expressed during development to establish cell–cell contact-mediated signaling. It is strongly down-regulated in normal adult human tissue and becomes re-expressed in certain diseases or damaged tissues to induce complex processes of tissue remodeling and regeneration (Noutsias et al. 2001). CAR-overexpression studies suggest that there is a threshold level above which CAR expression is detrimental to the muscle tissue. Since CAR expression is strictly regulated during development, it might be not surprising that its persistent, non-physiological overexpression may be deleterious in other tissues too.

2.5.6 CAR in Renal and Intestinal Tissues and in the Lymphatic System

Knockdown of CAR gene function by morpholino antisense oligonucleotides in zebrafish resulted in specific ultrastuctural defects in pronephric glomerular maturation and terminal epithelial differentiation. Although podocytes differentiate in CAR morphants they were not able to elaborate a regularly patterned architecture of foot processes. In the tubules loss of CAR resulted in a clear increase in distance between the neighboring membranes of epithelial cells but did not influence the formation of tight junctions. Additionally, in tubular epithelia lacking CAR apical microvilli were very much reduced in number and appear disorganized. These findings established a new role of CAR in the terminal differentiation of renal glomerular and tubular cell types (Raschperger et al. 2008).

Detailed analysis of internal organ morphology and physiology revealed several abnormalities in a recently developed conditional mouse model of CAR (Pazirandeh et al. 2011). CAR-depleted mice demonstrated a striking atrophy of the exocrine part of the pancreas and small but significant increase in the total number of thymocytes in the thymus. Furthermore, CAR-deficient mice also displayed dilated intestines along the whole intestinal system which showed normal length. This could be either due to exocrine dysfunction of the pancreas or due to an altered neuro/hormonal regulation of the gut motility.

CAR is expressed in neonatal lymphatic endothelial cells where it is found at cell–cell junctions (Vigl et al. 2009). Deletion of CAR by tamoxifen injections in mice containing a conditional allele of CAR at embryonic day 12.5—a developmental stage when CAR is no longer essential for cardiac development—resulted in dilated lymphatic vessels. These vessels were filled with erythrocytes and revealed gaps at lymphatic endothelial cell–cell junctions, indicating that CAR is also essential during the development of the lymphatic vasculature (Mirza et al. 2012).

2.5.7 Loss of CAR Expression Correlates with the Aggressiveness of Tumors

The application of adenoviral vectors in gene therapy depends strongly on the level of CAR expression in targeted tissue (Kim et al. 2002; Hemmi et al. 1998; Li et al. 1999). In this context many tumor samples have been exanimated for CAR expression. In addition, a number of cell adhesion proteins have been linked to cancer progression in that loss of cell–cell contacts allows malignant cells to detach from their neighbors and to escape (Okegawa et al. 2002). Interestingly, in several human cancer tissues CAR expression was significantly downregulated during the progression of the malignancy leading to the hypothesis of a tumor-suppressive role for CAR in human cancers. Furthermore, re-expression of CAR in highly tumorigenic CAR-deficient human prostate and bladder cancer cells suppressed tumor growth (Okegawa et al. 2000, 2001; Rauen et al. 2002). Similar results have been reported for glioblastoma cell line (Kim et al. 2003), malignant glioma, high-grade primary astrocytoma (Huang et al. 2005), thyroid tumor (Marsee et al. 2005), gastrointestinal cancer (esophageal, pancreatic, colorectal, and liver cancer) (Anders et al. 2003b, 2009; Korn et al. 2006), human endometrial adenocarcinoma (Anders et al. 2003b), skin cancer cell lines (Anders et al. 2003b), colon cancer cell line adenomas, primary colon cancers, and colon cancer metastases (Stecker et al. 2011). These observations suggest that loss of CAR expression increases migration and invasion of cancer cells and therefore leads to disease progression with an unfavorable clinical outcome (Korn et al. 2006; Rauen et al. 2002; Matsumoto et al. 2005; Buscarini et al. 2007; Okegawa et al. 2007; Anders et al. 2009; Sachs et al. 2002), whereas forced expression of CAR protein inhibits tumor growth in vitro and in vivo. A possible mechanism for the reduced expression of CAR in malignant cells is the activity of the Raf-MEK-ERK pathway, which is frequently deregulated in cancer. A series of studies have shown that activation of Raf-1 decreases CAR expression, and conversely, inhibition of ERK leads to increased accumulation of cell surface CAR (Huang et al. 2005; Korn et al. 2006; Anders et al. 2003a, 2009). Since CAR is a cell adhesion molecule, loss of CAR weakens intercellular adhesion and might increase proliferation and migration as well as invasion of cancer cells (Okegawa et al. 2000, 2001; Bruning and Runnebaum 2003, 2004; Huang et al. 2005; Wang et al. 2005). An impaired adhesion of cancer cells is considered as a crucial prerequisite for both invasion and metastatic spread (Buda and Pignatelli 2004; Kimura et al. 1997; Resnick et al. 2005; Grone et al. 2007). Moreover, CAR upregulation was associated with an accumulation of the cell cycle regulators p21 and hypophosphorylated retinoblastoma (pRb) protein (Okegawa et al. 2001) suggesting that CAR can inhibit cancer growth by behaving as a membrane receptor, which conveys its signal into the nucleus, suppressing the proliferative mechanisms (Okegawa et al. 2000, 2002; Kim et al. 2003; Huang et al. 2005).

2.6 CLMP Is Implicated in Congenital Short-Bowel Syndrome

Several loss-of-function mutations in the CLMP gene were found in patients with congenital short-bowel syndrome (CSBS) (Van Der Werf et al. 2012). Patients born with CSBS have substantially shorter small intestine, with an average length of 50 cm, compared to a normal length at birth of 190–280 cm. They also reveal intestinal malrotation. Because CSBS occurs in many consanguineous families, it is considered to be an autosomal-recessive disorder. Knockdown of CLMP in the zebrafish resulted in a similar but more severe phenotype including a significant reduction in intestinal length and lack of goblets cells in the mid-intestine. The discrete phenotype observed in human CSBS compared to the zebrafish CSBS model together with the broad expression of CLMP observed in many other tissues argues for functional redundancy of CLMP in human beings. Nevertheless, CLMP function is required for normal small intestine development in both fish and human beings, suggesting a potential evolutionary conservation in this gene's function and its loss of function has a pathological consequence in human beings causing CSBS (Van Der Werf et al. 2012). In addition, CLMP (known also as ACAM, adipocyte adhesion molecule, or ASAM, adipocyte-specific adhesion molecule) has been suggested to be involved in adipocyte differentiation and development of obesity (Eguchi et al. 2005). CLMP mRNA was upregulated in white adipose tissues (WATs) of Otsuka Long–Evans Tokushima fatty (OLETF) rats (an animal model for Type II diabetes and obesity) (Kawano et al. 1992) and in WATs of genetically obese db/db mice (Koya et al. 2000), diet-induced obese ICR mice, and human obese subjects. Also in primary cultured mouse and human adipocytes, CLMP mRNA expression was progressively upregulated during differentiation. These results indicated that CLMP mRNA expression is strongly correlated with accumulation of WATs in human and animal obesity state and that the intervening CLMP expression or activities may alter the adipocyte differentiation status and contribute to therapy of obesity and metabolic syndrome (Eguchi et al. 2005). A provisional analysis of a CLMP knockout mouse indicated that its absence causes hydronephrosis at adult stages (Tang et al. 2010).

2.7 BT-IgSF (Igsf11) Is Implicated in the Migration of Melanophores

BT-IgSF gene expression is highly restricted to testis and brain. In the brain BT-IgSF transcripts are found in both neurons and glial cells, with abundant expression especially in pyramidal cell layers of the dentate gyrus and hippocampus and in commissure fibers of the corpus callosum (Suzu et al. 2002). This expression pattern of BT-IgSF might suggest a role in spermatogenesis and the development or function of the nervous system. Interestingly, so far, three human patients have been reported to have an interstitial deletion of chromosomal 3q spanning the region 3q11–3q21—representing the location of the IgSF gene—and to have an agenesis of the corpus callosum (Genuardi et al. 1994; Mackie et al. 1998). Although the gene responsible for the corpus callosum agenesis in these patients has yet to be identified, the overlay of the BT-IgSF gene localization with the deleted region in these patients together with an abundant expression of BT-IgSF in the corpus callosum makes BT-IgSF a candidate gene responsible for this anomaly. Recent studies using zebrafish mutants indicated that BT-IgSF is implicated in pigment cell development and patterning (Eom et al. 2012). In the wild type melanophores form horizontal stripes during the larval-to-adult transformation while in the absence of BT-IgSF melanophores form an irregularly spotted pattern. BT-IgSF mediates cell–cell contact formation and promotes migration and survival of melanophores.

2.8 Perspectives

CAR and the CAR-related proteins CLMP and BT-IgSF have recently received increased attention due to their role in specific disease states. In particular, CAR has been shown to be essential during the development of the heart, renal, lymphatic, and intestine tissues. Although these three adhesion proteins are expressed in the developing nervous system their functions on neural cells are currently not fully understood. In vitro experiments suggest that CAR might be implicated in the formation of neuronal circuits. In contrast to many IgCAMs, CAR is preferentially expressed in the developing nervous system and becomes downregulated at early postnatal stages. This unusual timing and pattern of expression suggest that CAR is most likely not a structural protein; in contrast it appears to be a developmental factor that might be essential for the development of the brain. Mouse models that allow specific inactivation of CAR within the complete or in parts of the brain might help to understand the function of CAR on neural cells.

Acknowledgments The critical reading of the manuscript by Dr Alistair Garratt is greatly acknowledged. HL is recipient of stipend of the international PhD program of the HGF. The authors' work was supported by a grant from DFG (Ra424/5-1).

Compliance with Ethics Requirements The authors declare that they have no conflicts of interest.

References

Anders M, Christian C, McMahon M, McCormick F, Korn WM (2003a) Inhibition of the Raf/MEK/ERK pathway up-regulates expression of the coxsackievirus and adenovirus receptor in cancer cells. Cancer Res 63:2088–2095

Anders M, Hansen R, Ding RX, Rauen KA, Bissell MJ, Korn WM (2003b) Disruption of 3D tissue integrity facilitates adenovirus infection by deregulating the coxsackievirus and adenovirus receptor. Proc Natl Acad Sci USA 100:1943–1948

Anders M, Vieth M, Rocken C, Ebert M, Pross M, Gretschel S, Schlag PM, Wiedenmann B, Kemmner W, Hocker M (2009) Loss of the coxsackie and adenovirus receptor contributes to gastric cancer progression. Br J Cancer 100:352–359

Asher D, Finberg R (2005) CAR might provide a survival signal for myocardial cells. J Cell Sci 118:5679–5680

Asher DR, Cerny AM, Weiler SR, Horner JW, Keeler ML, Neptune MA, Jones SN, Bronson RT, DePinho RA, Finberg RW (2005) Coxsackievirus and adenovirus receptor is essential for cardiomyocyte development. Genesis 42:77–85

Bai M, Harfe B, Freimuth P (1993) Mutations that alter an Arg-Gly-Asp (RGD) sequence in the adenovirus type 2 penton base protein abolish its cell-rounding activity and delay virus reproduction in flat cells. J Virol 67:5198–5205

Bailey A, Mautner V (1994) Phylogenetic relationships among adenovirus serotypes. Virology 205:438–452

Bergelson JM, Cunningham JA, Droguett G, Kurt-Jones EA, Krithivas A, Hong JS, Horwitz MS, Crowell RL, Finberg RW (1997) Isolation of a common receptor for Coxsackie B viruses and adenoviruses 2 and 5. Science 275:1320–1323

Bergelson JM, Krithivas A, Celi L, Droguett G, Horwitz MS, Wickham T, Crowell RL, Finberg RW (1998) The murine CAR homolog is a receptor for coxsackie B viruses and adenoviruses. J Virol 72:415–419

Bewley MC, Springer K, Zhang YB, Freimuth P, Flanagan JM (1999) Structural analysis of the mechanism of adenovirus binding to its human cellular receptor, CAR. Science 286:1579–1583

Bowles NE, Richardson PJ, Olsen EG, Archard LC (1986) Detection of Coxsackie-B-virus-specific RNA sequences in myocardial biopsy samples from patients with myocarditis and dilated cardiomyopathy. Lancet 1:1120–1123

Bowles KR, Gibson J, Wu J, Shaffer LG, Towbin JA, Bowles NE (1999) Genomic organization and chromosomal localization of the human Coxsackievirus B-adenovirus receptor gene. Hum Genet 105:354–359

Brandt CD, Kim HW, Vargosko AJ, Jeffries BC, Arrobio JO, Rindge B, Parrott RH, Chanock RM (1969) Infections in 18,000 infants and children in a controlled study of respiratory tract disease. I. Adenovirus pathogenicity in relation to serologic type and illness syndrome. Am J Epidemiol 90:484–500

Bruning A, Runnebaum IB (2003) CAR is a cell-cell adhesion protein in human cancer cells and is expressionally modulated by dexamethasone, TNFalpha, and TGFbeta. Gene Ther 10:198–205

Bruning A, Runnebaum IB (2004) The coxsackie adenovirus receptor inhibits cancer cell migration. Exp Cell Res 298:624–631

Buda A, Pignatelli M (2004) Cytoskeletal network in colon cancer: from genes to clinical application. Int J Biochem Cell Biol 36:759–765

Buscarini M, Quek ML, Gilliam-Hegarich S, Kasahara N, Bochner B (2007) Adenoviral receptor expression of normal bladder and transitional cell carcinoma of the bladder. Urol Int 78:160–166

Carthy CM, Yang D, Anderson DR, Wilson JE, McManus BM (1997) Myocarditis as systemic disease: new perspectives on pathogenesis. Clin Exp Pharmacol Physiol 24:997–1003

Caruso L, Yuen S, Smith J, Husain M, Opavsky MA (2010) Cardiomyocyte-targeted overexpression of the coxsackie-adenovirus receptor causes a cardiomyopathy in association with beta-catenin signaling. J Mol Cell Cardiol 48:1194–1205

Chen JW, Ghosh R, Finberg RW, Bergelson JM (2003) Structure and chromosomal localization of the murine coxsackievirus and adenovirus receptor gene. DNA Cell Biol 22:253–259

Chen JW, Zhou B, Yu QC, Shin SJ, Jiao K, Schneider MD, Baldwin HS, Bergelson JM (2006) Cardiomyocyte-specific deletion of the coxsackievirus and adenovirus receptor results in hyperplasia of the embryonic left ventricle and abnormalities of sinuatrial valves. Circ Res 98:923–930

Chretien I, Robert J, Marcuz A, Garcia-Sanz JA, Courtet M, Du PL (1996) CTX, a novel molecule specifically expressed on the surface of cortical thymocytes in Xenopus. Eur J Immunol 26:780–791
Chretien I, Marcuz A, Courtet M, Katevuo K, Vainio O, Heath JK, White SJ, Du PL (1998) CTX, a Xenopus thymocyte receptor, defines a molecular family conserved throughout vertebrates. Eur J Immunol 28:4094–4104
Cohen CJ, Shieh JT, Pickles RJ, Okegawa T, Hsieh JT, Bergelson JM (2001) The coxsackievirus and adenovirus receptor is a transmembrane component of the tight junction. Proc Natl Acad Sci USA 98:15191–15196
Coyne CB, Bergelson JM (2005) CAR: a virus receptor within the tight junction. Adv Drug Deliv Rev 57:869–882
Coyne CB, Voelker T, Pichla SL, Bergelson JM (2004) The coxsackievirus and adenovirus receptor interacts with the multi-PDZ domain protein-1 (MUPP-1) within the tight junction. J Biol Chem 279:48079–48084
D'Ambrosio A, Patti G, Manzoli A, Sinagra G, Di LA, Silvestri F, Di SG (2001) The fate of acute myocarditis between spontaneous improvement and evolution to dilated cardiomyopathy: a review. Heart 85:499–504
Dietel M, Hafner N, Jansen L, Durst M, Runnebaum IB (2011) Novel splice variant CAR 4/6 of the coxsackie adenovirus receptor is differentially expressed in cervical carcinogenesis. J Mol Med (Berl) 89:621–630
Dorner A, Xiong D, Couch K, Yajima T, Knowlton KU (2004) Alternatively spliced soluble coxsackie-adenovirus receptors inhibit coxsackievirus infection. J Biol Chem 279:18497–18503
Dorner AA, Wegmann F, Butz S, Wolburg-Buchholz K, Wolburg H, Mack A, Nasdala I, August B, Westermann J, Rathjen FG, Vestweber D (2005) Coxsackievirus-adenovirus receptor (CAR) is essential for early embryonic cardiac development. J Cell Sci 118:3509–3521
Dudding BA, Wagner SC, Zeller JA, Gmelich JT, French GR, Top FH Jr (1972) Fatal pneumonia associated with adenovirus type 7 in three military trainees. N Engl J Med 286:1289–1292
Eguchi J, Wada J, Hida K, Zhang H, Matsuoka T, Baba M, Hashimoto I, Shikata K, Ogawa N, Makino H (2005) Identification of adipocyte adhesion molecule (ACAM), a novel CTX gene family, implicated in adipocyte maturation and development of obesity. Biochem J 387: 343–353
Eom DS, Inoue S, Patterson LB, Gordon TN, Slingwine R, Kondo S, Watanabe M, Parichy DM (2012) Melanophore migration and survival during zebrafish adult pigment stripe development require the immunoglobulin superfamily adhesion molecule Igsf11. PLoS Genet 8:e1002899
Excoffon KJ, Hruska-Hageman A, Klotz M, Traver GL, Zabner J (2004) A role for the PDZ-binding domain of the coxsackie B virus and adenovirus receptor (CAR) in cell adhesion and growth. J Cell Sci 117:4401–4409
Excoffon KJ, Gansemer ND, Mobily ME, Karp PH, Parekh KR, Zabner J (2010) Isoform-specific regulation and localization of the coxsackie and adenovirus receptor in human airway epithelia. PLoS One 5:e9909
Fechner H, Haack A, Wang H, Wang X, Eizema K, Pauschinger M, Schoemaker R, Veghel R, Houtsmuller A, Schultheiss HP, Lamers J, Poller W (1999) Expression of coxsackie adenovirus receptor and alphav-integrin does not correlate with adenovector targeting in vivo indicating anatomical vector barriers. Gene Ther 6:1520–1535
Fechner H, Noutsias M, Tschoepe C, Hinze K, Wang X, Escher F, Pauschinger M, Dekkers D, Vetter R, Paul M, Lamers J, Schultheiss HP, Poller W (2003) Induction of coxsackievirus-adenovirus-receptor expression during myocardial tissue formation and remodeling: identification of a cell-to-cell contact-dependent regulatory mechanism. Circulation 107:876–882
Feldman AM, McNamara D (2000) Myocarditis. N Engl J Med 343:1388–1398
Feuer R, Pagarigan RR, Harkins S, Liu F, Hunziker IP, Whitton JL (2005) Coxsackievirus targets proliferating neuronal progenitor cells in the neonatal CNS. J Neurosci 25:2434–2444
Figueiredo LT (2009) Viral pneumonia: epidemiological, clinical, pathophysiological and therapeutic aspects. J Bras Pneumol 35:899–906
Fischer R, Poller W, Schultheiss HP, Gotthardt M (2009) CAR-diology—a virus receptor in the healthy and diseased heart. J Mol Med (Berl) 87:879–884

Freimuth P, Philipson L, Carson SD (2008) The coxsackievirus and adenovirus receptor. Curr Top Microbiol Immunol 323:67–87

Genuardi M, Calvieri F, Tozzi C, Coslovi R, Neri G (1994) A new case of interstitial deletion of chromosome 3q, del(3q)(q13.12q21.3), with agenesis of the corpus callosum. Clin Dysmorphol 3:292–296

Godman GC, Bunting H, Melnick JL (1952) The histopathology of Coxsackie virus infection in mice. I. Morphologic observations with four different viral types. Am J Pathol 28:223–257

Grist NR, Reid D (1993) Epidemiology of viral infections of the heart. In: Banatvala JE (ed) Viral infections of the heart. Hodder and Stoughton Ltd, London, pp 23–30

Grist NR, Bell EJ, Reid D (1975) The epidemiology of enteroviruses. Scott Med J 20:27–31

Grone J, Weber B, Staub E, Heinze M, Klaman I, Pilarsky C, Hermann K, Castanos-Velez E, Ropcke S, Mann B, Rosenthal A, Buhr HJ (2007) Differential expression of genes encoding tight junction proteins in colorectal cancer: frequent dysregulation of claudin-1, -8 and -12. Int J Colorectal Dis 22:651–659

Guo YL, Bai R, Chen CX, Liu DQ, Liu Y, Zhang CY, Zen K (2009) Role of junctional adhesion molecule-like protein in mediating monocyte transendothelial migration. Arterioscler Thromb Vasc Biol 29:75–83

Harada H, Suzu S, Hayashi Y, Okada S (2005) BT-IgSF, a novel immunoglobulin superfamily protein, functions as a cell adhesion molecule. J Cell Physiol 204:919–926

Hattori M, Fujiyama A, Taylor TD, Watanabe H, Yada T, Park HS, Toyoda A, Ishii K, Totoki Y, Choi DK, Groner Y, Soeda E, Ohki M, Takagi T, Sakaki Y, Taudien S, Blechschmidt K, Polley A, Menzel U, Delabar J, Kumpf K, Lehmann R, Patterson D, Reichwald K, Rump A, Schillhabel M, Schudy A, Zimmermann W, Rosenthal A, Kudoh J, Schibuya K, Kawasaki K, Asakawa S, Shintani A, Sasaki T, Nagamine K, Mitsuyama S, Antonarakis SE, Minoshima S, Shimizu N, Nordsiek G, Hornischer K, Brant P, Scharfe M, Schon O, Desario A, Reichelt J, Kauer G, Blocker H, Ramser J, Beck A, Klages S, Hennig S, Riesselmann L, Dagand E, Haaf T, Wehrmeyer S, Borzym K, Gardiner K, Nizetic D, Francis F, Lehrach H, Reinhardt R, Yaspo ML (2000) The DNA sequence of human chromosome 21. Nature 405:311–319

Hayashi S, Hogg JC (2007) Adenovirus infections and lung disease. Curr Opin Pharmacol 7: 237–243

He Y, Chipman PR, Howitt J, Bator CM, Whitt MA, Baker TS, Kuhn RJ, Anderson CW, Freimuth P, Rossmann MG (2001) Interaction of coxsackievirus B3 with the full length coxsackievirus-adenovirus receptor. Nat Struct Biol 8:874–878

Heath JK, White SJ, Johnstone CN, Catimel B, Simpson RJ, Moritz RL, Tu GF, Ji H, Whitehead RH, Groenen LC, Scott AM, Ritter G, Cohen L, Welt S, Old LJ, Nice EC, Burgess AW (1997) The human A33 antigen is a transmembrane glycoprotein and a novel member of the immunoglobulin superfamily. Proc Natl Acad Sci USA 94:469–474

Hemmi S, Geertsen R, Mezzacasa A, Peter I, Dummer R (1998) The presence of human coxsackievirus and adenovirus receptor is associated with efficient adenovirus-mediated transgene expression in human melanoma cell cultures. Hum Gene Ther 9:2363–2373

Hirata K, Ishida T, Penta K, Rezaee M, Yang E, Wohlgemuth J, Quertermous T (2001) Cloning of an immunoglobulin family adhesion molecule selectively expressed by endothelial cells. J Biol Chem 276:16223–16231

Hofland CA, Eron LJ, Washecka RM (2004) Hemorrhagic adenovirus cystitis after renal transplantation. Transplant Proc 36:3025–3027

Honda T, Saitoh H, Masuko M, Katagiri-Abe T, Tominaga K, Kozakai I, Kobayashi K, Kumanishi T, Watanabe YG, Odani S, Kuwano R (2000) The coxsackievirus-adenovirus receptor protein as a cell adhesion molecule in the developing mouse brain. Brain Res Mol Brain Res 77:19–28

Horowitz MS (2001) Adenoviruses. In: Knipe DM, Howley PM (eds) Fields virology. Lippincott Williams & Wilkins, New York, NY, pp 2310–2326

Hotta Y, Honda T, Naito M, Kuwano R (2003) Developmental distribution of coxsackie virus and adenovirus receptor localized in the nervous system. Brain Res Dev Brain Res 143:1–13

Huang KC, Altinoz M, Wosik K, Larochelle N, Koty Z, Zhu L, Holland PC, Nalbantoglu J (2005) Impact of the coxsackie and adenovirus receptor (CAR) on glioma cell growth and invasion: requirement for the C-terminal domain. Int J Cancer 113:738–745

Imrie CW, Ferguson JC, Sommerville RG (1977) Coxsackie and mumpsvirus infection in a prospective study of acute pancreatitis. Gut 18:53–56

Ison MG (2006) Adenovirus infections in transplant recipients. Clin Infect Dis 43:331–339

Ito M, Kodama M, Masuko M, Yamaura M, Fuse K, Uesugi Y, Hirono S, Okura Y, Kato K, Hotta Y, Honda T, Kuwano R, Aizawa Y (2000) Expression of coxsackievirus and adenovirus receptor in hearts of rats with experimental autoimmune myocarditis. Circ Res 86:275–280

Kashimura T, Kodama M, Hotta Y, Hosoya J, Yoshida K, Ozawa T, Watanabe R, Okura Y, Kato K, Hanawa H, Kuwano R, Aizawa Y (2004) Spatiotemporal changes of coxsackievirus and adenovirus receptor in rat hearts during postnatal development and in cultured cardiomyocytes of neonatal rat. Virchows Arch 444:283–292

Katoh M, Katoh M (2003) IGSF11 gene, frequently up-regulated in intestinal-type gastric cancer, encodes adhesion molecule homologous to CXADR, FLJ22415 and ESAM. Int J Oncol 23:525–531

Kaufman HE (2011) Adenovirus advances: new diagnostic and therapeutic options. Curr Opin Ophthalmol 22:290–293

Kawano K, Hirashima T, Mori S, Saitoh Y, Kurosumi M, Natori T (1992) Spontaneous long-term hyperglycemic rat with diabetic complications. Otsuka Long-Evans Tokushima Fatty (OLETF) strain. Diabetes 41:1422–1428

Kim JS, Lee SH, Cho YS, Choi JJ, Kim YH, Lee JH (2002) Enhancement of the adenoviral sensitivity of human ovarian cancer cells by transient expression of coxsackievirus and adenovirus receptor (CAR). Gynecol Oncol 85:260–265

Kim M, Sumerel LA, Belousova N, Lyons GR, Carey DE, Krasnykh V, Douglas JT (2003) The coxsackievirus and adenovirus receptor acts as a tumour suppressor in malignant glioma cells. Br J Cancer 88:1411–1416

Kimura Y, Shiozaki H, Hirao M, Maeno Y, Doki Y, Inoue M, Monden T, Ando-Akatsuka Y, Furuse M, Tsukita S, Monden M (1997) Expression of occludin, tight-junction-associated protein, in human digestive tract. Am J Pathol 151:45–54

Klaile E, Vorontsova O, Sigmundsson K, Muller MM, Singer BB, Ofverstedt LG, Svensson S, Skoglund U, Obrink B (2009) The CEACAM1 N-terminal Ig domain mediates cis- and trans-binding and is essential for allosteric rearrangements of CEACAM1 microclusters. J Cell Biol 187:553–567

Korn WM, Macal M, Christian C, Lacher MD, McMillan A, Rauen KA, Warren RS, Ferrell L (2006) Expression of the coxsackievirus- and adenovirus receptor in gastrointestinal cancer correlates with tumor differentiation. Cancer Gene Ther 13:792–797

Kostrewa D, Brockhaus M, D'Arcy A, Dale GE, Nelboeck P, Schmid G, Mueller F, Bazzoni G, Dejana E, Bartfai T, Winkler FK, Hennig M (2001) X-ray structure of junctional adhesion molecule: structural basis for homophilic adhesion via a novel dimerization motif. EMBO J 20:4391–4398

Koya D, Haneda M, Nakagawa H, Isshiki K, Sato H, Maeda S, Sugimoto T, Yasuda H, Kashiwagi A, Ways DK, King GL, Kikkawa R (2000) Amelioration of accelerated diabetic mesangial expansion by treatment with a PKC beta inhibitor in diabetic db/db mice, a rodent model for type 2 diabetes. FASEB J 14:439–447

Law LK, Davidson BL (2005) What does it take to bind CAR? Mol Ther 12:599–609

Li Y, Pong RC, Bergelson JM, Hall MC, Sagalowsky AI, Tseng CP, Wang Z, Hsieh JT (1999) Loss of adenoviral receptor expression in human bladder cancer cells: a potential impact on the efficacy of gene therapy. Cancer Res 59:325–330

Lim BK, Xiong D, Dorner A, Youn TJ, Yung A, Liu TI, Gu Y, Dalton ND, Wright AT, Evans SM, Chen J, Peterson KL, McCulloch AD, Yajima T, Knowlton KU (2008) Coxsackievirus and adenovirus receptor (CAR) mediates atrioventricular-node function and connexin 45 localization in the murine heart. J Clin Invest 118:2758–2770

Lisewski U, Shi Y, Wrackmeyer U, Fischer R, Chen C, Schirdewan A, Juttner R, Rathjen F, Poller W, Radke MH, Gotthardt M (2008) The tight junction protein CAR regulates cardiac conduction and cell-cell communication. J Exp Med 205:2369–2379

Liu PP, Mason JW (2001) Advances in the understanding of myocarditis. Circulation 104: 1076–1082

Luissint AC, Lutz PG, Calderwood DA, Couraud PO, Bourdoulous S (2008) JAM-L-mediated leukocyte adhesion to endothelial cells is regulated in cis by alpha4beta1 integrin activation. J Cell Biol 183:1159–1173

Lynch JP III, Fishbein M, Echavarria M (2011) Adenovirus. Semin Respir Crit Care Med 32: 494–511

Mackie OC, Rooney SC, Hodgson SV, Berry AC (1998) Deletion of chromosome 3q proximal region gives rise to a variable phenotype. Clin Genet 53:220–222

Marsee DK, Vadysirisack DD, Morrison CD, Prasad ML, Eng C, Duh QY, Rauen KA, Kloos RT, Jhiang SM (2005) Variable expression of coxsackie-adenovirus receptor in thyroid tumors: implications for adenoviral gene therapy. Thyroid 15:977–987

Martin AB, Webber S, Fricker FJ, Jaffe R, Demmler G, Kearney D, Zhang YH, Bodurtha J, Gelb B, Ni J et al (1994) Acute myocarditis. Rapid diagnosis by PCR in children. Circulation 90: 330–339

Martino TA, Petric M, Weingartl H, Bergelson JM, Opavsky MA, Richardson CD, Modlin JF, Finberg RW, Kain KC, Willis N, Gauntt CJ, Liu PP (2000) The coxsackie-adenovirus receptor (CAR) is used by reference strains and clinical isolates representing all six serotypes of coxsackievirus group B and by swine vesicular disease virus. Virology 271:99–108

Matsumoto K, Shariat SF, Ayala GE, Rauen KA, Lerner SP (2005) Loss of coxsackie and adenovirus receptor expression is associated with features of aggressive bladder cancer. Urology 66:441–446

Melnick JL (1996) Enteroviruses: poliovirus, coxsacvkieciruses, echo- viruses, and newer enteroviruses. In: Fields BN (ed) Virology KDHP. Lippincott-Raven, Philadelphia, PA, pp 655–712

Mirza M, Raschperger E, Philipson L, Pettersson RF, Sollerbrant K (2005) The cell surface protein coxsackie- and adenovirus receptor (CAR) directly associates with the Ligand-of-Numb Protein-X2 (LNX2). Exp Cell Res 309:110–120

Mirza M, Hreinsson J, Strand ML, Hovatta O, Soder O, Philipson L, Pettersson RF, Sollerbrant K (2006) Coxsackievirus and adenovirus receptor (CAR) is expressed in male germ cells and forms a complex with the differentiation factor JAM-C in mouse testis. Exp Cell Res 312:817–830

Mirza M, Pang MF, Zaini MA, Haiko P, Tammela T, Alitalo K, Philipson L, Fuxe J, Sollerbrant K (2012) Essential role of the coxsackie- and adenovirus receptor (CAR) in development of the lymphatic system in mice. PLoS One 7:e37523

Mitchell LS, Taylor B, Reimels W, Barrett FF, Devincenzo JP (2000) Adenovirus 7a: a community-acquired outbreak in a children's hospital. Pediatr Infect Dis J 19:996–1000

Muckelbauer JK, Kremer M, Minor I, Diana G, Dutko FJ, Groarke J, Pevear DC, Rossmann MG (1995) The structure of coxsackievirus B3 at 3.5 A resolution. Structure 3:653–667

Nagai M, Yaoita E, Yoshida Y, Kuwano R, Nameta M, Ohshiro K, Isome M, Fujinaka H, Suzuki S, Suzuki J, Suzuki H, Yamamoto T (2003) Coxsackievirus and adenovirus receptor, a tight junction membrane protein, is expressed in glomerular podocytes in the kidney. Lab Invest 83:901–911

Nalbantoglu J, Pari G, Karpati G, Holland PC (1999) Expression of the primary coxsackie and adenovirus receptor is downregulated during skeletal muscle maturation and limits the efficacy of adenovirus-mediated gene delivery to muscle cells. Hum Gene Ther 10:1009–1019

Nasdala I, Wolburg-Buchholz K, Wolburg H, Kuhn A, Ebnet K, Brachtendorf G, Samulowitz U, Kuster B, Engelhardt B, Vestweber D, Butz S (2002) A transmembrane tight junction protein selectively expressed on endothelial cells and platelets. J Biol Chem 277:16294–16303

Noutsias M, Fechner H, de Jonge H, Wang X, Dekkers D, Houtsmuller AB, Pauschinger M, Bergelson J, Warraich R, Yacoub M, Hetzer R, Lamers J, Schultheiss HP, Poller W (2001) Human coxsackie-adenovirus receptor is colocalized with integrins alpha(v)beta(3) and alpha(v)beta(5) on the cardiomyocyte sarcolemma and upregulated in dilated cardiomyopathy: implications for cardiotropic viral infections. Circulation 104:275–280

Okegawa T, Li Y, Pong RC, Bergelson JM, Zhou J, Hsieh JT (2000) The dual impact of coxsackie and adenovirus receptor expression on human prostate cancer gene therapy. Cancer Res 60:5031–5036

Okegawa T, Pong RC, Li Y, Bergelson JM, Sagalowsky AI, Hsieh JT (2001) The mechanism of the growth-inhibitory effect of coxsackie and adenovirus receptor (CAR) on human bladder cancer: a functional analysis of car protein structure. Cancer Res 61:6592–6600

Okegawa T, Li Y, Pong RC, Hsieh JT (2002) Cell adhesion proteins as tumor suppressors. J Urol 167:1836–1843

Okegawa T, Sayne JR, Nutahara K, Pong RC, Saboorian H, Kabbani W, Higashihara E, Hsieh JT (2007) A histone deacetylase inhibitor enhances adenoviral infection of renal cancer cells. J Urol 177:1148–1156

Patzke C, Max KE, Behlke J, Schreiber J, Schmidt H, Dorner AA, Kroger S, Henning M, Otto A, Heinemann U, Rathjen FG (2010) The coxsackievirus-adenovirus receptor reveals complex homophilic and heterophilic interactions on neural cells. J Neurosci 30:2897–2910

Pazirandeh A, Sultana T, Mirza M, Rozell B, Hultenby K, Wallis K, Vennstrom B, Davis B, Arner A, Heuchel R, Lohr M, Philipson L, Sollerbrant K (2011) Multiple phenotypes in adult mice following inactivation of the Coxsackievirus and Adenovirus Receptor (Car) gene. PLoS One 6:e20203

Petrella J, Cohen CJ, Gaetz J, Bergelson JM (2002) A zebrafish coxsackievirus and adenovirus receptor homologue interacts with coxsackie B virus and adenovirus. J Virol 76:10503–10506

Poller W, Fechner H, Noutsias M, Tschoepe C, Schultheiss HP (2002) Highly variable expression of virus receptors in the human cardiovascular system. Implications for cardiotropic viral infections and gene therapy. Z Kardiol 91:978–991

Prota AE, Campbell JA, Schelling P, Forrest JC, Watson MJ, Peters TR, Aurrand-Lions M, Imhof BA, Dermody TS, Stehle T (2003) Crystal structure of human junctional adhesion molecule 1: implications for reovirus binding. Proc Natl Acad Sci USA 100:5366–5371

Raschperger E, Engstrom U, Pettersson RF, Fuxe J (2004) CLMP, a novel member of the CTX family and a new component of epithelial tight junctions. J Biol Chem 279:796–804

Raschperger E, Thyberg J, Pettersson S, Philipson L, Fuxe J, Pettersson RF (2006) The coxsackie- and adenovirus receptor (CAR) is an in vivo marker for epithelial tight junctions, with a potential role in regulating permeability and tissue homeostasis. Exp Cell Res 312:1566–1580

Raschperger E, Neve EP, Wernerson A, Hultenby K, Pettersson RF, Majumdar A (2008) The coxsackie and adenovirus receptor (CAR) is required for renal epithelial differentiation within the zebrafish pronephros. Dev Biol 313:455–464

Rauen KA, Sudilovsky D, Le JL, Chew KL, Hann B, Weinberg V, Schmitt LD, McCormick F (2002) Expression of the coxsackie adenovirus receptor in normal prostate and in primary and metastatic prostate carcinoma: potential relevance to gene therapy. Cancer Res 62: 3812–3818

Resnick MB, Gavilanez M, Newton E, Konkin T, Bhattacharya B, Britt DE, Sabo E, Moss SF (2005) Claudin expression in gastric adenocarcinomas: a tissue microarray study with prognostic correlation. Hum Pathol 36:886–892

Roelvink PW, Lizonova A, Lee JG, Li Y, Bergelson JM, Finberg RW, Brough DE, Kovesdi I, Wickham TJ (1998) The coxsackievirus-adenovirus receptor protein can function as a cellular attachment protein for adenovirus serotypes from subgroups A, C, D, E, and F. J Virol 72:7909–7915

Roelvink PW, Mi LG, Einfeld DA, Kovesdi I, Wickham TJ (1999) Identification of a conserved receptor-binding site on the fiber proteins of CAR-recognizing adenoviridae. Science 286: 1568–1571

Rux JJ, Burnett RM (2004) Adenovirus structure. Hum Gene Ther 15:1167–1176

Sachs MD, Rauen KA, Ramamurthy M, Dodson JL, De Marzo AM, Putzi MJ, Schoenberg MP, Rodriguez R (2002) Integrin alpha(v) and coxsackie adenovirus receptor expression in clinical bladder cancer. Urology 60:531–536

Sasse A, Wallich M, Ding Z, Goedecke A, Schrader J (2003) Coxsackie-and-adenovirus receptor mRNA expression in human heart failure. J Gene Med 5:876–882

Seiradake E, Lortat-Jacob H, Billet O, Kremer EJ, Cusack S (2006) Structural and mutational analysis of human Ad37 and canine adenovirus 2 fiber heads in complex with the D1 domain of coxsackie and adenovirus receptor. J Biol Chem 281:33704–33716

Shaw CA, Holland PC, Sinnreich M, Allen C, Sollerbrant K, Karpati G, Nalbantoglu J (2004) Isoform-specific expression of the Coxsackie and adenovirus receptor (CAR) in neuromuscular junction and cardiac intercalated discs. BMC Cell Biol 5:42

Shaw CA, Larochelle N, Dudley RW, Lochmuller H, Danialou G, Petrof BJ, Karpati G, Holland PC, Nalbantoglu J (2006) Simultaneous dystrophin and dysferlin deficiencies associated with high-level expression of the coxsackie and adenovirus receptor in transgenic mice. Am J Pathol 169:2148–2160

Shi Y, Chen C, Lisewski U, Wrackmeyer U, Radke M, Westermann D, Sauter M, Tschope C, Poller W, Klingel K, Gotthardt M (2009) Cardiac deletion of the Coxsackievirus-adenovirus receptor abolishes Coxsackievirus B3 infection and prevents myocarditis in vivo. J Am Coll Cardiol 53:1219–1226

Sinnreich M, Shaw CA, Pari G, Nalbantoglu J, Holland PC, Karpati G (2005) Localization of coxsackie virus and adenovirus receptor (CAR) in normal and regenerating human muscle. Neuromuscul Disord 15:541–548

Skevaki CL, Galani IE, Pararas MV, Giannopoulou KP, Tsakris A (2011) Treatment of viral conjunctivitis with antiviral drugs. Drugs 71:331–347

Sollerbrant K, Raschperger E, Mirza M, Engstrom U, Philipson L, Ljungdahl PO, Pettersson RF (2003) The Coxsackievirus and adenovirus receptor (CAR) forms a complex with the PDZ domain-containing protein ligand-of-numb protein-X (LNX). J Biol Chem 278:7439–7444

Stecker K, Vieth M, Koschel A, Wiedenmann B, Rocken C, Anders M (2011) Impact of the coxsackievirus and adenovirus receptor on the adenoma-carcinoma sequence of colon cancer. Br J Cancer 104:1426–1433

Suzu S, Hayashi Y, Harumi T, Nomaguchi K, Yamada M, Hayasawa H, Motoyoshi K (2002) Molecular cloning of a novel immunoglobulin superfamily gene preferentially expressed by brain and testis. Biochem Biophys Res Commun 296:1215–1221

Sze KL, Lee WM, Lui WY (2008) Expression of CLMP, a novel tight junction protein, is mediated via the interaction of GATA with the Kruppel family proteins, KLF4 and Sp1, in mouse TM4 Sertoli cells. J Cell Physiol 214:334–344

Tang T, Li L, Tang J, Li Y, Lin WY, Martin F, Grant D, Solloway M, Parker L, Ye W, Forrest W, Ghilardi N, Oravecz T, Platt KA, Rice DS, Hansen GM, Abuin A, Eberhart DE, Godowski P, Holt KH, Peterson A, Zambrowicz BP, de Sauvage FJ (2010) A mouse knockout library for secreted and transmembrane proteins. Nat Biotechnol 28:749–755

Tatrai E, Bedi K, Kovalszky I, Hartyanszky I, Laszik A, Acsady G, Sotonyi P, Hubay M (2011) No mutation but high mRNA expression of Coxsackie-Adenovirus Receptor was observed in both dilated and ischemic cardiomyopathy. Forensic Sci Int 212:47–50

Thoelen I, Keyaerts E, Lindberg M, Van RM (2001a) Characterization of a cDNA encoding the bovine coxsackie and adenovirus receptor. Biochem Biophys Res Commun 288:805–808

Thoelen I, Magnusson C, Tagerud S, Polacek C, Lindberg M, Van RM (2001b) Identification of alternative splice products encoded by the human coxsackie-adenovirus receptor gene. Biochem Biophys Res Commun 287:216–222

Tomko RP, Xu R, Philipson L (1997) HCAR and MCAR: the human and mouse cellular receptors for subgroup C adenoviruses and group B coxsackieviruses. Proc Natl Acad Sci USA 94:3352–3356

Tomko RP, Johansson CB, Totrov M, Abagyan R, Frisen J, Philipson L (2000) Expression of the adenovirus receptor and its interaction with the fiber knob. Exp Cell Res 255:47–55

Van Der Werf CS, Wabbersen TD, Hsiao NH, Paredes J, Etchevers HC, Kroisel PM, Tibboel D, Babarit C, Schreiber RA, Hoffenberg EJ, Vekemans M, Zeder SL, Ceccherini I, Lyonnet S, Ribeiro AS, Seruca R, Te Meerman GJ, van Ijzendoorn SC, Shepherd IT, Verheij JB, Hofstra RM (2012) CLMP is required for intestinal development, and loss-of-function mutations cause congenital short-bowel syndrome. Gastroenterology 142:453–462

van Raaij MJ, Chouin E, van der Zandt H, Bergelson JM, Cusack S (2000) Dimeric structure of the coxsackievirus and adenovirus receptor D1 domain at 1.7 A resolution. Structure Fold Des 8:1147–1155

Verdino P, Witherden DA, Havran WL, Wilson IA (2010) The molecular interaction of CAR and JAML recruits the central cell signal transducer PI3K. Science 329:1210–1214

Vigl B, Zgraggen C, Rehman N, Banziger-Tobler NE, Detmar M, Halin C (2009) Coxsackie- and adenovirus receptor (CAR) is expressed in lymphatic vessels in human skin and affects lymphatic endothelial cell function in vitro. Exp Cell Res 315:336–347

Volkmer H, Schreiber J, Rathjen FG (2013) Regulation of adhesion by flexible ectodomains of IgCAMs. Neurochem Res 38(6):1092–1099

Waldman M, Marshall V, Whitby D, Kopp JB (2008) Viruses and kidney disease: beyond HIV. Semin Nephrol 28:595–607

Walters RW, Freimuth P, Moninger TO, Ganske I, Zabner J, Welsh MJ (2002) Adenovirus fiber disrupts CAR-mediated intercellular adhesion allowing virus escape. Cell 110:789–799

Wang X, Bergelson JM (1999) Coxsackievirus and adenovirus receptor cytoplasmic and transmembrane domains are not essential for coxsackievirus and adenovirus infection. J Virol 73:2559–2562

Wang B, Chen G, Li F, Zhou J, Lu Y, Ma D (2005) Inhibitory effect of coxsackie adenovirus receptor on invasion and metastasis phenotype of ovarian cancer cell line SKOV3. J Huazhong Univ Sci Technolog Med Sci 25(85–7):93

Weber C, Fraemohs L, Dejana E (2007) The role of junctional adhesion molecules in vascular inflammation. Nat Rev Immunol 7:467–477

Wessely R, Klingel K, Santana LF, Dalton N, Hongo M, Jonathan LW, Kandolf R, Knowlton KU (1998) Transgenic expression of replication-restricted enteroviral genomes in heart muscle induces defective excitation-contraction coupling and dilated cardiomyopathy. J Clin Invest 102:1444–1453

Wickham TJ, Mathias P, Cheresh DA, Nemerow GR (1993) Integrins alpha v beta 3 and alpha v beta 5 promote adenovirus internalization but not virus attachment. Cell 73:309–319

Witherden DA, Verdino P, Rieder SE, Garijo O, Mills RE, Teyton L, Fischer WH, Wilson IA, Havran WL (2010) The junctional adhesion molecule JAML is a costimulatory receptor for epithelial gammadelta T cell activation. Science 329:1205–1210

Yoon JW, London WT, Curfman BL, Brown RL, Notkins AL (1986) Coxsackie virus B4 produces transient diabetes in nonhuman primates. Diabetes 35:712–716

Yuen S, Smith J, Caruso L, Balan M, Opavsky MA (2011) The coxsackie-adenovirus receptor induces an inflammatory cardiomyopathy independent of viral infection. J Mol Cell Cardiol 50:826–840

Zarraga AL, Kerns FT, Kitchen LW (1992) Adenovirus pneumonia with severe sequelae in an immunocompetent adult. Clin Infect Dis 15:712–713

Zen K, Liu Y, McCall IC, Wu T, Lee W, Babbin BA, Nusrat A, Parkos CA (2005) Neutrophil migration across tight junctions is mediated by adhesive interactions between epithelial coxsackie and adenovirus receptor and a junctional adhesion molecule-like protein on neutrophils. Mol Biol Cell 16:2694–2703

Chapter 3
GLIALCAM, A Glial Cell Adhesion Molecule Implicated in Neurological Disease

Alejandro Barrallo-Gimeno and Raúl Estévez

Abstract GlialCAM (also named HepaCAM) is a cell adhesion molecule expressed mainly in glial cells from the central nervous system and the liver. GlialCAM plays different roles according to its cellular context. In epithelial cell lines, overexpression of GlialCAM increases cell adhesion and motility but also inhibits cell growth in tumor cell lines, leading to senescence. In glial cells, however, its function is quite different. GlialCAM acts a regulator of subcellular traffic of MLC1, a protein with unknown function involved in the pathogenesis of megalencephalic leukoencephalopathy with subcortical cysts (MLC), a rare neurological condition. Moreover, GlialCAM itself has been found to be responsible for some of the cases of this disease. Additionally, GlialCAM also works as an auxiliary subunit of the chloride channel ClC-2, regulating its targeting to cell–cell junctions and modifying its functional properties. In summary, GlialCAM has different functions not only related to its adhesive nature, and defects in these functions lead to neurological disease.

3.1 Introduction

Glial cells are the most abundant cells in the nervous system. They form a heterogeneous group of cells that can be distinguished morphologically and functionally (Zhang and Barres 2010). Glial cells play numerous and diverse roles in the normal neural physiology, tightly linked to neuronal function, and therefore glial alteration may lead to neurological disease (Allaman et al. 2011; Parpura et al. 2012). For example, during embryonic development, radial glial cells act as a scaffold for the proper migration of neurons from their place of birth to their final location within

A. Barrallo-Gimeno, Ph.D. • R. Estévez, Ph.D. (✉)
Physiology Section, Department of Physiological Sciences II, School of Medicine,
University of Barcelona, Av. Feixa Llarga s/n, 08907 Barcelona, Spain
e-mail: restevez@ub.edu

V. Berezin and P.S. Walmod (eds.), *Cell Adhesion Molecules: Implications in Neurological Diseases*, Advances in Neurobiology 8, DOI 10.1007/978-1-4614-8090-7_3,

the brain. In this way, newborn neurons must recognize glial cells through extracellular matrix proteins or signaling cues (Molofsky et al. 2012). Astrocytes, together with endothelial cells and pericytes, constitute the blood–brain barrier, effectively isolating the brain from the rest of the organism. Astrocyte end feet surround blood vessels, taking up nutrients for the neurons and disposing metabolic waste, in order to keep brain metabolic homeostasis (Belanger et al. 2011). Oligodendrocytes form the myelin sheath around axons of the central nervous system, contributing to the fast propagation of action potentials along neural pathways (Piaton et al. 2010). In neural pathways, astrocytes are also fundamental for maintaining the ionic conditions required for neuronal electrical activity, mainly the buffering of the excess of extracellular potassium ions produced by action potential propagation, that are conducted away from the nodes of Ranvier (Rash 2010). These ionic fluxes from myelin to astrocytic end feet have an osmotically associated water flow, and therefore astrocytes are important for water homeostasis and osmotic balance in the brain. Astrocytes are highly sensitive to changes in extracellular osmolarity and can display prominent cell volume responses that may lead to macroscopic swelling of the brain.

Cell adhesion molecules (CAMs) may exhibit different properties depending on the context in which they are studied, and their functions may go well beyond the mere cell adhesion to other cells or to another substrates. Such is the case of the protein we are dealing with in this chapter, initially described as a tumor suppressor in the liver, then found to be expressed normally in glial cells, and finally involved in a rare neurological disease, interacting with and modulating the activity of a chloride channel.

3.2 Initial Identification of HepaCAM in the Liver

HepaCAM was first found in a screen for silenced genes in human hepatocellular carcinoma (HCC): its mRNA was expressed in normal tissue samples, but downregulated in many carcinoma nodules and in five HCC-derived cell lines (Moh et al. 2003; Chung Moh et al. 2005). HepaCAM turned out to be a 46 kDa transmembrane protein of 416 amino acids, comprising two immunoglobulin (Ig)-like domains (one V-set and another one of the C2-type) in the extracellular side and an intracellular low-complexity proline-rich tail that can be phosphorylated. The extracellular domain can be modified through N-glycosylation (Gaudry et al. 2008). HepaCAM has been conserved throughout vertebrate evolution, especially in its extracellular part, but no evidence of its presence in invertebrates could be found (our unpublished data). It was found that HepaCAM molecules are able to form homodimers on the cell surface, independently of the presence of the intracellular domain (Moh et al. 2005). Interestingly, HepaCAM intracellular domain can be cleaved, affecting all its proposed functions (see below), suggesting a mechanism of regulation (Zhang et al. 2010).

HepaCAM distribution within the cells depended on whether the cells were isolated or making cell–cell contacts: in the first case the majority of the protein would be diffusely localized in the cytoplasm or concentrated at plasma membrane protrusions, while in the latter case the protein would concentrate at cell junctions, suggesting a role in cell adhesion (Moh et al. 2005). We will return to this point later, when this distribution was found to be physiologically relevant.

The role in cell adhesion was confirmed when transfection of HepaCAM in a HCC cell line increased its adherence to the substrate, motility, and invasion properties (Chung Moh et al. 2005). The cell adhesion role may depend on the interaction of HepaCAM with the actin cytoskeleton through its intracellular domain, which may also be important for its localization. In addition, unknown interactions involving individual extracellular Ig-like domains were also found to be important for HepaCAM localization, adhesion, and motility properties (Moh et al. 2009a, b). Furthermore, it has been suggested that a fraction of HepaCAM molecules resides in lipid rafts thanks to its interaction with caveolin-1 (Moh et al. 2009a, b).

HepaCAM was downregulated not only in liver tumors, but also in tumors of different origin, as well as in several cancer cell lines (Moh et al. 2008). Expression of HepaCAM on breast cancer MCF7 cells had an anti-proliferative effect dependent on the increased expression of p53 and the cyclin-dependent kinase inhibitor p21 and also on the downregulation of cMyc in renal carcinoma cells (Zhang et al. 2011). Altogether, these data suggest that HepaCAM may act not only as a CAM, but also as a tumor suppressor, although the mechanism that mediates this role is not yet defined.

3.3 HepaCAM Is Also GlialCAM

HepaCAM was rediscovered twice more in different contexts. A screening for cell surface molecules expressed in myelinating nerves in the peripheral nervous system identified many novel CAMs, one of which was HepaCAM (Spiegel et al. 2006). Two years later, a bioinformatic screen targeting Ig-like domains identified a molecule identical to HepaCAM. Surprisingly, when its expression in an array of adult human normal tissues was analyzed, the highest levels were observed in the central nervous system (Favre-Kontula et al. 2008a, b), much higher than in the liver where it was initially discovered.

When the expression of HepaCAM was analyzed in mouse tissues, it was absent from the liver (Favre-Kontula et al. 2008a, b). These data were corroborated by protein expression analysis using specific antibodies, raising doubts whether the downregulation in tumors and cancer cell lines of epithelial origin was an unspecific consequence of the experimental model. Nevertheless, forced overexpression of HepaCAM in the poorly differentiated U373-MG glioblastoma cell line was able to induce the re-expression of glial fibrillary acid protein (GFAP), a marker of differentiated astrocytes. HepaCAM overexpression caused additional morphological signs of differentiation, as well as growth arrest and increased adhesion but decreased motility, as opposed to the epithelial cell lines, in which HepaCAM expression increased motility (Lee et al. 2009). Thus, in the context of glial cells, it was suggested that HepaCAM may function as a glial differentiation factor.

The nervous system expression profile of HepaCAM was refined by western blot of different areas using a specific antibody, finding higher levels in cerebellum, entorhinal cortex, pons, medulla, and spinal cord (Favre-Kontula et al. 2008a, b). These authors also generated a lacZ knock-in mouse, allowing the accurate identification of which cell types were precisely expressing HepaCAM. Then, it was found that HepaCAM was expressed specifically in glial cells: ependymal cells, astrocytic end feet around blood vessels, and oligodendrocytes in myelin sheaths of the white matter. Also, HepaCAM expression was found in rat primary astrocytes and oligodendrocytes in culture and concentrated in cell–cell contacts or at the tip of processes in isolated cells, in a similar manner as it was observed in cancer cell lines as we mentioned earlier (Favre-Kontula et al. 2008a, b). Similar results were found by our group (Lopez-Hernandez et al. 2011a, b; Jeworutzki et al. 2012). Therefore, HepaCAM was appropriately renamed as GlialCAM, and we will use this name from now on.

3.4 GlialCAM Has a Role in Neurological Disease

3.4.1 GlialCAM Mutations Originate a White Matter Disorder

The interest on GlialCAM increased when it was found responsible for a variant of a rare neurological disease: megalencephalic leukoencephalopathy with subcortical cysts (MLC), a syndrome characterized by the presence of edema in the white matter of the nervous system (van der Knaap et al. 2012).

MLC is an inherited disease, with infantile onset evident as the affected children show macrocephaly during the first year of life, but no neurological signs at this point. After several years, the patients develop progressive cerebellar ataxia and spasticity, mild cognitive deterioration, and autism in some cases. There is a broad spectrum in the symptoms presented by the patients: from mild to severe. The most evident pathological indication of MLC comes from magnetic resonance imaging (MRI) that shows diffuse cerebral white matter abnormality with swelling, compressing the ventricles and subarachnoid spaces. Subcortical cysts are present in the anterior temporal region and often also in the frontal and parietal regions. Over the years, the white matter swelling decreases, but cysts become larger and more abundant (van der Knaap et al. 1995a, b). Electron microscopy of an affected brain biopsy showed that vacuoles were in fact covered with myelin membranes and that the vacuoles were present in the outer part of myelin sheaths (van der Knaap et al. 1996), as well as in astrocytes (Duarri et al. 2011).

Most MLC patients carry two mutated alleles of the gene *MLC1*, which was identified by linkage analysis (Leegwater et al. 2001). *MLC1* codes for an eight transmembrane domain protein with very weak amino acid sequence identity to ion channels. However, up to now, its function is unknown. Experiments of RNA interference in astrocytes and with primary lymphocytic cell lines derived from MLC patients suggest that MLC1 function may be related to the volume-regulated anion channel (VRAC). VRAC channel activity is very important when cells are exposed to hypoosmotic medium, as this chloride activity coupled to water efflux is necessary to restore volume to its original size (Ridder et al. 2011). MLC1 is expressed at the plasma membrane of astroglial processes around blood vessels in the cortex and white matter, in Bergmann glia in the cerebellum, and also in blood white cells, but not in myelin-forming oligodendrocytes (Teijido et al. 2004, 2007; Boor et al. 2007; Ambrosini et al. 2008). Disease-causing mutations of different types (missense, null, and frame shift) are spread over the entire MLC1 protein (Leegwater et al. 2002; Ilja-Boor et al. 2006; Montagna et al. 2006), but there is no correlation between genotype and phenotype. In fact, all mutations have the same effect: a reduction on the MLC1 levels at the plasma membrane due to reduced protein stability leading to its degradation (Teijido et al. 2004; Duarri et al. 2008).

Approximately 20–30 % of the MLC patients carry no mutation on the *MLC1* gene and do not show genetic linkage to the genomic region where *MLC1* is localized, indicating that other gene or genes must be affected in these patients. Two different phenotypes could be described in these patients: the classical MLC presentation and a benign, remitting phenotype characterized by early macrocephaly, but no motor or cognitive deterioration later in life (van der Knaap et al. 2010).

This is the point when GlialCAM comes into the spotlight once more. A biochemical screen in search of MLC1 interacting proteins as potential candidates for a second MLC disease gene revealed that GlialCAM was the most abundant protein co-purified with MLC1 antibodies. Subsequent analysis of MLC patients without *MLC1* mutations showed that they did harbor mutations in *GLIALCAM*, and these mutations were found spread on the entire extracellular domain (Lopez-Hernandez et al. 2011a). We should note that patients with *GLIALCAM* mutations showed no sign of liver malfunction, reinforcing the notion that the first finding of this molecule was serendipitous.

GlialCAM met all the requirements to be a MLC1 partner: it not only co-purified by immunoprecipitation, interacted directly (as revealed by FRET and split-TEV methods) and modified the traffic of MLC1, but also co-localized in astrocytic end feet in the human brain and in cell junctions in rat primary astrocyte cultures (Lopez-Hernandez et al. 2011a, b). Strikingly, *GLIALCAM* mutations do not affect its own expression, but do interfere with the formation of homo-complexes and with its subcellular localization. Furthermore, as MLC1 localization depends on GlialCAM, expression of the mutant variants of GlialCAM induced a more diffused localization of both proteins, still at the plasma membrane, but not concentrated at cell–cell contact areas (Fig. 3.1). However, mutations in *MLC1* do not affect GlialCAM expression or targeting (Lopez-Hernandez et al. 2011b), as it was confirmed in siRNA MLC1 depleted astrocytes (Duarri et al. 2011).

In addition, recent results (Capdevila-Nortes et al. 2013) using RNA interference of GlialCAM in rat primary astrocytes have shown that the endoplasmatic reticulum exit of MLC1 depends on GlialCAM. Therefore, lack of GlialCAM leads to a reduced expression of MLC1 (Fig. 3.2). In summary, GlialCAM works as an obligate beta subunit of MLC1, stabilizing and escorting it to the plasma membrane in cell–cell junctions. As we believe that the localization of MLC1 in tissue at astrocytic end feet between astrocytes is crucial for its physiological role, defects in the localization of MLC1 caused by *GLIALCAM* mutations may explain why mutations in both genes cause the same disease.

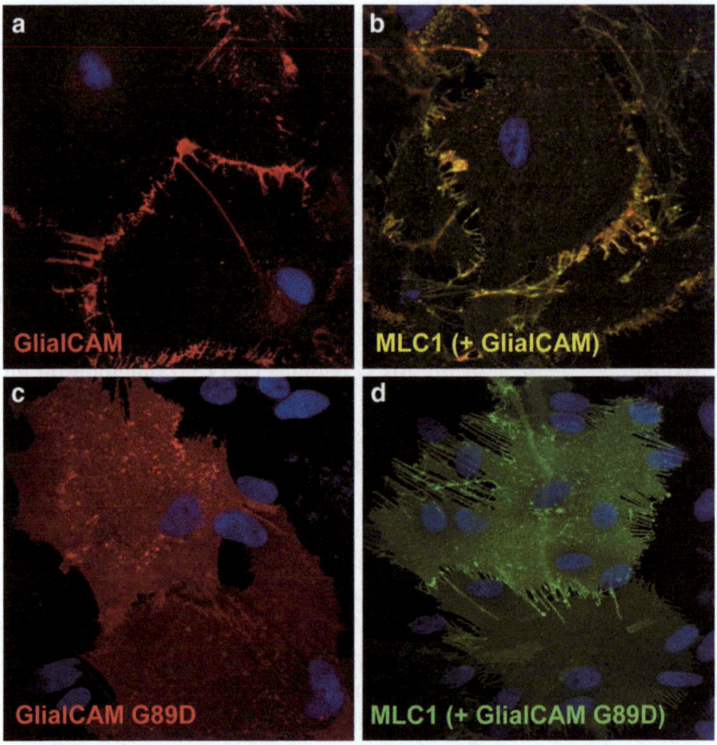

Fig. 3.1 MLC-related mutations in *GLIALCAM* abolish the targeting of GlialCAM and MLC1 to astrocyte cell junctions. Localization in astrocyte junctions of GlialCAM (**a**) and MLC1 (co-expressed with GlialCAM) (**b**) after adenoviral-mediated transduction. In contrast, the presence of an MLC-causing mutation (G89D) caused mislocalization of GlialCAM (**c**) and MLC1 (**d**)

3.4.2 A Dominant Role for GlialCAM in Disease

Interestingly, while some MLC patients carried two mutated *GLIALCAM* alleles, others showed only one mutated allele, suggesting dominant inheritance (Lopez-Hernandez et al. 2011a). This was correlated with the MLC benign, remitting phenotype, only present during early infancy. Dominant mutations were clustered on the V-set Ig-like extracellular domain of the protein, although it is not known why some mutations behave as dominant and others as recessive.

The dominant nature of these *GLIALCAM* mutations was confirmed with in vitro experiments in cell culture: when wild-type GlialCAM was co-transfected with the dominant mutated variant, normal MLC1 localization could not be rescued as it was the case of the classical recessive mutant GlialCAM variants (Lopez-Hernandez et al. 2011a).

Why dominant mutations in GlialCAM lead to a phenotype that remit during development? One possibility is that GlialCAM may have additional roles only necessary at early stages of development. In this direction, unpublished data from our

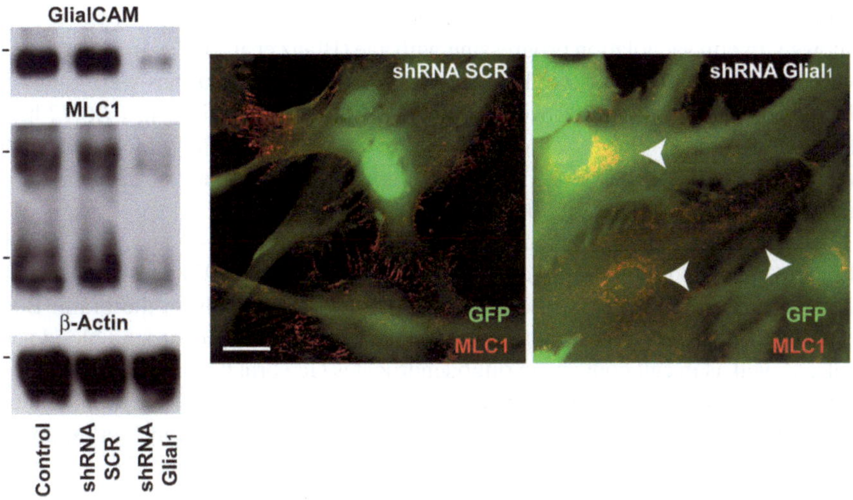

Fig. 3.2 Adenovector-mediated knockdown of GlialCAM caused MLC1 retention and degradation. Primary astrocytes were transduced with adenovirus expressing a shRNA control (SCR) or a shRNA that reduces GlialCAM expression, as detected by western blot. Reduction of GlialCAM expression leads to the reduction of MLC1 expression, as assessed by western blot. Actin was used as a loading control. *Small lanes* indicate the weight if the molecular markers are used. In an analogous manner, MLC1 localization after GlialCAM knockdown was detected in an intracellular localization, probably the endoplasmatic reticulum

laboratory show that GlialCAM is expressed during embryonic development, in both mice and zebrafish embryos. At these stages, GlialCAM is specifically expressed in radial glia, raising the possibility that it may help neuronal migration in a more conventional function according to its adhesion molecule identity. This function may not be needed in adult stages. Alternatively, it may be possible that the GlialCAM distribution defect caused in vivo by dominant mutations is not as strong as the one caused by two recessive mutations. Therefore, the function of GlialCAM, MLC1, and other associated proteins (see below) may be required at higher levels at early stages of development, and this fact may explain the reversibility of the phenotype as development progresses on.

3.4.3 GlialCAM and Ion Homeostasis

Nevertheless, without a physiological function for MLC1, the pathogenesis of the disease remained a mystery. Yet another biochemical screen in search for GlialCAM interacting proteins revealed an additional clue. Using specific GlialCAM antibodies, the most abundant proteins purified were in this order: GlialCAM itself, MLC1 as expected, and the chloride channel ClC-2 (Jeworutzki et al. 2012). It has been proposed that MLC pathogenesis is related to a dysfunction in the regulation of the water content of the brain, as the myelin and astrocyte vacuolation suggests, probably due to an impaired ion transport across cellular membranes and osmotic imbalance. ClC-2 has been involved in these processes, and *Clcn2* knockout mice showed myelin vacuolation progressing with age (Blanz et al. 2007). In fact, *CLCN2* was tested as a potential candidate gene for MLC, but no mutations were found in MLC patients (Scheper et al. 2010). However, recently mutations in *CLCN2* have been found in a novel form of leukodistrophy chacterized by the presence of white matter edema (Depienne et al. 2013), highlighting the relevance of this protein in ion and water homeostasis in the brain.

GlialCAM co-precipitates and interacts directly with ClC-2, and, as ClC-2 is broadly expressed in almost every tissue (Thiemann et al. 1992) and GlialCAM is only expressed in the brain, they are expected to interact only in the brain. Both proteins co-localized in Bergmann glia, at the astrocytic end feet around blood vessels, as well as in cell contacts in oligodendrocytes (Jeworutzki et al. 2012). Indeed, transfection of *GLIALCAM* was necessary to target ClC-2 to cell junctions; otherwise ClC-2 was observed along the entire plasma membrane. Moreover, although ClC-2 is functional on its own, interaction with GlialCAM modified ClC-2 activity. Co-expression of GlialCAM and ClC-2 in *Xenopus* oocytes, HEK epithelial cells, or cultured primary astrocytes greatly increased ClC-2-mediated currents and modified their voltage rectification and kinetics (Jeworutzki et al. 2012). It was suggested that the effect of GlialCAM on ClC-2 activity may result from the opening of the common gate that acts on the two pores of the homodimeric ClC-2 channel. Therefore, GlialCAM acts as an auxiliary subunit of ClC-2 in the brain, regulating its targeting and modulating its activity (Fig. 3.3).

The effect of GlialCAM on ClC-2 currents depends on its extracellular domain, as a mutant lacking the cytoplasmic tail has the same effect as the complete protein (Jeworutzki et al. 2012). However, although MLC-causing *GLIALCAM* mutants failed to target ClC-2 to cell junctions, they did modify ClC-2 activity as the wild-type protein, raising the possibility that the most important function of GlialCAM is to drive ClC-2 channels to cell–cell contacts, where their activity would be more relevant for glial and myelin physiology.

Since GlialCAM interacted with and drove to cell junctions both MLC1 and ClC-2, it could be possible that the three proteins formed a tripartite complex in the same cell. However, MLC1 could not be co-purified with ClC-2, making that possibility less plausible. However, it still could be possible that they could form a complex in different cells.

The unexpected role of GlialCAM as an auxiliary subunit of an ion channel is not uncommon for CAMs. Voltage-gated sodium channels interact with several beta subunits that have a single extracellular Ig-like domain. These beta subunits modulate sodium currents and enhance cell surface expression of the channel, but they are also needed for cellular migration, neurite outgrowth, and axonal fasciculation. These functions depend on extracellular interactions with other CAMs and also with intracellular effectors that initiate transduction mechanisms (Brackenbury and Isom 2010). Another example of this phenomenon is the neuronal adhesion protein AMIGO (amphoterin-induced gene and ORF) that contains leucine-rich motifs and an Ig-like domain and is able to interact in a homo- or heterotypical fashion (Chen et al. 2006). AMIGO acts as an auxiliary subunit of the potassium channel Kv2.1, increasing its

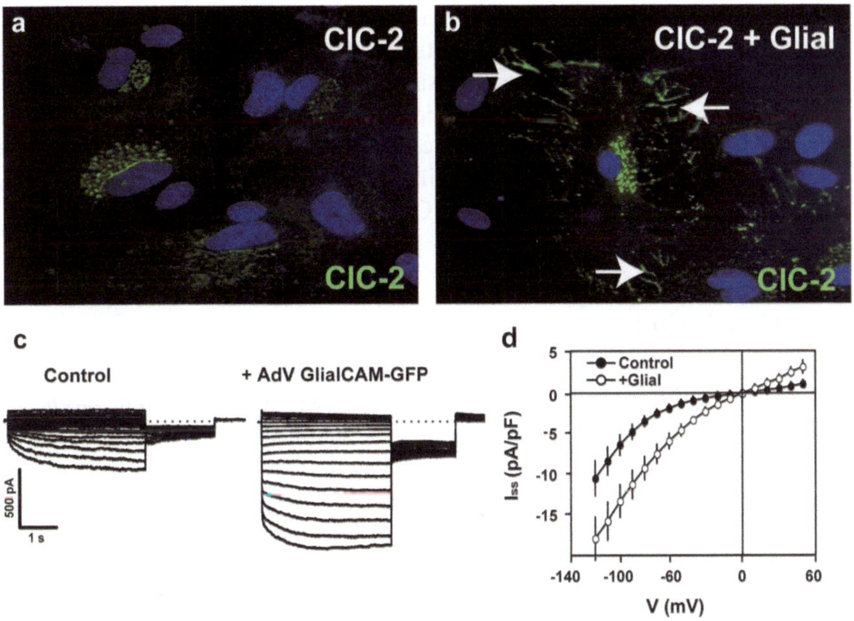

Fig. 3.3 GlialCAM modifies the targeting and the activity of the chloride channel ClC-2 in primary astrocytes. ClC-2 is detected in astrocyte junctions only after adenoviral-mediated transduction with GlialCAM (**b**), but not when expressed alone (**a**). (**c**) *Left*: representative trace of whole-cell inwardly rectifying chloride currents in dbcAMP-treated cultured rat astrocytes. *Right*: representative trace of chloride currents of dbcAMP-treated astrocytes transduced with adenoviruses expressing GlialCAM fused to GFP. The inset shows the voltage protocol used. (**d**) Average steady-state current–voltage relationship of dbcAMP-treated astrocytes (*circles*) or transduced with adenoviruses expressing GlialCAM-GFP (*filled circles*) in chloride medium. Recordings were performed in symmetrical chloride concentrations

conductance and helping the delayed rectifier current regulating neuronal excitability (Peltola et al. 2011). However, it is not known if AMIGO acts in a similar way to GlialCAM directing Kv2.1 cellular traffic. Other extracellular matrix proteins, like integrin receptors, interact physically and functionally with different types of ion channels, raising the possibility that ion channels have other functions than creating ion currents across the plasma membrane (Arcangeli and Becchetti 2006).

3.5 Perspectives

Many questions about GlialCAM biology remain to be answered. It is still not known if the adhesion properties shown in epithelial cells are relevant in glial cells. It is feasible that, besides being able to form homodimeric complexes in *cis* (in the same plasma membrane), GlialCAM could interact in *trans* (between adjacent cells) with itself or with other molecules, contributing to keep cell–cell junctions between astrocytes and/or oligodendrocytes. In addition, GlialCAM could act restraining MLC1 and ClC-2 movement in cell junctions, keeping them where their function in regulating ion fluxes to keep water osmosis is needed. Another possibility that needs to be explored is that GlialCAM is needed to keep neuronal activity, allowing the glial syncytium to restore normal ion concentrations to sustain action potential generation. Some of these questions will be properly addressed when a suitable animal model for the loss of function of GlialCAM is available.

3.6 Conclusion

In summary, GlialCAM is a protein that has shown us two different facets: the expected from a CAM, regulating adhesion and motility in epithelial cells, and the unexpected in glial cells, acting as a chaperone to direct the subcellular traffic of proteins MLC1 and ClC-2, concentrating them at cell–cell junctions, and modulating ClC-2-mediated chloride currents. Importantly, this latter role has been proven as physiologically relevant, as inactivating mutations in GlialCAM lead to a rare neurological disease, inducing the vacuolation of astrocytes and myelin sheaths.

Acknowledgments We thank all members of the lab and to our collaborators for their support. Studies in our lab are supported by SAF 2009-07014, PS09/02672-ERARE, ELA Foundation 2009-017C4 project, 2009 SGR 719, and an ICREA Academia prize.

Compliance with Ethics Requirements The authors declare that they have no conflicts of interest.

References

Allaman I, Belanger M, Magistretti PJ (2011) Astrocyte-neuron metabolic relationships: for better and for worse. Trends Neurosci 34:76–87

Ambrosini E, Serafini B, Lanciotti A, Tosini F, Scialpi F, Psaila R, Raggi C, Di Girolamo F, Petrucci TC, Aloisi F (2008) Biochemical characterization of MLC1 protein in astrocytes and its association with the dystrophin-glycoprotein complex. Mol Cell Neurosci 37:480–493

Arcangeli A, Becchetti A (2006) Complex functional interaction between integrin receptors and ion channels. Trends Cell Biol 16:631–639

Belanger M, Allaman I, Magistretti PJ (2011) Brain energy metabolism: focus on astrocyte-neuron metabolic cooperation. Cell Metab 14:724–738

Blanz J, Schweizer M, Auberson M, Maier H, Muenscher A, Hubner CA, Jentsch TJ (2007) Leukoencephalopathy upon disruption of the chloride channel ClC-2. J Neurosci 27:6581–6589

Boor I, Nagtegaal M, Kamphorst W, van der Valk P, Pronk JC, van Horssen J, Dinopoulos A, Bove KE, Pascual-Castroviejo I, Muntoni F et al (2007) MLC1 is associated with the dystrophin-glycoprotein complex at astrocytic endfeet. Acta Neuropathol 114:403–410

Brackenbury WJ, Isom LL (2010) Na channel beta subunits: overachievers of the Ion channel family. Front Pharmacol 2:53

Capdevila-Nortes X, Lopez-Hernandez T, Apaja PM, Lopez de Heredia M, Sirisi S, Callejo G, Arnedo T, Nunes V, Lukacs GL, Gasull X et al (2013). Insights into MLC pathogenesis: GlialCAM is an MLC1 chaperone required for proper activation of volume-regulated anion currents. Hum Mol Genet 22(21):4405–4416. doi:10.1093/hmg/ddt290

Chen Y, Aulia S, Li L, Tang BL (2006) AMIGO and friends: an emerging family of brain-enriched, neuronal growth modulating, type I transmembrane proteins with leucine-rich repeats (LRR) and cell adhesion molecule motifs. Brain Res Rev 51:265–274

Chung Moh M, Hoon Lee L, Shen S (2005) Cloning and characterization of hepaCAM, a novel Ig-like cell adhesion molecule suppressed in human hepatocellular carcinoma. J Hepatol 42:833–841

Depienne C, Bugiani M, Dupuits C, Galanaud D, Touitou V, Postma N, van Berkel C, Polder E, Tollard E, Darios F et al (2013) Brain white matter oedema due to ClC-2 chloride channel deficiency: an observational analytical study. Lancet Neurol 12:659–668

Duarri A, Teijido O, Lopez-Hernandez T, Scheper GC, Barriere H, Boor I, Aguado F, Zorzano A, Palacin M, Martinez A et al (2008) Molecular pathogenesis of megalencephalic leukoencephalopathy with subcortical cysts: mutations in MLC1 cause folding defects. Hum Mol Genet 17:3728–3739

Duarri A, Lopez de Heredia M, Capdevila-Nortes X, Ridder MC, Montolio M, Lopez-Hernandez T, Boor I, Lien CF, Hagemann T, Messing A et al (2011) Knockdown of MLC1 in primary astrocytes causes cell vacuolation: a MLC disease cell model. Neurobiol Dis 43:228–238

Favre-Kontula L, Rolland A, Bernasconi L, Karmirantzou M, Power C, Antonsson B, Boschert U (2008a) GlialCAM, an immunoglobulin-like cell adhesion molecule is expressed in glial cells of the central nervous system. Glia 56:633–645

Favre-Kontula L, Sattonnet-Roche P, Magnenat E, Proudfoot AE, Boschert U, Xenarios I, Vilbois F, Antonsson B (2008b) Detection and identification of plasma proteins that bind GlialCAM using ProteinChip arrays, SELDI-TOF MS, and nano-LC MS/MS. Proteomics 8:378–388

Gaudry JP, Arod C, Sauvage C, Busso S, Dupraz P, Pankiewicz R, Antonsson B (2008) Purification of the extracellular domain of the membrane protein GlialCAM expressed in HEK and CHO cells and comparison of the glycosylation. Protein Expr Purif 58:94–102

Ilja Boor PK, de Groot K, Mejaski-Bosnjak V, Brenner C, van der Knaap MS, Scheper GC, Pronk JC (2006) Megalencephalic leukoencephalopathy with subcortical cysts: an update and extended mutation analysis of MLC1. Hum Mutat 27:505–512

Jeworutzki E, Lopez-Hernandez T, Capdevila-Nortes X, Sirisi S, Bengtsson L, Montolio M, Zifarelli G, Arnedo T, Muller CS, Schulte U et al (2012) GlialCAM, a protein defective in a leukodystrophy, serves as a ClC-2 Cl(-) channel auxiliary subunit. Neuron 73:951–961

Lee LH, Moh MC, Zhang T, Shen S (2009) The immunoglobulin-like cell adhesion molecule hepaCAM induces differentiation of human glioblastoma U373-MG cells. J Cell Biochem 107:1129–1138

Leegwater PA, Yuan BQ, van der Steen J, Mulders J, Konst AA, Boor PK, Mejaski-Bosnjak V, van der Maarel SM, Frants RR, Oudejans CB et al (2001) Mutations of MLC1 (KIAA0027), encoding a putative membrane protein, cause megalencephalic leukoencephalopathy with subcortical cysts. Am J Hum Genet 68:831–838

Leegwater PA, Boor PK, Yuan BQ, van der Steen J, Visser A, Konst AA, Oudejans CB, Schutgens RB, Pronk JC, van der Knaap MS (2002) Identification of novel mutations in MLC1 responsible for megalencephalic leukoencephalopathy with subcortical cysts. Hum Genet 110:279–283

Lopez-Hernandez T, Ridder MC, Montolio M, Capdevila-Nortes X, Polder E, Sirisi S, Duarri A, Schulte U, Fakler B, Nunes V et al (2011a) Mutant GlialCAM causes megalencephalic leukoencephalopathy with subcortical cysts, benign familial macrocephaly, and macrocephaly with retardation and autism. Am J Hum Genet 88:422–432

Lopez-Hernandez T, Sirisi S, Capdevila-Nortes X, Montolio M, Fernandez-Duenas V, Scheper GC, van der Knaap MS, Casquero P, Ciruela F, Ferrer I et al (2011b) Molecular mechanisms of MLC1 and GLIALCAM mutations in megalencephalic leukoencephalopathy with subcortical cysts. Hum Mol Genet 20:3266–3277

Moh MC, Lee LH, Yang X, Shen S (2003) HEPN1, a novel gene that is frequently down-regulated in hepatocellular carcinoma, suppresses cell growth and induces apoptosis in HepG2 cells. J Hepatol 39:580–586

Moh MC, Zhang C, Luo C, Lee LH, Shen S (2005) Structural and functional analyses of a novel ig-like cell adhesion molecule, hepaCAM, in the human breast carcinoma MCF7 cells. J Biol Chem 280:27366–27374

Moh MC, Zhang T, Lee LH, Shen S (2008) Expression of hepaCAM is downregulated in cancers and induces senescence-like growth arrest via a p53/p21-dependent pathway in human breast cancer cells. Carcinogenesis 29:2298–2305

Moh MC, Lee LH, Zhang T, Shen S (2009a) Interaction of the immunoglobulin-like cell adhesion molecule hepaCAM with caveolin-1. Biochem Biophys Res Commun 378:755–760

Moh MC, Tian Q, Zhang T, Lee LH, Shen S (2009b) The immunoglobulin-like cell adhesion molecule hepaCAM modulates cell adhesion and motility through direct interaction with the actin cytoskeleton. J Cell Physiol 219:382–391

Molofsky AV, Krencik R, Ullian EM, Tsai HH, Deneen B, Richardson WD, Barres BA, Rowitch DH (2012) Astrocytes and disease: a neurodevelopmental perspective. Genes Dev 26:891–907

Montagna G, Teijido O, Eymard-Pierre E, Muraki K, Cohen B, Loizzo A, Grosso P, Tedeschi G, Palacin M, Boespflug-Tanguy O et al (2006) Vacuolating megalencephalic leukoencephalopathy with subcortical cysts: functional studies of novel variants in MLC1. Hum Mutat 27:292

Parpura V, Heneka MT, Montana V, Oliet SH, Schousboe A, Haydon PG, Stout RF Jr, Spray DC, Reichenbach A, Pannicke T et al (2012) Glial cells in (patho)physiology. J Neurochem 121:4–27

Peltola MA, Kuja-Panula J, Lauri SE, Taira T, Rauvala H (2011) AMIGO is an auxiliary subunit of the Kv2.1 potassium channel. EMBO Rep 12:1293–1299

Piaton G, Gould RM, Lubetzki C (2010) Axon-oligodendrocyte interactions during developmental myelination, demyelination and repair. J Neurochem 114:1243–1260

Rash JE (2010) Molecular disruptions of the panglial syncytium block potassium siphoning and axonal saltatory conduction: pertinence to neuromyelitis optica and other demyelinating diseases of the central nervous system. Neuroscience 168:982–1008

Ridder MC, Boor I, Lodder JC, Postma NL, Capdevila-Nortes X, Duarri A, Brussaard AB, Estevez R, Scheper GC, Mansvelder HD et al (2011) Megalencephalic leucoencephalopathy with cysts: defect in chloride currents and cell volume regulation. Brain 134:3342–3354

Scheper GC, van Berkel CG, Leisle L, de Groot KE, Errami A, Jentsch TJ, Van der Knaap MS (2010) Analysis of CLCN2 as candidate gene for megalencephalic leukoencephalopathy with subcortical cysts. Genet Test Mol Biomarkers 14:255–257

Spiegel I, Adamsky K, Eisenbach M, Eshed Y, Spiegel A, Mirsky R, Scherer SS, Peles E (2006) Identification of novel cell-adhesion molecules in peripheral nerves using a signal-sequence trap. Neuron Glia Biol 2:27–38

Teijido O, Martinez A, Pusch M, Zorzano A, Soriano E, Del Rio JA, Palacin M, Estevez R (2004) Localization and functional analyses of the MLC1 protein involved in megalencephalic leukoencephalopathy with subcortical cysts. Hum Mol Genet 13:2581–2594

Teijido O, Casaroli-Marano R, Kharkovets T, Aguado F, Zorzano A, Palacin M, Soriano E, Martinez A, Estevez R (2007) Expression patterns of MLC1 protein in the central and peripheral nervous systems. Neurobiol Dis 26:532–545

Thiemann A, Grunder S, Pusch M, Jentsch TJ (1992) A chloride channel widely expressed in epithelial and non-epithelial cells. Nature 356:57–60

van der Knaap MS, Barth PG, Stroink H, van Nieuwenhuizen O, Arts WF, Hoogenraad F, Valk J (1995a) Leukoencephalopathy with swelling and a discrepantly mild clinical course in eight children. Ann Neurol 37:324–334

van der Knaap MS, Valk J, Barth PG, Smit LM, van Engelen BG, Tortori Donati P (1995b) Leukoencephalopathy with swelling in children and adolescents: MRI patterns and differential diagnosis. Neuroradiology 37:679–686

van der Knaap MS, Barth PG, Vrensen GF, Valk J (1996) Histopathology of an infantile-onset spongiform leukoencephalopathy with a discrepantly mild clinical course. Acta Neuropathol 92:206–212

van der Knaap MS, Lai V, Kohler W, Salih MA, Fonseca MJ, Benke TA, Wilson C, Jayakar P, Aine MR, Dom L et al (2010) Megalencephalic leukoencephalopathy with cysts without MLC1 defect. Ann Neurol 67:834–837 van der Knaap MS, Boor I, Estevez R (2012) Megalencephalic leukoencephalopathy with subcortical cysts: chronic white matter edema due to a defect in brain ion and water homeostasis. Lancet Neurol 11(11):973–985

Zhang Y, Barres BA (2010) Astrocyte heterogeneity: an underappreciated topic in neurobiology. Curr Opin Neurobiol 20:588–594

Zhang T, Moh MC, Lee LH, Shen S (2010) The immunoglobulin-like cell adhesion molecule hepaCAM is cleaved in the human breast carcinoma MCF7 cells. Int J Oncol 37:155–165

Zhang QL, Luo CL, Wu XH, Wang CY, Xu X, Zhang YY, Liu Q, Shen SL (2011) HepaCAM induces G1 phase arrest and promotes c-Myc degradation in human renal cell carcinoma. J Cell Biochem 112:2910–2919

Chapter 4
The Neuroplastins: Multifunctional Neuronal Adhesion Molecules—Involvement in Behaviour and Disease

Philip Beesley, Michaela Kraus, and Nathalie Parolaro

Abstract The neuroplastins np65 and np55 are neuronal and synapse-enriched immunoglobulin (Ig) superfamily cell adhesion molecules that contain 3 and 2 Ig domains, respectively. Np65 is neuron specific whereas np55 is expressed in many tissues. They are multifunctional proteins whose physiological roles are defined by the partner proteins they bind to and the signalling pathways they activate. The neuroplastins are implicated in activity-dependent long-term synaptic plasticity. Thus neuroplastin-specific antibodies and a recombinant peptide inhibit long-term potentiation in hippocampal neurones. This is mediated by activation of the p38MAP kinase signalling pathway, resulting in the downregulation of the surface expression of GluR1 receptors. Np65, but not np55, exhibits *trans*-homophilic binding. Both np65 and np55 induce neurite outgrowth and both activate the FGF receptor and associated downstream signalling pathways. Np65 binds to and co-localises with $GABA_A$ receptor subtypes and may play a role in anchoring them to specific synaptic and extrasynaptic sites. Most recently the neuroplastins have been shown to chaperone and support the monocarboxylate transporter MCT2 in transporting lactate across the neuronal plasma membrane.

Thus the neuroplastins are multifunctional adhesion molecules which support neurite outgrowth, modulate long-term activity-dependent synaptic plasticity, regulate surface expression of GluR1 receptors, modulate $GABA_A$ receptor localisation, and play a key role in delivery of monocarboxylate energy substrates both to the

P. Beesley, B.Sc., Ph.D. (✉)
School of Biological Sciences, Royal Holloway University of London, Egham Hill, Egham, Surrey TW20 0EX, UK

Life Technologies, 3 Fountain Drive, Inchinnan Business Park, Paisley PA4 9RF, UK
e-mail: p.beesley@rhul.ac.uk

M. Kraus, B.Sc., Ph.D. • N. Parolaro, B.Sc., Ph.D.
School of Biological Sciences, Royal Holloway University of London, Egham Hill, Egham, Surrey TW20 0EX, UK

V. Berezin and P.S. Walmod (eds.), *Cell Adhesion Molecules: Implications in Neurological Diseases*, Advances in Neurobiology 8, DOI 10.1007/978-1-4614-8090-7_4,

synapse and to extrasynaptic sites. The diverse functions and range of signalling pathways activated by the neuroplastins suggest that they are important in modulating behaviour and in relation to human disease.

4.1 Introduction

The neuroplastins were originally identified as glycoprotein components of rat brain synaptic membrane (SM) and postsynaptic density (PSD) fractions (Hill et al. 1988). The two glycoproteins were detected by a single monoclonal antibody raised against SM glycoproteins and were originally termed gp65 and gp55 on the basis of their apparent molecular weights. These early studies showed several striking differences in the localisation of the two glycoproteins. Gp65 is brain specific and is enriched in PSD preparations. In contrast gp55 is not detectable in PSDs and was originally only detected in brain. However, later studies show that it is expressed in a wide range of tissues including skeletal muscle, heart, thymus, spleen, kidney, and liver (Langnaese et al. 1997). Furthermore, whereas gp55 is present in all brain regions gp65 is concentrated in subsets of forebrain neurons, is present only in low levels in cerebellum, is barely detectable in brain stem regions (pons and medulla), and is not present in spinal cord or peripheral nerve. As a result of studies implicating gp65 in particular in long-term activity-dependent synaptic plasticity the molecules were termed neuroplastin (np) 65 and 55 (Smalla et al. 2000). The neuroplastins were also identified using a signal sequence trap method to identify novel and secreted membrane proteins from a bone marrow stromal cell line (Shirozu et al. 1996) and termed stromal derived receptor 1, SDR1.

Sequencing studies (Langnaese et al. 1997) confirmed the close structural relationship of the neuroplastins and identified them as novel members of the Ig superfamily.

4.2 Expression and Structure of the Neuroplastins

4.2.1 Primary and Secondary Structure and X-Ray Crystallography

Np65 and np55 are members of the Ig superfamily comprising 3 and 2 Ig domains, respectively (Fig. 4.1), followed by a single 24-amino acid (aa) (leu308–tyr331) transmembrane-spanning sequence and a short (34 aa) intracellular domain (Langnaese et al. 1997). The intracellular domain is hydrophobic, but does contain several charged aas. A 28-aa signal peptide which is cleaved in the mature proteins is present at the N-terminus. The neuroplastins arise from a single gene by alternative splicing. In humans this is located on chromosome 15q22 and comprises nine exons (NCBI; Gene ID 27020).

The sequence data show that the peptide structure of np65 and np55 is identical except with respect to the additional 116 aa np65-specific N-terminal Ig1 domain. Np65 and np55 comprise 348 and 233 aas, respectively, and contain six sites for

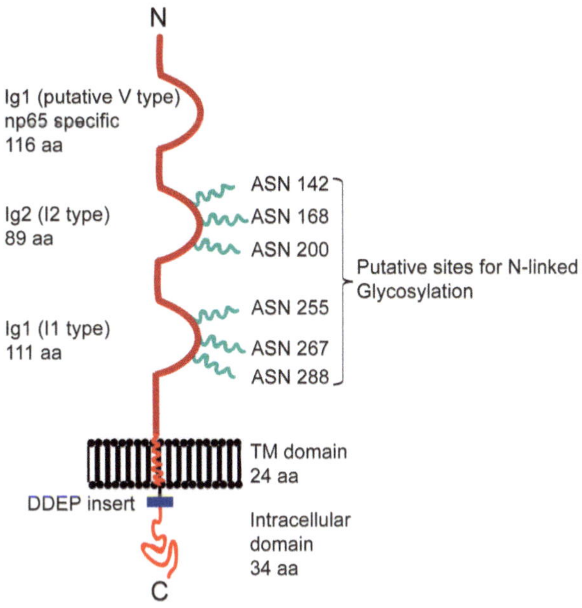

Fig. 4.1 Schematic diagram of the neuroplastins. The 3 Ig domains (Ig1–3), the transmembrane (TM), and intracellular domains are indicated, together with the six putative sites for N-glycosylation and the 4-aa (DDEP) alternative splice insert in the intracellular domain. The Ig1 domain is np65 specific. The Ig2 and Ig3 domains are common to both np65 and np55

N-linked glycosylation all located on the extracellular sequence common to both neuroplastins.

Splice variants (np +DDEP and np –DDEP) of np65 and np55 containing a 4-aa acidic 227-Asp-Asp-Glu-Pro-230 insert have been identified (Langnaese et al. 1997; Kreutz et al. 2001). The np65 and np55 +DDEP splice variants are expressed in brain and in retina, though they are expressed at much lower levels than the corresponding –DDEP splice variants. In general the pattern of expression of the +DDEP variants parallels that of the –DDEP isoforms (Kreutz et al. 2001). Since both np65 and np55 +DDEP splice variants are detected it is unlikely that they are synapse specific as neither neuroplastin shows a specific synaptic localisation in either brain or retina. However, it is plausible that the 4-aa insert is involved in mediating an intracellular binding interaction, though this remains to be established.

The X-ray crystal structure for Ig2 and Ig3 has been solved by analysis of a recombinant protein, ecto np55, comprising Ig domains 2 and 3 (aas1-193) of np55 (Owczarek et al. 2010). The Ig2 and Ig3 domains are in the extended conformation and are oriented at 45° to each other (Fig. 4.2a, b). The 89-aa Ig2 domain is of the intermediate 2 (I2) type comprising the classical two β sheets and eight β strands. The A, B, and E β strands form one sheet and the A′, C, C′, F, and G β strands form the other. The two β sheets are connected by a cysteine bridge between aas 25 and 73. Ig3 is linked to Ig2 by only 3 aas (91-Ala-Ala-Pro-93). The β strands are all antiparallel with the exception of the A′ strand which is parallel to the C-terminal sequence of the G strand.

The Ig3 domain is 12 aas longer than Ig2. It is an intermediate type 1 (I1) domain comprising nine β strands. Here the A, B, D, and E β strands form one β sheet while the A′, C, C′, F, and G β strands form the other. The two sheets are connected by the disulfide bond between Cys 114 and Cys 171. As for the Ig2 domain all strands run

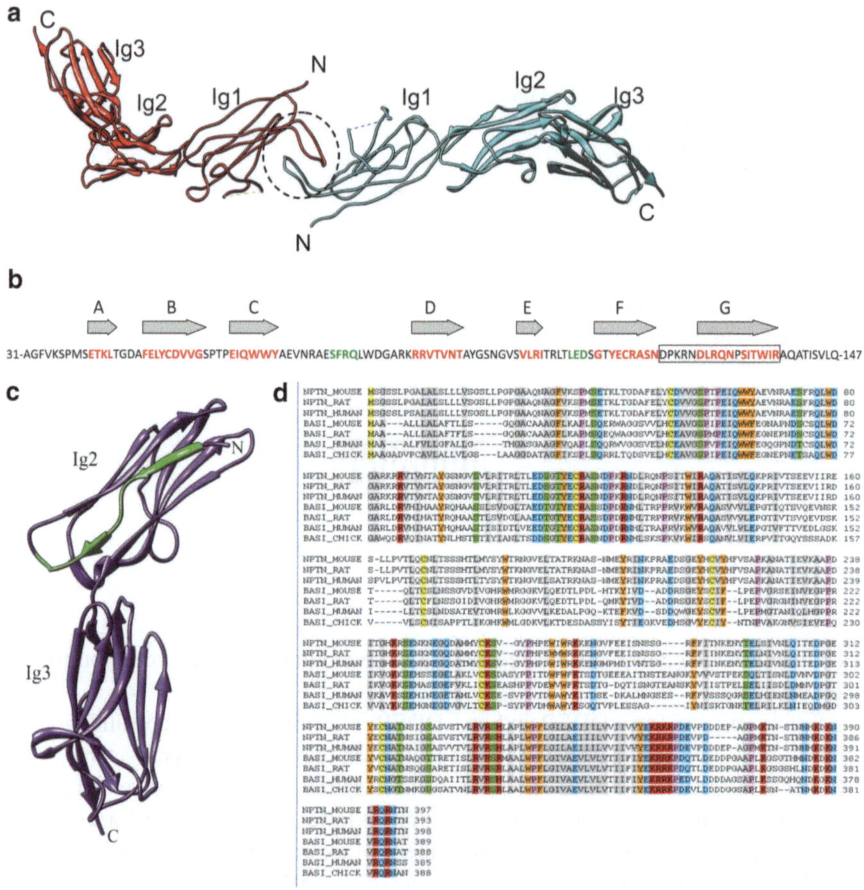

Fig. 4.2 The primary and secondary structure of the neuroplastins. (**a**) Backbone model of the secondary structure of np65 indicating the 3 Ig domains. Np65 exhibits *trans* homophilic binding and the *dotted circle* indicates the location of the binding site between np65 molecules in opposing cell surfaces. (**b**) Linear sequence for the Ig1 domain of np65. The *arrows* indicate the β strands with the amino acids predicted to form each strand shown in *red*. (**c**) Backbone model of the Ig2 and Ig3 domains of np55. The 2 Ig domains are oriented at 45° to each other. The putative FGF receptor binding site is shown in *green*. (**d**) Neuroplastin structure is highly conserved between species. This is shown by the sequence alignment of neuroplastin from human, mouse, and rat and that for basigin (the nearest Ig superfamily related protein) from human, mouse, rat, and chicken. Reprinted from Owczarek and Berezin (2012) with permission

antiparallel with the exception of the A′ strand which, as for the Ig2 domain, is in parallel with the C-terminal sequence of the G strand.

Subsequently Owczarek et al. (2011) have used X-ray crystallography to analyse the structure of ecto np65, a recombinant protein encoding all 3 Ig domains of np65. The Ig1 domain is predicted to be of the V type. Although part of the Ig2 domain and all of the Ig3 domain were disordered the data for the structure of the Ig1 domain were sufficiently interpretable to predict contact sites for np65 *trans-homophilic* binding (Fig. 4.2a). These data are discussed in more detail in Sect. 4.4.1.

4.2.2 *The Neuroplastins Are Members of the Basigin Group of the Ig Superfamily*

Comparison of sequence data (Fig. 4.2d) shows that the neuroplastins are most closely related to basigin/CD147 (also variously termed HT7 antigen, 5A11 antigen, or neurothelin in the chick, M6 antigen or EMPPRIN in the human, and MRC OX47 antigen in the rat) and are members of the basigin/CD147 group of the Ig superfamily (Langnaese et al. 1997). The basigin family comprises three members, basigin, embigin (gp70), and the neuroplastins (Huang et al. 1993; Muramatsu and Miyauchi 2003). Basigin, like neuroplastin, is expressed as 2 and 3 Ig domain isoforms termed basigin 1 and 2, respectively (Miyauchi et al. 1990; Kanekura et al. 1991; Ochrietor et al. 2003). Sequence homology between the neuroplastins and basigin ranges from 40 to 45 % although specific regions show much higher homology. The Ig2 domains share only 20 % homology whereas homology between the Ig3 domains is much higher, 40 %. The highest degree of homology is within and around the transmembrane and intracellular domains. Interestingly in this context Sarto-Jackson et al. (2012) have identified a neuroplastin intracellular amino acid sequence which can be recognized by proteins containing Src homology domain 3 binding domains though its functional significance remains to be established. Regions of highest homology are likely to reflect binding interactions common to both neuroplastins and basigin. Embigin shows much lower homology with basigin (28 % with the 2 Ig isoform). All three members of the family unusually have a conserved charged glutamate residue (aa 320 in neuroplastin) in the middle of the transmembrane domain (Muramatsu and Miyauchi 2003) and this may be important in mediating cis interactions with other membrane proteins, notably monocarboxylate transporters (see Sect. 4.4.3). As for neuroplastin the intracellular domains of both basigin and embigin are short (39 and 47 aas, respectively).

Species homology for neuroplastin is much higher than that for basigin. There is 94 % homology between the human and rodent neuroplastin genes, whereas for basigin this is only 58 %. However, the transmembrane region and intracellular domains of basigin are completely conserved between human, mouse, and chick, suggesting the functional importance of these regions of the molecule (Muramatsu and Miyauchi 2003).

4.2.3 *Oligosaccharide Structure of the Neuroplastins*

The neuroplastins contain six potential sites for N-glycosylation all located on the common Ig2 and 3 domains (Fig. 4.1). These are Asns 142, 168, and 200 in the Ig2 domain and Asns 255, 267, and 288 in the Ig3 domain (Langnaese et al. 1997). Most if not all six sites are glycosylated since the apparent molecular weights of np65 and 55 are reduced from 65 and 55 kDa to 40 and 28 kDa, respectively, following complete chemical or enzymatic deglycosylation (Willmott et al. 1992). Indeed X-ray crystallographic analysis of the recombinant protein encoding the Ig2 and Ig3 domains of np55 (ecto np55) shows clear electron density for *N*-acetylglucosamine at Asns142 and 255 with additional electron density indicating glycosylation at Asns 200 and 267 (Owczarek et al. 2010).

A full analysis of the oligosaccharide structure of the neuroplastins is not yet available. However, a number of np65 and np55 glycoforms have been identified. In most brain regions both neuroplastins occur as doublets of apparent molecular weights 63 and 67 kDa (np65) and 52 and 57 kDa (np55) (Willmott et al. 1992). A third novel np65 glycoform of apparent molecular weight 69 kDa is detected only in striatum. Both high mannose and complex/hybrid oligosaccharide structures are present on the brain glycoforms since they all bind concanavalin A. Neuraminidase treatment shows that np65 and np55 contain similar amounts of sialic acid. Importantly, using an antibody specific for Fucα(1–2) Gal saccharides (Smalla et al. 1998) show that np65 contains this glycan. It has been suggested that fucosylation of terminal galactose residues of brain glycoproteins in the α(1–2) position is crucial for neuronal plasticity, including phenomena such as long-term potentiation and long-term memory formation. Several lines of evidence support this suggestion (1) task-specific learning and long-term potentiation (LTP) induce fucosylation in hippocampal neurones; (2) injection of L-fucose or 2′-fucosyllactose enhances LTP in hippocampal slices; (3) injection of 2-deoxygalactose, an inhibitor of formation of Fucα(1–2) linkages, causes reversible amnesia and inhibits maintenance of LTP (Pohle et al. 1987; Matthies et al. 1996; Rose and Jork 1987; Krug et al. 1991).

There is considerable tissue variation in the number and molecular weight of np55 glycoforms, indicating that the pattern of glycosylation of np55 is dependent, at least in part, on cell type (Langnaese et al. 1998). Most tissues express a 44 kD glycoform. Liver and muscle also express a 61 kD glycoform with up to four separate glycoforms being expressed in some tissues, most notably heart. The functional significance of the various glycoforms is not clear at present.

4.3 Localisation and Developmental Expression of the Neuroplastins

4.3.1 Localisation in Rodent and Human Brain

Expression of np65, in contrast to np55, is specific to the brain with the exception that it is also expressed in the retina (Hill et al. 1988; Langnaese et al. 1997; Smalla et al. 2000; Kreutz et al. 2001). Strikingly and similar to the neuroplastins, the 3 Ig domain basigin 2 isoform exhibits a much more restricted expression than the 2 Ig basigin 1 isoform (Ochrietor et al. 2003) and is solely expressed by photoreceptors in the retina where it plays a key role in lactate transport (see Sect. 4.4.3).

Immunocytochemical, biochemical fractionation, and gene expression studies show that np65 levels exhibit an anterior–posterior axis of expression within the brain. Thus np65 is predominantly expressed by subsets of forebrain neurons in the cortex, hippocampus, and striatum. It is present in lower amounts in midbrain regions such as the thalamus and hypothalamus, is barely detectable in brain stem regions such as pons and medulla, and is not detectable in spinal cord or peripheral nerve. It is not detected in glial cells. In contrast, although there is variation in the expression of np55 between brain regions there is no anterior–posterior axis of expression and it is present in readily detectable levels in all brain regions. For example, in the adult mouse although comparable levels of np55 to np65 are detected in cortex and hippocampus, in cerebellum 95 % of total neuroplastin is np55 (Marzban et al. 2003). Immunocytochemical studies show that expression of np65 is laminar in both cerebral cortex and in hippocampus. Np65 immunoreactivity occurs as punctate deposits concentrated in neuropil regions, i.e., layers II, III, and Vb/VI in cerebral cortex and in the stratum radiatum and stratum oriens in hippocampus (Hill et al. 1988; Smalla et al. 2000). The barrel fields in layer 1V of the somatosensory cortex exhibit particularly strong immunoreactivity. In hippocampus punctate deposits of np65 immunoreactivity also surround the somata of the CA1 pyramidal neurones. Strikingly in rat brain the level of hippocampal np65 immunoreactivity is much lower in the CA3 compared to the CA1 region.

The punctate nature and localisation of np65 immunoreactivity suggested that it is located, at least in part, in postsynaptic structures. Indeed, the initial biochemical fractionation studies showed that np65, but not np55, is concentrated in forebrain PSDs. A postsynaptic localisation of a fraction of np65 has been confirmed by co-localisation of eGFP-tagged np65 with the postsynaptic density marker PSD95 on dendritic spines of transfected, cultured hippocampal neurons (Fig. 4.3). Although a portion of np65 is present in the postsynaptic region it is clear that punctate deposits of immunoreactivity are present in extrasynaptic regions of the neurone. The nature of these punctate deposits is unresolved, but they are not lipid rafts (Kraus and Beesley, unpublished data).

The localisation of np55 has been studied in detail in mouse cerebellum (Marzban et al. 2003) where immunocytochemical studies show that np55 immunoreactivity is primarily synaptic and is concentrated at parallel fibre synapses on Purkinje cells in the molecular layer and on synaptic glomeruli in the granule cell layer. This localisation suggests that np55, in addition to np65, plays important functional roles at the synapse. No glial staining is detected. Strikingly np55 is expressed in parasagittal (zebrin) stripes. These parasagittal bands have been described for numerous molecules (Armstrong and Hawkes 2000; Ozol et al. 1999) and most thoroughly investigated for Zebrin II/aldolase C (Ahn et al. 1994; Brochu et al. 1990). Np55 is enriched in the zebrin II negative stripes. Although the functional significance of this pattern of np55 expression is not clear it has been suggested that the differential expression of a suite of molecules associated with glutamatergic neurotransmission and energy production in parasagittal stripes in cerebellum reflects differences in synaptic efficacy at both parallel fibre and climbing fibre synapses and may be associated with LTD, long-term depression (e.g., Ito 2002). This in turn may be, at least in part, associated with a role of the neuroplastins as accessory proteins for monocarboxylate transporters (MCTs) as discussed in Sect. 4.4.3.

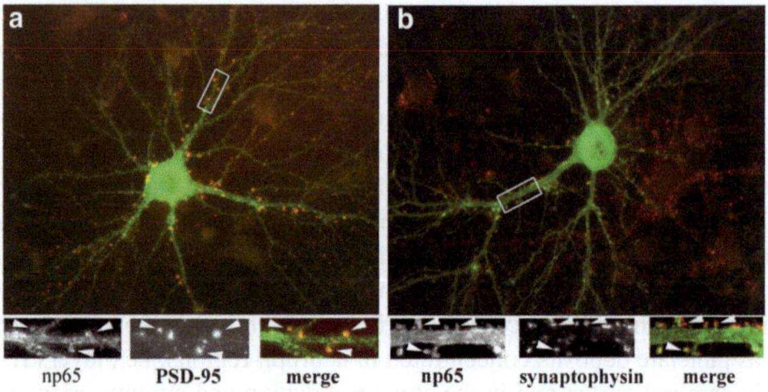

Fig. 4.3 Co-localisation of EGFPnp65 with PSD-95 and synaptophysin. Double immunofluorescent labelling of hippocampal neurones transfected with EGFP-tagged np65 lacking the DDEP insert and stained for EGFP (*green*) and either PSD-95 (*red*) (**a**) or synaptophysin (*red*) (**b**). A fraction of the EGFPnp65 co-localises with PSD-95 (*arrowheads*, **a**), but appears to be adjacent to presynaptic synaptophysin (*arrowheads*, **b**). Primary cultures of hippocampal neurons were prepared from 18-day-old embryonic rats and transfected using the calcium phosphate method. Scale bar 10 μm

While many features of neuroplastin expression are common to rodent and human brain some striking differences have been reported (Bernstein et al. 2007). In human brain np65 is expressed at comparable levels in human cerebral cortex and cerebellum. This is in contrast to mouse brain where expression of np65 in cerebellum is much lower than for cerebral cortex. Furthermore no parasagittal bands of neuroplastin immunoreactivity are detected in human cerebellum and the sharp demarcation in neuroplastin immunoreactivity observed between CA1 and CA3 regions of rat hippocampus is not observed in human brain. It is not clear at present whether another cell adhesion molecule substitutes for np65/np55 in human hippocampus or whether the lack of the CA1/CA3 boundary in neuroplastin expression reflects a functional difference between the two species. However, it is likely that these differences reflect species-specific differences in neuroplastin function and may, in part, be related to the role of neuroplastins as accessory proteins for the monocarboxylate transporter MCT2 (see Sect. 4.4.3).

4.3.2 Developmental Expression of Neuroplastins in Brain

Developmental studies show that np55 is expressed in brain earlier than np65. Np55 is expressed in low levels in embryonic brain, its level increases rapidly during the first 2 postnatal weeks, thereafter increasing gradually to reach the stable adult level between postnatal weeks 4 and 5 (Hill et al. 1989; Langnaese et al. 1997; Marzban et al. 2003). In contrast np65 is not detected in embryonic brain and its levels increase rapidly only during the second and third postnatal weeks, thereafter reaching the stable adult level. However, more recent studies have shown that np65 can be detected in hippocampal neurons prepared from embryonic day 18 rats after 24 h in culture (Owczarek et al. 2011).

In mouse cerebellum transient expression of punctate deposits of np55 mmuno reactivity is observed outlining the Purkinje cell somata at postnatal day (P)7, but this is not detected by P10 (Marzban et al. 2003). It is plausible that these correspond to the transient synapses made between the developing climbing fibres and the Purkinje cell somata which are eliminated at later developmental stages (Mason 1987). The parasagittal stripes of neuroplastin immunoreactivity are detected by P5. However, as cerebellar compartmentation and afferent topography are already established by P5 it is unlikely that the neuroplastins are involved in these processes although it is likely that they are important in later events leading to refinement of connectivity.

4.3.3 Localisation and Developmental Expression of the Neuroplastins in Retina

The localisation of the neuroplastins has been studied in detail in the rat retina (Kreutz et al. 2001). In situ hybridisation studies used four probes which distinguish between np65 without the DDEP insert in the intracellular domain (–DDEP), np65 +DDEP, np55 –DDEP, and np55 +DDEP. As for brain np55 was expressed earlier than np65. Np55 transcripts were already detected at P1 whereas np65 transcripts were not detected until P4. As for brain the level of expression of both neuroplastins increases through development to reach a stable adult level. The pattern of expression of the +DDEP and –DDEP isoforms is identical. The spatiotemporal pattern of np65 expression is consistent with role of np65 in synapse formation and is concomitant with synapse development in the outer plexiform layer (OPL). Strong np65 expression is detected in the outer nuclear layer (ONL) and the inner half of the inner nuclear layer (INL) at day 6. The ONL comprises the cell bodies of the photoreceptors, while the INL comprises the cell bodies of the retinal interneurones, i.e., horizontal cells, bipolar cells, and amacrine cells. The synapses between the retinal neurones are confined to the inner and outer plexiform layers, IPL and OPL, respectively. The developmental profile of np65 expression in the INL is consistent with the formation of ribbon synapses between photoreceptors and bipolar cells in the IPL which occurs between P 11–13 in albino rats. Immunocytochemical studies using antibodies specific for the np65-specific Ig1 domain and for the np65 and np55 common Ig2 and Ig3 domains show that immunoreactivity for both neuroplastins is confined to the IPL and OPL, the two synaptic layers of the retina. Significant co-localisation of np65 and synaptophysin immunoreactivity suggests a partial presynaptic localisation consistent with *trans*-homophilic binding of np65 at the synapse. However, it is clear that there is a perisynaptic localisation of some neuroplastin immunoreactivity and, surprisingly, evidence of some glial staining. Retina is the only tissue for which any glial localisation of neuroplastins has been reported. Western blot analysis of microdissected retinal sections showed abundant expression of both neuroplastins in sections comprising mainly the granule cell layer (GCL) and IPL, or INL and OPL or ONL. Strikingly np65 was found to be much more abundant than np55 in the OPL section again consistent with a role for np65 at ribbon synapses. Neuroplastins are also expressed in optic nerve tissue. Overall the results are consistent with a role for neuroplastins in retinal synapse development and synaptic function.

4.4 Binding Interactions and Functions of the Neuroplastins

4.4.1 Homophilic Binding Interactions

Ig superfamily adhesion molecules including the NCAMs, L1, and F11 typically exhibit both cis- and *trans*-homophilic and, in many cases, heterophilic binding interactions [for reviews see Brummendorf and Rathjen (1996), Hansen et al. (2008), Walsh and Doherty (1997)]. Binding experiments using covaspheres coated with the extracellular domains of np65 and np55 coupled to the human Ig Fc domain show that np65 but not np55 exhibits *trans*-homophilic binding (Smalla et al. 2000). Molecular modelling studies suggest that np65 extends some 13–16 nm into the synaptic cleft and thus homophilic binding between the Ig1 domains of opposing np65 molecules extending from the pre- and postsynaptic membranes can span the synaptic cleft which is some 20–30 nm (Gray 1987) and indeed evidence supports both a pre- and a postsynaptic location for a pool of np65.

The np65 *trans*-homophilic binding site has been identified (Owczarek et al. 2011). Analysis of the structure of the Ig1 domain of np65 for crystals of ecto-np65 shows that the F-G loop is oriented perpendicular to the surface and therefore is able to bind to the corresponding F-G loop of the Ig1 domain on an opposing np65 molecule (Fig. 4.2 a, b). The key interacting elements are proposed to be hydrogen bonds formed between the side chain amides of lysine, arginine, asparagine, and arginine and the backbone carboxyls of the opposing F-G loop. The modelling data were confirmed by surface plasmon resonance studies using ecto np65 and a synthetic peptide, enplastin, containing the putative np65 homophilic binding motif. Enplastin (121-DPKRNDLRQNPSITWIR-137) binds to ecto np65 with a Kd (1.20 μM) similar to that observed for ecto np65-ecto np65 binding (0. 52 μM). The np65 *trans*-homophilic binding interaction is central to the role of np65 in mediating neurite outgrowth and in LTP (Owczarek et al. 2011; Smalla et al. 2000).

Unlike np65, np55 does not exhibit *trans*-homophilic binding. In contrast to np55 the 2 Ig basigin 1 does exhibit *trans*-homophilic binding (Belton *et al.* 2008). However, as for basigin, it is likely that both np65 and np55 interact in cis to form homodimers. Chemical cross-linking studies (Fadool and Linser 1996) and co-expression of FLAG- and HA-tagged basigin show that basigin forms homodimers present at the plasma membrane (Yoshida et al. 2000). The interaction is mediated through the N-terminal Ig domain of the 2 Ig isoform. Similarly chemical cross-linking studies of SM preparations show that both np65 and np55 can exist as homodimers (Beesley and Langnaese, unpublished results) and are consistent with both neuroplastins being present in the membrane as homodimers. Recent studies (Sarto-Jackson et al. 2012) showing FRET between ECFP- and EYFP-tagged np65 when co-expressed in HEK cells support the formation of cis np65 homo-oligomers.

4.4.2 *The Neuroplastins Play Important Roles in Synaptic Plasticity and Activate P38 MAP Kinase*

As for a number of other cell adhesion molecules including L1, NCAM, and members of the cadherin family (Cremer et al. 1998: Luthi et al. 1994; Tang et al. 1998), the neuroplastins have been shown to play key roles in activity-dependent synaptic plasticity. Several neuroplastin antibodies including one that is specific for the np65 N-terminal Ig1 domain almost completely suppress the maintenance, but not the induction of LTP at CA1 synapses in hippocampal slices (Smalla et al. 2000). This effect is also observed with a recombinant protein encoding all three neuroplastin Ig domains fused to the human Ig Fc domain, np65Ig1–3Fc. The np65-induced inhibition of the maintenance phase of LTP has been shown to be mediated by activation of p38MAP kinase which in turn results in the loss of surface GluR1 containing glutamate AMPA receptors by internalisation (Empson et al. 2006). Treatment of organotypic hippocampal slice cultures with np65Ig1–3Fc resulted in an increase in phosphorylation of p38MAP kinase concomitant with the inhibition of LTP. P38MAP kinase is activated by dual phosphorylation on threonine 180 and tyrosine 182 residues in the Thr-Gly-Tyr motif in the activation loop of the kinase subdomain VIII (Ono and Han 2000; Mielke and Herdegen 2000; Paul et al. 1997). The np65Ig1–3Fc-induced block of LTP was reversed by the p38MAP kinase-specific inhibitor SB202190, showing that the inhibition of LTP by np65 is mediated by the kinase.

The availability of surface glutamate receptors is critical for the induction and maintenance of hippocampal LTP (Malinow and Malenka 2002) and multiple signal transduction pathways have been shown to regulate AMPA receptor incorporation into synapses during LTP including CAM kinase II (Lisman et al. 2002), MAP kinase (Zhu et al. 2002), protein kinase C (Boehm et al. 2006), and PI 3 kinase (Arendt et al. 2010). Strikingly the number of GluR1 receptors at the cell surface, but not the total GluR1 receptor number, was reduced following treatment of hippocampal slices with np65Ig1–3Fc. This response was blocked by SB202190 confirming the role of p38MAP kinase in the downregulation of surface GluR1 receptors. The mechanism by which np65 activates p38MAP kinase remains unclear. However, it is unlikely that np65 directly binds to the enzyme since there is no known kinase binding motif on the neuroplastin intracellular domain. Owczarek et al. (2011) report that application of ecto np65 to hippocampal neurons

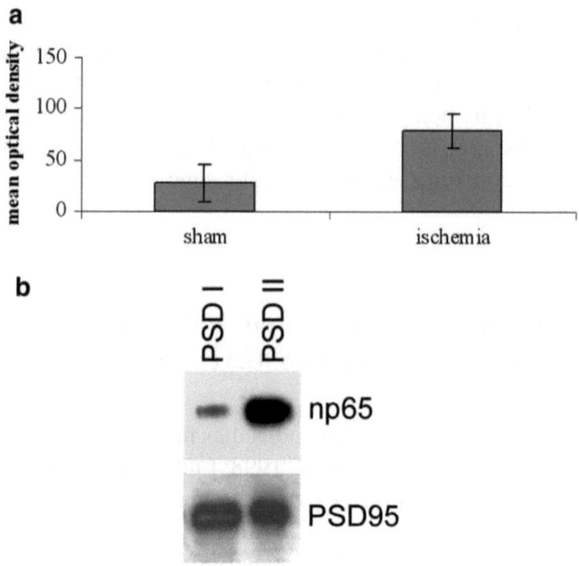

Fig. 4.4 (**a**) Ischemia results in increased association of np65 with the PSD. Forebrain PSDs were prepared from ischemic (15 min transient global ischemia induced by four-vessel occlusion) or sham-operated animals following 6 h recovery. PSDs were prepared by Triton X-100 extraction of SM. Western blotted samples were probed with a monoclonal antibody recognising both neuroplastin isoforms. The level of np65 in the PSD was analysed by densitometry and the results shown are the mean optical density ± SEM for three samples. The difference was statistically significant at the $P > 0.01$ level. (**b**) Np65 is a peripheral PSD protein. Samples of PSDs prepared by the double Triton X-100 method (Cho et al. 1992) or the phase partitioning method (Gurd et al. 1982), PSDs I and II, respectively, were separated by SDS PAGE and Western blotted. The blots were probed with antibodies specific for the neuroplastins or PSD95. The integral PSD protein PSD95 was present in high levels in both PSD preparations. However, only low amounts of np65 were detected in PSDs prepared using the Triton X-100 method compared to PSDs prepared using the phase partitioning method, suggesting that np65 is a peripheral PSD component. No np55 was detected in either PSD preparation

in culture induces an increase in intracellular calcium at hippocampal synapses. Thus it is plausible that this triggers a signalling cascade resulting in p38MAP kinase activation.

A number of paradigms which result in sustained increases in synaptic activity alter the level of np65 present in the PSD. Thus kainate seizures result in a significant increase in the amount of np65 present in PSDs prepared from seizured compared to kainate-treated non-seizured and control animals (Smalla et al. 2000). Similarly the level of np65 in a PSD-enriched fraction prepared from hippocampal slices showed a marked increase following induction of LTP. Transient global ischemia also results in a large increase in the level of PSD-associated np65 (Fig. 4.4a). A dynamic model for np65 translocation into the PSD is supported by recent data which suggest that np65 is a loosely bound rather than an integral PSD protein. PSDs can be prepared from synaptic membranes either using a double Triton X-100 procedure (Cho et al. 1992) or by phase partitioning with *n*-octyl glucopyranoside and polyethylene glycol (Gurd et al. 1982). The former method is more rigorous and solubilises a number of loosely bound PSD proteins. The level of np65 in PSDs prepared by the more rigorous Triton X-100 procedure is much lower than is observed using the milder phase partitioning protocol (Fig. 4.4b). In contrast the level of PSD95, an integral protein component of PSDs at glutamatergic synapses, is comparable in PSDs prepared using either method. Taken together these data suggest that np65 is a loosely bound PSD protein which can be rapidly translocated into the PSD and strongly support a dynamic model in which the level of np65 associated with the PSD changes in response to sustained changes in synaptic activity.

4.4.3 The Neuroplastins Are Accessory Proteins for the Proton-Linked Monocarboylate Transporter MCT2

Monocarboxylates such as lactate, pyruvate, and ketone bodies are important respiratory fuels for the developing and adult nervous system (Bergersen 2007). Specific proteins, the monocarboxylate transporters (MCTs), mediate the rapid uptake of these fuels across the cell membrane. The MCT family contains 14 members of which only MCT1 to MCT4, MCT8, and MCT10 have been functionally characterised. MCT1 to MCT4 mediate the proton-linked transport of monocarboxylates such as lactate, pyruvate, and ketone bodies across the plasma membrane [for review, see Halestrap and Wilson (2012)] whilst MCT8 is a thyroid hormone transporter (Friesema et al. 2003) and MCT10 (TAT1) is an aromatic amino acid transporter (Kim et al. 2002). MCT1, MCT2, and MCT4 are expressed in the brain. MCT1 is predominantly localised in glia and MCT4 in glia and cerebral vasculature while MCT2 is the major neuronal MCT in rodents, though not in humans (Debernardi et al. 2003; Halestrap and Wilson 2012; Karin et al. 2002; Koehler-Stec et al. 1998; Rafiki et al. 2003). Neurons have been shown to use lactate released by glial cells as an important respiratory fuel and MCT2 has been implicated as the major MCT isoform responsible for this uptake (Pierre and Pellerin 2005). Furthermore, immunocytochemical studies suggest that MCT2 is concentrated in the postsynaptic density (PSD) most probably by virtue of the C-terminal SXI motif that is suggested to interact with PDZ domains on PSD 93 (Bergersen et al. 2001). MCTs 1, 2, 3, and 4 all require ancillary proteins to be expressed in an active form at the plasma membrane and the interaction between the two proteins is essential for the transport function [for review, see Halestrap and Wilson (2012)]. Basigin is the preferred partner for MCT1 (Kirk et al. 2000; Wilson et al. 2002) while embigin is the preferred partner for MCT2 in some tissues and is important in modulating MCT2 activity (Ovens et al. 2010; Wilson et al. 2005). The critical importance of lactate transport for some tissues is well illustrated in homozygous basigin knockout mice. These mice are blind from the time of eye-opening with subsequent degeneration of the photoreceptors (Hori et al. 2000; Philp et al. 2003; Ochrietor and Linser 2004). This is most likely caused by the observed failure of MCT1 and MCT4 to integrate into the plasma membranes of the Muller glial cells and photoreceptors, although the level of MCT1 transcript is unaffected. As a result the export of lactate from the Muller cells, which depend primarily on glycolysis for energy production, and its subsequent import into the photoreceptors as a source for oxidative phosphorylation are disrupted resulting in photoreceptor degeneration.

Several observations provide conclusive evidence that the neuroplastins can act as the preferred partner to chaperone MCT2 to the cell surface and support its function in lactate transport across the cell membrane (Wilson et al. 2013). COS cells are a suitable model system as they do not express either neuroplastin or MCT2. When singly transfected with CFP-tagged MCT2 the COS cells show no staining of the plasma membrane, the expressed proteins remaining in the perinuclear compartment. Although COS cells singly transfected with EYFP-tagged neuroplastin do show staining at the plasma membrane a considerable amount remains within the cytoplasm. However, when co-transfected with either np55 or np65 the majority of MCT2 is co-expressed with neuroplastin at the plasma membrane. Thus it is clear that co-expression of MCT2 and np55 or np65 enables both proteins to be properly targeted to the cell surface consistent with their direct interaction. Fluorescence resonance energy transfer (FRET) confirms a direct binding interaction between the neuroplastins and MCT2 in doubly transfected COS cells.

These results are supported by the use of antisense technology to investigate the interaction of neuroplastins and MCT2 in *Xenopus laevis* oocytes. These oocytes do not express MCT2 (Broer et al. 1999), but do express basigin and neuroplastin. Immunofluorescence studies show, as expected, that MCT2 is concentrated at the plasma membrane of control oocytes injected with MCT2 cRNA. However, in oocytes treated with neuroplastin antisense RNA there is no significant MCT2 at the plasma membrane (Wilson et al. 2013).

The functional consequences of the reduced transport and localisation of MCT2 at the oocyte plasma membrane are shown by the fact that lactate transport is significantly increased in the oocytes injected with MCT2 cRNA. However, if the oocytes are also treated with neuroplastin antisense RNA lactate transport is reduced to the level observed in control uninjected oocytes. Thus neuroplastins can play a key role in delivering MCT2 to the cell surface and supporting lactate transport across the plasma membrane. However, a key question is do neuroplastins support MCT2 function in neurons? Immunocytochemical studies of cerebellum provide supporting evidence for this proposition. MCT2 and neuroplastin show parallel patterns of localisation on Purkinje cell somata and dendrites and most strikingly both are concentrated in the same parasagittal stripes in the cerebellar vermis. These data suggest that neuroplastins are the preferred accessory proteins for MCT2 in at least some neuronal populations (Wilson et al. 2013).

Previous electron microscope immunogold studies of MCT2 localisation in cerebellum showed a concentration of MCT2 immunoreactivity at the postsynaptic densities of parallel fibre-Purkinje cell synapses (Bergersen et al. 2001, 2002). Interestingly, the distribution of MCT2 immunoreactivity paralleled that of δ2-glutamate receptors. The intracellular domains of both MCT2 and of δ2-glutamate receptors have an SXI PDZ binding motif and it was suggested that both proteins could be anchored to the PSD by interaction with one of the PDZ domains of PSD93. Lactate is the main monocarboxylate found in the adult brain. Various reports show that lactate, mainly released by astrocytes, is used as an important energy substrate by neurons and sustains neuronal activity, including action potential propagation (Pierre and Pellerin 2005). MCT2 expression is upregulated in cultures of cortical neurons treated with noradrenaline (Pierre et al. 2003). Furthermore Suzuki et al. (2011) have demonstrated the importance of astrocyte-neuron lactate transport in long-term memory formation and in LTP. They also show that disrupting the expression of MCT1 and MCT4, both of which are expressed in astrocytes, leads to amnesia, but this can be rescued by lactate, but not glucose. While disruption of the neuronally expressed MCT2 also causes amnesia, significantly this cannot be rescued by lactate. Taken together, these data support a role for MCT2 to ensure adequate supply of lactate as a respiratory fuel to energise the neurone and the synapse after stimulation. This hypothesis is supported by the rapid translocation of np65 into the PSD under conditions of sustained synaptic activity as discussed in Sect. 4.4.2. At present details of all neuronal populations in which neuroplastin is the key accessory protein for MCT2 remain to be established. Furthermore there are no definitive studies of the detailed localization of embigin, the preferred partner for MCT2 in some tissues, in brain although it is expressed in neurons (Little 2011).

As mentioned in Sect. 4.3.1. it is tempting to speculate that the differences in neuroplastin expression and localisation between human and rodent brain may at least in part be related to the difference in MCT function, MCT2 being the major neuronal MCT in rodents, but not humans.

A significant role for np65 in ischemic insult is suggested by the threefold increase in np65 in forebrain PSDs following transient ischemia (Fig. 4.4a). It is plausible that this is at least in part related to its function as an accessory protein and chaperone for MCT2. The glucose and oxygen deprivation which results from the ischemia disrupts cerebral energy metabolism. Lactate is suggested to be an obligatory substrate for recovery following hypoxia (Schurr et al. 1997). Furthermore, blockade of lactate transport exacerbates delayed neuronal damage in rat brain following ischemia (Schurr et al. 2001). These results are consistent with an increased PSD level of MCT2 as a neuroprotective mechanism in the rodent brain.

The preferential localisation of neuroplastins and MCT2 to subsets of Purkinje cells in parasagittal stripes may be related to synaptic plasticity and LTD. LTD occurs at parallel fibre-Purkinje cell and mossy fibre-granule cell synapses. Although the function of the parasagittal stripes is not clearly understood at present the evidence suggests that parallel fibre synapses have different postsynaptic constituents in different Purkinje cell compartments. Certainly parallel fibres cross the boundaries of and form synapses with Purkinje cell dendrites in different parasagittal stripes. Indeed many molecules implicated in LTD are expressed differentially in parasagittal stripes in subsets of Purkinje cells (reviewed in Armstrong and Hawkes 2000). Furthermore, a suite of molecules associated with glutamatergic transmission and with LTD show differential expression between zebrin II positive and negative stripes. These include the metabotropic glutamate receptor (Mateos et al. 2001), excitatory amino acid transporter 4 (Dehnes et al. 1998; Wadiche and Jahr 2005), an inositol 1,4,5-trisphosphate (IP3) receptor (Furutama et al. 2010), phospholipase Cβ3/4 (Sarna et al. 2006), and protein kinase C (Barmack et al. 2000). Importantly molecules involved in energy generation notably zebrin II (aldolase C) and now MCT2 are present in parasagittal stripes. Taken together differential localisation of neuroplastins and MCT2 between subsets of Purkinje cells may at least in part be related to a differential requirement for lactate as an energy source, which in turn may relate to differences in LTD between the subsets of Purkinje cells.

4.4.4 Np65 and Np55 Support Neurite Outgrowth Via Homophilic Binding and Activation of the FGF Receptor Respectively

In common with many Ig superfamily cell adhesion molecules including L1 and NCAM [for reviews see Hansen et al. (2008), Walsh and Doherty (1997)] both neuroplastins have been shown to support neurite outgrowth (Owczarek et al. 2010, 2011; Owczarek and Berezin 2012). Whereas np65-induced neurite outgrowth is dependent on *trans*-homophilic binding, for np55 neurite outgrowth is mediated by activation of the FGF receptor. However, there are similarities in the signalling cascades that are activated. p38MAP kinase and ERK 1/2 together with calcium signalling are implicated in both np65- and np55-induced neurite outgrowth.

The expression of np55 prenatally with a rapid increase in level in postnatal weeks 1 and 2 in rat brain suggested that it may play a role in neurite outgrowth. Indeed, recombinant protein encoding the two np55 Ig domains, ecto np55, induces neurite outgrowth in cultures of dissociated hippocampal neurones. On the basis of the high homology between a 13 aa sequence in the Ig2 domain of np55 and a motif in the Ig1 domain of the FGFR1 structure Owczarek et al. (2010) predicted that np55 might bind to and signal through the FGF receptor tyrosine kinase as has been reported for other cell adhesion molecules.

The proposal by several groups that NCAMs, L1, and cadherins signal through the FGF receptor is supported by an overwhelming body of evidence [for reviews, see Hansen et al. (2008), Walsh and Doherty (1997)]. FGF receptors are composed of up to 3 Ig domains, D1–D3. The D2 and D3 domains are sufficient for FGF binding. FGF receptors contain a specific motif of acidic amino acids, the acid box, in the linker region between the D1 (Ig1) and D2 (Ig2) domains. Contiguous with the acid box is a D2 domain sequence of some 20 amino acids, the CAM homology domain, that shares sequence homology with the variable alternatively spliced exon (VASE) of NCAM, the sequence between Ig3 and Ig4 of L1, and an HAV containing sequence in N-cadherin, (Doherty and Walsh 1996). Antibodies raised against either the acid box or the CAM homology domain and peptide mimetics of the CAM homology domain inhibit NCAM- or cadherin-induced neurite outgrowth (Williams et al. 1994), leading to the suggestion that these CAMs interact directly in cis with this region of the FGF receptor. More recent data show that whereas the

CAM homology domain is not essential for CAM/FGF receptor interactions the acid box is an absolute requirement (Sanchez-Heras et al. 2006). The FGF receptor has been shown to interact with NCAM through the first and second NCAM fibronectin type 3 domain (Anderson et al. 2005; Kiselyov et al. 2003) in addition to the acid box and with N-cadherin through the fourth and fifth cadherin domains (Williams et al. 2001).

In the case of np55 a different binding motif to that for NCAM, L1, and N cadherin binding to the FGF receptor, namely, 121RIVTSEEVIIRDS134, corresponding to residues 121–134 (5–17 of ecto np55) has been identified (Fig. 4.2c). This sequence encompasses the A and A′ β strands together with a part of the A-B loop of the Ig2 domain of np55. This sequence shows 69 % identity and 77 % sequence similarity with residues 80–91 of the FGFR1 receptor Ig1 domain (RI-TGEEVEVRDS). Neither of the other two members of the basigin group has this homology, suggesting that the interaction with the FGFR is specific to the neuroplastins (Owczarek et al. 2010).

The hypothesis that np55 signals by binding to the FGFR was tested by surface plasmon resonance experiments. Ecto np55 (np55 Ig2 and Ig3) recombinant protein bound to sensor chips coated with recombinant protein comprising the D2 and D3 domains of the FGFR1 receptor (Owczarek et al. 2010) with a Kd of 11 μM. The D2 and D3 FGF receptor domains have been shown to be sufficient for the receptor dimerisation essential for activation of the tyrosine kinase domain (Mohammadi et al. 2005), while the D1 domain is involved in autoinhibition of ligand binding affinity (Kiselyov et al. 2006). Although the Kd value indicates relatively weak binding it is of the same order as thc Kds for NCAM (10 μm) and L1 (3.25 μM) binding to the FGF receptor (Kiselyov et al. 2003; Kulahin et al. 2008).

On the basis of these studies a synthetic peptide, narpin, comprising a tetramer of the np55 FGF receptor homology sequence, i.e., RIVTSEEVIIRDS, has been used as a tool to investigate np55-induced signalling. Treatment of Trex 93 cells expressing the FGFR1 IIIc splice variant with either ecto np55 or narpin induced FGFR autophosphorylation indicating receptor activation. Activation of the FGF receptor by np55 is at least in part responsible for mediating np55-induced neurite outgrowth since both the ecto np55- and narpin-induced neurite outgrowths in cultured hippocampal neurones are inhibited by the FGF receptor inhibitor SU5402. Furthermore, transfection of hippocampal neurones with a dominant negative kinase deleted FGFR1 blocked ecto np55- and narpin-induced neurite outgrowth. The binding site for np55 on the FGFR1 receptor overlaps with that for FGF2 since the latter inhibited np55-induced neurite outgrowth. This is in contrast to the NCAM and L1 binding site on the FGF receptor which shows no overlap with the FGF binding site.

Given the involvement of p38MAP kinase in np65-mediated inhibition of LTP and the report of p38MAP kinase activation by basigin (Lim et al. 1998) MAP kinase signalling pathways are prime candidates for mediating np55-induced neurite outgrowth. Indeed phosphorylation of p38MAP kinase and ERK 1/2 was observed in neurons treated with ecto np55 (Owczarek et al. 2010). FGF2, which can also induce neurite outgrowth, increased phosphorylation of ERK 1/2, but not p38MAP kinase.

The p38MAP kinase inhibitor SB202190 abolished the neurite outgrowth effect of np55 indicating its involvement in the mechanism of this response. Significantly ecto np55 induced a 30–40 % increase in synaptic calcium concentration which was significantly reduced by the FGFR1 receptor inhibitor SU5402 and by lavendustin A, a general protein tyrosine kinase antagonist, thus implicating calcium in the np55-induced signalling cascade(s) leading to increased neurite outgrowth.

Initially the relatively late expression of np65 suggested that it did not play a role in neurite outgrowth. Subsequently, however, its expression in hippocampal neurons from E19 rat embryos after 24 h in culture was confirmed (Owczarek et al. 2011). As for np55 recombinant protein encoding all three extracellular domains of np65 (ecto np65) induces neurite outgrowth in cultures of both hippocampal neurones and cerebellar granule cells. The mechanism involves np65 homophilic adhesion since it is blocked by enplastin, the peptide that specifically blocks the trans np65–np65 interaction. However, enplastin can also act as a partial agonist, supporting neurite outgrowth by hippocampal neurons grown on HEK cells, which do not express np65. Again as for np55, the np65 homophilic interaction appears to activate several signalling cascades. For these experiments hippocampal neurons were grown on a layer of HEK cells expressing np65 or on enplastin-coated culture dishes. Specific inhibitors of the FGFR1 receptor (SU5402), ERK 1/2 (PD98059), and p38MAP kinase (SB202190) all inhibited np65-induced neurite outgrowth. The involvement of ERK1/2 in this response is in contrast to the mechanism of np65-mediated inhibition of LTP since PD8059 did not have any effect on this response and no increase in phopho-ERK was detected (Empson et al. 2006). It is not clear at present whether np65 *trans*-homophilic binding and activation of the FGF receptor are synergistic or sequential events.

Ca^{2+} influx and subsequent activation of CaM kinase II are important in mediating np65-induced neurite outgrowth as has been shown for many other CAMS which exhibit homophilic binding (Hansen et al. 2008). Np65-mediated induction of neurite outgrowth was inhibited by KN-93, a compound which binds to the calcium binding site of CaM kinase II, thus inhibiting calmodulin binding to the enzyme. Application of either ecto np65 or enplastin to hippocampal neurons induced an increase in synaptic calcium. However, whereas the ecto-np65 induces a transient increase in calcium the enplastin-induced increase is sustained, suggesting differences in their mechanisms of activation of calcium influx.

4.4.5 Neuroplastin 65 Binds to $GABA_A$ Receptors

From a screen of potential neuroplastin binding partners Sarto-Jackson et al. (2012) have shown that the neuroplastins interact with $GABA_A$ receptor subtypes. $GABA_A$ receptors comprise five subunits most usually comprising one γ, two α, and two β subunits (Olsen and Sieghart 2008). Depending on subunit composition the various receptor subtypes exhibit different pharmacological properties and different localisation at synaptic and extrasynaptic sites. $GABA_A$ receptors containing $\alpha 1$, 2, or 3 subunits are localised mainly at synaptic sites and interact with the scaffolding protein gephyrin (Tretter et al. 2012; Kneussel and Loebrich 2007). Gephyrin anchors the receptor to the underlying postsynaptic complex and prevents lateral diffusion of the receptors. Receptors containing the α5 subunit are mainly extrasynaptic and link to the actin cytoskeleton via radixin (Loebrich et al. 2006). Co-purification experiments using affinity columns of antibody specific for the intracellular domain of neuroplastin or for the $GABA_A$ receptor β2 subunit demonstrate a specific interaction between neuroplastin and $GABA_A$ receptors. Co-immunoprecipitation experiments using antibodies directed against neuroplastin or the extracellular domain of $GABA_A$ receptor α-subunits confirm the interaction and that the np65-$GABA_A$ complex is present on the cell surface. Interestingly three times as much np65 as np55 co-immunoprecipates with $GABA_A$ receptors. This is consistent with the selective enrichment of np65 at postsynaptic structures (Hill et al. 1988; Smalla et al. 2000). A direct interaction between np65 and $GABA_A$ receptors is confirmed by significant FRET between ECFP-tagged np65, with either EYFP-tagged α1 subunit and wild-type β2 and γ2 subunits or EYFP-tagged β2 subunit and wild-type α1 and γ2 subunits. Immunocytochemical studies in hippocampal neuronal cultures and sections show that np 65 co-localises with α1 and α2, but not α3 subnunits at GABAergic synapses and α5 subunits at extrasynaptic sites in cultures. Strikingly the co-localisation is often observed at several synaptic sites along the same dendrite, but absent from others. The observation that only some 25 % in total of the neuroplastin clusters co-localise with $GABA_A$ receptors is consistent with co-localisation of a fraction of neuroplastin punctae with PSD95, a marker for glutamatergic postsynaptic structures (see Fig. 4.3). Furthermore, only a small proportion of gephyrin co-localised with np65 and synaptic clusters of np65 that co-localised with synaptic $GABA_A$ receptors but not with gephyrin were detected. A functional role for neuroplastin 65 in the subcellular localisation of specific $GABA_A$ receptor subtypes is suggested by the neuroplastin 65 shRNA-induced downregulation of α2 containing $GABA_A$ receptors at GABAergic synapses. In the shRNA-treated neurons α2-subunit staining was more diffuse and did not co-localise with vesicular inhibitory amino acid transporter, a presynaptic marker of GABAergic synapses. These data together with the involvement of neuroplastins in recycling of GluR1 receptors suggest multiple roles for these CAMs in modulating synaptic signalling through receptor localisation and recycling. Indeed the *trans*-homophilic binding of pre- and postsynaptic np65 localised in clusters may play important roles in anchoring key proteins (neurotransmitter receptors, monocarboxylate transporters, and enzymes involved in signalling cascades) in their correct synaptic and subcellular compartments.

4.5 Behavioural Effects of Neuroplastins and Linkage to Human Disease

Unlike several other CAMs including L1 (De Angelis et al. 2002) and basigin (Nabeshima et al. 2006), no direct role for neuroplastins in human disease and neurological disorders has yet been clearly defined. Basigin, the nearest Ig superfamily relative to the neuroplastins is directly linked to tumour metastasis through upregulation of several extracellular matrix metalloproteinases (Gabison, et al. 2005; Nabeshima et al. 2006; Yan et al. 2005). However, the diverse range of biological roles, variety of binding interactions, and the multiple signalling pathways activated by the neuroplastins all suggest that they are likely to play important roles both in regulating behaviour and in human disease.

4.5.1 Behavioural Effects of Np65 and Np55 Peptide Mimetics

Given the role of the neuroplastins in synaptic plasticity and in neurite outgrowth, the potential of peptide mimetics to modulate behavioural responses is of considerable interest. Np55-mediated activation of FGFR1 coupled with the observation that FGF2 decreases depressive behaviour in rats (Evans et al. 2004) led Owczarek et al. (2011) to investigate whether narpin has antidepressant activity using the forced swim test as a model paradigm. Treatment of rodents with classical antidepressants in this test results in decreased immobility and increases in active swimming and climbing behaviour (Porsolt et al. 1978). Narpin treatment resulted in antidepressant-like changes in behaviour, i.e., decreased floating behaviour and increased climbing behaviour. The possibility that these effects were due to enhanced locomotor activity was excluded by testing the effects of narpin in the open field test. Although the mechanism for the antidepressant effect of narpin is not clear an NCAM-derived synthetic peptide, FGL, and FGF2, both of which exert an antidepressant effect, have been shown to increase ERK phosphorylation (Aonurm-Helm et al. 2008; Garcia-Maya et al. 2006), thus leading to the suggestion that ligands for the FGF receptor may modulate behaviour via the ERK pathway (Owczarek et al. 2011). These results raise the question of whether narpin may be a useful therapeutic agent for treatment of mental disorders.

Np65 has been shown to play a key role in activity-dependent synaptic plasticity. This led to Owczarek et al. (2011) to test the effect of enplastin, the np65-specific peptide mimetic, on spatial learning behaviour using the Morris water maize paradigm. Enplastin treatment resulted in increased times for animals to find the hidden platform on day 1. However, no effects were observed on subsequent days, suggesting that the peptide has an inhibitory effect on the initial, but not later phases of the learning process.

4.5.2 Linkage of Neuroplastins to Human Disease

The neuroplastins have been linked to two very diverse disorders: schizophrenia and breast cancer.

Treatments of rodents with the psychotomimetics methamphetamine or phencyclidine are two commonly used animal models for schizophrenia. In humans chronic methamphetamine use induces a psychotic state closely resembling schizophrenia

(Sato et al. 1983) and phencyclidine usage mimics both the positive and negative symptoms of schizophrenia (Javitt and Zukin 1991). Gene expression profiling of either phencyclidine- or methamphetamine-treated rats revealed that neuroplastin and basigin are upregulated in both groups from a total of only 41 genes that were up- or downregulated by these treatments (Ouchi et al. 2005). Subsequent studies of human schizophrenic patients identified four single nucleotide polymorphisms (SNPs) in the 5′ upstream putative promoter and 5′ untranslated region (Saito et al. 2007) of the neuroplastin gene. Of these one (del-G-G) exhibited increased frequency in schizophrenia, while a second (T-G-T) exhibited decreased frequency and, from experiments with reporter constructs, significantly lowered neuroplastin transcription. Saito et al. (2007) propose that the T-G-T SNP, through its inhibitory role in neuroplastin transcription, could lower the onset risk for schizophrenia.

In a screen to select potential tumour antigens in B-cell actively proliferating regions of tumour-draining lymph nodes from human breast cancer patients, Rodriguez-Pinto et al. (2009) identified neuroplastin as a potential candidate. Subsequent analysis of human breast carcinoma tissue revealed the expression of both 55 and 45 kDa np55 glycoforms. Significant neuroplastin immunoreactivity was detected in some 20 % of breast tissue showing invasive carcinoma compared with only 2.5 % of control breast tissue. Interestingly a larger percentage of neuroplastin positive tumours was found in tissue showing distant (50 % neuroplastin positive) rather than lymph node (20 % neuroplastin positive) metastasis, leading to the suggestion that neuroplastin expression may promote tumour invasion. Strikingly neuroplastin overexpression in a breast cancer cell line transplanted into nude mice resulted in a significant increase in tumour growth and angiogenesis. The neuroplastin-induced increase in angiogenesis observed in vivo is mediated by an increase in production of vascular endothelial growth factor. Thus, if aberrantly expressed neuroplastin could promote breast tumour growth and metastasis. This raises the potential use of neuroplastin as a biomarker for breast tumour screening and for development of therapeutic approaches.

The pattern of neuroplastin immunoreactivity in cerebellum has been studied in relation to a mouse model of the Niemann-Pick type A lysosomal storage disease, the acid sphingomyelinase knockout (ASKMO) mouse. In this mouse a targeted disruption of the acid sphingomyelinase gene disrupts cholesterol metabolism and results in widespread axonal and dendritic abnormalities and Purkinje cell death (Otterbach and Stoffel 1995; Sarna et al. 2001). In homozygous wild-type mice neuroplastin immunoreactivity in Purkinje cells is confined to the dendrites whereas homozygous ASKMO mice frequently exhibit high levels of neuroplastin immunoreactivity in surviving Purkinje cell somata (Marzban et al. 2003). This altered cellular location is thought to be associated with a generalised defect in protein trafficking. Interestingly Purkinje cell neuroplastin expression and localisation are not dependent either upon the normal histotypic organisation of the cerebellum or formation of normal parallel fibre-Purkinje cell synaptic connections since neuroplastin is readily detected in the Purkinje cell dendrites of both the *disabled* (*dab*) and *cerebellar folia deficient* (*cdf*) mice (Marzban et al. 2003). In the *disabled* mouse there is a targeted disruption of the Reelin signalling pathway resulting in a profound Purkinje cell ectopia as these neurons do not migrate from their embryonic clusters, but remain as ectopic clusters among the deep cerebellar nuclei and receive little or no parallel fibre input (Howell et al. 1997; Gallagher et al. 1998). Similarly *cdf* mice also exhibit ectopic Purkinje cells which express abundant dendritic neuroplastin immunoreactivity (Marzban et al. 2003).

4.6 Conclusions and Future Perspectives

In common with many cell adhesion molecules the neuroplastins exhibit multiple homophilic and heterophilic binding interactions and activate several signalling pathways, and in vitro studies strongly suggest they play key roles in a wide variety of physiological functions (Fig. 4.5). A role in cell adhesion has been demonstrated only for np65, but not np55, and is mediated by *trans*-homophilic binding between opposing np65 Ig1 domains. This interaction in turn regulates surface localisation of GluR1 receptors so affecting synaptic plasticity. However, the in vivo physiological importance of np65 in long-term activity-dependent synaptic plasticity and in behavioural responses such as learning and memory is not clear at present.

Both np65 and np55 can induce neurite outgrowth in vitro. They can bind to and activate the FGF receptor together with a resultant regulation of signalling pathways including p38MAP kinase, ERK 1/2, and calcium/CaM kinase II. It seems likely that these are interrelated signalling networks rather than operating as individual discrete pathways. Np65 *trans*-homophilic binding also induces neurite outgrowth, but whether this is sequential or synergistic with FGF receptor activation is not clear. The physiological mechanism which initiates neuroplastin binding to the FGF

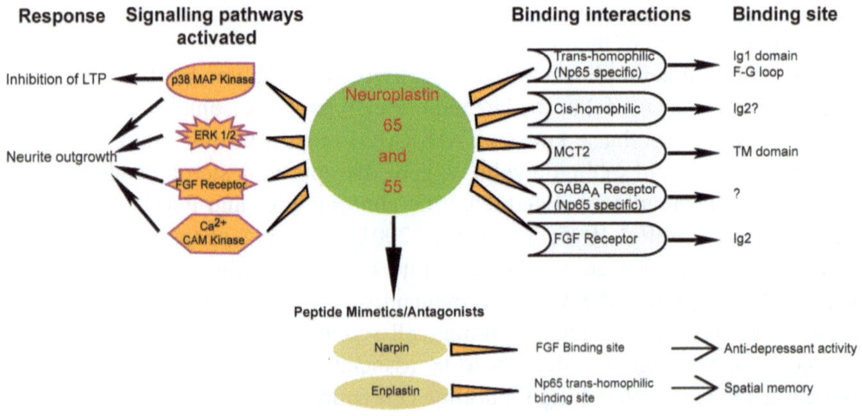

Fig. 4.5 Neuroplastin binding interactions, activation of signalling pathways, and physiological responses. The figure summarises homophilic and heterophilic neuroplastin binding interactions, the signalling pathways activated by *trans* homophilic (np65), and FGF binding together with the resultant physiological responses. The actions of neuroplastin peptide mimetics/antagonists are also summarised

receptor is also not known. Furthermore, given the large number of CAMs that can support neurite outgrowth the importance for this role of the neuroplastins in brain development and remodelling in vivo again remains to be established.

Both np65 and np55 are present at synapses formed by specific neuronal subpopulations. These are partially overlapping, but partially separate, e.g., np55 localisation at parallel fibre-Purkinje cell synapses. However, only np65 is found in PSD preparations and is rapidly translocated into the PSD in response to sustained increases in synaptic activity. The trafficking mechanism does not appear to be intrinsic to neuroplastin structure alone since eGFP-tagged np65 and np55 show identical localisation when expressed in hippocampal neurons in culture (Kraus and Beesley, unpublished observations).

The two neuroplastins have multiple but distinct roles at the synapse. For np65 these include regulation of GluR1 receptor internalisation in relation to synaptic plasticity and the recently reported direct binding to $GABA_A$ receptor subtypes with a putative role in $GABA_A$ receptor localisation. The functions of np55 at the synapse are less clear, but a role in LTD has been suggested.

Both np65 and np55 role can chaperone and act as the accessory protein for MCT2, thus supporting the import of lactate into neurons. This role is likely to be important at the synapse as well as along axons and dendrites. At present the neuronal populations where neuroplastin rather than embigin is the preferred partner have not been fully characterised, and the importance of the neuroplastins in MCT2 function and energy supply to neurons in vivo remains to be established. However, it seems likely that translocation of np65 into the PSD under conditions of sustained changes in synaptic activity is, in part, directly related to its role in supporting MCT2 function and energy supply to the synapse.

At present the functions of np55 in tissues other than brain have not been investigated, although it seems likely that in some tissues this may be related to MCT2 function.

Studies defining the physiological functions of the neuroplastins in vivo must be a prime focus for future research. Use of narpin and enplastin as tools to investigate the in vivo functions of the neuroplastins suggests roles in spatial learning and as targets for novel antidepressant therapies (Fig. 4.5), and a neuroprotective role for np65 is suggested by the upregulation of the protein in PSDs following recovery from transient global ischemia. Thus a key goal for future research is the development of further tools, crucially inducible, regional and np65/55-specific neuroplastin knockout mice, to gain further insights into the functions of the neuroplastins and to understand their roles in neurological disorders.

Acknowledgments The authors work cited here was supported by the Wellcome Trust and the BBSRC. We are grateful for critical discussion or reading of sections of the manuscript by E Gundelfinger, A Halestrap, R Hawkes, and K-H Smalla. We are grateful to S. Owczarek and V. Berezin for permission to reproduce Fig. 4.2 from Owczarek and Berezin (2012).

Compliance with Ethics Requirements The authors declare that they have no conflicts of interest.

References

Ahn AH, Dziennis S, Hawkes R, Herrup K (1994) The cloning of zebrin II reveals its identity with aldolase C. Development 120:2081–2090

Anderson AA, Kendal CE, Garcia-Maya M, Kenny AV, Morris-Triggs SA, Wu T, Reynolds R, Hohenester E, Saffell JL (2005) A peptide from the first fibronectin domain of NCAM acts as an inverse agonist and stimulates FGF receptor activation, neurite outgrowth and survival. J Neurochem 95:570–583

Aonurm-Helm A, Jurgenson M, Zharkovsky T, Sonn K, Berezin V, Bock E, Zharkovsky A (2008) Depression-like behaviour in neural cell adhesion molecule (NCAM)-deficient mice and its reversal by an NCAM-derived peptide, FGL. Eur J Neurosci 28:1618–1628

Arendt KL, Royo M, Fernandez-Moreal M, Knafo S, Petrok CN, Martens JR, Esteban JA (2010) PIP3 controls synaptic function by maintaining AMPA receptor clustering at the postsynaptic membrane. Nat Neurosci 13:36–44

Armstrong CL, Hawkes R (2000) Pattern formation in the cerebellar cortex. Biochem Cell Biol 78:551–562

Barmack NH, Qian Z, Yoshimura J (2000) Regional and cellular distribution of protein kinase C in rat cerebellar Purkinje cells. J Comp Neurol 427:235–254

Belton RJ Jr, Chen L, Mesquita FS, Nowak RA (2008) Basigin-2 is a cell surface receptor for soluble basigin ligands. J Biol Chem 283:17805–17814

Bergersen LH (2007) Is lactate food for neurons? Comparison of monocarboxylate transporter subtypes in brain and muscle. Neuroscience 145:11–19

Bergersen L, Waerhaug O, Helm J, Thomas M, Laake P, Davies AJ, Wilson MC, Halestrap AP, Ottersen OP (2001) A novel postsynaptic density protein: the monocarboxylate transporter MCT2 is co-localized with delta-glutamate receptors in postsynaptic densities of parallel fiber-Purkinje cell synapses. Exp Brain Res 136:523–534

Bergersen L, Rafiki A, Ottersen OP (2002) Immunogold cytochemistry identifies specialized membrane domains for monocarboxylate transport in the central nervous system. Neurochem Res 27:89–96

Bernstein HG, Small KH, Bogerts B, Gordon-Weeks PR, Beesley PW, Gundelfinger ED, Kreutz MR (2007) The immunolocalisation of the synaptic glycoprotein neuroplastin differs substantially between the human and rodent brain. Brain Res 1134:107–112

Boehm J, Kang MG, Johnson RC, Esteban J, Huganir R, Malinow R (2006) Synaptic incorporation of AMPA receptors during LTP is controlled by a PKC phosphorylation site on GluR1. Neuron 51:213–225

Brochu G, Maler L, Hawkes R (1990) Zebrin II: a polypeptide antigen expressed selectively by Purkinje cells reveals compartments in rat and fish cerebellum. J Comp Neurol 291:538–552

Broer S, Broer A, Schneider HP, Stegen C, Halestrap AP, Deitmer JW (1999) Characterization of the high-affinity monocarboxylate transporter MCT2 in Xenopus laevis oocytes. Biochem J 341:529–535

Brummendorf T, Rathjen F (1996) Structure/function relationships of axon-associated adhesion molecules of the immunoglobulin superfamily. Curr Opin Neurobiol 6:584–593

Cho K-O, Hunt CA, Kennedy MB (1992) The rat brain postsynaptic density fraction contains a homology of the Drosophila discs-large tumor suppressor protein. Neuron 9:929–942

Cremer H, Chazal G, Carleton A, Goridis C, Vincent JD, Lledo PM (1998) Long-term but not short-term plasticity at mossy fiber synapses is impaired in neural cell adhesion molecule-deficient mice. Proc Natl Acad Sci USA 95:13242–13247

De Angelis E, Watkins A, Schäfer M, Brümmendorf T, Kenwrick S (2002) Disease-associated mutations in L1 CAM interfere with ligand interactions and cell-surface expression. Hum Mol Genet 11:1–12

Debernardi R, Pierre K, Lengacher S, Magistretti PJ, Pellerin L (2003) Cell-specific expression pattern of monocarboxylate transporters in astrocytes and neurons observed in different mouse brain cortical cell cultures. J Neurosci Res 73:141–155

Dehnes Y, Chaudhry FA, Ullensvang K, Lehre KP, Storm-Mathisen J, Danbolt NC (1998) The glutamate transporter EAAT4 in rat cerebellar Purkinje cells: a glutamate-gated chloride channel concentrated near the synapse in parts of the dendritic membrane facing astroglia. J Neurosci 18:3606–3619

Doherty P, Walsh FS (1996) CAM-FGF receptor interactions: a model for axonal growth. Mol Cell Neurosci 8:99–111

Empson RM, Buckby LE, Kraus M, Bates KJ, Crompton MR, Gundelfinger ED, Beesley PW (2006) The cell adhesion molecule neuroplastin-65 inhibits hippocampal long-term potentiation via a mitogen-activated protein kinase p38-dependent reduction in surface expression of GluR1-containing glutamate receptors. J Neurochem 99:850–860

Evans SJ, Choudary PV, Neal CR, Li JZ, Vawter MP, Tomita H, Lopez JF, Thompson RC, Meng F, Stead JD, Walsh DM, Myers RM, Bunney WE, Watson SJ, Jones EG, Akil H (2004) Dysregulation of the fibroblast growth factor system in major depression. Proc Natl Acad Sci USA 101:15506–15511

Fadool JM, Linser P (1996) Evidence for the formation of multimeric forms of the 5A11/HT7 antigen. Biochem Biophys Res Commun 229:280–282

Friesema EC, Ganguly S, Abdalla A, Manning Fox JE, Halestrap AP, Visser TJ (2003) Identification of monocarboxylate transporter 8 as a specific thyroid hormone transporter. J Biol Chem 278:40128–40135

Furutama D, Morita N, Takano R, Sekine Y, Sadakata T, Shinoda Y, Hayashi K, Mishima Y, Mikoshiba K, Hawkes R, Furuichi T (2010) Expression of the IP3R1 promoter-driven nls-LacZ transgene in Purkinje cell parasagittal arrays of developing mouse cerebellum. J Neurosci Res 88:2810–2825

Gabison EE, Hoang-Xuan T, Mauviel A, Menashi S (2005) EMMPRIN/CD147, an MMP modulator in cancer, development and tissue repair. Biochimie 87:361–368

Gallagher E, Howell BW, Soriano P, Cooper JA, Hawkes R (1998) Cerebellar abnormalities in the disabled (mdab1-1) mouse. J Comp Neurol 402:238–251

Garcia-Maya M, Anderson AA, Kendal CE, Kenny AV, Edwards-Ingram LC, Holladay A, Saffell JL (2006) Ligand concentration is a driver of divergent signaling and pleiotropic cellular responses to FGF. J Cell Physiol 206:386–393

Gray G (1987) Synapse, morphology. In: Adelman G (ed) Encyclopaedia of neuroscience, vol 11. Birkhauser, Boston, MA, pp 1158–1162

Gurd JW, Gordon-Weeks P, Evans WH (1982) Biochemical and morphorlological comparison of postsynaptic densities prepared from rat, hamster and monkey brains by phase partitioning. J Neurochem 39:1117–1124

Halestrap AP, Wilson MC (2012) The monocarboxylate transporter family-role and regulation. IUBMB Life 64:109–119

Hansen SM, Berezin V, Bock E (2008) Signalling mechanisms of neurite-outgrowth induced by the cell adhesion molecules NCAM and N-cadherin. Cell Mol Life Sci 65:3809–3821

Hill IE, Selkirk CP, Hawkes RB, Beesley PW (1988) Characterization of novel glycoprotein components of synaptic membranes and postsynaptic densities, gp65 and gp55, with a monoclonal antibody. Brain Res 461:27–43

Hill I, Willmott T, Selkirk P, Gordon-Weeks P, Beesley PW (1989) Expression of two synapse-enriched glycoproteins gp65 and gp55 during rat brain development. Biochem Soc Trans 17:770

Hori K, Katayama N, Kachi S, Kondo M, Kadomatsu K, Usukura J, Muramatsu T, Mori S, Miyake Y (2000) Retinal dysfunction in basigin deficiency. Invest Ophthalmol Vis Sci 41: 3128–3133

Howell BW, Hawkes R, Soriano P, Cooper JA (1997) Neuronal position in the developing brain is regulated by mouse disabled-1. Nature 389:733–737

Huang RP, Ozawa M, Kadomatsu K, Muramatsu T (1993) Embigin, a member of the immunoglobulin superfamily expressed in embryonic cells, enhances cell-substratum adhesion. Dev Biol 155:307–314

Ito M (2002) The molecular organization of cerebellar long-term depression. Nat Rev Neurosci 3:896–902

Javitt DC, Zukin SR (1991) Recent advances in the phencyclidine model of schizophrenia. Am J Psychiatry 148:1301–1308

Kanekura T, Miyauchi T, Tashiro M, Muramatsu T (1991) Basigin, a new member of the immunoglobulin superfamily: genes in different mammalian species, glycosylation changes in the molecule from adult organs and possible variation in the N-terminal sequences. Cell Struct Funct 16:23–30

Karin P, Magistretti PJ, Pellerin L (2002) MCT2 is a major neuronal monocarboxylate transporter in the adult mouse brain. J Cereb Blood Flow Metab 22:586–595

Kim DK, Kanai Y, Matsuo H, Kim JY, Chairoungdua A, Kobayashi Y, Enomoto A, Cha SH, Goya T, Endou H (2002) The human T-type amino acid transporter-1: characterization, gene organization, and chromosomal location. Genomics 79:95–103

Kirk P, Wilson MC, Heddle C, Brown MH, Barclay AN, Halestrap AP (2000) CD147 is tightly associated with lactate transporters MCT1 and MCT4 and facilitates their cell surface expression. EMBO J 19:3896–3904

Kiselyov V, Skaldchikov G, Hinsby AM, Jensen PH, Kulahin N, Soroka V, Pedersen N, Tsetlin V, Poulsen FM, Berezin V, Bock E (2003) Structural basis for a direct interaction between FGFR1 and NCAM and evidence for a regulatory role of ATP. Structure 11:691–701

Kiselyov V, Kochoyan A, Poulsen FM, Bock E, Berezin V (2006) Elucidation of the mechanism of the regulatory function of the Ig1 module of the fibroblast growth factor receptor 1. Protein Sci 15:2318–2322

Kneussel M, Loebrich S (2007) Trafficking and synaptic anchoring of ionotropic inhibitory neurotransmitter receptors. Biol Cell 99:297–309

Koehler-Stec EM, Simpson IA, Vannucci SJ, Landschulz KT, Landschulz WH (1998) Monocarboxylate transporter expression in mouse brain. Am J Physiol 275:516–524

Kreutz MR, Langnaese K, Dieterich DC, Seidenbecher CI, Zuschratter W, Beesley PW, Gundelfinger ED (2001) Distribution of transcript and protein isoforms of the synaptic glycoprotein neuroplastin in rat retina. Invest Ophthalmol Vis Sci 42:1907–1914

Krug M, Jork R, Reymann K, Wagner M, Matthies H (1991) The amnesic substance 2-deoxygalactose suppresses the maintenance of hippocampal LTP. Brain Res 540:237–242

Kulahin N, Li S, Hinsby A, Kiselyov V, Berezin V, Bock E (2008) Fibronectin type III (FN3) modules of the neuronal cell adhesion molecule L1 interact directly with the fibroblast growth factor (FG) receptor. Mol Cell Neurosci 37:528–536

Langnaese K, Beesley PW, Gundelfinger ED (1997) Synaptic membrane glycoproteins gp65 and gp55 are new members of the immunoglobulin superfamily. J Biol Chem 272:821–827

Langnaese K, Mummery R, Gundelfinger ED, Beesley PW (1998) Immunoglobulin superfamily members gp65 and gp55: tissue distribution of glycoforms. FEBS Lett 429:284–288

Lim M, Martinez T, Jablons D, Cameron R, Guo H, Toole B, Li JD, Basbaum C (1998) Tumor-derived EMMPRIN (extracellular matrix metallo-proteinase inducer) stimulates collagenase transcription through MAPK p38. FEBS Lett 441:88–92

Lisman J, Schulman H, Cline H (2002) The molecular basis of CaMKII function in synaptic and behavioural memory. Nat Rev Neurosci 3:175–190

Little LN (2011) Characterisation of Basigin and the interaction between embigin and monocarboxylate transporter-1,-2 and -4 (MCT1, MCT2, MCT4) in the mouse brain. M.Sc. thesis, University of North Florida

Loebrich S, Bähring R, Katsuno T, Tsukita S, Kneussel M (2006) Activated radixin is essential for GABAA receptor alpha5 subunit anchoring at the actin cytoskeleton. EMBO J 25:987–999

Luthi A, Laurent JP, Figurov A, Muller D, Schachner M (1994) Hippocampal long-term potentiation and neural cell adhesion molecules L1 and NCAM. Nature 372:777–779

Malinow R, Malenka RC (2002) AMPA receptor trafficking and synaptic plasticity. Annu Rev Neurosci 25:103–126

Marzban H, Khanzada U, Shabir S, Hawkes R, Langnaese K, Smalla KH, Bockers TM, Gundelfinger ED, Gordon-Weeks PR, Beesley PW (2003) Expression of the immunoglobulin

superfamily neuroplastin adhesion molecules in adult and developing mouse cerebellum and their localisation to parasagittal stripes. J Comp Neurol 462:286–301
Mason CA (1987) The development of cerebellar mossy fibres and climbing fibres: embryonic and postnatal features. In: King JS (ed) New concepts in cerebellar neurobiology. Alan R. Liss, Inc., New York, NY, pp 57–88
Mateos JM, Osorio A, Azkue JJ, Benítez R, Elezgarai I, Bilbao A, Díez J, Puente N, Kuhn R, Knöpfel T, Hawkes R, Doñate-Oliver F, Grandes P (2001) Parasagittal compartmentalization of the metabotropic glutamate receptor mGluR1b in the cerebellar cortex. Eur J Anat 5:15–21
Matthies H, Staak S, Krug M (1996) Fucose and fucosyllactose enhance in-vitro hippocampal long-term potentiation. Brain Res 725:276–280
Mielke K, Herdegen T (2000) JNK and p38 stress kinases-degenerative effectors of signal-transduction cascades in the nervous system. Prog Neurobiol 61:45–60
Miyauchi T, Kanekura T, Yamaoka A, Ozawa M, Miyazawa S, Muramatsu T (1990) Basigin, a new, broadly distributed member of the immunoglobulin superfamily, has strong homology with both the immunoglobulin V domain and the beta-chain of major histocompatibility complex class II antigen. J Biochem 107:770–774
Mohammadi M, Olsen SK, Ibrahami OA (2005) Structural basis for fibroblast growth factor receptor activation. Cytokine Growth Factor Rev 16:107–137
Muramatsu T, Miyauchi T (2003) Basigin (CD147): a multifunctional transmembrane protein involved in reproduction, neural function, inflammation and tumor invasion. Histol Histopathol 18:981–987
Nabeshima K, Iwasaki H, Koga K, Hojo H, Suzumiya J, Kikuchi M (2006) Emmprin (basigin/CD147): matrix metalloproteinase modulator and multifunctional cell recognition molecule that plays a critical role in cancer progression. Pathol Int 56:359–367
Ochrietor JD, Linser PJ (2004) 5A11/Basigin gene products are necessary for proper maturation and function of the retina. Dev Neurosci 26:380–387
Ochrietor JD, Moroz TP, van Ekeris L, Clamp MF, Jefferson SC, deCarvalho AC, Fadool JM, Wistow G, Muramatsu T, Linser PJ (2003) Retina-specific expression of 5A11/Basigin-2, a member of the immunoglobulin gene superfamily. Invest Ophthalmol Vis Sci 44:4086–4096
Olsen RW, Sieghart W (2008) International Union of Pharmacology. LXX. Subtypes of gamma-aminobutyric acid(A) receptors: classification on the basis of subunit composition, pharmacology, and function. Update. Pharmacol Rev 60:243–260
Ono K, Han J (2000) The p38 signal transduction pathway. Activation and function. Cell Signal 12:1–13
Otterbach B, Stoffel W (1995) Acid sphingomyelinase-deficient mice mimic the neurovisceral form of human lysosomal storage disease (Niemann-Pick disease). Cell 81:1053–1061
Ouchi Y, Kubota Y, Kuramasu A, Watanabe T, Ito C (2005) Gene expression profiling in whole cerebral cortices of phencyclidine- or methamphetamine-treated rats. Mol Brain Res 140:142–149
Ovens MJ, Manoharan C, Wilson MC, Murray CM, Halestrap AP (2010) The inhibition of monocarboxylate transporter 2 (MCT2) by AR-C155858 is modulated by the associated ancillary protein. Biochem J 431:217–225
Owczarek S, Berezin V (2012) Neuroplastin: cell adhesion molecule and signaling receptor. Int J Biochem Cell Biol 44:1–5
Owczarek S, Kiryushko D, Larsen MH, Gajhede M, Sandi C, Berezin V, Bock E, Soroka V (2010) Neuroplastin-55 binds to and signals through the fibroblast growth factor receptor. FASEB J 24:1139–1150
Owczarek S, Soroka V, Kiryushko D, Yuan Q, Sandi C, Berezin V, Bock E (2011) Neuroplastin-65 and a mimetic peptide derived from its homophilic binding site modulate neuritogenesis and neuronal plasticity. J Neurochem 117:984–994
Ozol K, Hayden JM, Oberdick J, Hawkes R (1999) Transverse zones in the vermis of the mouse cerebellum. J Comp Neurol 412:95–111
Paul A, Wilson S, Belham CM, Robinson CJM, Scott PH, Gould G, Plevin R (1997) Stress-activated protein kinases: activation, regulation and function. Cell Singal 9:403–410

Philp NJ, Ochrietor JD, Rudoy C, Muramatsu T, Linser PJ (2003) Loss of MCT1, MCT3, and MCT4 expression in the retinal pigment epithelium and neural retina of the 5A11/basigin-null mouse. Invest Ophthalmol Vis Sci 44:1305–1311

Pierre K, Pellerin L (2005) Monocarboxylate transporters in the central nervous system: distribution, regulation and function. J Neurochem 94:1–14

Pierre K, Debernardi R, Magistretti PJ, Pellerin L (2003) Noradrenaline enhances monocarboxylate transporter 2 expression in cultured mouse cortical neurons via a translational regulation. J Neurochem 86:1468–1476

Pohle W, Acosta L, Rüthrich H, Krug M, Matthies H (1987) Incorporation of [3H]fucose in rat hippocampal structures after conditioning by perforant path stimulation and after LTP-producing tetanization. Brain Res 410:245–256

Porsolt RD, Anton G, Blavet N, Jalfre M (1978) Behavioural despair in rats: a new model sensitive to antidepressant treatments. Eur J Pharmacol 47:379–391

Rafiki A, Boulland JL, Halestrap AP, Ottersen OP, Bergersen L (2003) Highly differential expression of the monocarboxylate transporters MCT2 and MCT4 in the developing rat brain. Neuroscience 122:677–688

Rodriguez-Pinto D, Sparkowski J, Keough MP, Phoenix KN, Vumbaca F, Han D, Gundelfinger ED, Beesley P, Claffey KP (2009) Identification of novel tumor antigens with patient-derived immune-selected antibodies. Cancer Immunol Immunother 58:221–234

Rose SPR, Jork R (1987) Long-term-memory formation in chicks is blocked by 2-deoxygalactose, a fucose analogue. Behav Neural Biol 48:246–258

Saito A, Fujikura-Ouchi Y, Kuramasu A, Shimoda K, Akiyama K, Matsuoka H, Ito C (2007) Association study of putative promoter polymorphisms in the neuroplastin gene and schizophrenia. Neurosci Lett 411:168–173

Sanchez-Heras E, Howell FV, Williams G, Doherty P (2006) The fibroblast growth factor receptor acid box is essential for interactions with N-cadherin and all of the major isoforms of neural cell adhesion molecule. J Biol Chem 281:35208–35216

Sarna J, Miranda SR, Schuchman EH, Hawkes R (2001) Patterned cerebellar Purkinje cell death in a transgenic mouse model of Niemann Pick type A/B disease. Eur J Neurosci 13:1873–1880

Sarna JR, Marzban H, Watanabe M, Hawkes R (2006) Complementary stripes of phospholipase Cß3 and Cß4 expression by Purkinje cell subsets in the mouse cerebellum. J Comp Neurol 496:303–313

Sarto-Jackson I, Milenkovic I, Smalla KH, Gundelfinger ED, Kaehne T, Herrera-Molina R, Thomas S, Kiebler MA, Sieghart W (2012) The cell adhesion molecule neuroplastin-65 is a novel interaction partner of γ-aminobutyric acid type A receptors. J Biol Chem 287:14201–14214

Sato M, Chen C-C, Akiyama S, Otsuki S (1983) Acute exacerbation of paranoid psychotic state after long-term abstinence in patients with previous methamphetamine psychosis. Biol Psychiatry 18:429–440

Schurr A, Payne RS, Miller JJ, Rigor BM (1997) Brain lactate is an obligatory aerobic energy substrate for functional recovery after hypoxia: further in vitro validation. J Neurochem 69: 423–426

Schurr A, Payne RS, Miller JJ, Tseng MT, Rigor BM (2001) Blockade of lactate transport exacerbates delayed neuronal damage in a rat model of cerebral ischemia. Brain Res 895:268–272

Shirozu M, Tada H, Tashiro K, Nakamura T, Lopez ND, Nazarea M, Hamada T, Sato T, Nakano T, Honjo T (1996) Characterisation of novel and secreted membrane proteins isolated by the signal sequence trap method. Genomics 37:273–280

Smalla KH, Angenstein F, Richter K, Gundelfinger ED, Staak S (1998) Identification of fucose alpha(1-2) galactose epitope-containing glycoproteins from rat hippocampus. Neuroreport 9: 813–817

Smalla KH, Matthies H, Langnaese K, Shabir S, Bockers TM, Wyneken U, Staak S, Krug M, Beesley PW, Gundelfinger ED (2000) The synaptic glycoprotein neuroplastin is involved in long-term potentiation at hippocampal CA1 synapses. Proc Natl Acad Sci USA 97:4327–4332

Suzuki A, Stern SA, Bozdagi O, Huntley GW, Walker RH, Magistretti PJ, Alberini CM (2011) Astrocyte-neuron lactate transport is required for long-term memory formation. Cell 144:810–823
Tang L, Hung CP, Schuman EM (1998) A role for the cadherin family of cell adhesion molecules in hippocampal long-term potentiation. Neuron 20:1165–1175
Tretter V, Mukherjee J, Maric HM, Schindelin H, Sieghart W, Moss SJ (2012) Gephyrin, the enigmatic organizer at GABAergic synapses. Front Cell Neurosci 6:1–16
Wadiche JI, Jahr CE (2005) Patterned expression of Purkinje cell glutamate transporters controls synaptic plasticity. Nat Neurosci 8:1329–1334
Walsh FS, Doherty P (1997) Neural cell adhesion molecules of the immunoglobulin superfamily. Ann Rev Cell Dev Biol 13:425–456
Williams EJ, Furness J, Walsh FS, Doherty P (1994) Activation of the FGF receptor underlies neurite outgrowth stimulated by L1, N-CAM, and N-cadherin. Neuron 13:583–594
Williams EJ, Williams G, Howell FV, Skaper SD, Walsh FS, Doherty P (2001) Identification of an N-cadherin motif that can interact with the fibroblast growth factor receptor and is required for axonal growth. J Biol Chem 276:43879–43886
Willmott T, Skitsa I, Hill I, Mummery R, Beesley PW (1992) Molecular characterisation and structural relationship of the synapse-enriched glycoproteins gp65 and gp55. J Neurochem 58: 2037–2043
Wilson MC, Meredith D, Halestrap AP (2002) Fluorescence resonance energy transfer studies on the interaction between the lactate transporter MCT1 and CD147 provide information on the topology and stoichiometry of the complex in situ. J Biol Chem 277:3666–3672
Wilson MC, Meredith D, Fox JE, Manoharan C, Davies AJ, Halestrap AP (2005) Basigin (CD147) is the target for organomercurial inhibition of monocarboxylate transporter isoforms 1 and 4: the ancillary protein for the insensitive MCT2 is EMBIGIN (gp70). J Biol Chem 280:27213–27221
Wilson MC, Kraus M, Marzban H, Sarna JR, Wang Y, Hawkes R, Halestrap AP, Beesley PW (2013) The Neuroplastin adhesion molecules are accessory proteins that chaperone the monocarboxylate transporter MCT 2 to the neuronal cell surface. PLOSONE In Press
Yan L, Zucker S, Toole BP (2005) Roles of the multifunctional glycoprotein, emmprin (basigin; CD147), in tumour progression. Thromb Haemost 93:199–204
Yoshida S, Shibata M, Yamamoto S, Hagihara M, Asai N, Takahashi M, Mizutani S, Muramatsu T, Kadomatsu K (2000) Homo-oligomer formation by basigin, an immunoglobulin superfamily member, via its N-terminal immunoglobulin domain. Eur J Biochem 267:4372–4380
Zhu JJ, Qin Y, Zhao M, Van Aelst L, Malinow R (2002) Ras and Rap control AMPA receptor trafficking during synaptic plasticity. Cell 110:443–455

Chapter 5
Roles of Nectins and Nectin-Like Molecules in the Nervous System

Masahiro Mori, Yoshiyuki Rikitake, Kenji Mandai, and Yoshimi Takai

Abstract Nectins are immunoglobulin-like cell adhesion molecules (CAMs) constituting a family with four members. They exclusively localize at adherens junctions (AJs) between two neighboring cells. Nectins bind to afadin through their C-termini and are linked to the actin cytoskeleton. In addition to nectins, there are nectin-like molecules (Necls), which resemble nectins in their structures and constitute a family with five members. Nectins and Necls are involved in the formation of various kinds of cell–cell adhesion and diverse cellular functions including cell polarization, movement, proliferation, survival, and differentiation. In neuronal tissues, nectins and Necls functionally play crucial roles as CAMs at

M. Mori, M.D., Ph.D.
Division of Neurophysiology, Department of Physiology and Cell Biology, Kobe University Graduate School of Medicine, Kobe, Japan

Faculty of Health Sciences, Kobe University Graduate School of Health Sciences, Kobe, Japan

CREST, JST, Kobe, Japan
e-mail: morima@med.kobe-u.ac.jp

Y. Rikitake, M.D., Ph.D.
Division of Signal Transduction, Department of Biochemistry and Molecular Biology, Kobe University Graduate School of Medicine, Kobe, Japan

Division of Cardiovascular Medicine, Department of Internal Medicine, Kobe University Graduate School of Medicine, Kobe, Japan

CREST, JST, Kobe, Japan
e-mail: rikitake@med.kobe-u.ac.jp

K. Mandai, M.D., Ph.D. • Y. Takai, M.D., Ph.D. (✉)
Division of Pathogenetic Signaling, Department of Biochemistry and Molecular Biology, Kobe University Graduate School of Medicine, Kobe 650-0047, Japan

CREST, JST, Kobe, Japan
e-mail: mandai@med.kobe-u.ac.jp; ytakai@med.kobe-u.ac.jp

V. Berezin and P.S. Walmod (eds.), *Cell Adhesion Molecules: Implications in Neurological Diseases*, Advances in Neurobiology 8, DOI 10.1007/978-1-4614-8090-7_5,

neuron–neuron and neuron–glia interactions. For example, the members of the nectin and Necl families are involved in synapse formation and remodeling in the hippocampus, a key brain region for learning and memory. Nectins also play important roles in the auditory system. Moreover, nectins and Necls are associated with human neurological diseases when mutated or upregulated. Thus, nectins and Necls are crucial for physiology and pathology in the nervous system.

5.1 Introduction

Cells in multicellular organisms form cell–cell junctions and contacts that play essential roles in various cellular processes, including morphogenesis, differentiation, proliferation, and migration. Cell–cell junctions and contacts can be homotypic or heterotypic; for example, the former is formed between two neighboring epithelial cells and the latter is formed between differentiating germ cells and their supporter Sertoli cells in the testis. Cell–cell junctions are mediated by cell adhesion molecules (CAMs). Cadherins, which make up a superfamily with >100 members, serve as key Ca^{2+}-dependent CAMs in a variety of cell–cell junctions (Hirano and Takeichi 2012). The members of the immunoglobulin (Ig) superfamily also play important roles as Ca^{2+}-independent CAMs (Brummendorf and Lemmon 2001). Nectins have emerged as Ig-like CAMs that contribute to a variety of cell–cell junctions and contacts, acting cooperatively with or independently of cadherins (Takai et al. 2008a, b). In addition, nectin-like molecules (Necls), which have domain structures similar to those of nectins, have been intensively studied (Takai et al. 2008a, b). Nectins and Necls interact in *trans* with each other and in *cis* with growth factor receptors and integrins, and regulate a variety of cell functions, including polarization, movement, proliferation, survival, and differentiation in addition to cell adhesion and contacts. On the other hand, nectins and Necls serve as viral receptors and are also associated with human diseases, such as cancer, Alzheimer's disease (AD), and Margarita island ectodermal dysplasia (Hogle 2002; Takai et al. 2008a, b; Spear et al. 2000; Spear and Longnecker 2003). Here we first introduce the molecular and biological properties of nectins and Necls, then describe their roles in the nervous systems, and finally address neuropsychiatric diseases caused by dysfunction of nectins and Necls.

5.2 Nectins and Necls

5.2.1 *Molecular Properties of Nectins*

Nectins comprise a family with four members, nectin-1, nectin-2, nectin-3, and nectin-4, all of which have an extracellular region with three Ig-like loops, a single transmembrane region, and a cytoplasmic tail region (Fig. 5.1) (Takai et al. 2008a, b).

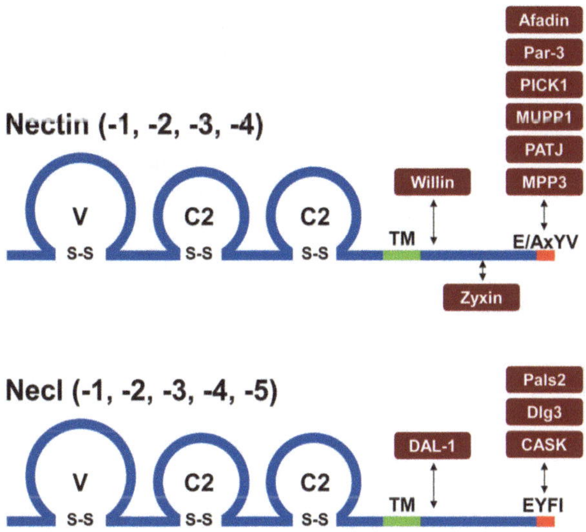

Fig. 5.1 Molecular structures and modes of *trans*-interactions of nectins, Necls, and their binding proteins. Nectins and Necls share three Ig-like extracellular domains comprising an N-terminal variable region-like (V) domain and two constant region-like (C2) domains, a transmembrane region (TM), and a cytoplasmic domain. The nectin C-terminus contains interaction motifs (E/AxYV, where x represents any amino acid) that allow interaction with afadin, Par-3, PICK1, MUPP1, PATJ, and MPP3. Willin and zyxin interact with the juxtamembrane and cytoplasmic regions of nectins, respectively. The C-terminus of Necl-1 and Necl-2 contains interaction motifs (EYFI) that allow interaction with Pals2, Dlg3, and CASK. DAL-1 interacts with the juxtamembrane regions of Necl-2. Bidirectional arrows indicate the binding

Nectins have several molecular properties. (1) They show a Ca^{2+}-independent cell–cell adhesion activity. Each nectin first forms homo-*cis*-dimers and then homo- or hetero-*trans*-dimers through the extracellular region, causing cell–cell adhesion. Heterophilic *trans*-interactions of nectins are stronger than their homophilic *trans-interactions* (Samanta et al. 2012; Satoh-Horikawa et al. 2000). By surface plasmon resonance analysis, the dissociation constants (*K*ds) for the interaction between nectin-1 and nectin-3 and between nectin-2 and nectin-3 are 2.3 and 360 nM, respectively (Ikeda et al. 2003), while those for homophilic binding of nectin-1, nectin-2, nectin-3, and nectin-4 are 17.5, 0.4, 228, and 153 μM, respectively (Harrison et al. 2012) (Fig. 5.2). Thus, among various combinations, the heterophilic *trans*-interaction between nectin-1 and nectin-3 is the strongest, followed by that between nectin-3 and nectin-2. (2) The *trans*-interactions of nectins induce the activation of Rap1, Cdc42, and Rac small G proteins (Takai et al. 2008a). (3) Nectins bind through their cytoplasmic tails to afadin, an actin filament (F-actin)-binding protein which connects nectins to the actin cytoskeleton (Mandai et al. 1997; Takai et al. 2008a, b) in a manner analogous to the way in which cadherins are connected to the cytoskeleton by binding through their cytoplasmic tails to the α- and β-catenin complex (Hirano and Takeichi 2012). (4) Nectins also bind through their cytoplasmic tails to partitioning defective three homologue (Par-3, also known as PARD3 in mammals), a cell

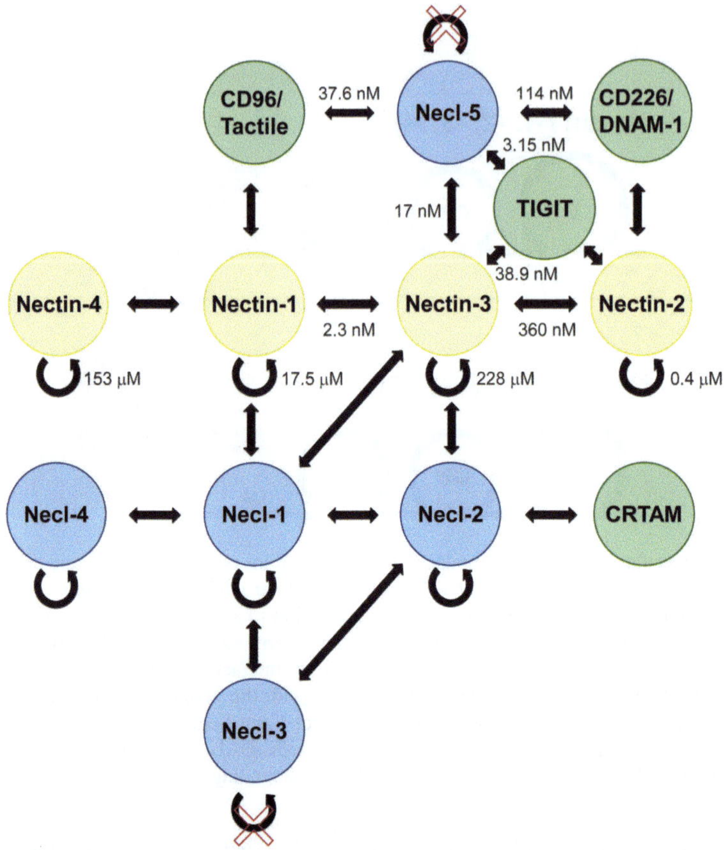

Fig. 5.2 *Trans*-interactions of nectins, Necls, and other Ig-like molecules. Homophilic (*looped arrows*) and heterophilic (*double-headed arrows*) *trans*-interactions are presented. Heterophilic *trans*-interaction between nectin-1 and nectin-3 is the strongest followed by that between nectin-3 and Necl-5 and that between nectin-2 and nectin-3. *Red crossing bars* indicate absence of homophilic interaction. DNAM1, DNAX accessory molecule 1; TIGIT, T cell immunoreceptor with Ig and ITIM domains; CRTAM, Class I-restricted T-cell associated molecule. Values besides *arrows* are *K*d for hemophilic and heterophilic interactions

polarity protein, which forms a complex with the other cell polarity proteins aPKC and Par-6 and regulates cell polarization (Ohno 2001; Takai et al. 2008a, b; Takekuni et al. 2003). (5) Nectins interact *in trans* heterophilically with Necls (Takai et al. 2008a, b; Ikeda et al. 2003; Shingai et al. 2003; Mueller and Wimmer 2003) and other Ig-like CAMs such as Tactile/CD96, DNAM-1/CD226, and TIGIT through their extracellular regions (Fig. 5.2) (Bottino et al. 2003; Fuchs et al. 2004; Pende et al. 2005, 2006; Stanietsky et al. 2009; Yu et al. 2009). The *K*ds for the heterophilic interaction between nectin-3 and Necl-5 and between nectin-3 and TIGIT are 17 and 38.9 nM, respectively (Ikeda et al. 2003; Yu et al. 2009). (6) Nectins interact with growth factor receptors. Nectin-3 interacts *in cis* with the platelet-derived growth factor (PDGF) receptor (Kanzaki et al. 2008), while nectin-1 interacts *in cis* with the fibroblast growth factor receptor (Bojesen et al. 2012). Nectin-1 and nectin-3, but not nectin-2, physically interact *in cis* with integrin $\alpha_v\beta_3$ at cell–cell adhesion sites (Sakamoto et al. 2006).

5.2.2 Molecular Properties of Necls

Necls, which are structurally similar to nectins, have five members and consist of an extracellular region with three Ig-like loops, a single transmembrane region, and a cytoplasmic region, but, unlike nectins, they do not bind afadin (Takai et al. 2008a, b). These include Necl-1/TSLL1/SynCAM3/CADM3, Necl-2/IGSF4/RA175/SgIGSF/TSLC1/SynCAM1/CADM1, Necl-3/SynCAM2/CADM2, Necl-4/TSLL2/SynCAM4/CADM4, and Necl-5/poliovirus receptor (PVR)/CD155/Tage4. Necl-1 homophilically interacts *in trans* and heterophilically interacts *in trans* with nectin-1, nectin-3, Necl-2, Necl-3, and Necl-4 (Fig. 5.2) (Kakunaga et al. 2005; Niederkofler et al. 2010; Shingai et al. 2003). Necl-2 also homophilically interacts *in trans* and heterophilically interacts *in trans* with nectin-3, Necl-1, and Necl-3, as well as another Ig-like CAM CRTAM (Boles et al. 2005; Niederkofler et al. 2010; Shingai et al. 2003). Necl-3 does not homophilically interact *in trans* but heterophilically interacts *in trans* with Necl-1 and Necl-2 (Niederkofler et al. 2010; Pellissier et al. 2007). Necl-4 homophilically interacts *in trans* and heterophilically interacts *in trans* with Necl-1 (Maurel et al. 2007; Spiegel et al. 2007). Necl-5 does not homophilically interact *in trans* but heterophilically interacts *in trans* with nectin-3 and other Ig-like CAMs such as Tactile/CD96 (Fuchs et al. 2004), DNAM-1/CD226 (Bottino et al. 2003; Pende et al. 2005, 2006), and TIGIT (Stanietsky et al. 2009; Yu et al. 2009). The *K*ds for the interactions between Necl-5 and TIGIT, between Necl-5 and Tactile/CD96, and between Necl-5 and DNAM-1/CD226 are 3.15, 37.6, and 114 nM, respectively (Yu et al. 2009). Moreover, Necl-1 and Necl-3 interact *in cis* with integrin $\alpha_6\beta_4$ (Mizutani et al. 2011). Necl-2 interacts *in cis* with the ErbB3 receptor, the ErbB4 receptor, integrin $\alpha_6\beta_4$, and integrin $\alpha_v\beta_3$ (Kawano et al. 2010; Mizutani et al. 2011). Necl-5 interacts *in cis* with integrin $\alpha_v\beta_3$ and growth factor receptors, such as the PDGF receptor and the vascular endothelial growth factor (VEGF) receptor (Minami et al. 2010; Kinugasa et al. 2012). Necl-4 has not been reported to bind any growth factor receptors.

5.2.3 Proteins Associating with Nectins and Necls

Nectin-Binding Proteins

Some, but not all, members of the nectin family are able to bind various cytoplasmic proteins in addition to afadin, including Par-3 (Takekuni et al. 2003), protein interacting with PRKCA1 (PICK1) (Reymond et al. 2005), multiple PDZ domain protein (MUPP1, also known as MPDZ), Pals1-associated tight junction protein (PATJ)

(Adachi et al. 2009), membrane palmitoylated protein 3 (MPP3) (Dudak et al. 2011), zyxin (Call et al. 2011), and willin (Ishiuchi and Takeichi 2012) (Fig. 5.1). Afadin has multiple domains: from the N-terminus to the C-terminus it has two Ras-associated domains, a forkhead-associated domain, a dilute domain, a PDZ domain, three proline-rich domains, and an F-actin-binding domain (Takai et al. 2008a, b). Afadin binds many proteins, including transmembrane proteins such as the Eph receptor tyrosine kinases, the synaptic transmembrane protein neurexins, the Notch receptor ligand jagged-1, the junction cell adhesion molecule JAM, and the gap junction protein connexin 36; peripheral membrane proteins such as the tight junction (TJ) protein ZO-1, the afadin- and α-actinin-binding protein ADIP, the actin-binding protein profilin, the vinculin-binding protein ponsin, the cadherin-binding protein α-catenin, and the Arg/Abl-interacting protein nArgBp2; and signaling molecules such as Rap1, Ras, Rit, and Rin small G proteins, Rap1 GTPase-activating protein SPA-1, Bcr and c-Src protein kinases, T-cell oncogene LIM domain only 2 (LMO2), and the tumor suppressor LIM domain only 7 (LMO7) (Kawabe et al. 1999; Begay-Muller et al. 2002; Shao et al. 1999; Takai et al. 2008a, b; Li et al. 2012). The tyrosine kinase Ryk is reported to interact with afadin, but this interaction is controversial (Trivier and Ganesan 2002). Thus, these proteins are indirectly associated with nectins through afadin. Moreover, the first Ras-associated domain of afadin mediates its self-association (Liedtke et al. 2010).

Necl-Binding Proteins

Necls do not directly bind afadin but binds many other proteins. Necl-1 and Necl-2 bind the membrane-associated guanylate kinase (MAGUK) family members, Pals2, Dlg3/MPP3, and calcium/calmodulin-associated Ser/Thr kinase (CASK), while Necl-2 additionally binds the tumor suppressor gene product DAL1, a band 4.1 family member protein, which connects Necl-2 to the actin cytoskeleton (Fukuhara et al. 2003; Shingai et al. 2003; Yageta et al. 2002). Pals2 is known to bind Lin-7, which is implicated in the proper localization of the Let-23 protein in *Caenorhabditis elegans*, the homologue of mammalian epidermal growth factor receptor (Kakunaga et al. 2004). Although PVR/CD155, a human homologue of Necl-5, was reported to bind Tctex-1, a subunit of the dynein motor complex (Mueller et al. 2002), the interaction between a mouse homologue of Necl-5 and Tctex-1 was negligible in our study (Minami et al. 2010).

5.3 Nectins and Necls in Non-neuronal Tissues

5.3.1 Nectins in a Variety of Cell–Cell Junctions of Non-neuronal Tissues

Nectins are widely expressed in many types of cells and play roles in the formation of cell–cell junctions or contacts, in some cases acting cooperatively with cadherins and in other cases independently of them. Mammalian tissues and organs are composed of two or more cell types that can adhere homotypically or heterotypically

(Rikitake et al. 2012). There are homotypic and heterotypic cell–cell adhesions. Nectins and cadherins are involved in the formation of both groups of cell–cell adhesion, but cadherins are important for the formation of homotypic cell–cell adhesion, whereas nectins are important for the formation of heterotypic cell–cell adhesions, because cadherins interact *in trans* almost exclusively homophilically between the same members, whereas nectins interact *in trans* both homophilically and heterophilically between the same and different members, and, more importantly, because heterophilic *trans*-interactions of nectins are much stronger than their homophilic *trans*-interactions.

The typical homotypic cell–cell adhesion is formed between neighboring epithelial cells and endothelial cells. In these cells, there is a junctional complex comprised of TJs and adherens junctions (AJs). Nectin-1, nectin-2, and nectin-3 and possibly nectin-4 symmetrically localize at AJs and are involved in the formation of AJs, acting cooperatively with E-cadherin, a cadherin superfamily member that serves as a key CAM at AJs in epithelial cells (Gumbiner 1996; Takeichi 1991), and in the subsequent formation of TJs (Takai et al. 2008a, b). The *trans*-interactions of nectins induce the activation of Rap1, Cdc42, and Rac (Takai et al. 2008a, b), while those of E-cadherins induce the activation of Rac (Yap and Kovacs 2003). The activation of these molecules regulates the reorganization of the actin cytoskeleton, which is required for the formation of cell–cell adhesion (Takai et al. 2008a, b). The formation of TJs is dependent on the formation of AJs (Tsukita and Furuse 1999). The symmetric homotypic cell–cell adhesion is also formed in fibroblasts. In these cells, both N-cadherin and nectins colocalize at AJs and are involved in the formation of AJs.

The typical heterotypic cell–cell adhesion is found between many types of cells, such as between Sertoli cells and spermatids during spermatid differentiation in the testis, between commissural axons and basal processes of floor plate cells in the neural tube, between the pigment cell and non-pigment cell layers of the ciliary epithelium, between ameloblasts and stratum intermedium cells in the developing tooth, and between hair cells and supporting cells in the cochlea in the inner ear. At the Sertoli cell–spermatid junctions, nectin-2 and nectin-3 reside specifically in Sertoli cells and spermatids, respectively, and serve as essential CAMs (Ozaki-Kuroda et al. 2002). At the junctions between the apical membranes of pigment and non-pigment epithelia in the ciliary body of the eye, nectin-1 and nectin-3 localize at both sides and P-cadherin symmetrically localizes at both sides (Inagaki et al. 2005). Nectins and P-cadherin mediate the apex–apex adhesion between the pigment and non-pigment epithelia of the ciliary body. At these junctions, nectins, but not cadherins, are major CAMs. The roles of nectins in cell–cell adhesion between commissural axons and basal processes of floor plate cells in the neural tube and between hair cells and supporting cells in the cochlea in the inner ear are described below.

In addition to the stable adhesion for cell–cell junctions, such as AJs and TJs, weak and transient cell–cell adhesions are found between blood cells and between blood cells and vascular endothelial cells. Nectin-2 and nectin-3 are expressed in blood cells that lack cadherins. The exact roles of nectins in blood cells remain unknown, but they may play roles in transiently formed cell–cell contacts, such as those between macrophages and lymphocytes and between leukocytes and vascular endothelial cells during *trans*-endothelial migration.

5.3.2 Necls in Cell Adhesion, Migration, and Proliferation

Necl-2 is widely expressed in various tissues and localizes at the basolateral plasma membrane of epithelial cells, although it is not at the specialized cell–cell junctions such as AJs, TJs, and desmosomes (Shingai et al. 2003; Takai et al. 2008a, b). The ability of nectin-3 to interact *in cis* with Necl-2 suggests that Necl-2 is recruited to the nectin-3-based cell–cell adhesion sites during the formation of AJs (Takai et al. 2008a, b). After Necl-2 is assembled to the primordial cell–cell adhesion sites, it may be translocated from there to the extrajunctional region of the basolateral plasma membrane. The mechanism of segregation of Necl-2 from nectin-3 at the plasma membrane is currently unknown.

Necl-5 physically and functionally interacts *in cis* with integrin $\alpha_V\beta_3$ and the PDGF receptor and stimulates cell movement by enhancing both integrin $\alpha_V\beta_3$- and PDGF receptor-induced signalings. Necl-5 co-localizes with integrin $\alpha_V\beta_3$ and the PDGF receptor at peripheral ruffles and with integrin $\alpha_V\beta_3$ at focal complexes (Minami et al. 2010). Necl-5 facilitates the integrin $\alpha_V\beta_3$-dependent, PDGF receptor-induced activation of Rac, which regulates the formation of peripheral ruffles and focal complexes. Necl-5 is also involved in the contact inhibition of cell movement. When two moving cells collide with each other, Necl-5 on the surface of one cell interacts *in trans* heterophilically with nectin-3, which may be diffusely distributed along the adjacent cell surface, initiating the formation of cell–cell junctions (Ikeda et al. 2003). This *trans*-interaction induces the activation of Cdc42 and Rac (Sato et al. 2004), which enhances the reorganization of the actin cytoskeleton and increases the number of cell–cell adhesion sites. However, because the *trans-interaction* of Necl-5 with nectin-3 is transient, Necl-5 is downregulated and endocytosed from the plasma membrane in a clathrin-dependent manner (Fujito et al. 2005), which reduces cell movement. On the other hand, nectin-3 dissociated from Necl-5 is retained on the plasma membrane and subsequently interacts *in trans* with nectin-1, which most feasibly interacts *in trans* with nectin-3 (Ikeda et al. 2003). Then, cadherin is recruited to the nectin-based adhesion sites, eventually establishing AJs. Hence, the cell–cell contact-induced *trans*-interaction of Necl-5 with nectin-3 and the subsequent downregulation of Necl-5 are at least one of the mechanisms of the contact inhibition of cell movement (Fujito et al. 2005).

In addition, Necl-5 enhances the PDGF-induced cell proliferation by shortening the period of the G_1 phase of the cell cycle (Kakunaga et al. 2004). Necl-5 enhances the PDGF-induced activation of the Ras-Raf-MEK-ERK pathway and consequently upregulates cyclins D2 and E and downregulates p27Kip1. Necl-5 regulates the VEGF-induced angiogenesis by controlling the interaction of VEGF receptor 2 with integrin $\alpha_V\beta_3$ and the VEGF receptor 2-mediated activation of downstream proangiogenic and survival signals, including Rap1, Akt, and endothelial nitric oxide synthase (Kinugasa et al. 2012).

Necl-5 also heterophilically interacts *in trans* with DNAM-1/CD226, which is expressed in human natural killer cells (Bottino et al. 2003). DNAM-1 has one extracellular region with two Ig-like loops, one transmembrane region, and one cytoplasmic region. Heterophilic *trans*-interactions of CD155/hNecl-5 with DNAM-1, poliovirus, and an anti-CD155 monoclonal antibody stimulate the phosphorylation of Necl-5 by Src kinases and recruit SH2-domain-containing tyrosine phosphatase-2 (Oda et al. 2004).

5.4 Nectins and Necls in Neuronal Tissues

5.4.1 Nectins and Necls at Neuron–Neuron Interactions

Nectins and Necls in Synapse Formation

Interneuronal synapses are asymmetric homotypic cell–cell adhesions. At the synapses, at least two types of intercellular junctions with different functions have been recognized: synaptic junctions (SJs) and puncta adherentia junctions (Fig. 5.3). Synaptic

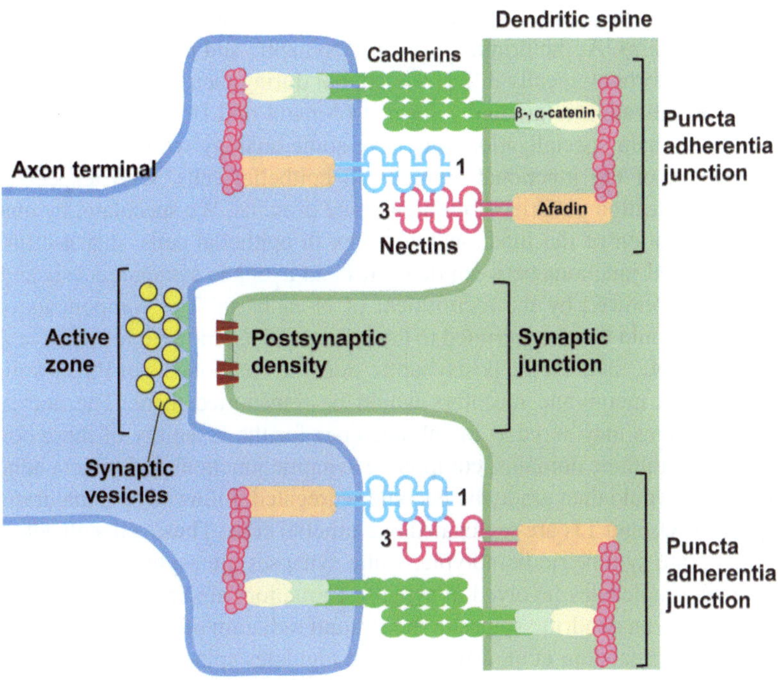

Fig. 5.3 Puncta adherentia junctions in neurons. Synapse between a mossy fiber terminal of a granule cell and a dendrite of a pyramidal cell in the CA3 region of the hippocampus contains two types of junctions: synaptic junctions and puncta adherentia junctions. Nectin-1 and nectin-3 asymmetrically localize at the mossy fiber terminal (presynaptic side) and at the dendrite of pyramidal cell (postsynaptic side), respectively, and form the puncta adherentia junctions in cooperation with cadherins

junctions are associated with synaptic vesicles that are docked at the presynaptic active zone where Ca^{2+} channels localize and with postsynaptic densities (PSDs) that are regarded as sites of specific receptors to which the neurotransmitter binds. Puncta adherentia junctions are not associated with synaptic vesicles or PSDs and appear to be similar in ultrastructure to the AJs of epithelial cells. They are regarded as mechanical adhesion sites between presynaptic axon terminals and PSDs. At the mossy fiber synapses, synapses between the mossy fiber terminals and the dendrites of pyramidal cells in the CA3 area of the hippocampus, both synaptic and puncta adherentia junctions, are highly specialized and are actively remodeled in an activity-dependent manner (Amaral and Dent 1981). N-Cadherin and αN- and β-catenins localize symmetrically at both the presynaptic and postsynaptic sides of puncta adherentia junctions, whereas nectin-1 and nectin-3 localize asymmetrically at the presynaptic and postsynaptic sides of puncta adherentia junctions, respectively (Mizoguchi et al. 2002) (Fig. 5.3). Puncta adherentia junctions have been regarded as symmetrical junctions on the basis of the morphological symmetry and symmetrical distribution of N-cadherin (Mueller and Wimmer 2003; Shingai et al. 2003), but their molecular architecture in this region is asymmetrical, at least in part, with regard to nectins (Mizoguchi et al. 2002).

The molecular mechanism of synapse formation is thought to be analogous, in part, with that of the epithelial junctions in terms of the localization patterns of the junctional proteins. At the primitive synapse, synaptic and puncta adherentia junctions are not morphologically differentiated, but during their maturation membrane domain specialization is gradually formed (Amaral and Dent 1981). This neural membrane domain specialization may have some analogy with that found during the formation of the junctional complex in epithelial cells, with respect to the dynamic localization patterns of the junctional proteins. We speculate, by analogy with the formation of the junctional complex in epithelial cells, that nectins first form primordial junctions between dendrites and axons in synaptogenesis and that this event is followed by the recruitment of N-cadherin. The components of the active zones would then be recruited to the primordial junctions to form active zones at the presynaptic side. At the postsynaptic side, the components of PSDs would be assembled and membrane receptors would be transported there. The nectin and cadherin systems may serve as membrane cues for the assembly of these components. The membrane domains, comprising synaptic junctions and puncta adherentia junctions, would then gradually become segregated, followed by a maturation of synapses as AJs and TJs are segregated in epithelial cells. Thus, cell–cell adhesions in epithelia are symmetric homotypic, while synapses are asymmetric homotypic. Of the many molecules involved in synapse formation, afadin is required for synapse formation on dendritic spines in the stratum radiatum of the CA1 region of the hippocampus (Beaudoin et al. 2012). Afadin regulates spine morphology in cooperation with Rap1, which is activated by NMDA receptors (Xie et al. 2005). Afadin is recruited to the plasma membrane by activated Rap1 and induces spine neck elongation, while afadin is dissociated from the membrane by inactive Rap1 and induces spine enlargement, suggesting that afadin could be involved in activity-dependent synaptic plasticity. However, it remains unclear whether these functions of afadin are involved in the action of nectins in synapse formation.

ZO-1 associates with TJs in epithelial cells (Stevenson et al. 1986) and binds to F-actin. ZO-1 belongs to the membrane-associated guanylate kinase-like homologues (MAGUKs) family (Itoh et al. 1993; González-Mariscal et al. 2000) and plays a key role in the formation and maintenance of TJs in epithelial cells and endothelial cells (Hartsock and Nelson 2008; Wolburg and Lippoldt 2002). In neurons, ZO-1 co-localizes with nectins and cadherins at puncta adherentia junctions (Inagaki et al. 2003), which suggests that ZO-1 plays a role in the segregation of the components of synaptic junctions and puncta adherentia junctions, as is described for the role of ZO-1 in epithelial cells (Hogle 2002).

In addition to nectins, afadin, cadherins, and catenins, neuroligins and neurexins have been implicated in synapse formation (Biederer et al. 2002; Missler et al. 1998). Neuroligins and neurexins localize at the presynaptic and postsynaptic sides of SJs, respectively (Ushkaryov et al. 1992; Song et al. 1999). Neuroligins induce stable junctions with presynapse-like properties between neurons and neuroligin-expressing fibroblasts that are co-cultured with dissociated hippocampal neurons (Dean et al. 2003; Scheiffele et al. 2002). N-Cadherin and neuroligin-1 cooperate to control vesicle clustering at nascent synapses (Stan et al. 2010). They also in concert regulate the formation of glutamatergic synapses (Aiga et al. 2011). The relationship between the nectin–afadin complex and the neurexin–neuroligin complex in synaptogenesis is not known. However, there are at least afadin-dependent and/or neuroligin-dependent signaling pathways in synaptogenesis (unpublished observation).

Necls have been reported as Ig-like CAMs at synapses and named SynCAM1-3 (Biederer et al. 2002). Biederer et al. reported that SynCAM1/Necl-2 was specifically synthesized in mouse brain (Biederer et al. 2002), whereas we found that Necl-2 was ubiquitously expressed (Kakunaga et al. 2005) as reported elsewhere (Wakayama et al. 2001; Fukami et al. 2002; Shingai et al. 2003). Presumably, the reason caused this inconsistency is that the anti-SynCAM1/Necl-2 Ab used by Biederer and coworkers (Biederer et al. 2002) may recognize Necl-1 but not Necl-2. Although they reported that SynCAM1 co-localized with synaptophysin and localized at synaptic junctions (Biederer et al. 2002), we could not repeat these results and the reason for this inconsistency remains unknown. SynCAM1/Necl-2 in particular has been shown to be involved in synapse formation and remodeling. Glutamatergic synaptic transmission is reconstituted between cultured neurons and non-neuronal cells co-expressing glutamate receptors with SynCAM1/Necl-2, suggesting that a single type of SynCAM1/Necl-2 as well as the glutamate receptor is sufficient for a functional postsynaptic response (Biederer et al. 2002; Sara et al. 2005). SynCAM1/Necl-2 acts in developing neurons to shape migrating growth cones and contributes to the adhesive differentiation of their axo-dendritic contacts (Stagi et al. 2010). In addition to the involvement in the organization of synapses SynCAM1/Necl-2 may recruit both the AMPA receptors and the NMDA receptors during synapse formation (Hoy et al. 2009). Moreover, Necl-2 may be involved in neuronal migration, axon growth, path finding, and fasciculation on the axons of differentiating neurons in addition to cell adhesion in the neuroepithelium and the synapses (Fujita et al. 2005). The functions of SynCAM1/Necl-2 are modulated by polysialic acid during integration of proteoglycan NG2-positive glial cells into neural networks (Galuska et al. 2010). Overexpression of Necl-2 leads to the upregulation of CASK and increased Ca^{2+}-independent cell adhesion (Giangreco et al. 2009). CASK is recruited to developing axon terminals by Necl-2 and neurexin/neuroligin (Kakunaga et al. 2005).

Nectins and Necls in Synapse Remodeling

Spines are dynamic structures that undergo rapid remodeling and experience-dependent spine remodeling provides a structural basis for learning and memory (Yuste and Bonhoeffer 2001). Synaptic activity that induces long-term potentiation, a long-lasting enhancement of synaptic strength, promotes spine enlargement and new spine formation (Matsuzaki et al. 2004). Spine structure and synaptic function are closely related (Kasai et al. 2003). The mechanisms that control the development and remodeling of spiny synapses under normal and pathological conditions need to be studied. Immature spines are often thin and elongated with filopodia; during their maturation, spine length decreases and the proportion of mushroom spines increases. The molecular details of how the filopodia are formed are still unknown, but they might be formed by the Cdc42 activated by the *trans*-interactions of nectins (Bottino et al. 2003; Takai et al. 2008a, b). N-Cadherin is involved in the formation of dendritic spines (Amaral and Dent 1981). It has been reported that scatter factor/hepatocyte growth factor and 12-*O*-tetradecanoylphorbol-13-acetate induce ectodomain shedding of nectin-1, which results in the formation of an extracellular fragment of nectin-1 (Tsukita et al. 2001; Yamada et al. 2004). In addition, nectin-1 serves as a substrate for presenilin/γ-secretase in the brain (Kim et al. 2002). The extracellular fragment of nectin-1 formed by this shedding may bind to dendritic nectin-3 and induce the formation of filopodia, which would result in changes to spine morphology. In the *afadin* conditionally deficient mice crossed with camk2a-Cre mice, the active zone protein, bassoon, and the postsynaptic density protein, PSD-95, are accumulated at mossy fiber-CA3 pyramidal cell synapses, while perforated PSDs tend to be more frequently observed than in control mice (Majima et al. 2009). Perforated PSDs are observed in synapses that undergo remodeling (Yuste and Bonhoeffer 2001). Thus, afadin is likely to regulate the remodeling of synapses. Whereas previous studies have advanced our understanding of molecular mechanisms of synapse formation, molecular mechanisms underlying synaptic remodeling remain largely unknown.

As the components of heterophilic *trans*-synaptic adhesion complexes such as a SynCAM1/Necl-2–SynCAM2/Necl-3 complex and a SynCAM3/Necl-1–SynCAM4/Necl-4 complex, Necls contribute to synapse organization and function (Fogel et al. 2007). SynCAM1/Necl-2 is also involved in synapse remodeling (Robbins et al. 2010). Necl-2 contributes to the regulation of synapse number and plasticity and impacts how neuronal networks undergo activity-dependent changes. Lateral self-assembling of SynCAM1/Necl-2 within the synaptic cleft promotes

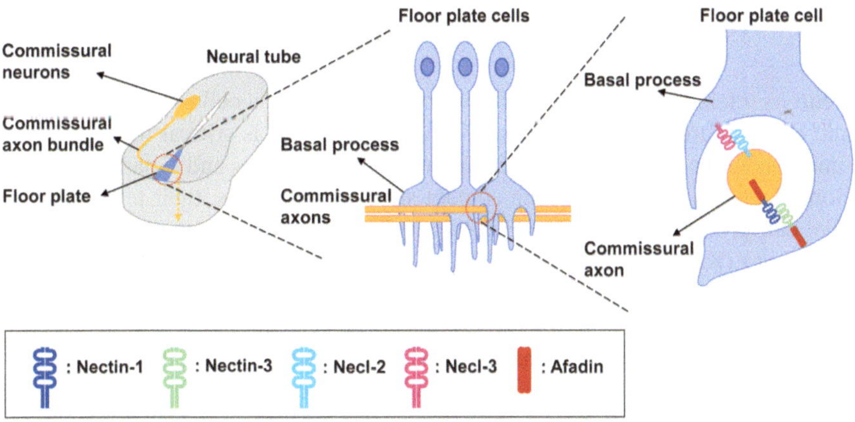

Fig. 5.4 Localization and roles of nectins and Necls at the contacts between commissural axons and floor plate cells in the neural tube. When the commissural axons make cell contacts with the dendrites of the floor plate cells, they extend across the central canal to make shift either to the rostral side or to the caudal side. Nectin-3 and Necl-3 on extending axons interact with nectin-1 and Necl-2 on dendrites of the floor plate cells, respectively. Cadherins do not localize at the contact sites

synapse induction and modulates their structure (Fogel et al. 2011). N-Glycosylation of SynCAM1/Necl-2 and SynCAM2/Necl-3 differentially affects their binding interface and implicates posttranslational modification as a mechanism to regulate *trans*-synaptic adhesion (Fogel et al. 2010).

Nectins and Necls at Contacts Between Commissural Axons and Floor Plate Cells

In the neural tube, commissural axons grow toward the ventral midline, cross the floor plate, and then abruptly change their trajectory from the circumferential to the longitudinal axis (Fig. 5.4). This axon guidance is mediated by the contacts between commissural axons and the basal processes of floor plate cells. Nectin-1 and nectin-3 asymmetrically localize at the commissural axon side and the floor plate cell side, respectively, of the plasma membranes at their contact sites and play an important role in the trajectory of the commissural axons (Okabe et al. 2004a). In addition to the nectin-1 and nectin-3 system, Necls are also involved in the trajectory of commissural axons. Necl-3 that is expressed by floor plate cells interacts with Necl-2 that is expressed by commissural axons to mediate a turning response in post-crossing commissural axons in the developing chick spinal cord *in vivo* (Niederkofler et al. 2010). Cadherins do not localize at the contact sites, while nectins and Necls localize there and may serve as CAMs. The weak *trans*-interaction between nectins and/or Necls, instead of the strong adhesion mediated by cadherins, might be advantageous when commissural axons continuously elongate while they are attached to floor plate cells.

5.4.2 Necls at Neuron–Glia Interactions

Neurons interact not only with neurons but also with glial cells, such as astrocytes and oligodendrocytes. Interaction of neurons with glia is critical for a variety of functions in the nervous system, including neural activities and synapse transmission. Necl-1 is expressed at the contact sites among axons, their terminals, and glial cell processes that cooperatively form axon bundles, synapses, and myelinated axons (Kakunaga et al. 2005). Necl-1 is likely to serve as a CAM at the non-junctional cell–cell contact sites of the nervous tissues. In fact, Necl-1 plays an important role in the initial axon-oligodendrocyte recognition and adhesion in central nervous system myelination (Park et al. 2008). Necl-4 in Schwann cells plays an important role in initiating peripheral nervous system myelination as the glial binding partner for Necl-1 on the axon (Fig. 5.5) (Maurel et al. 2007; Spiegel et al. 2007). Necl-2-mediated glia cell adhesiveness is affected by erbB4 receptor activation (Sandau et al. 2011). Necl-3 also acts as an adhesion molecule between different cell types, interacting with other Necls in the central and peripheral nervous systems (Pellissier et al. 2007). Thus, in both the central and peripheral nervous systems, Necls are involved in myelination by mediating adhesion among different cell type such as neuron and glial cells.

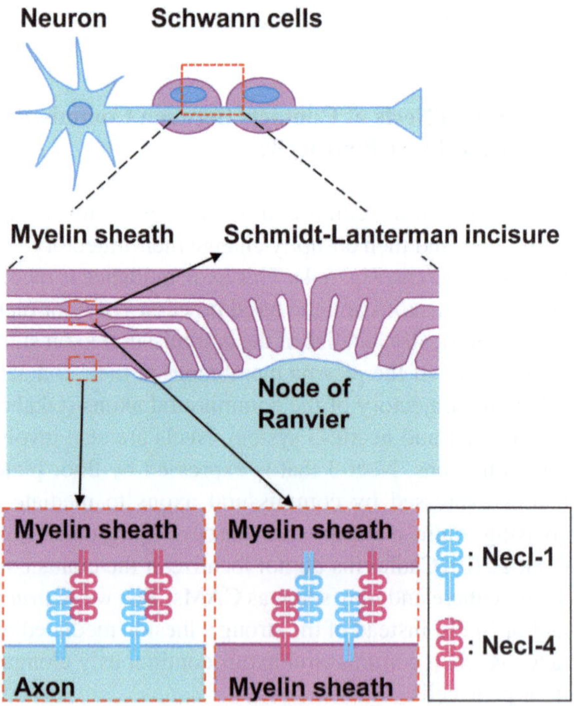

Fig. 5.5 Myelin sheath of the peripheral nerve. Necl-1 is specifically expressed at the contact sites among axons and glia cell processes that form the myelin sheath. Necl-4 in the Schwann cells plays an important role in initiating peripheral nervous system myelination as the glial binding partner for Necl-1 on the axon. At the Schwann cell–axon contact, Necl-1 on the axon interacts *in trans* with Necl-4 on the Schwann cell, while at the autotypic junctions formed between the myelin lamellae at the Schmidt–Lanterman incisure, Necl-1 interacts *in trans* with Necl-4

5.5 Nectins in the Auditory Epithelium

In the organ of Corti, sensory hair cells and supporting cells are observed. Hair cells convert sounds into electrical signals, which are transmitted to the brain. Hair cells and supporting cells are highly organized to form a checkerboard-like pattern (Kelley 2006). However, molecular mechanisms that regulate this characteristic pattern had remained unknown. In the mouse organ of Corti, hair cells and supporting cells express nectin-1 and nectin-3, respectively, but both cells possess nectin-2. The *trans*-interaction between nectin-1 and nectin-3 mediates the heterotypic adhesion between these two cell types, as the fine mosaic pattern is lost in *nectin-1–/–* mice and *nectin-3–/–* mice (Togashi et al. 2011). Moreover, in these mutant mice, the position of the kinocilium and the orientation of stereociliary bundles in hair cells are altered (unpublished observation). Thus, the *trans*-interaction between nectin-1 and nectin-3 is critical not only for checkerboard-like pattern formation, but also positioning of the kinocilium and stereociliary bundle orientation in hair cells.

5.6 Nectins and Necls in Diseases

5.6.1 Nectins and Necls as Viral Receptors

Virus interaction with cellular receptors is an essential step for recognition of the host cell and for commitment of the virus to initiate infection. Some viruses such as herpes virus and poliovirus show a tropism for neurons. Upon peripheral infection such viruses may enter the central nervous system and cause massive damage, either by direct virus-conferred effects or by immunopathology. Nectin-1 was originally isolated as one of the PVR-related proteins and named PRR-1 (Lopez et al. 1995). Nectin-2 was originally isolated as the murine homologue of human PVR, but turned out to be another PVR-related protein and was named PRR-2 (Eberlé et al. 1995). Neither PRR-1 nor PRR-2 has thus far been shown to serve as a PVR. They were later shown to serve as receptors for α-herpes viruses, facilitating their entry and intercellular spreading, and renamed HveC and HveB, respectively (Table 5.1) (Geraghty et al. 1998; Spear et al. 2000). Human nectin-1 allows entry of all α-herpes viruses tested so far, including herpes simplex virus (HSV) types 1 and 2, pseudorabies virus, and bovine herpes virus type 1 (Geraghty et al. 1998). Human nectin-2 can mediate entry of a restricted number of α-herpes viruses (Warner et al. 1998). The interaction of nectin-1 or nectin-2 with one of the HSV envelope glycoproteins recruits other viral glycoproteins to initiate fusion between the viral envelope and a cell membrane, thereby mediating the entry of the viral nucleocapsid into the cell (Spear and Longnecker 2003). The usual manifestations of HSV disease are mucocutaneous lesions. HSV establishes latent infection of neurons in sensory ganglia and causes recurrent lesions at the sites of primary infection. In HSV disease, the intercellular spreading significantly contributes to the pathogenesis. The interaction of nectin-1 with afadin increases the efficiency of intercellular spreading, but

Table 5.1 Nectins and Necls as viral receptors

Nectins/Necls	Viruses
Nectin-1	HSV-1, HSV-2, PVR, BHV-1
Nectin-2	HSV-2, PVR
Nectin-3	Not identified
Nectin-4	MV
Necl-1, -2, -3, -4	Not identified
Necl-5	PVR, BHV-1, PV

not the entry, of HSV-1. The E-cadherin–catenin system increases the efficiency of both the entry and intercellular spreading of HSV-1 (Sakisaka et al. 2001). Nectin-4 was recently identified as the epithelial cell receptor for the measles virus (Mühlebach et al. 2011; Noyce et al. 2011). Coupled with recent observations made in measles virus-infected macaques, this discovery has led to a new paradigm for how the virus accesses the respiratory tract and exits the host. Human Necl-5 (hNecl-5) was originally isolated as a receptor for poliovirus and was named hPVR (Koike et al. 1990; Mendelsohn et al. 1989). Poliovirus infects susceptible cells through hNecl-5/hPVR. It is thought that binding of hNecl-5/hPVR to poliovirus, the outer coat of which is an icosahedral protein shell, initiates conformational changes that enable the altered virion to bind to membranes and to invade cells even in the absence of the receptor (Hogle 2002). It is not clear whether the target membrane for entry is the plasma membrane or an endosomal membrane. Poliovirus is the causative agent of poliomyelitis. The usual manifestations of poliomyelitis disease are the spread and replication of virus in the central nervous system, particularly in the motor neurons. The cytoplasmic domain of hNecl-5/hPVR on the surface of endosomes that might enclose an intact poliovirion could interact with cytoplasmic dynein and the endosomes could be transported in a retrograde direction along microtubules through the axon to the neural cell body where replication of poliovirus occurs. It remains unknown whether other nectins and Necls serve as viral receptors. Thus, nectins and Necls are not only CAMs but also viral receptors and play a critical role in the pathogenesis of neurotrophic viral infections. Therefore, nectins and Necls could be therapeutic targets or probes as viral receptors. For example, nectin-4 is also a tumor cell marker that is highly expressed on the apical surface of many adenocarcinoma cell lines, making it a potential target for the oncolytic therapy by measles virus (Noyce and Richardson 2012).

5.6.2 Nectins in Neurological Diseases

Many lines of evidence suggest the association of nectins with pathogenesis of various neurological diseases. Mutations in the nectin-1 gene are responsible for cleft lip/palate ectodermal dysplasia, Margarita island ectodermal dysplasia and Zlotogora-Ogür syndrome, which are characterized by cleft lip/palate, syndactyly, mental retardation, and ectodermal dysplasia (Sozen et al. 2001; Suzuki et al. 2000).

Impairment of the function of nectin-1 in synapse formation would explain the cause of the mental retardation. These phenotypes mainly affect the places where nectin-1 is specifically expressed (Okabe et al. 2004b). Mutations in human nectin-4 cause an ectodermal dysplasia–syndactyly syndrome that is characterized by the combination of hair and tooth abnormalities, alopecia, and cutaneous syndactyly (Brancati et al. 2010). Recent genome-wide association studies of various populations, including Japanese and African Americans, have shown a genetic association between single nucleotide polymorphisms (SNPs) in *NECTIN-2* and late-onset AD (Harold et al. 2009; Logue et al. 2011; Takei et al. 2009), and mutations in *NECTIN-3* are associated with human ocular disease and congenital ocular defects (Lachke et al. 2012). A SNP in the 3′UTR region of *NECTIN-2* is one of the 13 genome-wide significant SNPs that map within or close to the *APOE* (Apolipoprotein E) locus on chromosome 19, whose polymorphic expression is widely associated with AD (Harold et al. 2009). These results suggest that together with a known association of APOE with AD, genetic variations in the *NECTIN-2* gene may have implications for predisposition to this disease. At the synapses in the CA3 area of the hippocampus, the number of puncta adherentia junctions is decreased in both *nectin-1–/–* mice and *nectin-3–/–* mice (Honda et al. 2006). Furthermore, in the *nectin-1*-deficient mice, there is an abnormal trajectory of mossy fibers at the stratum lucidum of the hippocampus, possibly as a result of impaired puncta adherentia junctions. Both the *nectin-1–/–* mice and *nectin-3–/–* mice show microphthalmia and display a separation of the apex–apex adhesion between the pigment and non-pigment epithelia of the ciliary body (Inagaki et al. 2005).

5.6.3 Possible Involvement of Nectins and Necls in Other Diseases

Besides neurological diseases, nectins and Necls may be involved in the pathogenesis of various other diseases. Both *nectin-2–/–* mice and *nectin-3–/–* mice exhibit the male-specific infertility phenotype and have defects in the later steps of sperm morphogenesis, exhibiting distorted nuclei and abnormal distribution of mitochondria (Bouchard et al. 2000; Inagaki et al. 2005; Ozaki-Kuroda et al. 2002). The structure of Sertoli cell–spermatid junctions is severely impaired, and the localization of afadin and nectin-3 or nectin-2 is disorganized in the *nectin-2–/–* mice and *nectin-3–/–* mice, respectively. In all the cases, the impaired phenotypes occur at cell–cell junctions and contacts where the functions of the two nectins are not redundant. The heterophilic *trans*-interaction of nectins plays a particularly important role in maintaining the specialized junctions and contacts between different types of cells. Mice deficient in nectin-1, nectin-2, or nectin-3 do not apparently show impaired organization of AJs and TJs in tissues where multiple types of nectins are expressed, which might indicate that the nectins have overlapping functions in these tissues. *Afadin–/–* mice are embryonic lethal and show developmental defects at stages around gastrulation, including disorganization of the ectoderm, impaired migration of the mesoderm, and loss of somites and other structures that are derived from both the ectoderm and the mesoderm (Ikeda et al. 1999). In the ectoderm of the mutant mice, the organization of AJs and TJs is highly impaired. One reason why *afadin–/–* mice show more severe phenotypes than nectin-deficient mice is because afadin does not have family members. Necl-2 and Necl-5 are likely to contribute to tumorigenesis. Necl-2 serves as a tumor suppressor in human non-small cell lung cancer (Kuramochi et al. 2001). Rodent Necl-5 was identified as a product of a gene overexpressed in rat and mouse colon carcinoma (Chadeneau et al. 1994, 1996). Necl-5 is expressed at low levels in many cells, but its expression level is upregulated in many carcinomas (Ikeda et al. 2004; Chadeneau et al. 1994, 1996). Necl-5 is overexpressed in human colorectal carcinoma and malignant glioma (Masson et al. 2001). Upregulated Necl-5 in cancer is responsible at least partly for the enhanced motility and proliferation of cancer cells (Ikeda et al. 2003, 2004).

5.7 Conclusions and Perspectives

Evidence has been accumulated that nectins and Necls are important for various aspects of the nervous system physiology, such as synapse formation and remodeling, the trajectory of the commissural axons in the neural tube, myelination, and development of the auditory epithelium. However, questions about nectins and Necls still remain that include their roles in neuronal circuit formation, synaptic plasticity, neuronal cell differentiation, establishment of planar cell polarity in the auditory epithelium, and formation of heterotypic cell adhesions in the nervous system. Functional analysis combined with fine molecular and biological manipulations will answer these questions. For example, the *trans*-interaction between nectin-1 and nectin-3 may affect the output of the hippocampal mossy fiber circuit by changing the balance of excitatory and inhibitory synaptic transmission. Live imaging of hippocampal mossy fibers by means of a fluorescent dye will enable us to clarify the involvement of nectins and Necls in synapse formation and remodeling after induction of long-term potentiation. Further studies of the relationship between structural remodeling and change in functional parameter such as action potential firing rate and amplitude of synaptic response will provide valuable information to help answer how changes in synaptic structures contribute to changes in function of neuronal circuits. Moreover, conditional inactivation of afadin, a nectin-binding protein, also will help understand further the molecular mechanisms of synapse formation and remodeling.

On the other hand, nectins and Necls have been implicated in pathophysiology of neurological disorders. Several members of nectins and Necls have been identified as virus receptors. Mutations in the nectin genes can be the causes of hereditary neurological disorders and SNPs in the nectin genes are associated with neurodegenerative diseases. To assess the contribution of nectins and Necls in the pathogenesis of neuropsychiatric diseases, their significance should be studied at molecular, cellular, and *in vivo* levels. In particular, *in vivo* analysis is important to clarify the initiation and progression of disease processes precisely. Fortunately, *nectin-1−/−*, *nectin-2−/−*, *nectin-3−/−*, *Necl-2/SynCAM1−/−*, *Necl-4−/−*, *Necl-5−/−* mice have been already generated. Such mouse models would be powerful tools to advance our understanding of the significance of nectins and Necls in pathophysiology of neurological disorders.

Acknowledgments We thank our colleagues and collaborators, for their great contributions and excellent achievements. This work was supported by grants-in-aid for Scientific Research from the Ministry of Education, Culture, Sports, Science and Technology, Japan.

Compliance with Ethics Requirements The authors declare that they have no conflicts of interest.

References

Adachi M, Hamazaki Y, Kobayashi Y, Itoh M, Tsukita S, Furuse M, Tsukita S (2009) Similar and distinct properties of MUPP1 and Patj, two homologous PDZ domain-containing tight-junction proteins. Mol Cell Biol 29:2372–2389

Aiga M, Levinson JN, Bamji SX (2011) N-cadherin and neuroligins cooperate to regulate synapse formation in hippocampal cultures. J Biol Chem 286:851–858

Amaral DG, Dent JA (1981) Development of the mossy fibers of the dentate gyrus: I. A light and electron microscopic study of the mossy fibers and their expansions. J Comp Neurol 195:51–86

Beaudoin GM III, Schofield CM, Nuwal T, Zang K, Ullian EM, Huang B, Reichardt LF (2012) Afadin, a Ras/Rap effector that controls cadherin function, promotes spine and excitatory synapse density in the hippocampus. J Neurosci 32:99–110

Begay-Muller V, Ansieau S, Leutz A (2002) The LIM domain protein Lmo2 binds to AF6, a translocation partner of the MLL oncogene. FEBS Lett 521:36–38

Biederer T, Sara Y, Mozhayeva M, Atasoy D, Liu X, Kavalali ET, Südhof TC (2002) SynCAM, a synaptic adhesion molecule that drives synapse assembly. Science 297:1525–1531

Bojesen KB, Clausen O, Rohde K, Christensen C, Zhang L, Li S, Kohler L, Nielbo S, Nielsen J, Gjorlund MD, Poulsen FM, Bock E, Berezin V (2012) Nectin-1 binds and signals through the fibroblast growth factor receptor. J Biol Chem 287(44):37420–37433

Boles KS, Barchet W, Diacovo T, Cella M, Colonna M (2005) The tumor suppressor TSLC1/NECL-2 triggers NK-cell and CD8+ T-cell responses through the cell-surface receptor CRTAM. Blood 106:779–786

Bottino C, Castriconi R, Pende D, Rivera P, Nanni M, Carnemolla B, Cantoni C, Grassi J, Marcenaro S, Reymond N, Vitale M, Moretta L, Lopez M, Moretta A (2003) Identification of PVR (CD155) and Nectin-2 (CD112) as cell surface ligands for the human DNAM-1 (CD226) activating molecule. J Exp Med 198:557–567

Bouchard MJ, Dong Y, McDermott BM Jr, Lam DH, Brown KR, Shelanski M, Bellve AR, Racaniello VR (2000) Defects in nuclear and cytoskeletal morphology and mitochondrial localization in spermatozoa of mice lacking nectin-2, a component of cell–cell adherens junctions. Mol Cell Biol 20:2865–2873

Brancati F, Fortugno P, Bottillo I, Lopez M, Josselin E, Boudghene-Stambouli O, Agolini E, Bernardini L, Bellacchio E, Iannicelli M, Rossi A, Dib-Lachachi A, Stuppia L, Palka G, Mundlos S, Stricker S, Kornak U, Zambruno G, Dallapiccola B (2010) Mutations in PVRL4, encoding cell adhesion molecule nectin-4, cause ectodermal dysplasia-syndactyly syndrome. Am J Hum Genet 87:265–273

Brummendorf T, Lemmon V (2001) Immunoglobulin superfamily receptors: cis-interactions, intracellular adaptors and alternative splicing regulate adhesion. Curr Opin Cell Biol 13:611–618

Call GS, Chung JY, Davis JA, Price BD, Primavera TS, Thomson NC, Wagner MV, Hansen MD (2011) Zyxin phosphorylation at serine 142 modulates the zyxin head-tail interaction to alter cell-cell adhesion. Biochem Biophys Res Commun 404:780–784

Chadeneau C, LeMoullac B, Denis MG (1994) A novel member of the immunoglobulin gene superfamily expressed in rat carcinoma cell lines. J Biol Chem 269:15601–15605

Chadeneau C, LeCabellec M, LeMoullac B, Meflah K, Denis MG (1996) Over-expression of a novel member of the immunoglobulin superfamily in Min mouse intestinal adenomas. Int J Cancer 68:817–821

Dean C, Scholl FG, Choih J, DeMaria S, Berger J, Isacoff E, Scheiffele P (2003) Neurexin mediates the assembly of presynaptic terminals. Nat Neurosci 6:708–716

Dudak A, Kim J, Cheong B, Federoff HJ, Lim ST (2011) Membrane palmitoylated proteins regulate trafficking and processing of nectins. Eur J Cell Biol 90:365–375

Eberlé F, Dubreuil P, Mattei MG, Devilard E, Lopez M (1995) The human PRR2 gene, related to the human poliovirus receptor gene (PVR), is the true homolog of the murine MPH gene. Gene 159:267–272

Fogel AI, Akins MR, Krupp AJ, Stagi M, Stein V, Biederer T (2007) SynCAMs organize synapses through heterophilic adhesion. J Neurosci 27:12516–12530

Fogel AI, Li Y, Giza J, Wang Q, Lam TT, Modis Y, Biederer T (2010) N-glycosylation at the SynCAM (synaptic cell adhesion molecule) immunoglobulin interface modulates synaptic adhesion. J Biol Chem 285:34864–34874

Fogel AI, Stagi M, Perez de Arce K, Biederer T (2011) Lateral assembly of the immunoglobulin protein SynCAM 1 controls its adhesive function and instructs synapse formation. EMBO J 30:4728–4738

Fuchs A, Cella M, Giurisato E, Shaw AS, Colonna M (2004) Cutting edge: CD96 (tactile) promotes NK cell-target cell adhesion by interacting with the poliovirus receptor (CD155). J Immunol 172:3994–3998

Fujita E, Urase K, Soyama A, Kouroku Y, Momoi T (2005) Distribution of RA175/TSLC1/SynCAM, a member of the immunoglobulin superfamily, in the developing nervous system. Brain Res Dev Brain Res 154:199–209

Fujito T, Ikeda W, Kakunaga S, Minami Y, Kajita M, Sakamoto Y, Monden M, Takai Y (2005) Inhibition of cell movement and proliferation by cell-cell contact-induced interaction of Necl-5 with nectin-3. J Cell Biol 171:165–173

Fukami T, Satoh H, Fujita E, Maruyama T, Fukuhara H, Kuramochi M, Takamoto S, Momoi T, Murakami Y (2002) Identification of the Tslc1 gene, a mouse orthologue of the human tumor suppressor TSLC1 gene. Gene 295:7–12

Fukuhara H, Masuda M, Yageta M, Fukami T, Kuramochi M, Maruyama T, Kitamura T, Murakami Y (2003) Association of a lung tumor suppressor TSLC1 with MPP3, a human homologue of Drosophila tumor suppressor Dlg. Oncogene 22:6160–6165

Galuska SP, Rollenhagen M, Kaup M, Eggers K, Oltmann-Norden I, Schiff M, Hartmann M, Weinhold B, Hildebrandt H, Geyer R, Mühlenhoff M, Geyer H (2010) Synaptic cell adhesion molecule SynCAM 1 is a target for polysialylation in postnatal mouse brain. Proc Natl Acad Sci USA 107:10250–10255

Geraghty RJ, Krummenacher C, Eisenberg RJ, Cohen GH, Spear PG (1998) Entry of aherpesviruses mediated by poliovirus receptor related protein 1 and poliovirus receptor. Science 280: 1618–1620

Giangreco A, Jensen KB, Takai Y, Miyoshi J, Watt FM (2009) Necl 2 regulates epidermal adhesion and wound repair. Development 136:3505–3514

González-Mariscal L, Betanzos A, Avila-Flores A (2000) MAGUK proteins: structure and role in the tight junction. Semin Cell Dev Biol 11:315–324

Gumbiner BM (1996) Cell adhesion: the molecular basis of tissue architecture and morphogenesis. Cell 84:345–357

Harold D, Abraham R, Hollingworth P, Sims R, Gerrish A, Hamshere ML, Pahwa JS, Moskvina V, Dowzell K, Williams A, Jones N, Thomas C, Stretton A, Morgan AR, Lovestone S, Powell

J, Proitsi P, Lupton MK, Brayne C, Rubinsztein DC, Gill M, Lawlor B, Lynch A, Morgan K, Brown KS, Passmore PA, Craig D, McGuinness B, Todd S, Holmes C, Mann D, Smith AD, Love S, Kehoe PG, Hardy J, Mead S, Fox N, Rossor M, Collinge J, Maier W, Jessen F, Schurmann B, Van Den Bussche H, Heuser I, Kornhuber J, Wiltfang J, Dichgans M, Frolich L, Hampel H, Hull M, Rujescu D, Goate AM, Kauwe JS, Cruchaga C, Nowotny P, Morris JC, Mayo K, Sleegers K, Bettens K, Engelborghs S, De Deyn PP, Van Broeckhoven C, Livingston G, Bass NJ, Gurling H, McQuillin A, Gwilliam R, Deloukas P, Al-Chalabi A, Shaw CE, Tsolaki M, Singleton AB, Guerreiro R, Muhleisen TW, Nothen MM, Moebus S, Jockel KH, Klopp N, Wichmann HE, Carrasquillo MM, Pankratz VS, Younkin SG, Holmans PA, O'Donovan M, Owen MJ, Williams J (2009) Genome-wide association study identifies variants at CLU and PICALM associated with Alzheimer's disease. Nat Genet 41:1088–1093

Harrison OJ, Vendome J, Brasch J, Jin X, Hong S, Katsamba PS, Ahlsen G, Troyanovsky RB, Troyanovsky SM, Honig B, Shapiro L (2012) Nectin ectodomain structures reveal a canonical adhesive interface. Nat Struct Mol Biol 19:906–915

Hartsock A, Nelson WJ (2008) Adherens and tight junctions: structure, function and connections to the actin cytoskeleton. Biochim Biophys Acta 1778:660–669

Hirano S, Takeichi M (2012) Cadherins in brain morphogenesis and wiring. Physiol Rev 92:597–634

Hogle JM (2002) Poliovirus cell entry: common structural themes in viral cell entry pathways. Annu Rev Microbiol 56:677–702

Honda T, Sakisaka T, Yamada T, Kumazawa N, Hoshino T, Kajita M, Kayahara T, Ishizaki H, Tanaka-Okamoto M, Mizoguchi A, Manabe T, Miyoshi J, Takai Y (2006) Involvement of nectins in the formation of puncta adherentia junctions and the mossy fiber trajectory in the mouse hippocampus. Mol Cell Neurosci 31:315–325

Hoy JL, Constable JR, Vicini S, Fu Z, Washbourne P (2009) SynCAM1 recruits NMDA receptors via protein 4.1B. Mol Cell Neurosci 42:466–483

Ikeda W, Nakanishi H, Miyoshi J, Mandai K, Ishizaki H, Tanaka M, Togawa A, Takahashi K, Nishioka H, Yoshida H, Mizoguchi A, Nishikawa S, Takai Y (1999) Afadin: a key molecule essential for structural organization of cell–cell junctions of polarized epithelia during embryogenesis. J Cell Biol 146:1117–1132

Ikeda W, Kakunaga S, Itoh S, Shingai T, Takekuni K, Satoh K, Inoue Y, Hamaguchi A, Morimoto K, Takeuchi M, Imai T, Takai Y (2003) Tage4/nectin-like molecule-5 heterophilically *trans-interacts* with cell adhesion molecule nectin-3 and enhances cell migration. J Biol Chem 278:28167–28172

Ikeda W, Kakunaga S, Takekuni K, Shingai T, Satoh K, Morimoto K, Takeuchi M, Imai T, Takai Y (2004) Nectin-like molecule-5/Tage4 enhances cell migration in an integrin-dependent, nectin-3-independent manner. J Biol Chem 279:18015–18025

Inagaki M, Irie K, Deguchi-Tawarada M, Ikeda W, Ohtsuka T, Takeuchi M, Takai Y (2003) Nectin-dependent localization of ZO-1 at puncta adhaerentia junctions between the mossy fiber terminals and the dendrites of the pyramidal cells in the CA3 area of adult mouse hippocampus. J Comp Neurol 460:514–524

Inagaki M, Irie K, Ishizaki H, Tanaka-Okamoto M, Morimoto K, Inoue E, Ohtsuka T, Miyoshi J, Takai Y (2005) Roles of cell-adhesion molecules nectin 1 and nectin 3 in ciliary body development. Development 132:1525–1537

Ishiuchi T, Takeichi M (2012) Nectins localize Willin to cell-cell junctions. Genes Cells 17:387–397

Itoh M, Nagafuchi A, Yonemura S, Kitani-Yasuda T, Tsukita S, Tsukita S (1993) The 220-kD protein colocalizing with cadherins in non-epithelial cells is identical to ZO-1, a tight junction-associated protein in epithelial cells: cDNA cloning and immunoelectron microscopy. J Cell Biol 121:491–502

Kakunaga S, Ikeda W, Shingai T, Fujito T, Yamada A, Minami Y, Imai T, Takai Y (2004) Enhancement of serum- and platelet-derived growth factor-induced cell proliferation by Necl-5/Tage4/PVR/CD155 through the Ras-Raf-MEK-ERK signaling. J Biol Chem 279: 36419–36425

Kakunaga S, Ikeda W, Itoh S, Deguchi-Tawarada M, Ohtsuka T, Mizoguchi A, Takai Y (2005) Nectin-like molecule-1/TSLL1/SynCAM3: a neural tissue-specific immunoglobulin-like cell-cell adhesion molecule localizing at non-junctional contact sites of presynaptic nerve terminals, axons and glia cell processes. J Cell Sci 118:1267–1277

Kanzaki N, Ogita H, Komura H, Ozaki M, Sakamoto Y, Majima T, Ijuin T, Takenawa T, Takai Y (2008) Involvement of the nectin-afadin complex in PDGF-induced cell survival. J Cell Sci 121:2008–2017

Kasai H, Matsuzaki M, Noguchi J, Yasumatsu N, Nakahara H (2003) Structure-stability-function relationships of dendritic spines. Trends Neurosci 26:360–368

Kawabe H, Hata Y, Takeuchi M, Ide N, Mizoguchi A, Takai Y (1999) nArgBP2, a novel neural member of ponsin/ArgBP2/vinexin family that interacts with synapse-associated protein 90/postsynaptic density-95-associated protein (SAPAP). J Biol Chem 274:30914–30918

Kawano S, Mizutani K, Miyata M, Ikeda W, Takai Y (2010) Interaction of integrin $\alpha_6\beta_4$ with ErbB3 and implication in heregulin-induced ErbB3/ErbB2-mediated DNA synthesis. Genes Cells 15:995–1001

Kelley MW (2006) Regulation of cell fate in the sensory epithelia of the inner ear. Nat Rev Neurosci 7:837–849

Kim DY, Ingano LA, Kovacs DM (2002) Nectin-1alpha, an immunoglobulin-like receptor involved in the formation of synapses, is a substrate for presenilin/gamma-secretase-like cleavage. J Biol Chem 277:49976–49981

Kinugasa M, Amano H, Satomi-Kobayashi S, Nakayama K, Miyata M, Kubo Y, Nagamatsu Y, Kurogane Y, Kureha F, Yamana S, Hirata K, Miyoshi J, Takai Y, Rikitake Y (2012) Necl-5/poliovirus receptor interacts with VEGFR2 and regulates VEGF-induced angiogenesis. Circ Res 110:716–726

Koike S, Horie H, Ise I, Okitsu A, Yoshida M, Iizuka N, Takeuchi K, Takegami T, Nomoto A (1990) The poliovirus receptor protein is produced both as membrane-bound and secreted forms. EMBO J 9:3217–3224

Kuramochi M, Fukuhara H, Nobukuni T, Kanbe T, Maruyama T, Ghosh HP, Pletcher M, Isomura M, Onizuka M, Kitamura T, Sekiya T, Reeves RH, Murakami Y (2001) TSLC1 is a tumor-suppressor gene in human non-small-cell lung cancer. Nat Genet 27:427–430

Lachke SA, Higgins AW, Inagaki M, Saadi I, Xi Q, Long M, Quade BJ, Talkowski ME, Gusella JF, Fujimoto A, Robinson ML, Yang Y, Duong QT, Shapira I, Motro B, Miyoshi J, Takai Y, Morton CC, Maas RL (2012) The cell adhesion gene PVRL3 is associated with congenital ocular defects. Hum Genet 131:235–250

Li X, Lynn BD, Nagy JI (2012) The effector and scaffolding proteins AF6 and MUPP1 interact with connexin36 and localize at gap junctions that form electrical synapses in rodent brain. Eur J Neurosci 35:166–181

Liedtke M, Ayton PM, Somervaille TC, Smith KS, Cleary ML (2010) Self-association mediated by the Ras association 1 domain of AF6 activates the oncogenic potential of MLL-AF6. Blood 116:63–70

Logue MW, Schu M, Vardarajan BN, Buros J, Green RC, Go RC, Griffith P, Obisesan TO, Shatz R, Borenstein A, Cupples LA, Lunetta KL, Fallin MD, Baldwin CT, Farrer LA, Multi-Institutional Research on Alzheimer Genetic Epidemiology (MIRAGE) Study Group (2011) A comprehensive genetic association study of Alzheimer disease in African Americans. Arch Neurol 68:1569–1579

Lopez M, Eberlé F, Mattei MG, Gabert J, Birg F, Bardin F, Maroc C, Dubreuil P (1995) Complementary DNA characterization and chromosomal localization of a human gene related to the poliovirus receptor-encoding gene. Gene 155:261–265

Majima T, Ogita H, Yamada T, Amano H, Togashi H, Sakisaka T, Tanaka-Okamoto M, Ishizaki H, Miyoshi J, Takai Y (2009) Involvement of afadin in the formation and remodeling of synapses in the hippocampus. Biochem Biophys Res Commun 385:539–544

Mandai K, Nakanishi H, Satoh A, Obaishi H, Wada M, Nishioka H, Itoh M, Mizoguchi A, Aoki T, Fujimoto T, Matsuda Y, Tsukita S, Takai Y (1997) Afadin: a novel actin filament-binding protein with one PDZ domain localized at cadherin-based cell-to-cell adherens junction. J Cell Biol 139:517–528

Masson D, Jarry A, Baury B, Blanchardie P, Laboisse C, Lustenberger P, Denis MG (2001) Overexpression of the CD155 gene in human colorectal carcinoma. Gut 49:236–240

Matsuzaki M, Honkura N, Ellis-Davies GC, Kasai H (2004) Structural basis of long-term potentiation in single dendritic spines. Nature 429:761–766

Maurel P, Einheber S, Galinska J, Thaker P, Lam I, Rubin MB, Scherer SS, Murakami Y, Gutmann DH, Salzer JL (2007) Nectin-like proteins mediate axon Schwann cell interactions along the internode and are essential for myelination. J Cell Biol 178:861–874

Mendelsohn CL, Wimmer E, Racaniello VR (1989) Cellular receptor for poliovirus: molecular cloning, nucleotide sequence, and expression of a new member of the immunoglobulin superfamily. Cell 56:855–865

Minami A, Mizutani K, Waseda M, Kajita M, Miyata M, Ikeda W, Takai Y (2010) Necl-5/PVR enhances PDGF-induced attraction of growing microtubules to the plasma membrane of the leading edge of moving NIH3T3 cells. Genes Cells 15:1123–1135

Missler M, Fernandez-Chacon R, Südhof TC (1998) The making of neurexins. J Neurochem 71:1339–1347

Mizoguchi A, Nakanishi H, Kimura K, Matsubara K, Ozaki-Kuroda K, Katata T, Honda T, Kiyohara Y, Heo K, Higashi M, Tsutsumi T, Sonoda S, Ide C, Takai Y (2002) Nectin: an adhesion molecule involved in formation of synapses. J Cell Biol 156:555–565

Mizutani K, Kawano S, Minami A, Waseda M, Ikeda W, Takai Y (2011) Interaction of nectin-like molecule 2 with integrin $\alpha_6\beta_4$ and inhibition of disassembly of integrin $\alpha_6\beta_4$ from hemidesmosomes. J Biol Chem 286:36667–36676

Mueller S, Wimmer E (2003) Recruitment of nectin-3 to cell–cell junctions through *trans-heterophilic* interaction with CD155, a vitronectin and poliovirus receptor that localizes to $\alpha_v\beta_3$-integrin-containing membrane microdomains. J Biol Chem 278:31251–31260

Mueller S, Cao X, Welker R, Wimmer E (2002) Interaction of the poliovirus receptor CD155 with the dynein light chain Tctex-1 and its implication for poliovirus pathogenesis. J Biol Chem 277:7897–7904

Mühlebach MD, Mateo M, Sinn PL, Prüfer S, Uhlig KM, Leonard VH, Navaratnarajah CK, Frenzke M, Wong XX, Sawatsky B, Ramachandran S, McCray PB Jr, Cichutek K, von Messling V, Lopez M, Cattaneo R (2011) Adherens junction protein nectin-4 is the epithelial receptor for measles virus. Nature 480:530–533

Niederkofler V, Baeriswyl T, Ott R, Stoeckli ET (2010) Nectin-like molecules/SynCAMs are required for post-crossing commissural axon guidance. Development 137:427–435

Noyce RS, Richardson CD (2012) Nectin 4 is the epithelial cell receptor for measles virus. Trends Microbiol 20:429–439

Noyce RS, Bondre DG, Ha MN, Lin LT, Sisson G, Tsao MS, Richardson CD (2011) Tumor cell marker PVRL4 (nectin 4) is an epithelial cell receptor for measles virus. PLoS Pathog 7:e1002240

Oda T, Ohka S, Nomoto A (2004) Ligand stimulation of CD155a inhibits cell adhesion and enhances cell migration in fibroblasts. Biochem Biophys Res Commun 319:1253–1264

Ohno S (2001) Intercellular junctions and cellular polarity: the PAR–aPKC complex, a conserved core cassette playing fundamental roles in cell polarity. Curr Opin Cell Biol 13:641–648

Okabe N, Shimizu K, Ozaki-Kuroda K, Nakanishi H, Morimoto K, Takeuchi M, Katsumaru H, Murakami F, Takai Y (2004a) Contacts between the commissural axons and the floor plate cells are mediated by nectins. Dev Biol 273:244–256

Okabe N, Ozaki-Kuroda K, Nakanishi H, Shimizu K, Takai Y (2004b) Expression patterns of nectins and afadin during epithelial remodeling in the mouse embryo. Dev Dyn 230:174–186

Ozaki-Kuroda K, Nakanishi H, Ohta H, Tanaka H, Kurihara H, Mueller S, Irie K, Ikeda W, Sasaki T, Wimmer E, Nishimune Y, Takai Y (2002) Nectin couples cell–cell adhesion and the actin scaffold at heterotypic testicular junctions. Curr Biol 12:1145–1150

Park J, Liu B, Chen T, Li H, Hu X, Gao J, Zhu Y, Zhu Q, Qiang B, Yuan J, Peng X, Qiu M (2008) Disruption of Nectin-like 1 cell adhesion molecule leads to delayed axonal myelination in the CNS. J Neurosci 28:12815–12819

Pellissier F, Gerber A, Bauer C, Ballivet M, Ossipow V (2007) The adhesion molecule Necl-3/SynCAM-2 localizes to myelinated axons, binds to oligodendrocytes and promotes cell adhesion. BMC Neurosci 8:90

Pende D, Spaggiari GM, Marcenaro S, Martini S, Rivera P, Capobianco A, Falco M, Lanino E, Pierri I, Zambello R, Bacigalupo A, Mingari MC, Moretta A, Moretta L (2005) Analysis of the receptor-ligand interactions in the natural killer-mediated lysis of freshly isolated myeloid or lymphoblastic leukemias: evidence for the involvement of the Poliovirus receptor (CD155) and Nectin-2 (CD112). Blood 105:2066–2073

Pende D, Castriconi R, Romagnani P, Spaggiari GM, Marcenaro S, Dondero A, Lazzeri E, Lasagni L, Martini S, Rivera P, Capobianco A, Moretta L, Moretta A, Bottino C (2006) Expression of the DNAM-1 ligands, Nectin-2 (CD112) and poliovirus receptor (CD155), on dendritic cells: relevance for natural killer-dendritic cell interaction. Blood 107:2030–2036

Reymond N, Garrido-Urbani S, Borg JP, Dubreuil P, Lopez M (2005) PICK-1: a scaffold protein that interacts with Nectins and JAMs at cell junctions. FEBS Lett 579:2243–2249

Rikitake Y, Mandai K, Takai Y (2012) The role of nectins in different types of cell–cell adhesion. J Cell Sci 125:3713–3722

Robbins EM, Krupp AJ, Perez de Arce K, Ghosh AK, Fogel AI, Boucard A, Südhof TC, Stein V, Biederer T (2010) SynCAM 1 adhesion dynamically regulates synapse number and impacts plasticity and learning. Neuron 68:894–906

Sakamoto Y, Ogita H, Hirota T, Kawakatsu T, Fukuyama T, Yasumi M, Kanzaki N, Ozaki M, Takai Y (2006) Interaction of integrin alpha(v)beta3 with nectin. Implication in cross-talk between cell-matrix and cell-cell junctions. J Biol Chem 281:19631–19644

Sakisaka T, Taniguchi T, Nakanishi H, Takahashi K, Miyahara M, Ikeda W, Yokoyama S, Peng YF, Yamanishi K, Takai Y (2001) Requirement of interaction of nectin-1a/HveC with afadin for efficient cell–cell spread of herpes simplex virus type 1. J Virol 75:4734–4743

Samanta D, Ramagopal UA, Rubinstein R, Vigdorovich V, Nathenson SG, Almo SC (2012) Structure of Nectin-2 reveals determinants of homophilic and heterophilic interactions that control cell-cell adhesion. Proc Natl Acad Sci U S A 109:14836–14840

Sandau US, Mungenast AE, McCarthy J, Biederer T, Corfas G, Ojeda SR (2011) The synaptic cell adhesion molecule, SynCAM1, mediates astrocyte-to-astrocyte and astrocyte-to-GnRH neuron adhesiveness in the mouse hypothalamus. Endocrinology 152:2353–2363

Sara Y, Biederer T, Atasoy D, Chubykin A, Mozhayeva MG, Südhof TC, Kavalali ET (2005) Selective capability of SynCAM and neuroligin for functional synapse assembly. J Neurosci 25:260–270

Sato T, Irie K, Ooshio T, Ikeda W, Takai Y (2004) Involvement of heterophilic *trans*-interaction of Necl-5/Tage4/PVR/CD155 with nectin-3 in formation of nectin-and cadherin-based adherens junctions. Genes Cells 9:791–799

Satoh-Horikawa K, Nakanishi H, Takahashi K, Miyahara M, Nishimura M, Tachibana K, Mizoguchi A, Takai Y (2000) Nectin-3, a new member of immunoglobulin-like cell adhesion molecules that shows homophilic and heterophilic cell-cell adhesion activities. J Biol Chem 275:10291–10299

Scheiffele P, Fan J, Choih J, Fetter R, Serafini T (2002) Neuroligin expressed in nonneuronal cells triggers presynaptic development in contacting axons. Cell 101:657–669

Shao H, Kadono-Okuda K, Finlin BS, Andres DA (1999) Biochemical characterization of the Ras-related GTPases Rit and Rin. Arch Biochem Biophys 371:207–219

Shingai T, Ikeda W, Kakunaga S, Morimoto K, Takekuni K, Itoh S, Satoh K, Takeuchi M, Imai T, Monden M, Takai Y (2003) Implications of nectin-like molecule-2/IGSF4/RA175/SgIGSF/TSLC1/SynCAM1 in cell–cell adhesion and transmembrane protein localization in epithelial cells. J Biol Chem 278:35421–35427

Song JY, Ichtchenko K, Sudhof TC, Brose N (1999) Neuroligin 1 is a postsynaptic cell-adhesion molecule of excitatory synapses. Proc Natl Acad Sci U S A 96:1100–1105

Sozen MA, Suzuki K, Tolarova MM, Bustos T, Fernandez Iglesias JE, Spritz RA (2001) Mutation of PVRL1 is associated with sporadic, non-syndromic cleft lip/palate in northern Venezuela. Nat Genet 29:141–142

Spear PG, Longnecker R (2003) Herpesvirus entry: an update. J Virol 77:10179–10185

Spear PG, Eisenberg RJ, Cohen GH (2000) Three classes of cell surface receptors for aherpesvirus entry. Virology 275:1–8

Spiegel I, Adamsky K, Eshed Y, Milo R, Sabanay H, Sarig-Nadir O, Horresh I, Scherer SS, Rasband MN, Peles E (2007) A central role for Necl4 (SynCAM4) in Schwann cell-axon interaction and myelination. Nat Neurosci 10:861–869

Stagi M, Fogel AI, Biederer T (2010) SynCAM 1 participates in axo-dendritic contact assembly and shapes neuronal growth cones. Proc Natl Acad Sci USA 107:7568–7573

Stan A, Pielarski KN, Brigadski T, Wittenmayer N, Fedorchenko O, Gohla A, Lessmann V, Dresbach T, Gottmann K (2010) Essential cooperation of N-cadherin and neuroligin-1 in the transsynaptic control of vesicle accumulation. Proc Natl Acad Sci USA 107:11116–11121

Stanietsky N, Simic H, Arapovic J, Toporik A, Levy O, Novik A, Levine Z, Beiman M, Dassa L, Achdout H, Stern-Ginossar N, Tsukerman P, Jonjic S, Mandelboim O (2009) The interaction of TIGIT with PVR and PVRL2 inhibits human NK cell cytotoxicity. Proc Natl Acad Sci USA 106:17858–17863

Stevenson BR, Siliciano JD, Mooseker MS, Goodenough DA (1986) Identification of ZO-1: a high molecular weight polypeptide associated with the tight junction (zonula occludens) in a variety of epithelia. J Cell Biol 103:755–766

Suzuki K, Hu D, Bustos T, Zlotogora J, Richieri-Costa A, Helms JA, Spritz RA (2000) Mutations of PVRL1, encoding a cell–cell adhesion molecule/herpesvirus receptor, in cleft lip/palate-ectodermal dysplasia. Nat Genet 25:427–430

Takai Y, Ikeda W, Ogita H, Rikitake Y (2008a) The immunoglobulin-like cell adhesion molecule nectin and its associated protein afadin. Annu Rev Cell Dev Biol 24:309–342

Takai Y, Miyoshi J, Ikeda W, Ogita H (2008b) Nectins and nectin-like molecules: roles in contact inhibition of cell movement and proliferation. Nat Rev Mol Cell Biol 9:603–615

Takei N, Miyashita A, Tsukie T, Arai H, Asada T, Imagawa M, Shoji M, Higuchi S, Urakami K, Kimura H, Kakita A, Takahashi H, Tsuji S, Kanazawa I, Ihara Y, Odani S, Kuwano R (2009) Genetic association study on in and around the APOE in late-onset alzheimer disease in Japanese. Genomics 93:441–448

Takeichi M (1991) Cadherin cell adhesion receptors as a morphogenetic regulator. Science 251:1451–1455

Takekuni K, Ikeda W, Fujito T, Morimoto K, Takeuchi M, MondenM TY (2003) Direct binding of cell polarity protein PAR-3 to cell–cell adhesion molecule nectin at neuroepithelial cells of developing mouse. J Biol Chem 278:5497–5500

Togashi H, Kominami K, Waseda M, Komura H, Miyoshi J, Takeichi M, Takai Y (2011) Nectins establish a checkerboard-like cellular pattern in the auditory epithelium. Science 333: 1144–1147

Trivier E, Ganesan TS (2002) RYK, a catalytically inactive receptor tyrosine kinase, associates with EphB2 and EphB3 but does not interact with AF-6. J Biol Chem 277:23037–23043

Tsukita S, Furuse M (1999) Occludin and claudins in tight-junction strands: leading or supporting players? Trends Cell Biol 9:268–273

Tsukita S, Furuse M, Itoh M (2001) Multifunctional strands in tight junctions. Nat Rev Mol Cell Biol 4:285–293

Ushkaryov YA, Petrenko AG, Geppert M, Sudhof TC (1992) Neurexins: synaptic cell surface proteins related to the alpha-latrotoxin receptor and laminin. Science 257:50–56

Wakayama T, Ohashi K, Mizuno K, Iseki S (2001) Cloning and characterization of a novel mouse immunoglobulin superfamily gene expressed in early spermatogenic cells. Mol Reprod Dev 60:158–164

Warner MS, Geraghty RJ, Martinez WM, Montgomery RI, Whitbeck JC, Xu R, Eisenberg RJ, Cohen GH, Spear PG (1998) A cell surface protein with herpesvirus entry activity (HveB) confers susceptibility to infection by mutants of herpes simplex virus type 1, herpes simplex virus type 2 and pseudorabies virus. Virology 246:179–189

Wolburg H, Lippoldt A (2002) Tight junctions of the blood-brain barrier: development, composition and regulation. Vascul Pharmacol 38:323–337

Xie Z, Huganir RL, Penzes P (2005) Activity-dependent dendritic spine structural plasticity is regulated by small GTPase Rap1 and its target AF-6. Neuron 48:605–618

Yageta M, Kuramochi M, Masuda M, Fukami T, Fukuhara H, Maruyama T, Shibuya M, Murakami Y (2002) Direct association of TSLC1 and DAL-1, two distinct tumor suppressor proteins in lung cancer. Cancer Res 62:5129–5133

Yamada A, Irie K, Fukuhara A, Ooshio T, Takai Y (2004) Requirement of the actin cytoskeleton for the association of nectins and other cell adhesion molecules at adherens and tight junctions in MDCK cells. Genes Cells 9:843–855

Yap AS, Kovacs EM (2003) Direct cadherin-activated cell signaling: a view from the plasma membrane. J Cell Biol 160:11–16

Yu X, Harden K, Gonzalez LC, Francesco M, Chiang E, Irving B, Tom I, Ivelja S, Refino CJ, Clark H, Eaton D, Grogan JL (2009) The surface protein TIGIT suppresses T cell activation by promoting the generation of mature immunoregulatory dendritic cells. Nat Immunol 10:48–57

Yuste R, Bonhoeffer T (2001) Morphological changes in dendritic spines associated with long-term synaptic plasticity. Annu Rev Neurosci 24:1071–1089

Chapter 6
ICAM-5: A Neuronal Dendritic Adhesion Molecule Involved in Immune and Neuronal Functions

Carl G. Gahmberg, Lin Ning, and Sonja Paetau

Abstract The neuron-specific intercellular adhesion molecule-5 (ICAM-5, telencephalin) is a member of the ICAM family of adhesion proteins. It has a complex structure with nine external immunoglobulin domains followed by a transmembrane and a cytoplasmic domain. The external part binds to β1- and β2-integrins and the matrix protein vitronectin, whereas its transmembrane domain binds to presenilins and the cytoplasmic domain to α-actinin and the ERM family of cytoplasmic proteins. In neurons it is confined to the soma and dendrites and it is enriched in dendritic filopodia with less expression in more mature dendritic spines. ICAM-5 strongly stimulates neurite outgrowth. ICAM-5 is cleaved by matrix metalloproteases upon activation of glutamate receptors or degraded through endocytosis resulting in increased spine maturation. Ablation of ICAM-5 expression increases functional synapse formation. The cleaved soluble fragment of ICAM-5 is immunosuppressive, which may be important in neuronal inflammatory diseases.

6.1 Introduction

Cell adhesion is necessary for the distribution of cells during development, for the formation of different organs and for the circulation of blood cells. It is absolutely pivotal for the development of the brain, but it is also equally important that the adhesion is regulated to enable the generation of dynamical changes in synapses and plasticity.

C.G. Gahmberg, M.D. (✉) • L. Ning, M.Sc. • S. Paetau, B.Sc.
Faculty of Bio- and Environmental Sciences, Division of Biochemistry and Biotechnology, University of Helsinki, Viikinkaari 5, 00014 Helsinki, Finland
e-mail: carl.gahmberg@helsinki.fi

V. Berezin and P.S. Walmod (eds.), *Cell Adhesion Molecules: Implications in Neurological Diseases*, Advances in Neurobiology 8, DOI 10.1007/978-1-4614-8090-7_6,

Four major families of adhesion proteins are known in mammals. These include the integrin family, the selectins, the immunoglobulin superfamily and the cadherins. The integrins are heterodimeric molecules composed of an α-chain and a β-chain. They are type I membrane proteins with the N-terminal domains on the outside of cells, followed by a transmembrane domain and a cytoplasmic part. There are 24 members in mammals which can be divided into different families according to their β-chains (Hynes 2002; Gahmberg et al. 2009; Luo et al. 2007). The β1-integrins are expressed in several tissues and they bind to extracellular matrix proteins such as collagens and fibronectin, but also to membrane proteins such as vascular adhesion protein-1 (VCAM-1). The β2-integrins are confined to leukocytes and bind to intercellular adhesion molecules (ICAMs) (Gahmberg et al. 2009; Luo et al. 2007). The β3-integrins are expressed for example in endothelial cells and platelets and bind to several different types of ligands. β4-integrins on endothelial cells bind mainly to laminins, but also to other ligands.

Importantly, the integrins need to become activated to be able to bind ligands, and it has become evident that conformational changes occur in their structures upon activation. Although there may exist exceptions, it is generally agreed that in the resting state the integrin ligand binding head is turned back towards the membrane, but when the integrin becomes activated, it straightens out, but the binding region remains closed. Further activation results in opening of the ligand binding domain with an increase in the affinity for ligands (Xiong et al. 2001; Luo et al. 2007). The integrins can be activated by ligand binding to non-integrin receptors on the cells followed by intracellular signalling to the integrins (inside-out activation). For example, different chemokines, which bind to chemokine cell surface receptors, are efficient activators of integrins (Gahmberg et al. 2009; Luo et al. 2007). But the integrins can also be activated by direct binding of ligands to the integrins on the outside of the cell followed by signalling (outside-in activation). How the communication over the membrane takes place is still incompletely understood, but we know that at least for β2-integrins, specific phosphorylations of the integrin cytoplasmic domains are essential. The phosphorylations result in turn in binding of intracellular 14-3-3 proteins, followed by binding of the G protein exchange factor Tiam1 and activation of the small G protein Rac-1 (Fagerholm et al. 2002, 2005, 2006; Hilden et al. 2003; Nurmi et al. 2007; Takala et al. 2008; Grönholm et al. 2011). Interactions with several cytoskeletal proteins such as filamin, talin, cytohesin-1 and the kindlins 1–3 are also important (Gahmberg et al. 2009).

The selectins are carbohydrate binding proteins recognizing sialyl Le^x-oligosaccharides on endothelial cells and leukocytes. These oligosaccharides can occur on several different glycoproteins and glycolipids. The selectins are active in the blood vessels and induce the rolling of leukocytes along the endothelial cell surfaces. This is often a prerequisite for the subsequent activation of the integrins.

The cadherins are expressed in epithelia and are important in organ formation and various adhesion events. They need Ca^{++} for activity and interact in a homophilic manner. Different cadherins can also bind to each other. They span the plasma membrane and bind to catenins in the cytoplasm and also here phosphorylation reactions may be important in the regulation of their activity (Takeichi 1991).

The large immunoglobulin superfamily of adhesion receptors is composed of molecules with one or more immunoglobulin (Ig) domains, which are involved in the interactions with integrins or act in homophilic binding events. A subfamily is formed by the five ICAM molecules (Gahmberg 1997).

In this review we focus on ICAM-5 (telencephalin), which is a neuronal cell surface protein present on the cell soma and dendrites. It exhibits properties important in synapse development and plasticity, but it is also involved in immunological reactions in the brain. Recent work has shown that this molecule may serve as a model protein in studies on synapse structure and development. It should be pointed out, however, that much of our present understanding of ICAM-5 is based on previous studies of the ICAM family of adhesion proteins and their integrin receptors.

No doubt it is necessary to characterize neuronal and glial molecules in detail to get an adequate understanding of how cognitive functions arise and develop. We realize that we are in the very beginning of this endeavour, but it is already apparent that it is possible to obtain informative results using well-defined experimental setups.

6.2 Expression, Structural Properties, and Binding Activities of ICAM-5

ICAM-5 (telencephalin) was originally identified using a monoclonal antibody, which showed an antigenic distribution in the telencephalon region of the brain including the hippocampus (Yoshihara and Mori 1994). Subsequent isolation and cloning of the molecule showed that it belongs to the intercellular adhesion molecule (ICAM) family (Yoshihara et al. 1994). Since their discovery (Rothlein et al. 1986; Patarroyo et al. 1987), these molecules were known to bind to the β2-integrin family of adhesion receptors, which are confined to leukocytes (Gahmberg et al. 2009; Takeichi 1991). The ICAMs are type I membrane proteins and are schematically drawn in Fig. 6.1. The external parts are formed by Ig domains, followed by a single transmembrane domain and a cytoplasmic part. ICAM-1 and ICAM-3 contain five Ig domains and ICAM-2 and ICAM-4 two Ig domains, whereas ICAM-5 has nine Ig domains. ICAM-1 is expressed in leukocytes but also in many other organs and it is often upregulated by various cytokines (Gahmberg et al. 1998), whereas ICAM-2 shows a more stable expression (Nortamo et al. 1991). It is confined to leukocytes, platelets, and endothelial cells. ICAM-3 is leukocyte specific and it is important in the regulation of immune responses. It binds to the DC-SIGN protein (Geijtenbeek et al. 2000) through the carbohydrate on DC-SIGN (van Kooyk and Geijtenbeek 2003; Bogoevska et al. 2007). ICAM-4 (Landsteiner–Wiener antigen) is erythroid cell specific and it may be important in the turnover of senescent red cells (Bailly et al. 1995; Toivanen et al. 2008). In contrast to these, ICAM-5 is neuron specific. The much more complex structure of ICAM-5 as compared to the other ICAMs indicated already early that it can express functions not associated with the other ICAM molecules.

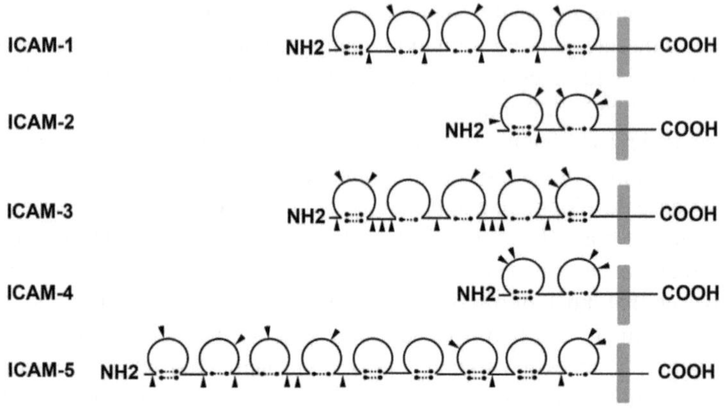

Fig. 6.1 Schematic structures of the ICAM family proteins. The immunoglobulin domains are depicted as *half circles* and the potential N-glycosylation sites are marked (*inverted triangle*)

In humans the ICAM-5 gene is on chromosome 19 (19p13.2) between the ICAM-1 and ICAM-3 genes and close to that of ICAM-4 (Sugino et al. 1997; Kilgannon et al. 1998). In contrast, the ICAM-2 gene is on chromosome 17 (17q23–25). The mouse ICAM-5 gene is on chromosome 9.

ICAM-5 is highly conserved in different mammalian species (Yang 2012). Most work on the protein has been done with the human, murine, rabbit and rat proteins, and the results from one species can in most instances be applied to others (Gahmberg et al. 2008). ICAM-5 shows a low expression in embryos, but the expression rapidly increases after birth. This is the time when large numbers of synapses are formed, and as discussed below, ICAM-5 is important during synapse formation and maturation.

The LFA-1 leukocyte integrin (αLβ2, CD11a/CD18) binds to ICAM-5 and the binding region in ICAM-5 is in its first Ig domain where glutamate-37 plays a key role (Mizuno et al. 1997; Tian et al. 1997; Zhang et al. 2008). The structure of the complex formed by this domain with the binding domain in LFA-1 (I-domain) has been determined by X-ray diffraction (Zhang et al. 2008). Upon binding, the α7-helix of the I-domain is moving away from its normal position enabling the binding of the corresponding helix from a neighbouring I-domain. Whether this occurs in cells is not known, but if so, such a mechanism could facilitate the clustering of LFA-1 molecules and in this way increase their avidity. Very recent work shows that β1-integrins, notably the α5β1-integrin, also bind to ICAM-5 through the two first Ig domains (Conant et al. 2011; Ning et al. 2013) (Fig. 6.2). Furthermore, it was recently found that ICAM-5 can bind to the cell matrix protein vitronectin (Furutani et al. 2012). Vitronectin is a soluble protein, which itself is a β1-integrin ligand through its arginine-glycine-aspartic acid (RGD) sequence. This binding site is located in the NH_2-terminal part of the protein, whereas the binding site for ICAM-5 is in the hemopexin domain COOH-terminal of the RGD sequence. The binding of vitronectin to ICAM-5 induced phosphorylation of ezrin–radixin–moesin (ERM) proteins resulting in the recruitment of actin filaments (Furutani et al. 2012).

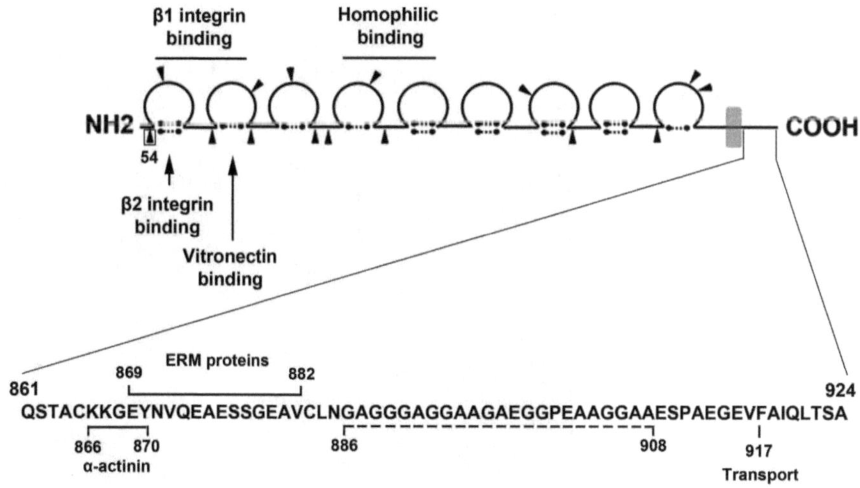

Fig. 6.2 Schematic structure of human ICAM-5 and important regions indicated. The binding sites of ligands in the external part are shown above. The potential N-glycosylation sites are marked (*inverted triangle*) and the site important in intracellular transport (Ohgomori et al. 2012). Below is the sequence of the cytoplasmic domain with the binding sites of α-actinin and ERM proteins indicated. The glycine–alanine-rich region is indicated (*dashed line*) and the phenylalanine important in the transport to dendrites (F917). The numbering is from the beginning of the signal sequence

ICAM-5 is heavily N-glycosylated and studies of rat brain ICAM-5 show a high proportion of high-mannose oligosaccharides (Ohgomori et al. 2009). Ohgomori et al. (2012) mutated all 15 N-glycosylation sites, one at a time, and found that the oligosaccharide at asparagine-54 was necessary for the transport of ICAM-5 to the cell surface, whereas mutation of the other sites had no effect. The asparagine-54 mutated molecule could not induce the formation of filopodia, which is a characteristic feature of ICAM-5.

The cytoplasmic domain of ICAM-5 is relatively long (Fig. 6.2). The proximal region contains the KKGEY sequence, which is important for the interaction with α-actinin (Nyman-Huttunen et al. 2006). It has been shown that ICAM-5-transfected Paju neuronal cells exhibit a co-distribution of α-actinin and ICAM-5. Introduction of the KKGEY peptide in the Paju neuronal cell line rounded up the cells and transfection with an ICAM-5 KK/AA mutant resulted in loss of the α-actinin/ICAM-5 co-distribution, indicating that the ICAM-5–α-actinin interaction is needed for cell adhesion. The ICAM-5/α-actinin co-distribution was most evident in dendritic shafts and filopodia (Nyman-Huttunen et al. 2006).

The cytoplasmic domain of ICAM-5 also interacts with the ERM family of cytoplasmic proteins (Furutani et al. 2007), and an acidic region in the domain distal to the KKGEY sequence was important in ERM protein binding (Fig. 6.2).

The ICAM-5/ERM co-distribution was most prevalent in filopodia (Furutani et al. 2007). The cytoplasmic domain also contains a long glycine/alanine-rich segment, but whether this is functionally important is not known. It could, however, give a flexible character to the cytoplasmic domain, which could facilitate

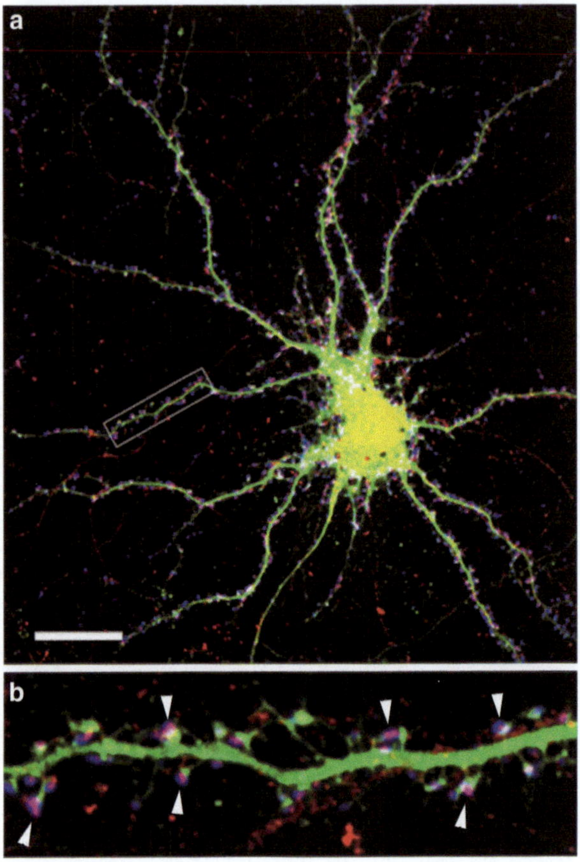

Fig. 6.3 An EGFP-transfected mouse neuron at 15 DIV. (**a**) The neuron was stained with an antibody to β1-integrins (*red*) and the presynaptic marker Synapsin I (*blue*). (**b**) Zoomed view of the area marked in (**a**). *Arrow heads* indicate β1-integrin immunoreactivity co-localized with Synapsin I, and opposite to EGFP-labelled spines

interactions with cytoplasmic molecules. The cytoplasmic domain is important for the transport of ICAM-5 to dendrites. Mutation of phenylalanine-905 in murine ICAM-5, which corresponds to phenylalanine-917 in human ICAM-5 (Fig. 6.2), abrogated the specific transport to dendrites (Mitsui et al. 2005) (Fig. 6.3).

6.3 ICAM-5 Regulates the Maturation of Dendritic Spines

ICAM-5 expression is strictly confined to dendrites and the soma, and axons do not express the protein. It stimulates the formation of neurites and their arborisation (Tamada et al. 1998; Tian et al. 2000). Filopodia and thin spines protruding from the dendritic shafts express high levels of ICAM-5, and these are considered to be precursors of mature spines (Tian et al. 2007). During spine maturation ICAM-5 expression is gradually decreased. Therefore, one could anticipate that ICAM-5 is most important early during dendritic development.

Knock-out of the ICAM-5 gene does not result in any observable changes in the gross anatomy of the brain (Nakamura et al. 2001). This is in contrast to the results obtained with several other adhesion molecules. Ablation of NCAM, cadherin-11, contactin and L1 resulted in morphological abnormalities in the brain and loss of central functions. However, more detailed studies of neurons showed that knock-out of ICAM-5 resulted in a decrease of filopodia and an increase in mature spines (Ning et al. 2013; Matsuno et al. 2006). Overexpression, on the other hand, increased the number of filopodia and thin spines and slowed down the development of mature mushroom spines. These results indicate that ICAM-5 acts as an inhibitor of rapid spine maturation, which results in the development of more mature spines and the formation of functional synapses. An important finding was made when it was shown that activation of glutamate (NMDA, AMPA) receptors resulted in ICAM-5 cleavage by matrix metalloproteases (MMPs). Addition of agonists to hippocampal neuron cultures activated MMP-2 and -9, which in turn cleaved ICAM-5 close to the lipid bilayer on the external side of the membrane (Tian et al. 2007). This resulted in an increase in the size of the spine heads, characteristic of mature spines. Disruption of the cytoskeletal association of ICAM-5 by treatment with agents disrupting the actin cytoskeleton, such as cytochalasin b, promoted the cleavage. Furthermore, the resulting soluble ICAM-5 (sICAM-5) increased the number and length of filopodia in wt neurons but not in ICAM-$5^{-/-}$ neurons. This effect could be due to homophilic binding or binding to β1-integrins in dendrites, but the exact mechanism is not known.

An additional mechanism of down-regulation of ICAM-5 resulting in an effect on the differentiation of spines was recently reported. The small G protein ADP-ribosylation factor 6 (ARF6) and its activator EFA6A induced endocytosis of ICAM-5 in flotillin-positive membrane subdomains (Raemaekers et al. 2012). This required dephosphorylation/release of ERM protein binding to ICAM-5. The reduction in ICAM-5, through activation of endocytosis (autophagy), resulted in a decrease in filopodia and an increase in spines.

6.4 Functional Synapse Formation Depends on ICAM-5–β1-Integrin Interactions

The above-mentioned studies showed that a decrease in ICAM-5 levels resulted in increased amounts of mushroom spines. Detailed studies have now shown that the α5β1-integrin is expressed in presynaptic sites and that it is juxtaposed to immature filopodia and covers mature mushroom spines (Ning et al. 2013). This finding indicated that the β1-integrin/ICAM-5 interaction is functionally important. The presynaptic and postsynaptic markers Synapsin I and PSD-95, respectively, were used to estimate the presence of pre- and postsynaptic contacts. We found that in wt neurons, Synapsin I largely distributed along the dendritic shafts, whereas in ICAM-$5^{-/-}$ neurons the marker was enriched in spine heads. The overlap between Synapsin I and PSD-95 was significantly increased in ICAM-$5^{-/-}$ neurons as compared to wt neurons.

To study if this increased interaction between presynaptic and postsynaptic elements in ICAM-5$^{-/-}$ neurons is functionally important, miniature excitatory postsynaptic currents (mEPSC) were recorded in wt and ICAM-5$^{-/-}$ neurons. Importantly, the ICAM-5$^{-/-}$ neurons showed an increase in the frequency of mEPSCs as compared to wt neurons whereas the amplitudes remained unchanged. The results show that in ICAM-5$^{-/-}$ neurons, the ICAM-5 deficiency increased the release of transmitters by affecting the contacts with the presynaptic sites.

A recombinant ICAM-5 molecule consisting of the two N-terminal domains binds to the α5β1-integrin (Ning et al. 2013). When the interaction between ICAM-5 and α5β1 was inhibited in neuron cultures using either ICAM-5- or β1-blocking antibodies, the release of the cleaved ICAM-5 molecule (sICAM-5) was increased. Opposite to this, strengthening the interaction between ICAM-5 and α5β1 using integrin activating antibodies reduced the amount of sICAM-5, and a similar effect was obtained by adding a construct consisting of domains 1–2 of ICAM-5 (Ning et al. 2013). Incubation of neurons with the ICAM-5 adhesion blocking antibodies resulted in a relative increase of mature spines and a similar effect was obtained with a β1-integrin blocking antibody. In contrast, when β1-integrin activating antibodies were used an opposite effect was seen. Now the relative proportion of filopodia increased as compared to mushroom spines (Ning et al. 2013). MMP-2 and MMP-9 were shown to be responsible for the generation of sICAM-5 (Tian et al. 2007). Evidently, a tight contact between the presynaptic and postsynaptic sites inhibited the action of the MMPs on ICAM-5.

Further proof of the importance of β1-integrins in synapse formation was obtained by down-regulation of their expression using small hairpin RNA. Co-culture of transfected neurons with non-transfected ones showed that dendrites in contact with axons with down-regulated β1-integrins expressed an increased proportion of mushroom spines.

The maturation of synapses, induced by a decrease in ICAM-5-involved adhesion, is associated with LTP (Sakurai et al. 1998). Interaction of sICAM-5 with β1-integrins resulted in increased LTP-evoked changes in action potential probability (Niederingshaus et al. 2012). An increase in neuronal excitability was also seen after up-regulation of MMP activity by activation of NMDA or AMPA receptors (Niederingshaus et al. 2012). Interestingly, also reference memory became better in ICAM-5$^{-/-}$ mice (Nakamura et al. 2001), and in experiments on auditory thalamocortical connectivity, it was found that deletion of the ICAM-5 gene increased plasticity (Barkat Rinaldi et al. 2011).

6.5 sICAM-5: A Functionally Important Molecule and a Marker of Degradation

The release of sICAM-5 by MMPs and the resulting stimulation of the development of dendritic spines is a physiological event. However, different drugs may act on ICAM-5. It has been shown that the highly addictive drug methamphetamine induces an increased cleavage of ICAM-5 by activating MMPs resulting in sICAM-5. An MMP inhibitor prevented the drug-induced cleavage, and administration of methamphetamine in vivo increased ICAM-5 cleavage (Conant et al. 2011). This is a very exciting result and may explain the changes seen in synaptic plasticity induced by methamphetamine.

Furthermore, sICAM-5 is immunosuppressive. The T lymphocyte receptor mediated activation of lymphocytes was prevented by sICAM-5, which was shown by decreased expression of the activation markers CD69, CD42 and CD25 (Tian et al. 2008). sICAM-5 seems to act by regulating the synthesis of cytokines. Thus sICAM-5 stimulated the synthesis of interferon-γ and transforming growth factor-β1, but not that of tumour necrosis factor-α. In contrast, ICAM-1 strongly activated the expression of all studied cytokines. The results indicate that ICAM-5 may be important in the regulation of various brain inflammatory conditions.

Presenilins-1 and -2 are components of the γ-secretase complex, which interacts with the amyloid precursor protein (APP) and liberates the amyloid β-peptide, which is thought to cause Alzheimer's disease (Annaert and de Strooper 2002; Selkoe and Kopan 2003). Presenilin-1 has been found to bind both to APP and ICAM-5 through the transmembrane domains of these proteins (Annaert et al. 2001). It does not, however, induce the down-regulation of ICAM-5 by the γ-secretase activity but by an autophagic mechanism (Esselens et al. 2004; Raemaekers et al. 2005). The transmembrane domains of presenilins are often mutated in patients with Alzheimer's disease, and deficiency of presenilin-1 resulted in the accumulation of ICAM-5 (Raemaekers et al. 2012). Evidently, the ARF6-mediated transport of ICAM-5 to autophagic vesicles was inhibited (Raemaekers et al. 2012; Esselens et al. 2004).

sICAM-5 production is increased in various clinical conditions affecting the brain. In hypoxic–ischemic injury induced by ligation of the carotid arteries in mice, there was a strong increase in the serum levels of sICAM-5 (Guo et al. 2000). In patients with acute encephalitis, sICAM-5 increased in the cerebrospinal fluid (Lindsberg et al. 2002). Herpes simplex virus (HSV-1) is a common cause of encephalitis and up to 90 % of the human population is exposed to the virus (Whitley and Roizman 2001). The virus infection is usually latent, but it is periodically reactivated, often during stress. The HSV-1 gene product UOL is involved in the neurovirulence, because deletion of the gene improved the outcome of encephalitis (Chan et al. 2005). It has been shown that the UOL protein binds to ICAM-5 and in wt HSV-1-infected mice there was a decrease in ICAM-5 expression (Tse et al. 2009). The mice infected with wt virus expressed higher levels of cytokines/chemokines than mice infected with UOL-deleted virus. The results indicate that sICAM-5 generated by virus infection increases cytokine production, which would be beneficial, but binding to the UOL protein compromises the outcome.

6.6 ICAM-5 and Microglia

Microglia are macrophage-like cells in the brain, which easily can become activated by a variety of stimuli. They are important for the innate immune response in the brain (Skaper et al. 2012). An important function of these cells is to kill infected

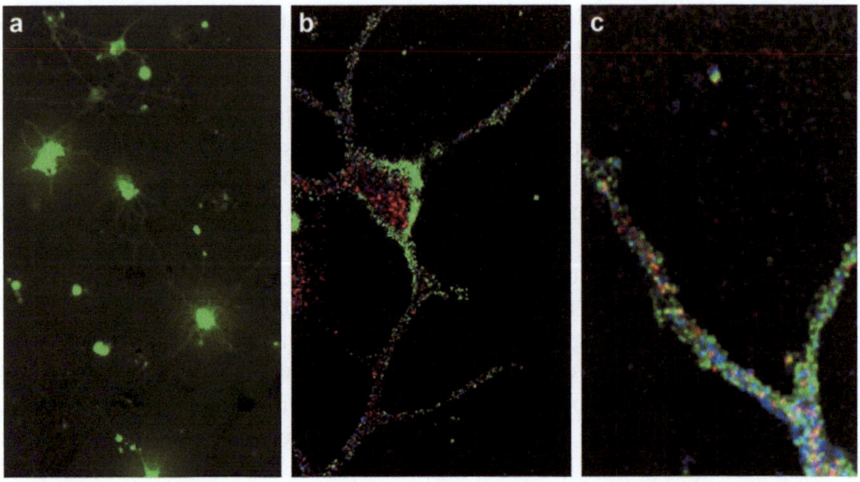

Fig. 6.4 Primary microglia culture from postnatal day 5 C57BL/6j mouse brain. Fixed cells were incubated with ICAM-5/D1-D2-Fc. The cells were stained with anti-IgG (*green*), anti-β1-integrin antibody (*red*) and β2-integrin antibody (*blue*). (**a**) Microglia culture, (**b**) a microglia in higher magnification, (**c**) a segment enlarged from (**b**). Some overlap of ICAM-5 with β1-integrins and β2-integrins is seen, but most ICAM-5 staining is separate from the integrin stainings

cells and phagocytize cellular and microbial remnants. But recent work shows that in addition, microglia are involved in the regulation of synapse plasticity and regeneration of neurons (Hughes 2012; Pascual et al. 2011).

Macrophages are known to contain β2-integrins and they are enriched in the αXβ2- and αDβ2-integrins (Gahmberg et al. 1997). Microglia are known to express the β2-integrins LFA-1 and Mac-1 and most probably they also express the other β2-integrins. Microglia could bind by their β2-integrins to endothelial cell ICAM-1 and ICAM-2 and astrocyte-expressed ICAM-1 and use their integrins for migration into the brain parenchyma. On the other hand, LFA-1 expressed in microglia could bind to ICAM-5 in dendrites and synapses. Macrophages also express β1-integrins, and the recent finding that β1-integrins bind ICAM-5 implies that the interaction between microglia and ICAM-5 may include both β1-integrin and β2-integrin family members. Furthermore, it has been shown that ICAM-5 binding results in spreading of microglia. Interestingly, this effect is not seen with ICAM-1 (Mizuno et al. 1999). ICAM-1 and ICAM-5 binding to the LFA-1 I-domain is similar, but in contrast to ICAM-1, ICAM-5 can bind to both LFA-1 and α5β1 (Ning et al. 2013).

Figure 6.4 shows ICAM-5 (domains 1–2) bound to mouse microglia stained with β1- and β2-integrin antibodies. The distribution of β1-integrins is patchy and seen in protrusions (Fig. 6.4b). Higher magnification shows some overlap of bound ICAM-5 with β1- and β2-integrins, but most of the added ICAM-5 seems to have a different distribution than the integrins. The results indicate that there may exist additional, presently unknown receptors for ICAM-5.

Recent work has shown that microglia have an influence on synaptic functions and eliminate or weaken functional synapses during development. This process must be important in brain plasticity underlying memory and learning. There is an obvious possibility that microglia use ICAM-5 as receptor when interacting with synapses. This could disrupt the binding to axonal β1-integrins affecting synapse functions. It would now be important to study these interactions in detail to be able to understand the mechanisms of synaptic pruning and the development of cognitive functions.

6.7 Perspectives

The CNS and the immune systems have much in common. Both exhibit memory functions, although they are of course very different. Neurons and immune cells share many cell surface molecules and have partially similar signalling systems. NCAM, CD200, CD22, CD47, the cadherins, RAGE and Thy-1 are examples of cell surface proteins present in both types of cells. In addition, neurons contain MHC class I transplantation antigens.

In a recent review we put forward the potential importance of neuronal regulation of immune responses in the brain (Tian et al. 2009). A number of cytokines are secreted by neurons and inflammatory cells, and direct interactions between immune cells such as T cells and microglia with neurons exist.

It is in fact highly rewarding to compare classical neuronal synapses with the more recently described immunological synapse (Fig. 6.5) (Dustin and Colman 2002). In both cases we have the pivotal specific interactions in the centre, i.e. MHC class I proteins interacting with the T cell receptor (TCR) in immune cells and glutamate binding to its receptors in neurons. The interaction between MHC proteins and TCR is relatively weak and there is of course no direct strengthening by glutamate and its receptors in neuronal synapses. Therefore, a number of additional molecules are needed to build up and make the synapses stronger (Dityatev et al. 2008). In neuronal synapses these include NCAM, cadherins, neurexin/neurogilin, synCAM and CD200. NCAM and cadherins exhibit homophilic binding, whereas CD200 and its receptor CD200R and ICAM-5 and β1-integrins show heterophilic binding. The immune synapses contain binding pairs such as ICAM-1/LFA-1, CD22/CD45 and B7/CD28.

The presence of similar or identical molecules in the neuronal and immunological synapses could make extensive crosstalks possible between neurons and immune cells. The crosstalks could take place in both directions, not only from immune cells to neurons, but also from neurons to immune cells. Several molecular interactions could be important here, but we want to put forward the possibility that ICAM-5 is of pivotal and rather unique importance. ICAM-5 is able to interact with both neuronal and leukocyte receptors. It is functionally active as a regulator of dendritic outgrowth and maturation, and it has an important role in the formation of axonal/dendritic contacts. On the other hand, it is able to bind to microglia, it may

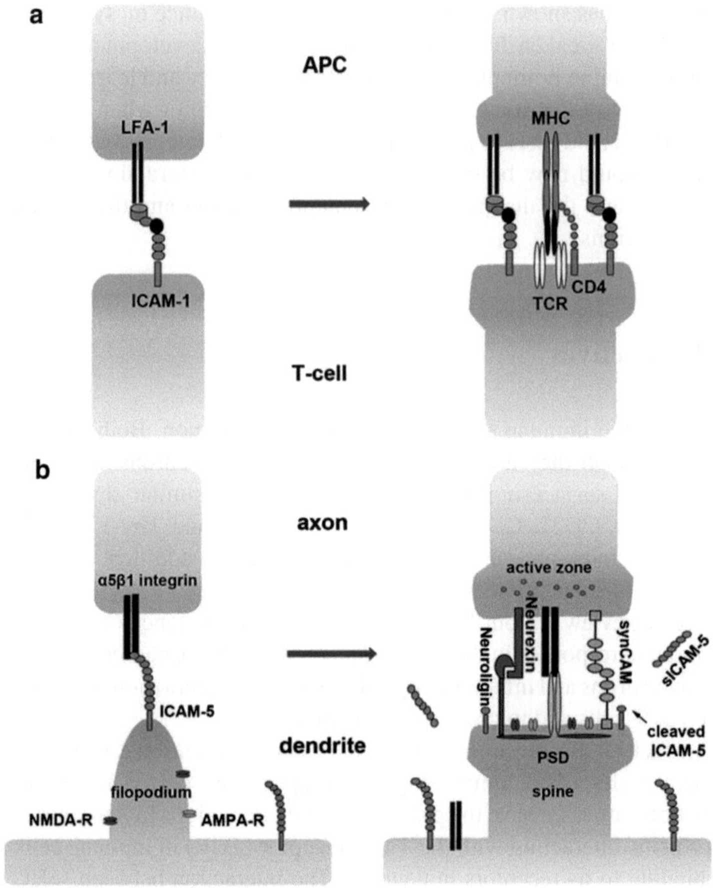

Fig. 6.5 Schematic comparison of the immunological and neuronal synapses. (**a**) In the antigen-presenting cell (APC), the LFA-1 integrin first interacts with ICAM-1 on T cells, but later the MHC transplantation antigen binds to the T cell receptor (TCR) and the LFA-1/ICAM-1 complex moves to the periphery. (**b**) In the neuronal synapse the initial contact is between axonal α5β1-integrin and filopodial ICAM-5. In the mature mushroom spines several molecules may bind the presynaptic and postsynaptic regions together. The amount of ICAM-5 is decreased and upon activation, ICAM-5 is cleaved by metalloproteases generating sICAM-5. The presynaptic β1-integrins remain bound to spines probably through the interactions with other adhesion molecules from the postsynaptic membrane

suppress T cell activation and it can act both as a membrane molecule and as a soluble mediator. These functional features are challenging objects of study in order to understand how the immunological and neuronal systems could cooperate in the CNS, but also for their potential importance in various pathological conditions affecting the brain. These may include Alzheimer's disease, stroke, encephalitis and multiple sclerosis.

Acknowledgements We thank Katarina Fellman for expert secretarial assistance and Dr. Mikaela Grönholm for comments on the manuscript. The original studies were supported by the Academy of Finland, the Sigrid Jusélius Foundation, the Finnish Medical Association, the Magnus Ehrnrooth Foundation, the Finnish Society of Science and Letters, the Wilhelm and Else Stockmann Foundation and the Liv och Hälsa Foundation.

Compliance with Ethics Requirements Carl G. Gahmberg declares that he has no conflict of interest. Lin Ning declares that she has no conflict of interest. Sonja Paetau declares that she has no conflict of interest. For studies with animals: all institutional and national guidelines for the care and use of laboratory animals were followed.

References

Annaert W, de Strooper B (2002) A cell biological perspective on Alzheimer's disease. Annu Rev Cell Dev Biol 18:25–51

Annaert W, Esselens C, Baert V, Boeve C, Snellings G, Cupers P, Craessaerts K, de Strooper B (2001) Interaction with telencephalin and the amyloid precursor protein predicts a ring structure for presenilins. Neuron 32:579–589

Bailly P, Tontti E, Hermand P, Cartron J-P, Gahmberg CG (1995) The red cell LW blood group protein is an intercellular adhesion molecule which binds to CD11/CD18 leukocyte integrins. Eur J Immunol 25:3316–3320

Barkat Rinaldi T, Polley D, Hensch T (2011) A critical period for auditory thalamocortical connectivity. Nat Neurosci 14:1189–1196

Bogoevska V, Nollau P, Lucka L, Grunow D, Klampe B, Uotila LM, Samsen A, Gahmberg CG, Wagener C (2007) DC-SIGN binds ICAM-3 isolated from pheripheral human leukocytes through Lewis x residues. Glycobiology 17:324–333

Chan D, Cohen J, Naito J, Mott KR, Osorio N, Jin L, Fraser N, Jones C, Wechsler SL, Perng GC (2005) A mutant deleted for most of the herpes simplex virus type 1 (HSV-1) UOL gene does not affect the spontaneous reactivation phenotype in rabbits. J Neurovirol 12:5–16

Conant K, Lonskaya I, Szklarczyk A, Krall C, Steiner J, Maquire-Zeiss K, Lim ST (2011) Methamphetamine-associated cleavage of the synaptic adhesion molecule intercellular adhesion molecule-5. J Neurochem 118:521–532

Dityatev A, Bukalo O, Schachner M (2008) Modulation of synaptic transmission and plasticity by cell adhesion and repulsion molecules. Neuron Glia Biol 4:197–209

Dustin M, Colman D (2002) Neural and immunological synaptic relations. Science 298: 785–789

Esselens C, Oorschot V, Baert V, Raemaekers T, Spittaels K, Serneels L, Zheng H, Saftig P, de Strooper B, Klumperman J, Annaert W (2004) Presenilin 1 mediates the turnover of telecephalin in hippocampal neurons via an autophagic degradative pathway. J Cell Biol 166:1041–1054

Fagerholm S, Morrice N, Gahmberg CG, Cohen P (2002) Phosphorylation of the cytoplasmic domain of the integrin CD18 chain by protein kinase C isoforms in leukocytes. J Biol Chem 277:1728–1738

Fagerholm SC, Hilden TJ, Nurmi SM, Gahmberg CG (2005) Specific integrin α and β chain phosphorylations regulate LFA-1 activation through affinity-dependent and -independent mechanisms. J Cell Biol 171:705–715

Fagerholm SC, Varis M, Stefandakis M, Hilden TJ, Gahmberg CG (2006) α-chain phosphorylation of the human leukocyte CD116b/CD18(Mac-1) integrin is pivotal for integrin activation to bind ICAMs and leukocyte extravasation in vivo. Blood 108:3379–3386

Furutani Y, Matsuno H, Kawasaki M, Sasaki T, Mori K, Yoshihara Y (2007) Interaction between telencephalin and ERM family proteins mediates dendritic filopodia formation. J Neurosci 27:8866–8876

Furutani Y, Kawasaki M, Matsuno H, Mitsui S, Mori K, Yoshihara Y (2012) Vitronectin induces phosphorylation of ezrin/radixin/moesin acting-binding proteins through binding to its novel neuronal receptor telencephalin. J Biol Chem 287:39041–39049. doi:10.1074/jbc.M112.383851

Gahmberg CG (1997) Leukocyte adhesion CD11/CD18 integrins and intercellular adhesion molecules. Curr Opin Cell Biol 9:643–650

Gahmberg CG, Tolvanen M, Kotovuori P (1997) Leukocyte adhesion. Structure and function of human leukocyte β_2-integrins and their cellular ligands. Eur J Biochem 245:215–232

Gahmberg CG, Valmu L, Tian L, Kotovuori P, Ihanus E, Tian L, Pessa-Morikawa T (1998) Leukocyte integrins and inflammation. Cell Mol Life Sci 54:549–555

Gahmberg CG, Tian L, Ning L, Nyman-Huttunen H (2008) ICAM-5-a novel two facetted adhesion molecule in the mammalian brain. Immunol Lett 117:131–135

Gahmberg CG, Fagerholm SC, Nurmi SM, Chavakis T, Marchesan S, Grönholm M (2009) Regulation of integrin activity and signalling. Biochim Biophys Acta 1790:431–444

Geijtenbeek TBH, Torensma SJ, van Vliet G, van Duijnhoven Gosse JA, van Kooyk Y, Figdor CG (2000) Identification of DC-SIGN, a novel dendritic cell-specific ICAM-3 receptor that supports primary immune responses. Cell 100:578–585

Grönholm M, Jahan F, Marchesan S, Karvonen U, Aatonen M, Narumanchi S, Gahmberg CG (2011) TCR-induced activation of LFA-1 involves signaling through Tiam1. J Immunol 187: 3613–3619

Guo H, Tong N, Turner T, Epstein LG, McDermott MP, Kilgannon P, Gelbard H (2000) Release of the neuronal glycoprotein ICAM-5 in serum after hypoxic-ischemic injury. Ann Neurol 48: 590–602

Hilden TJ, Valmu L, Kärkkäinen S, Gahmberg CG (2003) Threonine phosphorylation sites in the β_2 and β_7 leukocyte integrin polypeptides. J Immunol 170:4170–4177

Hughes V (2012) The constant gardener. Nature 485:570–571

Hynes RO (2002) Integrins: bidirectional, allosteric signaling machines. Cell 110:673–687

Kilgannon P, Turner T, Meyer J, Wisdom W, Gallatin WM (1998) Mapping of the ICAM-5 (telencephalin) gene, a neuronal member of the ICAM family, to a location between ICAM-1 and ICAM-3 on human chromosome 19p13.2. Genomics 54:328–330

Lindsberg PJ, Launes J, Tian L, Mikkola H, Subramanian V, Siren J, Hokkanen L, Hyypiä T, Carpen O, Gahmberg CG (2002) Release of soluble ICAM-5, a neuronal adhesion molecule, in acute encephalitis. Neurology 58:446–451

Luo B-H, Carman CV, Springer TA (2007) Structural basis of integrin regulation and signaling. Annu Rev Immunol 25:619–647

Matsuno H, Okabe S, Mishina M, Yanagida T, Mori K, Yoshihara Y (2006) Telencephalin slows spine maturation. J Neurosci 26:1776–1786

Mitsui S, Saito M, Hayashi K, Mori K, Yoshihara Y (2005) A novel phenylalanine-based targeting signal directs telencephalin to neuronal dendrites. J Neurosci 25:1122–1131

Mizuno T, Yoshihara Y, Inazawa J, Kagamiyama H, Mori K (1997) cDNA cloning and chromosomal localization of the human telencephalin and its distinctive interaction with lymphocyte function-associated antigen-1. J Biol Chem 272:1156–1163

Mizuno T, Yoshihara Y, Kagamiyama H, Ohsawa K, Ilmai Y, Kohsaka S, Mori K (1999) Neuronal adhesion molecule telencephalin induces rapid cell spreading of microglia. Brain Res 849:58–99

Nakamura K, Manabe T, Watanabe M, Mamiya T, Ichikawa R, Kiyama Y, Sanbo M, Yagi T, Inoue Y, Nabeshima T, Mori H, Mishina M (2001) Enhancement of hippocampal LTP, reference memory and sensorimotor gating in mutant mice lacking a telencephalon-specific cell adhesion molecule. Eur J Neurosci 13:179–189

Niederingshaus M, Chen X, Dzakpasu R, Conant K (2012) MMPs and soluble ICAM-5 increase neuronal excitability within in vitro networks of hippocampal neurons. PLoS One 10.1371 journal.pone.0042631

Ning L, Tian L, Smirnov S, Vihinen H, Llano O, Vick K, Davis RL, Rivera C, Gahmberg CG (2013) Interactions between intercellular adhesion molecule-5 (ICAM-5) and β_1 integrins regulate neuronal synapse information. J Cell Sci 126:77–89

Nortamo P, Salcedo R, Timonen T, Patarroyo M, Gahmberg CG (1991) A monoclonal antibody to the human leukocyte adhesion molecule ICAM-2. Cellular distribution and molecular characterization of the antigen. J Immunol 146:2530–2535

Nurmi SM, Autero M, Raunio AK, Gahmberg CG, Fagerholm SC (2007) Phosphorylation of the LFA-1 integrin β_2-chain on Thr-758 leads to adhesion, Rac-1/Cdc42 activation and stimulation of CD69 expression in human T cells. J Biol Chem 282:968–975

Nyman-Huttunen H, Tian L, Ning L, Gahmberg CG (2006) α-actinin-dependant cytoskeletal anchorage is important for ICAM-5 mediated neuritic outgrowth. J Cell Sci 119:3057–3066

Ohgomori T, Funatsu O, Nakaya S, Morita A, Ikekita M (2009) Structural study of the N-glycans of intercellular adhesion molecule-5(telencephalin). Biochim Biophys Acta 1790:1611–1623

Ohgomori T, Nanao T, Morita A, Ikekita M (2012) Asn(54)-linked glycan is critical for functional folding of intercellular adhesion molecule-5. Glycoconj J 29:47–55

Pascual O, Achour S, Rostaing P, Triller A, Bessis A (2011) Microglia activation triggers astrocyte-mediated modulation of excitatory neurotransmission. Proc Natl Acad Sci USA 109:197–205

Patarroyo M, Clark EA, Prieto J, Kantor C, Gahmberg CG (1987) Identification of a novel adhesion molecule in human leukocytes by monoclonal antibody LB-2. FEBS Lett 210:127–131

Raemaekers T, Esselens C, Annaert W (2005) Presenilin 1: more than just a γ-secretase. Biochem Soc Trans 33:559–561

Raemaekers T, Peric A, Baatsen P, Sannerud R, Declerck I, Baert V, Michiels C, Annaert W (2012) ARF6-mediated endosomal transport of telencephalin affects dendritic filopodia-to-spine maturation. EMBO J 31:3252–3269

Rothlein R, Dustin ML, Marlin SD, Springer TA (1986) A human intercellular adhesion molecule (ICAM-1) distinct from LFA-1. J Immunol 137:1270–1274

Sakurai E, Hashikawa T, Yoshihara Y, Kaneko S, Satoh M, Mori K (1998) Involvement of dendritic adhesion molecule telencephalin in hippocampal long-term potentiation. Neuroreport 9:881–886

Selkoe D, Kopan R (2003) Notch and presenilin: regulated intramembrane protelolysis links development and degeneration. Annu Rev Neurosci 26:565–597

Skaper S, Giusti P, Facci L (2012) Microglia and mast cells: two tracks on the road to neuroinflammation. FASEB J 26:3103–3117

Sugino H, Yoshihara Y, Copeland NG, Gilbert DJ, Jenkins NA, Mori K (1997) Genomic organization and chromosomal localization of the mouse telencephalin gene, a neuronal member of the ICAM family. Genomics 43:209–215

Takala H, Nurminen E, Nurmi SM, Aatonen M, Strandin T, Takatalo M, Kiema T, Gahmberg CG, Ylänne J, Fagerholm SC (2008) β_2 integrin phosphorylation on Thr758 acts as a molecular switch to regulate 14-3-3 and filamin binding. Blood 112:1853–1862

Takeichi M (1991) Cadherin cell adhesion receptors as a morphogenetic regulator. Science 251: 1451–1455

Tamada A, Yoshihara Y, Mori K (1998) Dendrite-associated cell adhesion molecule, telencephalin, promotes neurite outgrowth in mouse embryo. Neurosci Lett 240:163–166

Tian L, Yoshihara Y, Mizuno T, Mori K, Gahmberg CG (1997) The neuronal glycoprotein, telencephalin, is a cellular ligand for the CD11a/CD18 leukocyte integrin. J Immunol 158: 928–936

Tian L, Nyman H, Kilgannon P, Yoshihara Y, Mori K, Andersson LC, Kaukinen S, Rauvala H, Gallatin WM, Gahmberg CG (2000) Intercellular adhesion molecule-5 induces dendritic outgrowth by homophilic adhesion. J Cell Biol 150:243–252

Tian L, Stefanidakis M, Ning L, van Lint P, Nyman-Huttunen H, Libert C, Itohara S, Mishina M, Rauvala H, Gahmberg CG (2007) Activation of NMDA receptors promotes dendritic spine development through MMP-mediated ICAM-5 cleavage. J Cell Biol 178:687–700

Tian L, Lappalainen J, Autero M, Hänninen S, Rauvala H, Gahmberg CG (2008) Shredded neuronal ICAM-5 suppresses T cell activation. Blood 111:3615–3625

Tian L, Rauvala H, Gahmberg CG (2009) Neuronal regulation of immune responses in the central nervous system. Trends Immunol 30:91–99
Toivanen A, Ihanus E, Mattila M, Lutz HU, Gahmberg CG (2008) Importance of molecular studies on major blood groups—intercellular adhesion molecule-4, a blood group antigen involved in multiple cellular interactions. Biochim Biophys Acta 1780:456–466
Tse M, Lane C, Mott K, Onlamoon N, Hsiao HM, Peng GC (2009) ICAM-5 modulates cytokine/chemokine production in the CNS during the course of herpes simplex virus type 1 infection. J Neuroimmunol 213:12–19
van Kooyk Y, Geijtenbeek TBH (2003) DC-SIGN: escape mechanism for pathogens. Nat Rev Immunol 3:697–709
Whitley RJ, Roizman B (2001) Herpes simplex virus infections. Lancet 357:1513–1518
Xiong J-P, Stehle T, Diefenbach B, Zhang R, Dunker R, Scott DL, Joachimiak A, Goodman SL, Arnaout MA (2001) Crystal structure of the extracellular segment of integrin $\alpha v\beta 3$. Science 294:339–345
Yang H (2012) Structure, expression and function of ICAM-5. Comp Funct Genomics 2012: 11p
Yoshihara Y, Mori K (1994) Telencephalin: a neuronal area code molecule? Neurosci Res 21:119–124
Yoshihara Y, Oka S, Nemoto Y, Watanabe Y, Nagata S, Kagamiyama H, Mori K (1994) An ICAM-related neurone glycoprotein telencephalin, with brain segment specific expression. Neuron 12:541–553
Zhang H, Casasnovas JM, Jin M, Liu J, Gahmberg CG, Springer TA, Wang J (2008) An unusual allosteric mobility of the C-terminal helix of a high-affinity α_L integrin I domain variant bound to ICAM-5. Mol Cell 31:432–437

Chapter 7
ROUNDABOUT Receptors

Athéna R. Ypsilanti and Alain Chedotal

Abstract Roundabout receptors (Robo) and their Slit ligands were discovered in the 1990s and found to be key players in axon guidance. Slit was initially described as an extracellular matrix protein that was expressed by midline glia in *Drosophila*. A few years later, it was shown that, in vertebrates and invertebrates, Slits acted as chemorepellents for axons crossing the midline. Robo proteins were originally discovered in *Drosophila* in a mutant screen for genes involved in the regulation of midline crossing. This ligand–receptor pair has since been implicated in a variety of other neuronal and non-neuronal processes ranging from cell migration to angiogenesis, tumourigenesis and even organogenesis of tissues such as kidneys, lungs and breasts.

7.1 Robo and Slit Structures

7.1.1 Robo Structure

The Robo family belongs to the immunoglobulin (Ig) superfamily of cell adhesion molecules (CAM) (Kidd et al. 1998a; Sundaresan et al. 1998b). Three *robo* genes (*robo1-3*) have been identified in most species including *Drosophila* and birds, while a fourth *robo* gene (*robo4*) was also found in zebrafish and in mammals (Hao et al. 2001; Kidd et al. 1998a; Lee et al. 2001; Rajagopalan et al. 2000; Simpson et al. 2000; Vargesson et al. 2001; Yuan et al. 1999). In mammals, Robo1-3 are expressed in many tissues during development and particularly in the

A.R. Ypsilanti, Ph.D. (✉) • A. Chedotal, Ph.D., DR1 INSERM.
Institut de la Vision, INSERM UMRS 968, CNRS UMR 7210, 17 Rue Moreau, Paris 75012, France
e-mail: athena.ypsilanti@inserm.fr; alain.chedotal@inserm.fr

V. Berezin and P.S. Walmod (eds.), *Cell Adhesion Molecules: Implications in Neurological Diseases*, Advances in Neurobiology 8, DOI 10.1007/978-1-4614-8090-7_7,

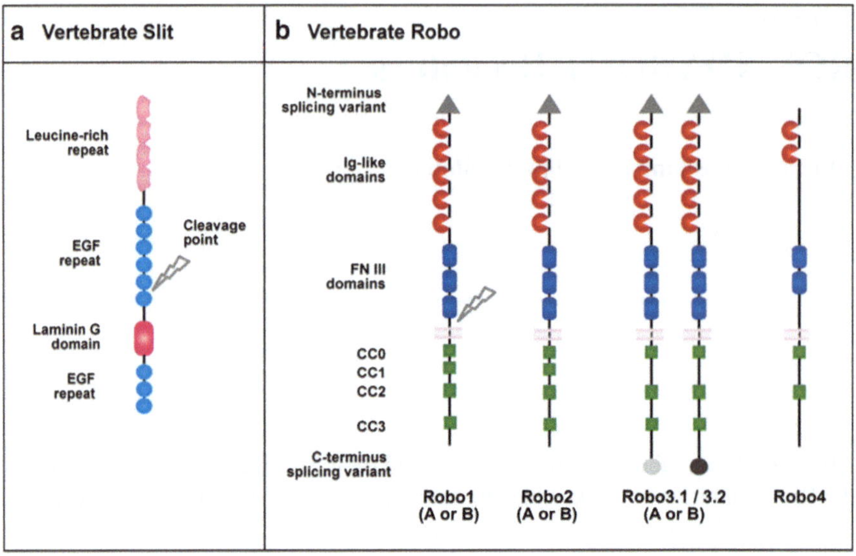

Fig. 7.1 Structure of Slits and Robos. (**a**) Structure of the prototypical vertebrate Slit protein containing four leucine-rich repeat domains, nine EGF repeat domains and a laminin G domain. Slit2 can be cleaved proteolytically to yield an inactive C-terminal and a biologically active N-terminal. (**b**) Structure of the vertebrate Robo receptors. The archetypal Robo receptor possesses five Ig-like domains, three fibronectin type III motifs, a transmembrane spanning region and four C-terminal domains (CC0–3). Robo3 lacks the CC1 domain and Robo4 lacks three Ig-like domains, one FN3 domain and both the CC1 and the CC3 domains. Alternative isoforms exist for Robo1-3: isoforms with variations at the N-terminal are referred to as Robo1-3 A or B whereas isoforms differing at the C-terminus are referred to as Robo3.1 or Robo3.2. Finally, Robo1 can be proteolytically cleaved to yield a soluble receptor

nervous system (Kidd et al. 1998b; Sundaresan et al. 1998a; Yuan et al. 1999). Robo4 also known as Magic Roundabout was originally thought to be expressed mainly by endothelial cells (Bedell et al. 2005; Park et al. 2003), but recent evidence indicates that it is also expressed in the central nervous system (CNS) of zebrafish (Bedell et al. 2005) and of mice (Zheng et al. 2011).

The archetypical Robo receptor possesses an extracellular region containing five Ig domains, three fibronectin type III motifs (FN3), a single and short transmembrane region, and a long intracellular tail containing four conserved cytoplasmic (CC0–CC3) domains (see Fig. 7.1). Still, in the mammalian Robo family, the structure of Robo3 and Robo4 presents some divergences from the prototypical Robo structure. In fact, Robo3 lacks the CC1 domain (Yuan et al. 1999; Jen et al. 2004). Moreover, Robo4 possesses only two conserved Ig domains, contains variants of the CC0 and CC2 motifs and lacks the CC1 and CC3 domain (Huminiecki et al. 2002). In addition, alternative isoforms or splice variants have been described for all Robo receptors to date (Camurri et al. 2005; Chen et al. 2008; Huminiecki et al. 2002; Jen et al. 2004; Kidd et al. 1998a; Yuan et al. 1999). In the majority of cases the biological implication of the alternative splicing remains undetermined, although alternative splicing of Robo3 into Robo3.1 and Robo3.2 isoforms was proposed to play an important role in the regulation of axonal midline crossing (Chen et al. 2008). In fact, Robo3.1 and Robo3.2 have opposite functions for commissural axons in the spinal cord: whereas Robo3.1 promotes midline crossing, Robo3.2 suppresses midline crossing.

The cytoplasmic domains of Robo receptors do not possess any catalytic activities and thus exert their effect by interacting with different signalling molecules (Chedotal 2007). A recent insightful study has suggested that, although specific Robo receptors have distinct cytoplasmic domains, the specific responses encoded by different Robos are instructed by differences in the extracellular Ig domain structures of the receptors (Evans and Bashaw 2010). Moreover, specific Ig domains in Robo receptor subtypes were found to encode different functions such as the Ig1 and Ig3 of Robo2 which were important in specifying lateral positioning of longitudinal axons whereas the Ig2 domain functions mostly in promoting midline crossing.

7.1.2 *Slit Structure*

Members of the Slit family have been found in a variety of species including *Drosophila*, *Caenorhabditis elegans*, *Xenopus*, chicks and mammals (Hao et al. 2001; Holmes et al. 1998; Kidd et al. 1999). In mammals, three Slit proteins (Slit1–3) are expressed in the nervous system and in other tissues (Holmes et al. 1998; Itoh et al. 1998; Li et al. 1999; Marillat et al. 2002). Slit proteins are generally large (around 200 kDa) secreted glycoproteins which are composed of the following domains: an N-terminal signal peptide, four leucine-rich repeats (LRR), seven to nine (in *Drosophila* and vertebrates, respectively) EGF-like sequences, a laminin-G domain containing agrin, laminin and perlecan homologies (ALP) and a C-terminal cysteine-rich region (see Fig. 7.1). Slit2 can be cleaved proteolytically in its EGF domain, thus generating an inactive C-terminal Slit fragment and a long N-terminal fragment (Slit-N) which retains the ability to bind Robo receptors and mediate biological activity (Battye et al. 2001; Brose et al. 1999; Chen et al. 2001; Nguyen-Ba-Charvet et al. 2001; Wang et al. 1999).

The active site of Slit2 was identified as being located on the concave surface of the LRR2 domain which mediates binding to the Ig1 domain of Robo1 (Howitt et al. 2004; Morlot et al. 2007). The residues on the binding sites of both the Slit and Robo proteins are highly conserved throughout evolution, thus accounting for the promiscuous binding of Slit ligands to Robo receptors (Brose et al. 1999; Sabatier et al. 2004). That being said, the aptitude of Robo3 to bind Slits has been inconsistently reported in the literature with some groups detecting binding and others not (Sabatier et al. 2004; Camurri et al. 2005; Mambetisaeva et al. 2005). Moreover, recent studies suggest that Robo4 is not a Slit receptor although this remains controversial (Koch et al. 2011; Park et al. 2003; Suchting et al. 2005; Verissimo et al. 2009; Wang et al. 2003; Zheng et al. 2011).

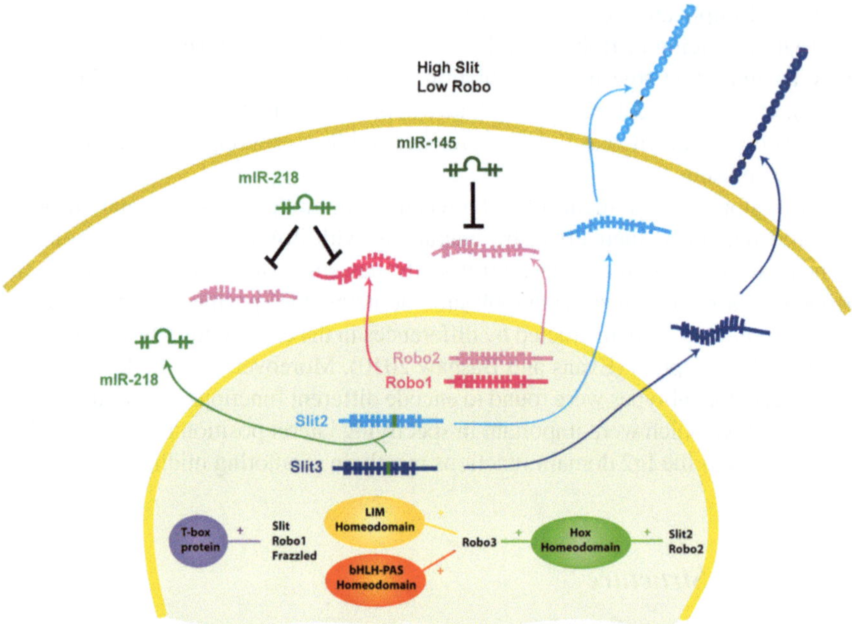

Fig. 7.2 Transcriptional and translational regulation of Slit–Robo. Several families of transcription factors such as the T-box, the bHLH-PAS, the LIM homeodomains or the Hox homeodomain transcription factors regulate the transcription of *Slits* and *Robos*. In addition, microRNAs like miR-218.1 (which is localised intronically in the *Slit2* gene) negatively regulate translation of Robo1 and Robo2. High expression of Slit2 thus leads to low levels of Robo1 and Robo2. Likewise, miR-145 is a negative regulator of Robo2 translation. Adapted from Chedotal (2011)

7.1.3 Regulation of Robo Expression

Membrane-bound proteins can be regulated in several ways including transcriptional or translational regulation and the control of receptor transport to the membrane. Robo receptors have been shown to be regulated in a variety of manners which will be outlined below.

Transcriptional and Translational Regulation

At a transcriptional level, Robo expression is regulated by a variety of proteins from different families of transcription factors (TFs) (see Fig. 7.2). Proteins of the LIM homeodomain family such as Lhx2 or Lhx9 have been shown to activate Robo3 in a direct manner (Marcos-Mondejar et al. 2012; Wilson et al. 2008). Moreover, Sim1, Sim2 and Arnt2 of the basic-helix-loop-helix-PAS transcription factor family are also able to regulate Robo3 expression (Marion et al. 2005; Schweitzer et al. 2013). Likewise, Robo2 and Robo3 are activated by Hoxa2, a member of the Hox family of TFs (Bravo-Ambrosio et al. 2012; Di Bonito et al. 2013; Geisen et al. 2008). And finally, a few members of the T-box family of TFs such as Midline protein (an orthologue of Tbx20) or Tbx1 are also known to regulate Robo and/or Slit expression (Calmont et al. 2009; Liu et al. 2009).

Another level of regulation of Robo expression is achieved by regulating Robo mRNAs at the translational level. MicroRNAs (miRNAs) are short non-coding RNA sequences (roughly 21–22 nucleotides in length) that bind to mRNAs to negatively modulate their post-transcriptional expression [Fig. 7.2 and reviewed by Carrington and Ambros (2003) and Carthew (2006)]. It was first shown in tumour cells that miR-218, which is downregulated in certain cancer cell lines, can repress Robo1 expression (Alajez et al. 2011; Tie et al. 2010). Intriguingly, this study also demonstrated that the miR-218 is synthesised from two precursors, mir-218-1 and mir-218-2, located in the intronic sequence of Slit2 and Slit3 genes, respectively. Mature mir-218 is mostly derived from the mir-218-2 precursor whose expression is negatively correlated with Slit3 expression, thus creating a negative feedback loop. In zebrafish heart tube formation, miR-218 negatively regulates both Robo1 and Robo2 (Fish et al. 2011). Furthermore, after sciatic nerve injury, several miRNAs are downregulated including miR-145 which can potentially target the 3′-UTR of Robo2 and one of its downstream signalling partners, srGAP1 (Zhang et al. 2011). Robo2 is upregulated prior to sciatic nerve injury and it was shown in vitro that miR-145 can downregulate Robo2 expression. Further, miRNAs are strongly downregulated following nerve injury, which could therefore facilitate Robo receptor upregulation by decreasing their translational regulation. Many miRNAs are specifically enriched in neuronal tissues and have been shown to be potent regulators of gene expression during development [reviewed by Klein et al. (2005)], thus presaging future discoveries in the translational control of Robos, their ligands and their downstream partners by miRNAs. Apart from miRNAs, post-transcriptional regulation can be achieved by RNA binding proteins such as Musashi1 which was shown to regulate Robo3 expression at the protein level in precerebellar neurons (Kuwako et al. 2010).

Control of the Subcellular Localisation of Robo

Robo function can be modulated by regulating its expression at the plasma membrane (see Fig. 7.3). In *Drosophila* the transmembrane protein Comm negatively regulates Slit repulsion by preventing the transport of Robo to the cell surface and instead favouring the routing of Robo to the endosome for degradation (Keleman et al. 2002). Interestingly Comm function is not conserved in all insects, and in particular, in *Tribolium castaneum* (the flour beetle), Comm does not downregulate the expression of Robo2/3 (Evans and Bashaw 2012). Furthermore, recently, a scaffolding protein Canoe (Cno) was also demonstrated to form a complex with Robo and stabilise its expression in growth cone filopodia (Slováková et al. 2012). A comparable regulation of the subcellular regulation of Robo proteins appears to exist in mammals. In fact, Robo1 binds directly to the ubiquitin-specific protease 33 (USP33) and this interaction is necessary for commissural neurons to respond to Slit and for their

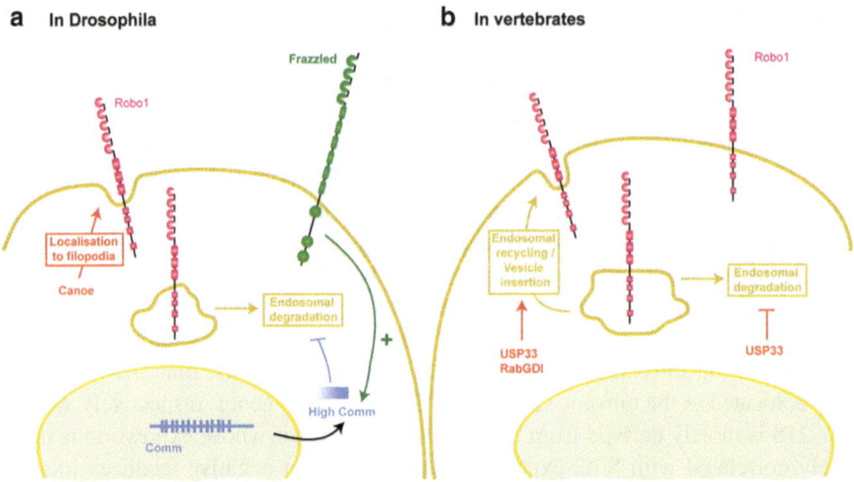

Fig. 7.3 Regulation of Robo at the membrane. (**a**) In *Drosophila*, Comm. acts at the midline as a negative regulator of Robo by downregulating Robo during midline crossing and thus suppressing Robo-induced midline repulsion (Keleman et al. 2002). The Netrin receptor, Frazzled, is a positive regulator of Comm transcription. In addition, Canoe stabilises Robo expression in growth cone filopodia. (**b**) Comm has no mammalian homologues, but the ubiquitin-specific protease 33 (USP33) acts to mediate slit responsiveness and ensure commissural axon crossing by either preventing Robo1 from being targeted for degradation or by facilitating its endosomal recycling. Moreover, RabGDI acts to promote Robo1 expression at the membrane

axons to cross the midline (Yuasa-Kawada et al. 2009a, b). USP33 appears to mediate these functions by either preventing Robo1 from being targeted for degradation or facilitating its endosomal recycling. Moreover, a component of the vesicular machinery, RabdGI, was also shown to be important for the post-translational regulation of Robo1 expression at the membrane on chick commissural axons (Philipp et al. 2012).

7.2 How Do Robo Proteins Transduce Their Effect?

As previously described, Slits bind to Robos through evolutionarily conserved domains and they constitute Robos' primary ligands. Robo activation at the cell surface is then translated within the cell by a number of signalling cascade events. Further, Slit–Robo signalling can be modulated by co-receptors, by homodimeric or heterodimeric receptor binding or by other ligands.

7.2.1 Downstream Signalling of Slit and Robo

Axon guidance receptors sense the environment around the leading process or the growth cone, and based on the attractiveness or repulsiveness of their surroundings, they act upon the cell membrane/cytoskeleton in order to mediate the appropriate movement either towards or away from the outside milieu. In order to accomplish this requirement, Robo–Slit signalling relies on two distinct processes: the modulation of calcium and cyclic nucleotides by Slit and the recruitment of downstream effectors.

Slit2 is able to induce an elevation of intracellular concentration of calcium (Ca^{2+}) in the growth cone of migrating cerebellar granule neurons (Xu et al. 2004) and leads to a reversal of migration direction in granular cells and other types of migrating neurons (Ward et al. 2005; Xu et al. 2004). Slit2-dependent direction change in cerebellar granule neurons was subsequently shown to cause a redistribution of RhoA activity towards the new leading front and to require RhoA activation (Guan et al. 2007). Cyclic nucleotides modulation of Slit signalling can also lead to cytoskeletal changes. For instance, the outgrowth that occurs when exposing dorsal root ganglia neurons to Slit2 can be inhibited by lowering cGMP concentrations using pharmacological inhibitors (Nguyen-Ba-Charvet et al. 2001).

In addition, Robo receptors rely on a number of proteins involved in cytoskeletal dynamics such as the small GTPases of the Rho family which are regulated by the guanine nucleotide exchange factors (GEFs) (Bustelo et al. 2007; Heasman and Ridley 2008). A family of RhoGAP proteins, designed as the slit–robo GAPS (srGAPS), have been shown to interact with the intracellular domain (CC3) of Robo1 especially in the presence of Slit2 (Wong et al. 2001). Recent data point to the involvement of srGAPs in neuronal migration, neuronal development, cell adhesion, neuronal morphogenesis, axonal regeneration and brain evolution (Bacon et al. 2009; Charrier et al. 2012; Coutinho-Budd et al. 2012; Guo and Bao 2010; Madura et al. 2004; Yao et al. 2008). Moreover, several Rho-GTPases such as RhoA, Rac1 and Cdc42 have been shown to be modulated by Slit–Robo signalling (Bashaw et al. 2000; Fan et al. 2003; Hu et al. 2005; Lundström et al. 2004; Wong et al. 2001). Finally, GEFs such as Son of Sevenless (SOS) have been shown to bind to the cytoplasmic domains of Robo proteins and this binding is upregulated in the presence of Slit (Yang and Bashaw 2006).

Active research is currently being pursued to understand more completely the signalling pathways downstream of Slit–Robo binding. The discovery of new effector molecules will undoubtedly uncover new functions for Robo–Slit signalling.

7.2.2 Modulation of Robo–Slit Signalling

Both Robos and Slits have other binding partners which can serve to modulate Robo–Slit signalling. One of the main modulators of the Robo–Slit signalling is another major axon guidance signalling pathway: the Netrin–DCC pathway. Interplay between these guidance-signalling pathways has been shown to have major functional impacts on axon guidance [reviewed by Killeen and Sybingco (2008)]. In addition, heparan sulphate proteoglycans (HSPG) have been identified as co-receptors for Slit (Conway et al. 2011; Lee et al. 2004; Pratt et al. 2006; Van Vactor et al. 2006). Finally, Robos are capable of homophilic and heterophilic interactions, thus modulating one another's functions.

Slit–Robo and Netrin–DCC pathways Intersect at Several Levels

Slit–Robo and Netrin–DCC signalling are key components in midline crossing. The Robo and DCC pathways modulate one another at several levels.

It is known that, in *Xenopus* spinal axons, Robo1 forms heterodimers with the Netrin receptor DCC in the presence of Slit in order to silence Netrin1 attraction at the midline (Stein et al. 2001). Moreover, in *C. elegans,* UNC40/DCC participates in SLT-SAX3/Robo signalling in a Netrin-independent manner and this signalling is mediated by UNC34/Ena (Yu et al. 2002). In addition, the RhoGEF protein UNC73/Trio (in conjunction with VAB-8 L protein) positively modulates SAX3/Robo as well as UNC40/DCC expression and function (Levy-Strumpf and Culotti 2007; Watari-Goshima et al. 2007). In turn, the negative regulation of UNC73/Trio leads to a decrease in SAX3/Robo levels (Vanderzalm et al. 2009). In addition, in mouse neurons, Trio modulates DCC signalling (Briançon-Marjollet et al. 2008). Finally, another modulator affecting the interplay of the Slit/Robo and Netrin/DCC pathways is Presenilin-1, a major component of the γ-secretase complex (see Fig. 7.4) (Bai et al. 2011). Indeed, in motorneurons, DCC is proteolytically cleaved by metalloproteases leading to the formation of a cytoplasmic DCC form known as DCC stubs which are then normally processed by γ-secretase to release them from the membrane. Netrin stimulation leads to metalloprotease activity and inhibits γ-secretase, thus leading to the accumulation of DCC stubs. These stubs retain the ability to mediate Netrin attraction but cannot interact with Robo receptors. With the loss of binding between Robo and DCC, Slit silencing of the Netrin–DCC signalling pathway is therefore reduced. In this way, motorneurons, which are normally unresponsive to Netrin, can over time gain responsiveness to Netrin through the release of Slit silencing of Netrin–DCC signalling. Finally, very recent data demonstrate that Netrin/DCC signalling modulates Slit/Robo repulsion of callosal axons (Fothergill et al. 2013). In fact, it was shown that DCC/Netrin dampens Slit/Robo repulsion of pre-crossing neocortical axons as they approach the midline and that this effect is progressively lost as DCC becomes downregulated in the cortex from E16 to P0. Interestingly, in this system, Netrin1 does not function as an attractant for these callosal axons.

Other Robo Co-receptors

The chemokine SDF1 (CXCL12) and its receptor CXCR4 were originally described as regulators of cell–cell interactions in the immune system (Gonzalo et al. 2000; Nagasawa et al. 1996). Recently, they have also been shown to play a role in neuronal cell migration [reviewed by Tiveron and Cremer (2008)]. Slit2 inhibits SDF1-induced chemotaxis and transendothelial migration of T lymphocytes and monocytes (Prasad et al. 2007). This inhibitory effect occurs through the interaction of CXCR4 to the CC3 domain of Robo1, a binding that is strengthened in the presence of Slit2. Slit2 also acts to modulate the SDF1–CXCR4 signalling pathway by decreasing phosphorylation of Akt as well as of Src kinase, Lck kinase and Rac activation.

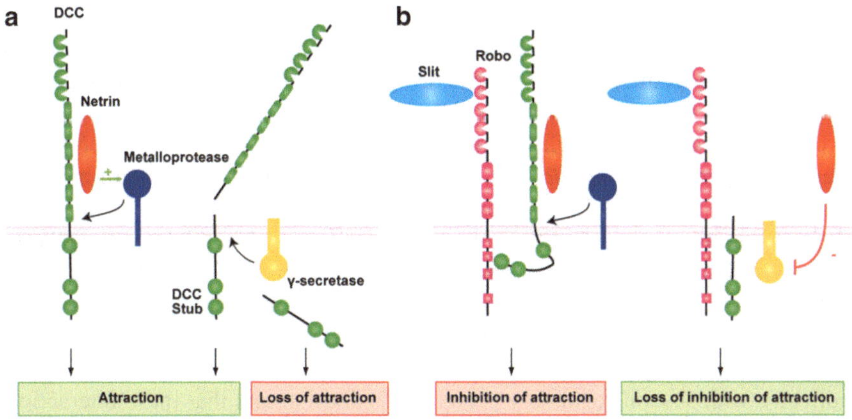

Fig. 7.4 Interplay of Slit–Robo and Netrin-DCC. (**a**) In commissural axons, Netrin1 binding to DCC acts to promote attraction. DCC can be proteolytically cleaved by metalloproteases to yield DCC stubs which retain their ability to mediate Netrin1 attraction. The stubs are then further processed by γ-secretase and thus lose their ability to mediate attraction. Netrin1 positively regulates metalloproteases and negatively regulates Presenilin-1, a major component of the γ-secretase complex. (**b**) In spinal cord motorneurons, DCC interaction with Robo1 leads to the inhibition of Netrin1-mediated attraction. After metalloprotease cleavage of DCC, DCC stubs are no longer able to interact with Robo1 but retain the ability to mediate Netrin1 attraction. Thus motorneurons gain responsiveness to Netrin1

In breast cancer, Slit2 blocks a whole host of SDF1-induced signalling involved in motility such as the activation MAP kinase or focal adhesion components (Prasad et al. 2004). Similarly, overexpression of Slit–Robo in breast cancer leads to a down-regulation of CXCR4 and a suppression of tumour growth (Marlow et al. 2008). In retinal ganglion cell axons, SDF1 reduces Slit-mediated axonal repulsion by modulating cyclic nucleotide signalling intermediates (Chalasani et al. 2003). In vivo, reducing SDF1 signalling rescues retinal axon pathfinding errors in mutants with a partial loss of Robo2 but not in full knockouts (Chalasani et al. 2007). Last, Slit2 through Robo1/CXCR4 might be used to modulate HIV infection (Anand et al. 2011, 2013).

Mounting evidence suggests that Slit/Robo interact with cell adhesion proteins, such as members of the cadherin family (N-cadherin, E-cadherin and P-cadherin), during development to modulate cell adhesion in different contexts. For instance, in chicken, placode and neural crest cells must interact intimately in order to produce proper assembly of the trigeminal ganglion. Here, gangliogenesis relies on the interaction of Slit1 expressed by the neural crest cells and Robo2 expressed by the incoming placodal ganglion cells (Shiau et al. 2008). N-cadherin, found on placode-derived sensory neurons, was further required to mediate the aggregation of placode cells and Slit2–Robo1 signalling was shown to positively modulate N-cadherin expression and function (Shiau and Bronner-Fraser 2009). Further, in an oral squamous cell carcinoma cell line (OSCC cells), it was shown that P-cadherin stabilises Slit2 expression and that P-cadherin binds to Robo3 with increasing affinity in the presence of increasing Slit2 concentrations (Bauer et al. 2011). In addition, Slit2 was shown to have a dose-dependent anti-migratory effect in the presence of P-cadherin. Altogether, these data provide evidence that Slit2–Robo3 signalling in combination with P-cadherin inhibits cell migration possibly by increasing adhesion of neighbouring cells through P-cadherin heterophilic interactions. In contrast, in heart formation in *Drosophila*, Slit–Robo signalling inhibits E-cadherin-mediated cell adhesion, thus enabling lumen formation (Santiago-Martínez et al. 2008). Similarly, in an in vitro study, Slit–Robo signalling negatively modulated N-cadherin/β-catenin-mediated cell adhesion through a complex series of secondary signalling events leading to the phosphorylation of β-catenin by Cables through a Robo–Abl–Cables interaction (Rhee et al. 2007). A general take-home message in terms of Slit–Robo signalling in cell adhesion is that Slit–Robo is able to interact directly or indirectly with several regulatory proteins of adhesion and that these interactions can lead to either positive or negative regulation of the adhesion machinery depending on tissue specificity or the availability of particular secondary effectors.

A recent study has shown that Robo modulates the signalling of another major family of axon guidance proteins, namely the Semaphorin3–Neuropilin ligand–receptor couple (Hernández-Miranda et al. 2010). Semaphorins are a large family of secreted or transmembrane proteins characterised by the presence of a conserved "Sema" domain. The first vertebrate Semaphorin discovered was a member of the Sema3 family which could induce growth cone collapse in sensory ganglion neurons (Kolodkin et al. 1993; Luo et al. 1993). Its actions were subsequently found to be transduced by the Neuropilin (Nrp) receptors (He and Tessier-Lavigne 1997; Kolodkin et al. 1997). Since then, more than 30 Semaphorins have been characterised and a large family of membrane-bound proteins, the Plexins, have been described as their principal receptors or co-receptors [reviewed by Jackson and Eickholt (2009)]. However, direct evidence was recently provided, supporting the interaction between Slit/Robo and Semaphorin/Plexin signalling (Hernández-Miranda et al. 2010). In rodents, cortical interneurons originate ventrally from the ganglionic eminence and migrate tangentially to the cortical plate. In Robo1 knockout mice, the number of interneurons is increased in the cortex and they take a shortcut through the normally repulsive striatum (Andrews et al. 2008). The striatum secretes chemorepulsive Semaphorin 3 proteins which repel Nrp1- and 2-expressing cortical interneurons. *Robo1*-deficient cortical interneurons were found to have lost responsiveness to Sema3A and Sema3F and this effect was found to be due to a reduction of Nrp1 expression on the cell surface of *Robo1* knockout cells (Hernández-Miranda et al. 2010). Moreover, it showed that Robo1 can directly bind to Nrp1 through its first two IgG domains.

Finally, Robos can bind to themselves and other Robo isoforms to form homodimers and heterodimers. For instance, Robo1 binds with both Robo1 and Robo4 (Sheldon et al. 2009), Robo2 with both Robo1 and Robo2 (Hivert et al. 2002) and Robo3 with both Robo3 and Robo1 (Camurri et al. 2005; Liu et al. 2004). Many cell types co-express several Robo receptors on their membrane and, depending on the context, their ability to dimerise could serve to potentiate or counteract one another's effects.

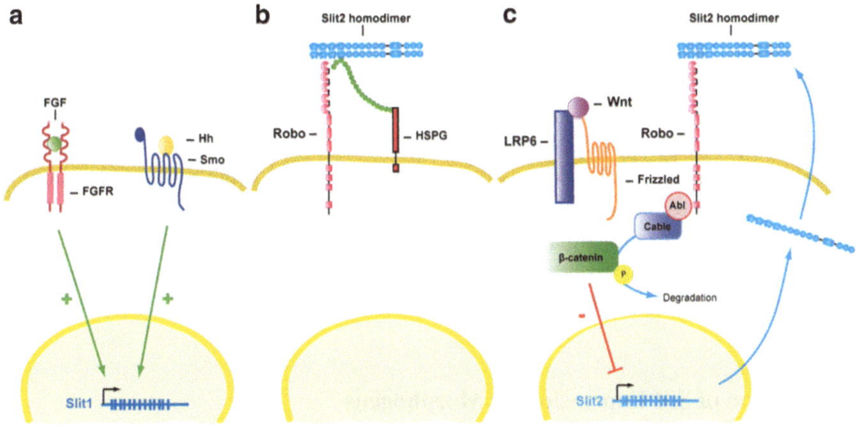

Fig. 7.5 Modulation of Slit expression. (**a**) In *Xenopus*, FGF/FGFR signalling and, in *zebrafish*, Hedgehog/Smoothened signalling positively modulate transcriptional regulation of *Slit1*. (**b**) HSPGs, consisting of a proteoglycan core protein with heparan sulphate chains, stabilise Robo and Slit2 binding by forming ternary complexes with Robo receptors and Slit through their HS polymers. (**c**) The canonical Wnt signalling pathway forms a feed-forward regulatory loop with Slit/Robo signalling. Robo phosphorylates β-catenin through the recruitment of Cables to Abl. This leads to β-catenin degradation which relieves β-catenin-mediated inhibition of transcriptional regulation on *Slit2*. Therefore in the absence of β-catenin Slit2 expression is increased which promotes the formation of a Robo/Abl complex which ultimately leads to further inhibition of the Wnt canonical signalling pathway

Modulation of Ligand Presentation

Slit is not the only ligand for Robo. Indeed, studies have identified HSPG as co-receptors of Slit (see Fig. 7.5b). HSPGs are proteoglycan core proteins attached to heparan sulphates (HS). They are polysaccharides that can be divided into different categories: the syndecans (Sdc) family and the glypicans family that are associated with the membrane, as well as secreted forms like perlecan or agrin. Investigations into the structural complex formed by Slits and Robos showed that the LRR2 domain of Slit and the Ig1–Ig2 domain of Robo could form a ternary complex with a heparin/heparan sulphate in order to stabilise Slit–Robo interaction (Fukuhara et al. 2008; Howitt et al. 2004; Hussain et al. 2006). Furthermore, structural work confirmed that another domain of Slit, the LRR4, binds HSPGs and is important for Slit homodimerisation (Seiradake et al. 2009).

Disrupting or removing HSPGs abolishes the repulsive action of Slit2 on *Xenopus* retinal axons, on olfactory axons and on olfactory precursor interneurons (Hu 2001; Piper et al. 2006). Moreover, in vivo ablation of enzymes involved in HSPGs synthesis, Ext1 and Ext2, induces defects at the optic chiasm similarly to those observed in the *Slit1/2*-deficient mice (Inatani et al. 2003; Plump et al. 2002). Further, knocking down *Hs6st1* or *Hs2st*, enzymes that are involved in HS sulphonation, leads to telencephalic phenotypes reminiscent of Slit/Robo mutants (Conway et al. 2011; Pratt et al.

2006). Hs6st1 was first shown to be necessary for Slit2-mediated repulsion of RGC axons (Pratt et al. 2006) and the genetic interaction between Hs6St1 and Slit2 was then confirmed by creating an Hs6st1 and Slit2 compound mutants (Conway et al. 2011).

In addition to HSPGs, Slit2 has recently been shown to bind to glycosylated α-dystroglycan (Wright et al. 2012). In the absence of glycosylation by β-1,3-N-acetyl-glucosaminyltransferase-1 (B3gnt1) and isoprenoid synthase domain containing (ISPD) proteins, several severe developmental phenotypes including axon guidance defects in the spinal cord are detected in mutant mice. It appears, therefore, that the glycosylated extracellular matrix protein acts to regulate the distribution of Slit protein in the floor plate of the spinal cord.

Modulation of Slit Expression by Morphogens

In addition to the tight regulation of Robo receptor expression, it was found that Slit expression itself was modulated by different morphogens and this regulation was linked to specific phenotypes in several cell types (see Fig. 7.5).

For example, in zebrafish, hedgehog (Hh) is required for correct Slit expression at the optic chiasm as well as for specifying positioning of Slit1a-expressing midline glia cells (Barresi et al. 2005). In mammals Hh signalling relies on Smoothened (Smo), and in a conditional knockout of Smo, cells in which Hh signalling was reduced also showed a marked reduction in the expression of Slit1 mRNA (Balordi and Fishell 2007).

Moreover, in *Xenopus*, FGF signalling positively regulates the expression of Slit1 (and Sema3A) in the optic tract (Atkinson-Leadbeater et al. 2010). Inhibiting the actions of the FGF receptor (FGFR) leads to a decrease in *Slit1* mRNA and phenocopies the retinal axon guidance errors seen in the absence of Slit1, thus suggesting that FGF signalling at the optic tract functions to maintain Slit1 (and Sema3A) expression.

Finally, a very complex signalling feedback loop appears to exist between Slit2 and members of the Wnt signalling pathway. Wnts are a large family of secreted glycoproteins that play a role in many stages of telencephalon development [reviewed by Bowerman (2008) and Killeen and Sybingco (2008)]. Wnt signalling has a canonical pathway whereby Wnt binds to Frizzled receptors and an additional receptor, LRP6, to stabilise β-catenin which is translocated into the nucleus and acts with the family of Tcf/Lef transcription factors to modulate gene expression. In an OSCC cell line, an in vitro experiment showed that after inhibiting β-catenin binding to the Tcf/Lef transcription factors, Slit2 expression was increased (Bauer et al. 2011). Slit2 was previously shown to modulate β-catenin phosphorylation through the recruitment of Cables to the Robo/Abl complex leading to β-catenin degradation (Rhee et al. 2007). In this feed-forward regulatory loop, Slit2 expression promotes inhibition of β-catenin, which leads to an increase in Slit2 expression.

In this section, we have established that both Robo receptors and Slit ligands are tightly regulated at several levels including transcriptional, translational and subcellular localisation. Moreover, we have shown that Slit/Robo signalling does not occur in a bubble, but rather that Slits and Robos interact with a variety of other signalling systems which allow them to mediate particular effects and which add another layer of regulation to Slit/Robo signalling.

7.3 Roles of Slit and Robo

By acting alone or in concert with other molecules, Slits and Robos are involved in a multitude of processes. In the CNS, Slit–Robo signalling has been implicated in numerous functions such as axon guidance, cell adhesion, axon and dendritic branching, cell proliferation and axonal targeting to name but a few. This section will focus on describing some of the known functions of Slit/Robo in the CNS using a specific case study to illustrate the role of Slit–Robo in each instance. For a more comprehensive overview of the roles of Slit and Robo signalling, please refer to the following reviews [reviewed by Chedotal (2007), Dickinson and Duncan (2010), Legg et al. (2008), and Ypsilanti et al. (2010)].

7.3.1 Slit–Robo Signalling Regulates the Guidance of Axons

The function of Slit and Robo in axon guidance was first discovered in the *Drosophila* ventral nerve cord and has since been extensively studied in the vertebrate hindbrain and in the spinal cord [reviewed by Dickson and Gilestro (2006), Garbe and Bashaw (2004), and O'Donnell et al. (2009)]. In the vertebrate forebrain, Slits and Robos have been shown to contribute to the guidance of several commissural tracts including the corpus callosum, the optic chiasm, the anterior commissure and the hippocampal commissure (Bagri et al. 2002; Plump et al. 2002). Moreover, longitudinal tracts are also regulated by Robo receptors such as the postoptic commissure in zebrafish or the lateral olfactory tract in mice (Devine and Key 2008; Fouquet et al. 2007). An interesting example of how Robo/Slit signalling regulates axon guidance can be illustrated by taking a look at callosal axon guidance.

In placental mammals, the corpus callosum (CC) is the main tract that connects right and left cerebral hemispheres and allows for the transmission and coordination of information originating from both sides of the cortex [reviewed by Chedotal (2011), Lindwall et al. (2007), and Plachez and Richards (2005)]. The CC is formed by axons of cortical pyramidal projection neurons which connect homotypic regions of the two hemispheres. As the first cortical axons extend towards the midline, they encounter several populations of glial cells that serve as guideposts to steer their growth cones across the midline including the glial wedge (GW), the indusium griseum glia (IGG) and the midline zipper glia (Shu et al. 2003b; Silver et al. 1982). A last crucial population of cells encountered by callosal axons at the midline is the subcallosal sling that is formed by a transient population of neurons and that is required for proper callosal development (Shu et al. 2003a; Silver and Ogawa 1983; Silver et al. 1982).

In the vicinity of the CC, Slit2 is expressed by the GW and the IGG, thus surrounding the Robo1- and Robo2-expressing callosal axons (Shu et al. 2003c). In *Slit2* knockout (KO) animals, a large proportion of callosal fibres misprojected ventrally prior to the midline and formed large swirls of axons known as Probst bundles (Bagri et al. 2002). *Slit1* KOs displayed a normal CC, whereas *Slit1;Slit2* double KOs, which die at birth, exhibited complete agenesis (Unni et al. 2012). Similar defects could also be observed in *Robo1* KO and *Robo1;Robo2* double KO mice in which callosal fibres were misrouted at the midline into huge fasciculated bundles that ectopically extend into the septum (Andrews et al. 2006; Lopez-Bendito et al. 2007). It is worth noting that in the absence of Robo1, axons showed an increase in fasciculation whereas in the absence of Slit2 the converse phenotype (i.e. defasciculation) was observed. This indicates that additional mechanisms are at play in this system.

Interestingly, several Robo and Slit partners have also been shown to regulate the guidance of callosal axons. The interaction between Slits and HSPGs is likely to be important at the corpus callosum. Mutants for Hs6st1, a structural modifier of heparan sulphates, showed a callosal phenotype with Probst bundles and growth of axons into the septum which was reminiscent of that of *Slit/Robo* mutants albeit more severe (Conway et al. 2011). Moreover, neither *Hs6st1*$^{+/-}$ nor *Slit2*$^{+/-}$ cortical axons exhibited any growth of cortical axons into the septum, but *Hs6st1*$^{+/-}$;*Slit2*$^{+/-}$ compound mutants exhibit the same phenotype as *Slit2* or *Hs6st1* full knockouts, thus indicating a synergistic genetic interaction between Slit2 and Hs6st1. Similarly, knocking down *EXT1* in mouse leads to profound axon misguidance of the main forebrain axon commissures reminiscent of phenotypes found in *Slit/Robo* mutants (Inatani et al. 2003). Moreover, the CC is totally absent in a percentage of *Mena* KOs as well as in all double mutants for *Mena* and *VASP*, which are downstream effector proteins of Slit–Robo signalling (Lanier et al. 1999; Menzies et al. 2004).

7.3.2 Cell Migration and Motility

In most developing organs, cells migrate in a precise direction, following stereotyped pathways under the influence of repulsive and attractive factors. Slit–Robo signalling has been shown to control the migration of several neuronal subtypes in mice including cortical interneurons (Andrews et al. 2006; Hernández-Miranda et al. 2010), cerebellar granule neurons (Guan et al. 2007; Xu et al. 2004) and inferior olivary neurons (Di Meglio et al. 2008). Studies suggest that Slit/Robo can control two properties in migrating cells, namely their directionality as well as their motility. This is best exemplified in migrating subventricular zone-derived precursors.

The subventricular zone (SVZ) located along the walls of the lateral ventricle is composed of mitotically active cells that can migrate over long distances along the rostral migratory stream (RMS) to the olfactory bulb (OB) to generate two types of olfactory interneurons (Lois and Alvarez-Buylla 1994). Simplistically, the SVZ contains three types of cells: type A, type B and type C cells as well as ependymal cells which line the ventricle (Doetsch et al. 1997). Type B cells are the neural stem cells that give rise to the rapidly dividing "transit-amplifying" cells or type C cells which in turn generate migratory postmitotic neuroblasts known as type A cells (Doetsch et al. 2002). Type A cells migrate a long distance from the walls of the lateral ventricle to reach the olfactory bulb along the rostral migratory stream (RMS) in chain formation (Jankovski and Sotelo 1996; Lois et al. 1996). During this long migration, the neuroblasts are protected from the surrounding brain parenchyma by an astrocytic sheath which acts as a sort of tunnel to channel them to the OB. Upon reaching the OB, the neuroblasts disperse and start migrating radially as they differentiate into two types of OB interneurons which integrate into the existing circuitry (Carleton et al. 2003; Kelsch et al. 2007).

The tangential migration of neuroblasts along the RMS depends mostly on repulsive cues. One of the main families of axon guidance molecules implicated in regulating the migration of SVZ-derived progenitors is the Slit–Robo ligand-receptor couple. Type A cells express Robo2 and Robo3 receptors (Kaneko et al. 2010), and in vitro, molecules secreted by the septum and the choroid plexus repel SVZ-derived neural progenitors (Wu et al. 1999). This repulsive effect was shown to be dependent on Slit1 and Slit2 which are produced in brain regions which neighbour the SVZ and the RMS and this repellent activity is lost in *Slit1;Slit2* knockouts animals (Nguyen-Ba-Charvet et al. 2004; Wu et al. 1999). Moreover, Slit1 is expressed by migrating neuroblasts along the RMS and appears to have both a cell-autonomous and cell-non-autonomous role in their migration (Nguyen-Ba-Charvet et al. 2004). Indeed, in *Slit1*-deficient mice, a portion of the SVZ-derived neuroblasts migrates caudally and migration defects are observed in *Slit1*-deficient explant and neurosphere cultures. In fact, both in culture and in situ, Slits are capable of reversing the direction of migrating SVZ neurons (Ward et al. 2003). In addition, these neurons turn away from Slit sources through recurring rounds of extending and retracting their processes leading to the selection of a dominant process that formed preferentially on the side of the cell body farthest away from the Slit source (Ward et al. 2005). Interestingly, the application of Slit to the leading edge of the migrating cells leads to a repositioning of the centrosome apparatus which is accompanied by a switch in cell directionality (Higginbotham et al. 2006). Moreover, RMS astrocytes of the glial tube express Robo receptors and are able to respond to repulsive Slits secreted by migrating neuroblasts by forming and maintaining glial tubes (Kaneko et al. 2010). In these ex vivo experiments, as well as in vivo, the astrocytic glial sheath surrounding neuroblasts showed a perturbed organisation. Slit1 expressed by neuroblasts is actually thought to act by repelling the Robo-expressing processes of neighbouring astrocytes, thus creating corridors within the brain parenchyma which can facilitate neuroblast migration. In the lateral ventricles, Slit2 is produced by the choroid plexus located caudally (Sawamoto et al. 2006). The beating of ependymal cilia on the walls of the lateral ventricle creates a high-caudal low-rostral concentration gradient of CSF molecules (including Slit2), which is postulated to drive the migration of SVZ neuroblasts rostrally towards the OB. In the absence of ependymal flow of the CSF, directional migration of neuroblasts is actually compromised. Slit/Robo not only controls the directionality of SVZ-derived neuroblasts but also their motility. In fact, *Slit1*-deficient neuroblasts in organotypic cultures migrate slower than control neuroblasts (Kaneko et al. 2010). This same effect is observed when placing wild-type neuroblasts on an organotypic brain slice derived from a *Slit1* KO brain, demonstrating that Slit1 also acts in a cell non-autonomous manner.

7.3.3 Dendritic and Axonal Arbourisation/Branching

Slit and Robo play a role in regulating the dendritic and axonal branching in a variety of systems. It was found that Slit2N, the amino terminal fragment of Slit2, was first characterised as a factor capable of inducing elongation and promoting branch formation of rat sensory axons (Wang et al. 1999). Slit–Robo signalling in axonal branching has since been studied in other models. For instance, applying ectopic Slit2 to trigeminal axonal tracts caused rapid axon branching and arbourisation (Ozdinler and Erzurumlu 2002). Moreover, an interesting study explored the role of Slit2 addition on cortical interneurons derived from an E13.5 medial ganglion eminence (MGE) explant (Sang et al. 2002). This study uncovered a dual role for Slit2 depending on the age of the cortical interneurons. In fact, newly generated interneurons were repelled by Slit2 addition and showed no branching effect, whereas 5-day-old interneurons responded to Slit2 addition by neurite branching and elongation. This switch in responsiveness to Slit2 underlines the fact that as neurons mature, their signalling pathways evolve to match their current environment and the new stage in their lives. Indeed, new interneurons must migrate away from the MGE towards the cortex, while old interneurons must send out neurites in order to integrate into the cortex. Moreover, adding Slit1 to cultured cortical neurons stimulated axon growth in wild-type but not *Robo1*$^{-/-}$;*Robo2*$^{-/-}$ cultures (Round and Sun 2011). This effect was mediated by the adaptor protein Nck2 which was previously shown to be an effector in EphB1 signalling (Stein et al. 1998).

The differential effects of Robo–Slit signalling on dendritic growth and branching, in different cell types and species, illustrate the specificity of Robo–Slit signalling.

7.4 Slits and Robos in Neurological Disease

Together with basic research implicating Robo/Slit signalling in a variety of functions in proper brain development, recent genetic screens conducted on pools of patients suffering from neurological diseases have uncovered a few mutations in Robo genes which suggest that Slit–Robo signalling may be important for proper brain function. The study of Slit–Robo in human diseases is in its infancy, but indications suggest that these proteins may play a role in several diseases.

7.4.1 Horizontal Gaze Palsy and Progressive Scoliosis

In mammals, Robo3 was discovered as a receptor upregulated in retinoblastoma-deficient mice (Yuan et al. 1999). Sequence analysis showed that it belongs to the Robo family of receptors although it lacks some amino acids and domains conserved in other Robos (see Sect. 7.1.1). Interestingly, parallel studies in human revealed that ROBO3 is mutated in patients suffering from a rare neurological syndrome named horizontal gaze palsy and progressive scoliosis (HGPPS) (Jen et al. 2004). HGPPS patients are unable to perform conjugate lateral eye movements and develop after birth a very severe scoliosis (Abu-Amero et al. 2009, 2011; Amouri et al. 2009; Jen et al. 2004). They do not have any other neurological symptoms. However, diffusion tensor imaging and the measure of evoked potentials revealed that in HGPPS patients, the corticospinal tract (which connects the motor cortex to neurons in the spinal cord) fails to cross the CNS midline at the pyramid level and projects to the ipsilateral side of the spinal cord instead. Likewise, ascending sensory projections activate the ipsilateral cortex instead of the contralateral one. Patients have a normal corpus callosum but otherwise seem to lack commissural connections at the level of the hindbrain and spinal cord. Likewise, Robo3 knockout mice lack all commissural connections in the caudal brain but have normal commissures in the forebrain (Marillat et al. 2004; Sabatier et al. 2004).

Although neurological deficits are similar in all HGPPS patients, the mutations (which are autosomal recessive) are distinct in all families identified so far and affect amino acids localised in various domains of the protein, from the most N-terminal region to the cytoplasmic domain. Most are just missense and their consequence on receptor function or expression is still unknown.

The neuronal defects underlying the horizontal gaze palsy is unclear as the abducens nerve, which plays a key role in lateral eye movement, is present in HGPPS patients. However, the eye movement defects can be mimicked in a Robo3 conditional mouse where the brainstem commissural projection which connects the abducens nucleus to the contralateral oculomotor nucleus is suppressed, suggesting that this also contributes to the defects in humans (Renier et al. 2010). Likewise, the severe hypoplasia of the basilar pons, one of the key feature of HGPPS brain, is most likely due to an abnormal migration of pontine neurons as is the case in Robo3 knockout mice (Geisen et al. 2008; Marillat et al. 2004). By contrast, the cause of the scoliosis is still unknown. How Robo3 functions to control axon guidance at the midline of the nervous system is still debated, but the current model suggests that Robo3 interacts with Robo1/Robo2 receptors and thereby blocks Slit-mediated repulsion before commissural axons have reached the CNS midline (Jaworski et al. 2010; Sabatier et al. 2004).

7.4.2 Diabetic Retinopathy

Diabetic retinopathy (DR) remains the first cause of new cases of blindness in adults in America (Zhang 2010). In fact, no fewer than 28.5 % of adults with diabetes present with DR. The disease is thought to be triggered by damage of small blood vessels in the retina (probably due to hyperglycemia) which results in ischemia in the retina and ultimately leads to the neovascularisation of the retinal surface. The proliferation of endothelial cells and the associated vascular growth further damage the retina as well as cause a decrease in visual acuity. In fact retinal damage leads to the neurodegeneration of several neuronal and glial cell types in the retina and forms a crucial component of the aetiology of DR (Barber 2003). Moreover, neurodegeneration contributes partially to the loss of vision that occurs in diabetes. Loss of vision resulting from neovascularisation occurs in several diseases other than DR and several lines of evidence suggest that Robo4 is a strong therapeutic candidate to modulate pathogenesis.

Axonal guidance molecules including Slit and their receptors have been demonstrated to regulate the growth and branching of blood vessels (Klagsbrun and Eichmann 2005). Robo4 and to a lesser extent Robo1 are the main Robo proteins expressed by blood vessels (Park et al. 2003). Moreover, Robo4, Robo1 and Slit2 are highly expressed by retinal pigment epithelium cells in patients with proliferative diabetic retinopathy (Huang et al. 2009b; Zhou et al. 2011a). Studies have demonstrated that Robo4 contributes to stabilising the vasculature in the retina whereas Robo1 is associated with neovascularisation (Huang et al. 2009a). That being said, a soluble form of the extracellular portion of Robo4 was shown to inhibit angiogenesis and block the migration of endothelial cells in vivo (Suchting et al. 2005). In addition, activation of Robo4 by Slit2 was shown to inhibit pathologic angiogenesis while its deletion promoted the pathobiology of vascular diseases (Jones et al. 2008, 2009). Moreover, Robo4 binds to the vascular Netrin receptor, UNC5B, which counteracts VEGF signalling, thus maintaining vascular integrity (Koch et al. 2011). In this case, Robo4 seems to act as a ligand for UNC5 and not as a receptor.

7.4.3 Dyslexia

ROBO1 was lately identified as a candidate gene for developmental dyslexia (Hannula-Jouppi et al. 2005; Mascheretti et al. 2012). Developmental dyslexia is a hereditary neurological disorder characterised by impairments in reading which cannot be explained by other factors such as lack of intelligence, motivation or opportunity. Interestingly, dyslexia is often associated in post-mortem studies with a mild disorganisation of cortical layering (Ramus 2004). In dyslexics, polymorphisms in the *ROBO1* gene have also been associated with language deficits (Bates et al. 2010).

Robo/Slit signalling has been implicated in several aspects of cortical development. As indicated previously, recent studies have shown that in the absence of

Robo1, the proliferation and migration of cortical interneurons are perturbed (Andrews et al. 2006, 2008; Hernández-Miranda et al. 2010). This defect was in part attributed to a Slit-independent mechanism whereby *Robo1*-deficient neurons lose responsiveness to Sema3A and Sema3F given that Robo1 can bind directly to the Sema3 receptor, Nrp1. In addition, knocking down Robo1, by in utero *electroporation* of an shRNA construct during embryogenesis, was recently found to lead to an accumulation of layer II/III cortical neurons in the marginal zone of the cortex just below the pia (Gonda et al. 2012). Interestingly this defect in cortical lamination occurred when Robo1 was silenced at E15.5 and E16.5 but not at E14.5, leading the authors to conclude that different cortical subtypes are more or less responsive to Robo signalling. Robo1 was previously shown to play a role in dendritic growth and branching in cortical neurons in vivo (Whitford et al. 2002) and Slit protein was shown to induce branching of mature cortical interneurons but not projection neurons (Sang et al. 2002). In addition, Robo1 was previously found to bind to Robo4 (Sheldon et al. 2009). Interestingly, a recent study reported that Robo4 was expressed in the developing cortex and that acute removal of Robo4 by in utero electroporation of a Robo4 siRNA led to the suppression of cortical migration (Sundaresan et al. 1998b; Zheng et al. 2011). This implies that timing of cortical radial migration might be dependent on Robo1–Robo4 signalling. Robo4 also caused a reduction of migration in an in vitro model of cell migration, but no defects were obvious in a Robo4 knockout mouse. The authors argued that perhaps other Robo receptors had a compensating effect. Finally, it was also recently shown that Slit–Robo signalling is implicated in the modulation of progenitor cell proliferation during brain development (Borrell et al. 2012). In fact, in the absence of Robo1/2 or Slit1/2, the secondary progenitor pool of intermediate precursor cells is expanded in the ventricular zone of mutant mice.

It will be interesting to determine whether the cortical phenotypes discovered in *Slit/Robo* mutant mice are contributing factors in developmental dyslexia. Another hypothesis for the role of Robo1 in dyslexia is based on a clinical study which found that dyslexic family members in a Finnish family were carriers of a weakly expressing haplotype of *ROBO1* (Lamminmaki et al. 2012). Moreover, the auditory pathway of the dyslexic family members showed some impairment upon testing of binaural suppression. Here, the strength of the interaural commissures was significantly correlated to the expression levels of ROBO1 in the blood of the dyslexics. One of the hypotheses put forward by the authors of this clinical investigation was that, Robo1 being important for axon commissural crossing, dyslexic patients most likely presented with defects in the crossing of axons along the auditory pathway.

7.4.4 Involvement in Brain Cancer

Angiogenesis is an important component of cancer metastasis and as discussed in Sect. 7.4.2, there is evidence that Slit–Robo signalling might be implicated in this component of cancer progression (Ballard and Hinck 2012; Mehlen et al. 2011).

Further, there is evidence that Slits and Robos are differentially regulated in many types of cancers in the body (Beggs et al. 2012; Biankin et al. 2012; Ito et al. 2006; Sundaresan et al. 1998b). The evidence does not point to a clear role of Slits and Robos in cancer where they appear to function as tumour suppressors in certain types of cancer (Alajez et al. 2011; Bauer et al. 2011; Marlow et al. 2008; Stella et al. 2009; Yuasa-Kawada et al. 2009a) and oncogenes in others (Schmid et al. 2007; Zhou et al. 2011b). A few studies point to a role of Slit–Robo in different types of brain cancer. For instance, in vitro, breast cancer cell lines were strongly attracted to Slit2, leading the researchers to hypothesise that Slit2 in the brain acts an attractant to metastatic breast cancer cells (Schmid et al. 2007). However, an in vitro study of medulloblastoma cancer cells demonstrated that Slit2 inhibited medulloblastoma cell invasion (Werbowetski-Ogilvie et al. 2006). In this study, although glioma tumours overexpress Slit and Robo, Slit2 had no effect on glioma cell invasion. Nonetheless, another study showed that Slit2 was overexpressed by glioma cells which mediated Slit2-induced chemorepulsion through Robo1 (Mertsch et al. 2008). Further studies are therefore required to support or not a role for Slit–Robo signalling in brain tumours.

7.4.5 Conclusion: Prospects of Slit–Robo Signalling in Neurological Diseases

Considering the brain phenotypes observed in various Slit/Robo mutant mice, one would expect Slit and Robo mutations to be present in a large variety of neurological diseases. It is therefore quite unexpected that, as of now, these molecules have not been implicated in more pathologies in the CNS. It is possible that, as is the case for mutant mice, humans with mutations in either Robos or Slits have decreased viability. However, as genetic screens of patients become more routine and cost-effective, we can speculate that an increasing number of mutations in SLIT and ROBO genes will be uncovered in diseases. These genetic screens will undoubtedly yield false positives and therefore further analysis will be warranted to ensure that Slit and Robo actually play a function in CNS regions affected in the disease using transgenic animals or that Slit–Robo signalling is affected in patients.

Analysis of ROBO receptors in autism, a severe neurodevelopmental disorder characterised by social and behavioural deficits, demonstrated an association between SNPs in ROBO3 and ROBO4 and autism as well as a reduction of ROBO1 and ROBO2 mRNA in autistic patients (Anitha et al. 2008). The decrease in ROBO2 mRNA and protein levels in autistic patients was subsequently confirmed in a post-mortem study (Suda et al. 2011). Moreover, a recent genetic screen of neurons differentiated from induced pluripotent stem cells obtained from schizophrenic patients showed that Slit–Robo genes were deregulated in schizophrenia (Brennand et al. 2011). In addition, ROBO1 and ROBO2 were picked up as risk genes in a study which integrated genetic data with brain imaging to identify new risk factors for schizophrenia (Potkin et al. 2010). This would indicate a new avenue of research to pursue in understanding the aetiology of schizophrenia, a severe and debilitating psychiatric disorder affecting approximately 1 % of the population. While its causes remains poorly understood, the prevailing theories point to schizophrenia as being a developmental disorder or a disorder linked to a dysregulation of dopamine (Murray et al. 2008). The neurodevelopmental hypothesis postulates that the specific pathological hallmarks of the disease are laid out during brain development but that the onset of the associated symptoms manifests in early adulthood.

This new research prospect is further supported by the fact that several other Slit–Robo binding partners have also been implicated in schizophrenia. For instance, a DCC single nucleotide polymorphism was shown to be implicated in schizophrenia (Grant et al. 2012). Moreover, a duplication in the *srGAP3* gene was recently found to segregate with psychosis in the family of a child with an early-onset heritable form of the disease (Wilson et al. 2011). This finding was further confirmed in a mouse knockout for *srGAP3* which revealed schizophrenia-related phenotypes (Waltereit et al. 2012). Interestingly, srGAP3 can bind to both Robo1 and Robo2 and a knockout mouse for srGAP3 showed mild guidance defects of spinal cord commissural axons at the floor plate as in the Robo2 KO (though ventral funiculus axons do not express high levels Robo2) (Bacon et al. 2011). Finally and quite interestingly DISC1 mutants have similar callosal defects as seen in Slit–Robo mutants and DISC1 has been known to be implicated in psychiatric disorders ranging from bipolar disorder to schizophrenia (Millar et al. 2000; Osbun et al. 2011).

Finally, further basic research into the role of Slit–Robo in the development of the CNS will not only serve to gain insights into pathologies but will undoubtedly also have a significant therapeutic impact. Upon brain injury or damage, such as in stroke or Huntington's disease, some neuroblasts of the SVZ are coopted out of their normal migratory pathway towards the olfactory bulb, and instead, they appear to be able to migrate to the zone of injury (Saha 2012). In a mouse model of brain injury, known as cryoinjury, Slit2 mRNA was upregulated surrounding the lesion (Hagino et al. 2003). In Sect. 7.3.2, we have highlighted the fact that neuroblasts are responsive to Slit–Robo signalling. In fact, in the presence of a dominant-negative Robo2 fusion protein, SVZ-derived cells were no longer repelled by Slit molecules (Lu et al. 2007). Such experiments are required in vivo to lessen the chemorepulsive upregulation of *Slits* upon brain injury, thus enabling neuroblasts to migrate to the area of brain injury and potentially repair the lesions. In other systems, controlling Slit–Robo signalling has already shown promises of therapeutic benefits, for instance, in reducing the migration of melanoma cells (Denk et al. 2011).

Compliance with Ethics Requirements The authors declare that they have no conflicts of interest.

References

Abu-Amero KK, Al Dhalaan H, Al Zayed Z, Hellani A, Bosley TM (2009) Five new consanguineous families with horizontal gaze palsy and progressive scoliosis and novel ROBO3 mutations. J Neurol Sci 276:22–26

Abu-Amero KK, Faletra F, Gasparini P, Parentin F, Pensiero S, Alorainy IA, Hellani AM, Catalano D, Bosley TM (2011) Horizontal gaze palsy and progressive scoliosis without ROBO3 mutations. Ophthalmic Genet 32:212–216

Alajez NM, Lenarduzzi M, Ito E, Hui ABY, Shi W, Bruce J, Yue S, Huang SH, Xu W, Waldron J et al (2011) MiR-218 suppresses nasopharyngeal cancer progression through downregulation of survivin and the SLIT2-ROBO1 pathway. Cancer Res 71:2381–2391

Amouri R, Nehdi H, Bouhlal Y, Kefi M, Larnaout A, Hentati F (2009) Allelic ROBO3 heterogeneity in Tunisian patients with horizontal gaze palsy with progressive scoliosis. J Mol Neurosci 39:337–341

Anand AR, Nagaraja T, Ganju RK (2011) A novel role for Slit2/Robo1 axis in modulating HIV-1 replication in T cells. AIDS 25:2105–2111

Anand AR, Zhao H, Nagaraja T, Robinson LA, Ganju RK (2013) N-terminal Slit2 inhibits HIV-1 replication by regulating the actin cytoskeleton. Retrovirology 10:2

Andrews W, Liapi A, Plachez C, Camurri L, Zhang J, Mori S, Murakami F, Parnavelas JG, Sundaresan V, Richards LJ (2006) Robo1 regulates the development of major axon tracts and interneuron migration in the forebrain. Development 133:2243–2252

Andrews W, Barber M, Hernadez-Miranda LR, Xian J, Rakic S, Sundaresan V, Rabbitts TH, Pannell R, Rabbitts P, Thompson H et al (2008) The role of Slit-Robo signaling in the generation, migration and morphological differentiation of cortical interneurons. Dev Biol 313:648–658

Anitha A, Nakamura K, Yamada K, Suda S, Thanseem I, Tsujii M, Iwayama Y, Hattori E, Toyota T, Miyachi T et al (2008) Genetic analyses of Roundabout(ROBO) axon guidance receptors in autism. Am J Med Genet 147B:1019–1027

Atkinson-Leadbeater K, Bertolesi GE, Hehr CL, Webber CA, Cechmanek PB, McFarlane S (2010) Dynamic expression of axon guidance cues required for optic tract development is controlled by fibroblast growth factor signaling. J Neurosci 30:685–693

Bacon C, Endris V, Rappold G (2009) Dynamic expression of the Slit-Robo GTPase activating protein genes during development of the murine nervous system. J Comp Neurol 513:224–236

Bacon C, Endris V, Andermatt I, Niederkofler V, Waltereit R, Bartsch D, Stoeckli ET, Rappold G (2011) Evidence for a role of srGAP3 in the positioning of commissural axons within the ventrolateral funiculus of the mouse spinal cord. PLoS ONE 6:e19887

Bagri A, Marín O, Plump AS, Mak J, Pleasure SJ, Rubenstein JLR, Tessier-Lavigne M (2002) Slit proteins prevent midline crossing and determine the dorsoventral position of major axonal pathways in the mammalian forebrain. Neuron 33:233–248

Bai G, Chivatakarn O, Bonanomi D, Lettieri K, Franco L, Xia C, Stein E, Ma L, Lewcock JW, Pfaff SL (2011) Presenilin-dependent receptor processing is required for axon guidance. Cell 144:106–118

Ballard MS, Hinck L (2012) A roundabout way to cancer. Adv Cancer Res 114:187–235

Balordi F, Fishell G (2007) Hedgehog signaling in the subventricular zone is required for both the maintenance of stem cells and the migration of newborn neurons. J Neurosci 27:5936–5947

Barber AJ (2003) A new view of diabetic retinopathy: a neurodegenerative disease of the eye. Prog Neuropsychopharmacol Biol Psychiatry 27:283–290

Barresi MJF, Hutson LD, Chien C-B, Karlstrom RO (2005) Hedgehog regulated Slit expression determines commissure and glial cell position in the zebrafish forebrain. Development 132: 3643–3656

Bashaw GJ, Kidd T, Murray D, Pawson T, Goodman CS (2000) Repulsive axon guidance: Abelson and Enabled play opposing roles downstream of the roundabout receptor. Cell 101:703–715

Bates TC, Luciano M, Medland SE, Montgomery GW, Wright MJ, Martin NG (2010) Genetic variance in a component of the language acquisition device: robo1 polymorphisms associated with phonological buffer deficits. Behav Genet 41:50–57

Battye R, Stevens A, Perry RL, Jacobs JR (2001) Repellent signaling by Slit requires the leucine-rich repeats. J Neurosci 21:4290–4298
Bauer K, Dowejko A, Bosserhoff A-K, Reichert TE, Bauer R (2011) Slit-2 facilitates interaction of P-cadherin with Robo-3 and inhibits cell migration in an oral squamous cell carcinoma cell line. Carcinogenesis 32:935–943
Bedell VM, Yeo S-Y, Park KW, Chung J, Seth P, Shivalingappa V, Zhao J, Obara T, Sukhatme VP, Drummond IA et al (2005) Roundabout4 is essential for angiogenesis in vivo. Proc Natl Acad Sci USA 102:6373–6378
Beggs AD, Jones A, El-Bahwary M, Abulafi M, Hodgson SV, Tomlinson IP (2013) Whole-genome methylation analysis of benign and malignant colorectal tumours. J Pathol 229:697–704
Biankin AV, Waddell N, Kassahn KS, Gingras M-C, Muthuswamy LB, Johns AL, Miller DK, Wilson PJ, Patch A-M, Wu J et al (2012) Pancreatic cancer genomes reveal aberrations in axon guidance pathway genes. Nature 491:399–405
Borrell V, Cárdenas A, Ciceri G, Galcerán J, Flames N, Pla R, Nóbrega-Pereira S, García-Frigola C, Peregrín S, Zhao Z et al (2012) Slit/Robo signaling modulates the proliferation of central nervous system progenitors. Neuron 76:338–352
Bowerman B (2008) Cell signaling. Wnt moves beyond the canon. Science 320:327–328
Bravo-Ambrosio A, Mastick G, Kaprielian Z (2012) Motor axon exit from the mammalian spinal cord is controlled by the homeodomain protein Nkx2.9 via Robo-Slit signaling. Development 139:1435–1446
Brennand KJ, Simone A, Jou J, Gelboin-Burkhart C, Tran N, Sangar S, Li Y, Mu Y, Chen G, Yu D et al (2011) Modelling schizophrenia using human induced pluripotent stem cells. Nature 473:221–225
Briançon-Marjollet A, Ghogha A, Nawabi H, Triki I, Auziol C, Fromont S, Piché C, Enslen H, Chebli K, Cloutier J-F et al (2008) Trio mediates netrin-1-induced Rac1 activation in axon outgrowth and guidance. Mol Cell Biol 28:2314–2323
Brose K, Bland KS, Wang KH, Arnott D, Henzel W, Goodman CS, Tessier-Lavigne M, Kidd T (1999) Slit proteins bind Robo receptors and have an evolutionarily conserved role in repulsive axon guidance. Cell 96:795–806
Bustelo XR, Sauzeau V, Berenjeno IM (2007) GTP-binding proteins of the Rho/Rac family: regulation, effectors and functions in vivo. Bioessays 29:356–370
Calmont A, Ivins S, Van Bueren KL, Papangeli I, Kyriakopoulou V, Andrews WD, Martin JF, Moon AM, Illingworth EA, Basson MA et al (2009) Tbx1 controls cardiac neural crest cell migration during arch artery development by regulating Gbx2 expression in the pharyngeal ectoderm. Development 136:3173–3183
Camurri L, Mambetisaeva E, Davies D, Parnavelas J, Sundaresan V, Andrews W (2005) Evidence for the existence of two Robo3 isoforms with divergent biochemical properties. Mol Cell Neurosci 30:485–493
Carleton A, Petreanu LT, Lansford R, Alvarez-Buylla A, Lledo P-M (2003) Becoming a new neuron in the adult olfactory bulb. Nat Neurosci 6:507–518
Carrington JC, Ambros V (2003) Role of microRNAs in plant and animal development. Science 301:336–338
Carthew RW (2006) Gene regulation by microRNAs. Curr Opin Genet Dev 16:203–208
Chalasani SH, Sabelko KA, Sunshine MJ, Littman DR, Raper JA (2003) A chemokine, SDF-1, reduces the effectiveness of multiple axonal repellents and is required for normal axon pathfinding. J Neurosci 23:1360–1371
Chalasani SH, Sabol A, Xu H, Gyda MA, Rasband K, Granato M, Chien C-B, Raper JA (2007) Stromal cell-derived factor-1 antagonizes slit/robo signaling in vivo. J Neurosci 27:973–980
Charrier C, Joshi K, Coutinho-Budd J, Kim J-E, Lambert N, de Marchena J, Jin W-L, Vanderhaeghen P, Ghosh A, Sassa T et al (2012) Inhibition of SRGAP2 function by its human-specific paralogs induces neoteny during spine maturation. Cell 149:923–935
Chedotal A (2007) Slits and their receptors. Adv Exp Med Biol 621:65–80
Chedotal A (2011) Further tales of the midline. Curr Opin Neurobiol 21:68–75

Chen JH, Wen L, Dupuis S, Wu JY, RAO Y (2001) The N-terminal leucine-rich regions in Slit are sufficient to repel olfactory bulb axons and subventricular zone neurons. J Neurosci 21:1548–1556

Chen Z, Gore BB, Long H, Ma L, Tessier-Lavigne M (2008) Alternative splicing of the Robo3 axon guidance receptor governs the midline switch from attraction to repulsion. Neuron 58: 325–332

Conway CD, Howe KM, Nettleton NK, Price DJ, Mason JO, Pratt T (2011) Heparan sulfate sugar modifications mediate the functions of slits and other factors needed for mouse forebrain commissure development. J Neurosci 31:1955–1970

Coutinho-Budd J, Ghukasyan V, Zylka MJ, Polleux F (2012) The F-BAR domains from srGAP1, srGAP2, and srGAP3 differentially regulate membrane deformation. J Cell Sci 125: 3390–3401

Denk AE, Braig S, Schubert T, Bosserhoff AK (2011) Slit3 inhibits activator protein 1-mediated migration of malignant melanoma cells. Int J Mol Med 28:721–726

Devine CA, Key B (2008) Robo-Slit interactions regulate longitudinal axon pathfinding in the embryonic vertebrate brain. Dev Biol 313:371–383

Di Bonito M, Narita Y, Avallone B, Sequino L, Mancuso M, Andolfi G, Franzè AM, Puelles L, Rijli FM, Studer M (2013) Assembly of the auditory circuitry by a hox genetic network in the mouse brainstem. PLoS Genet 9:e1003249

Di Meglio T, Nguyen-Ba-Charvet KT, Tessier-Lavigne M, Sotelo C, Chedotal A (2008) Molecular mechanisms controlling midline crossing by precerebellar neurons. J Neurosci 28:6285–6294

Dickinson RE, Duncan WC (2010) The SLIT-ROBO pathway: a regulator of cell function with implications for the reproductive system. Reproduction 139:697–704

Dickson BJ, Gilestro GF (2006) Regulation of commissural axon pathfinding by slit and its Robo receptors. Annu Rev Cell Dev Biol 22:651–675

Doetsch F, García-Verdugo JM, Alvarez-Buylla A (1997) Cellular composition and three-dimensional organization of the subventricular germinal zone in the adult mammalian brain. J Neurosci 17:5046–5061

Doetsch F, Petreanu L, Caillé I, Garcia-Verdugo JM, Alvarez-Buylla A (2002) EGF converts transit-amplifying neurogenic precursors in the adult brain into multipotent stem cells. Neuron 36:1021–1034

Evans TA, Bashaw GJ (2010) Functional diversity of Robo receptor immunoglobulin domains promotes distinct axon guidance decisions. Curr Biol 20:567–572

Evans TA, Bashaw GJ (2012) Slit/Robo-mediated axon guidance in *Tribolium* and *Drosophila*: divergent genetic programs build insect nervous systems. Dev Biol 363:266–278

Fan X, Labrador J-P, Hing H, Bashaw GJ (2003) Slit stimulation recruits Dock and Pak to the roundabout receptor and increases Rac activity to regulate axon repulsion at the CNS midline. Neuron 40:113–127

Fish JE, Wythe JD, Xiao T, Bruneau BG, Stainier DYR, Srivastava D, Woo S (2011) A Slit/miR-218/Robo regulatory loop is required during heart tube formation in zebrafish. Development 138:1409–1419

Fothergill T, Donahoo A-LS, Douglass A, Zalucki O, Yuan J, Shu T, Goodhill GJ, Richards LJ (2013) Netrin-DCC signaling regulates corpus callosum formation through attraction of pioneering axons and by modulating Slit2-mediated repulsion. Cereb Cortex. Epub ahead of print

Fouquet C, Di Meglio T, Ma L, Kawasaki T, Long H, Hirata T, Tessier-Lavigne M, Chedotal A, Nguyen-Ba-Charvet KT (2007) Robo1 and robo2 control the development of the lateral olfactory tract. J Neurosci 27:3037–3045

Fukuhara N, Howitt JA, Hussain S-A, Hohenester E (2008) Structural and functional analysis of slit and heparin binding to immunoglobulin-like domains 1 and 2 of Drosophila Robo. J Biol Chem 283:16226–16234

Garbe DS, Bashaw GJ (2004) Axon guidance at the midline: from mutants to mechanisms. Crit Rev Biochem Mol Biol 39:319–341

Geisen MJ, Di Meglio T, Pasqualetti M, Ducret S, Brunet J-F, Chedotal A, Rijli FM (2008) Hox paralog group 2 genes control the migration of mouse pontine neurons through slit-robo signaling. PLoS Biol 6:e142

Gonda Y, Andrews WD, Tabata H, Namba T, Parnavelas JG, Nakajima K, Kohsaka S, Hanashima C, Uchino S (2012) Robo1 regulates the migration and laminar distribution of upper-layer pyramidal neurons of the cerebral cortex. Cereb Cortex 23:1495–1508

Gonzalo JA, Lloyd CM, Peled A, Delaney T, Coyle AJ, Gutierrez-Ramos JC (2000) Critical involvement of the chemotactic axis CXCR4/stromal cell-derived factor-1 alpha in the inflammatory component of allergic airway disease. J Immunol 165:499–508

Grant A, Fathalli F, Rouleau G, Joober R, Flores C (2012) Association between schizophrenia and genetic variation in DCC: a case–control study. Schizophr Res 137:26–31

Guan C-B, Xu H-T, Jin M, Yuan X-B, Poo M-M (2007) Long-range Ca2+ signaling from growth cone to soma mediates reversal of neuronal migration induced by slit-2. Cell 129:385–395

Guo S, Bao S (2010) srGAP2 arginine methylation regulates cell migration and cell spreading through promoting dimerization. J Biol Chem 285:35133–35141

Hannula-Jouppi K, Kaminen-Ahola N, Taipale M, Eklund R, Nopola-Hemmi J, Kääriäinen H, Kere J (2005) The axon guidance receptor gene ROBO1 is a candidate gene for developmental dyslexia. PLoS Genet 1:e50

Hao JC, Yu TW, Fujisawa K, Culotti JG, Gengyo-Ando K, Mitani S, Moulder G, Barstead R, Tessier-Lavigne M, Bargmann CI (2001) *C. elegans* slit acts in midline, dorsal-ventral, and anterior-posterior guidance via the SAX-3/Robo receptor. Neuron 32:25–38

Hagino S, Iseki K, Mori T, Zhang Y, Hikake T, Yokoya S, Takeuchi M, Hasimoto H, Kikuchi S, Wanaka A (2003) Slit and glypican-1 mRNAs are coexpressed in the reactive astrocytes of the injured adult brain. Glia 42:130–138

He Z, Tessier-Lavigne M (1997) Neuropilin is a receptor for the axonal chemorepellent Semaphorin III. Cell 90:739–751

Heasman SJ, Ridley AJ (2008) Mammalian Rho GTPases: new insights into their functions from in vivo studies. Nat Rev Mol Cell Biol 9:690–701

Hernández-Miranda LR, Parnavelas JG, Chiara F (2010) Molecules and mechanisms involved in the generation and migration of cortical interneurons. ASN Neuro 2:e00031

Higginbotham H, Tanaka T, Brinkman BC, Gleeson JG (2006) GSK3beta and PKCzeta function in centrosome localization and process stabilization during Slit-mediated neuronal repolarization. Mol Cell Neurosci 32:118–132

Hivert B, Liu Z, Chuang C-Y, Doherty P, Sundaresan V (2002) Robo1 and Robo2 are homophilic binding molecules that promote axonal growth. Mol Cell Neurosci 21:534–545

Holmes GP, Negus K, Burridge L, Raman S, Algar E, Yamada T, Little MH (1998) Distinct but overlapping expression patterns of two vertebrate slit homologs implies functional roles in CNS development and organogenesis. Mech Dev 79:57–72

Howitt JA, Clout NJ, Hohenester E (2004) Binding site for Robo receptors revealed by dissection of the leucine-rich repeat region of Slit. EMBO J 23:4406–4412

Hu H (2001) Cell-surface heparan sulfate is involved in the repulsive guidance activities of Slit2 protein. Nat Neurosci 4:695–701

Hu H, Li M, Labrador J-P, McEwen J, Lai EC, Goodman CS, Bashaw GJ (2005) Cross GTPase-activating protein (CrossGAP)/Vilse links the Roundabout receptor to Rac to regulate midline repulsion. Proc Natl Acad Sci USA 102:4613–4618

Huang L, Yu W, Li X, Niu L, Li K, Li J (2009a) Robo1/robo4: different expression patterns in retinal development. Exp Eye Res 88:583–588

Huang L, Yu W, Li X, Xu Y, Niu L, He X, Dong J, Yan Z (2009b) Expression of Robo4 in the fibrovascular membranes from patients with proliferative diabetic retinopathy and its role in RF/6A and RPE cells. Mol Vis 15:1057–1069

Huminiecki L, Gorn M, Suchting S, Poulsom R, Bicknell R (2002) Magic roundabout is a new member of the roundabout receptor family that is endothelial specific and expressed at sites of active angiogenesis. Genomics 79:547–552

Hussain S-A, Piper M, Fukuhara N, Strochlic L, Cho G, Howitt JA, Ahmed Y, Powell AK, Turnbull JE, Holt CE et al (2006) A molecular mechanism for the heparan sulfate dependence of slit-robo signaling. J Biol Chem 281:39693–39698

Inatani M, Irie F, Plump AS, Tessier-Lavigne M, Yamaguchi Y (2003) Mammalian brain morphogenesis and midline axon guidance require heparan sulfate. Science 302:1044–1046

Ito H, Funahashi S-I, Yamauchi N, Shibahara J, Midorikawa Y, Kawai S, Kinoshita Y, Watanabe A, Hippo Y, Ohtomo T et al (2006) Identification of ROBO1 as a novel hepatocellular carcinoma antigen and a potential therapeutic and diagnostic target. Clin Cancer Res 12: 3257–3264

Itoh A, Miyabayashi T, Ohno M, Sakano S (1998) Cloning and expressions of three mammalian homologues of *Drosophila* slit suggest possible roles for Slit in the formation and maintenance of the nervous system. Brain Res Mol Brain Res 62:175–186

Jackson RE, Eickholt BJ (2009) Semaphorin signalling. Curr Biol 19:R504–R507

Jankovski A, Sotelo C (1996) Subventricular zone-olfactory bulb migratory pathway in the adult mouse: cellular composition and specificity as determined by heterochronic and heterotopic transplantation. J Comp Neurol 371:376–396

Jaworski A, Long H, Tessier-Lavigne M (2010) Collaborative and specialized functions of Robo1 and Robo2 in spinal commissural axon guidance. J Neurosci 30:9445–9453

Jen JC, Chan W-M, Bosley TM, Wan J, Carr JR, Rüb U, Shattuck D, Salamon G, Kudo LC, Ou J et al (2004) Mutations in a human ROBO gene disrupt hindbrain axon pathway crossing and morphogenesis. Science 304:1509–1513

Jones CA, London NR, Chen H, Park KW, Sauvaget D, Stockton RA, Wythe JD, Suh W, Larrieu-Lahargue F, Mukouyama Y-S et al (2008) Robo4 stabilizes the vascular network by inhibiting pathologic angiogenesis and endothelial hyperpermeability. Nat Med 14:448–453

Jones CA, Nishiya N, London NR, Zhu W, Sorensen LK, Chan AC, Lim CJ, Chen H, Zhang Q, Schultz PG et al (2009) Slit2-Robo4 signalling promotes vascular stability by blocking Arf6 activity. Nat Cell Biol 11:1325–1331

Kaneko N, Marín O, Koike M, Hirota Y, Uchiyama Y, Wu JY, Lu Q, Tessier-Lavigne M, Alvarez-Buylla A, Okano H et al (2010) New neurons clear the path of astrocytic processes for their rapid migration in the adult brain. Neuron 67:213–223

Keleman K, Rajagopalan S, Cleppien D, Teis D, Paiha K, Huber LA, Technau GM, Dickson BJ (2002) Comm sorts robo to control axon guidance at the *Drosophila* midline. Cell 110:415–427

Kelsch W, Mosley CP, Lin C-W, Lois C (2007) Distinct mammalian precursors are committed to generate neurons with defined dendritic projection patterns. PLoS Biol 5:e300

Kidd T, Brose K, Mitchell KJ, Fetter RD, Tessier-Lavigne M, Goodman CS, Tear G (1998a) Roundabout controls axon crossing of the CNS midline and defines a novel subfamily of evolutionarily conserved guidance receptors. Cell 92:205–215

Kidd T, Russell C, Goodman CS, Tear G (1998b) Dosage-sensitive and complementary functions of roundabout and commissureless control axon crossing of the CNS midline. Neuron 20: 25–33

Kidd T, Bland KS, Goodman CS (1999) Slit is the midline repellent for the robo receptor in *Drosophila*. Cell 96:785–794

Killeen MT, Sybingco SS (2008) Netrin, Slit and Wnt receptors allow axons to choose the axis of migration. Dev Biol 323:143–151

Klagsbrun M, Eichmann A (2005) A role for axon guidance receptors and ligands in blood vessel development and tumor angiogenesis. Cytokine Growth Factor Rev 16:535–548

Klein ME, Impey S, Goodman RH (2005) Role reversal: the regulation of neuronal gene expression by microRNAs. Curr Opin Neurobiol 15:507–513

Koch AW, Mathivet T, Larrivée B, Tong RK, Kowalski J, Pibouin-Fragner L, Bouvrée K, Stawicki S, Nicholes K, Rathore N et al (2011) Robo4 maintains vessel integrity and inhibits angiogenesis by interacting with UNC5B. Dev Cell 20:33–46

Kolodkin AL, Matthes DJ, Goodman CS (1993) The semaphorin genes encode a family of transmembrane and secreted growth cone guidance molecules. Cell 75:1389–1399

Kolodkin AL, Levengood DV, Rowe EG, Tai YT, Giger RJ, Ginty DD (1997) Neuropilin is a semaphorin III receptor. Cell 90:753–762

Kuwako K-I, Kakumoto K, Imai T, Igarashi M, Hamakubo T, Sakakibara S-I, Tessier-Lavigne M, Okano HJ, Okano H (2010) Neural RNA-binding protein Musashi1 controls midline crossing of precerebellar neurons through posttranscriptional regulation of Robo3/Rig-1 expression. Neuron 67:407–421

Lamminmaki S, Massinen S, Nopola-Hemmi J, Kere J, Hari R (2012) Human ROBO1 regulates interaural interaction in auditory pathways. J Neurosci 32:966–971
Lanier LM, Gates MA, Witke W, Menzies AS, Wehman AM, Macklis JD, Kwiatkowski D, Soriano P, Gertler FB (1999) Mena is required for neurulation and commissure formation. Neuron 22:313–325
Lee JS, Ray R, Chien CB (2001) Cloning and expression of three zebrafish roundabout homologs suggest roles in axon guidance and cell migration. Dev Dyn 221:216–230
Lee J-S, von der Hardt S, Rusch MA, Stringer SE, Stickney HL, Talbot WS, Geisler R, Nüsslein-Volhard C, Selleck SB, Chien C-B et al (2004) Axon sorting in the optic tract requires HSPG synthesis by ext2 (dackel) and extl3 (boxer). Neuron 44:947–960
Legg JA, Herbert JMJ, Clissold P, Bicknell R (2008) Slits and Roundabouts in cancer, tumour angiogenesis and endothelial cell migration. Angiogenesis 11:13–21
Levy-Strumpf N, Culotti JG (2007) VAB-8, UNC-73 and MIG-2 regulate axon polarity and cell migration functions of UNC-40 in *C. elegans*. Nat Neurosci 10:161–168
Li HS, Chen JH, Wu W, Fagaly T, Zhou L, Yuan W, Dupuis S, Jiang ZH, Nash W, Gick C et al (1999) Vertebrate slit, a secreted ligand for the transmembrane protein roundabout, is a repellent for olfactory bulb axons. Cell 96:807–818
Lindwall C, Fothergill T, Richards LJ (2007) Commissure formation in the mammalian forebrain. Curr Opin Neurobiol 17:3–14
Liu Z, Patel K, Schmidt H, Andrews W, Pini A, Sundaresan V (2004) Extracellular Ig domains 1 and 2 of Robo are important for ligand (Slit) binding. Mol Cell Neurosci 26:232–240
Liu Q-X, Hiramoto M, Ueda H, Gojobori T, Hiromi Y, Hirose S (2009) Midline governs axon pathfinding by coordinating expression of two major guidance systems. Genes Dev 23: 1165–1170
Lois C, Alvarez-Buylla A (1994) Long-distance neuronal migration in the adult mammalian brain. Science 264:1145–1148
Lois C, García-Verdugo JM, Alvarez-Buylla A (1996) Chain migration of neuronal precursors. Science 271:978–981
Lopez-Bendito G, Flames N, Ma L, Fouquet C, Di Meglio T, Chedotal A, Tessier-Lavigne M, Marín O (2007) Robo1 and Robo2 cooperate to control the guidance of major axonal tracts in the mammalian forebrain. J Neurosci 27:3395–3407
Lu W, van Eerde AM, Fan X, Quintero-Rivera F, Kulkarni S, Ferguson H, Kim H-G, Fan Y, Xi Q, Li Q-G et al (2007) Disruption of ROBO2 is associated with urinary tract anomalies and confers risk of vesicoureteral reflux. Am J Hum Genet 80:616–632
Lundström A, Gallio M, Englund C, Steneberg P, Hemphälä J, Aspenström P, Keleman K, Falileeva L, Dickson BJ, Samakovlis C (2004) Vilse, a conserved Rac/Cdc42 GAP mediating Robo repulsion in tracheal cells and axons. Genes Dev 18:2161–2171
Luo Y, Raible D, Raper JA (1993) Collapsin: a protein in brain that induces the collapse and paralysis of neuronal growth cones. Cell 75:217–227
Madura T, Yamashita T, Kubo T, Tsuji L, Hosokawa K, Tohyama M (2004) Changes in mRNA of Slit-Robo GTPase-activating protein 2 following facial nerve transection. Brain Res Mol Brain Res 123:76–80
Marcos-Mondejar P, Peregrín S, Li JY, Carlsson L, Tole S, Lopez-Bendito G (2012) The lhx2 transcription factor controls thalamocortical axonal guidance by specific regulation of robo1 and robo2 receptors. J Neurosci 32:4372–4385
Marillat V, Cases O, Nguyen-Ba-Charvet KT, Tessier-Lavigne M, Sotelo C, Chedotal A (2002) Spatiotemporal expression patterns of slit and robo genes in the rat brain. J Comp Neurol 442:130–155
Marillat V, Sabatier C, Failli V, Matsunaga E, Sotelo C, Tessier-Lavigne M, Chedotal A (2004) The slit receptor Rig-1/Robo3 controls midline crossing by hindbrain precerebellar neurons and axons. Neuron 43:69–79
Marion J-F, Yang C, Caqueret A, Boucher F, Michaud JL (2005) Sim1 and Sim2 are required for the correct targeting of mammillary body axons. Development 132:5527–5537

Marlow R, Strickland P, Lee JS, Wu X, Pebenito M, Binnewies M, Le EK, Moran A, Macias H, Cardiff RD et al (2008) SLITs suppress tumor growth in vivo by silencing Sdf1/Cxcr4 within breast epithelium. Cancer Res 68:7819–7827

Mascheretti S, Bureau A, Battaglia M, Simone D, Quadrelli E, Croteau J, Cellino MR, Giorda R, Beri S, Maziade M et al (2012) An assessment of gene-by-environment interactions in developmental dyslexia-related phenotypes. Genes Brain Behav 12:47–55

Mambetisaeva ET, Andrews W, Camurri L, Annan A, Sundaresan V (2005) Robo family of proteins exhibit differential expression in mouse spinal cord and Robo-Slit interaction is required for midline crossing in vertebrate spinal cord. Dev Dyn 233:41–51

Mehlen P, Delloye-Bourgeois C, Chedotal A (2011) Novel roles for Slits and netrins: axon guidance cues as anticancer targets? Nat Rev Cancer 11:188–197

Menzies AS, Aszodi A, Williams SE, Pfeifer A, Wehman AM, Goh KL, Mason CA, Fassler R, Gertler FB (2004) Mena and vasodilator-stimulated phosphoprotein are required for multiple actin-dependent processes that shape the vertebrate nervous system. J Neurosci 24:8029–8038

Mertsch S, Schmitz N, Jeibmann A, Geng JG, Paulus W, Senner V (2008) Slit2 involvement in glioma cell migration is mediated by Robo1 receptor. J Neurooncol 87:1–7

Millar JK, Wilson-Annan JC, Anderson S, Christie S, Taylor MS, Semple CA, Devon RS, St. Clair DM, Muir WJ, Blackwood DH et al (2000) Disruption of two novel genes by a translocation co-segregating with schizophrenia. Hum Mol Genet 9:1415–1423

Morlot C, Thielens NM, Ravelli RBG, Hemrika W, Romijn RA, Gros P, Cusack S, McCarthy AA (2007) Structural insights into the Slit-Robo complex. Proc Natl Acad Sci USA 104: 14923–14928

Murray RM, Lappin J, Di Forti M (2008) Schizophrenia: from developmental deviance to dopamine dysregulation. Eur Neuropsychopharmacol 18(Suppl 3):S129–S134

Nagasawa T, Hirota S, Tachibana K, Takakura N, Nishikawa S, Kitamura Y, Yoshida N, Kikutani H, Kishimoto T (1996) Defects of B-cell lymphopoiesis and bone-marrow myelopoiesis in mice lacking the CXC chemokine PBSF/SDF-1. Nature 382:635–638

Nguyen-Ba-Charvet KT, Brose K, Marillat V, Sotelo C, Tessier-Lavigne M, Chedotal A (2001) Sensory axon response to substrate-bound Slit2 is modulated by laminin and cyclic GMP. Mol Cell Neurosci 17:1048–1058

Nguyen-Ba-Charvet KT, Picard-Riera N, Tessier-Lavigne M, Baron-Van Evercooren A, Sotelo C, Chedotal A (2004) Multiple roles for slits in the control of cell migration in the rostral migratory stream. J Neurosci 24:1497–1506

O'Donnell M, Chance RK, Bashaw GJ (2009) Axon growth and guidance: receptor regulation and signal transduction. Annu Rev Neurosci 32:383–412

Osbun N, Li J, O'Driscoll MC, Strominger Z, Wakahiro M, Rider E, Bukshpun P, Boland E, Spurrell CH, Schackwitz W et al (2011) Genetic and functional analyses identify DISC1 as a novel callosal agenesis candidate gene. Am J Med Genet A 155A:1865–1876

Ozdinler PH, Erzurumlu RS (2002) Slit2, a branching-arborization factor for sensory axons in the mammalian CNS. J Neurosci 22:4540–4549

Park KW, Morrison CM, Sorensen LK, Jones CA, Rao Y, Chien C-B, Wu JY, Urness LD, Li DY (2003) Robo4 is a vascular-specific receptor that inhibits endothelial migration. Dev Biol 261: 251–267

Philipp M, Niederkofler V, Debrunner M, Alther T, Kunz B, Stoeckli ET (2012) RabGDI controls axonal midline crossing by regulating Robo1 surface expression. Neural Dev 7:36

Piper M, Anderson R, Dwivedy A, Weinl C, van Horck F, Leung KM, Cogill E, Holt C (2006) Signaling mechanisms underlying Slit2-induced collapse of *Xenopus* retinal growth cones. Neuron 49:215–228

Plachez C, Richards LJ (2005) Mechanisms of axon guidance in the developing nervous system. Curr Top Dev Biol 69:267–346

Plump AS, Erskine L, Sabatier C, Brose K, Epstein CJ, Goodman CS, Mason CA, Tessier-Lavigne M (2002) Slit1 and Slit2 cooperate to prevent premature midline crossing of retinal axons in the mouse visual system. Neuron 33:219–232

Potkin SG, Macciardi F, Guffanti G, Fallon JH, Wang Q, Turner JA, Lakatos A, Miles MF, Lander A, Vawter MP et al (2010) Identifying gene regulatory networks in schizophrenia. Neuroimage 53:839–847
Prasad A, Fernandis AZ, Rao Y, Ganju RK (2004) Slit protein-mediated inhibition of CXCR4-induced chemotactic and chemoinvasive signaling pathways in breast cancer cells. J Biol Chem 279:9115–9124
Prasad A, Qamri Z, Wu J, Ganju RK (2007) Slit-2/Robo-1 modulates the CXCL12/CXCR4-induced chemotaxis of T cells. J Leukoc Biol 82:465–476
Pratt T, Conway CD, Tian NMM-L, Price DJ, Mason JO (2006) Heparan sulphation patterns generated by specific heparan sulfotransferase enzymes direct distinct aspects of retinal axon guidance at the optic chiasm. J Neurosci 26:6911–6923
Rajagopalan S, Nicolas E, Vivancos V, Berger J, Dickson BJ (2000) Crossing the midline: roles and regulation of Robo receptors. Neuron 28:767–777
Ramus F (2004) Neurobiology of dyslexia: a reinterpretation of the data. Trends Neurosci 27: 720–726
Renier N, Schonewille M, Giraudet F, Badura A, Tessier-Lavigne M, Avan P, De Zeeuw CI, Chedotal A (2010) Genetic dissection of the function of hindbrain axonal commissures. PLoS Biol 8:e1000325
Rhee J, Buchan T, Zukerberg L, Lilien J, Balsamo J (2007) Cables links Robo-bound Abl kinase to N-cadherin-bound beta-catenin to mediate Slit-induced modulation of adhesion and transcription. Nat Cell Biol 9:883–892
Round JE, Sun H (2011) The adaptor protein Nck2 mediates Slit1-induced changes in cortical neuron morphology. Mol Cell Neurosci 47:265–273
Sabatier C, Plump AS, Le M, Brose K, Tamada A, Murakami F, Lee EY-HP, Tessier-Lavigne M (2004) The divergent Robo family protein rig-1/Robo3 is a negative regulator of slit responsiveness required for midline crossing by commissural axons. Cell 117:157–169
Saha B, Jaber M, Gaillard A (2012) Potentials of endogenous neural stem cells in cortical repair. Front Cell Neurosci 6:14
Sang Q, Wu J, Rao Y, Hsueh Y-P, Tan S-S (2002) Slit promotes branching and elongation of neurites of interneurons but not projection neurons from the developing telencephalon. Mol Cell Neurosci 21:250–265
Santiago-Martínez E, Soplop NH, Patel R, Kramer SG (2008) Repulsion by slit and roundabout prevents shotgun/E-cadherin-mediated cell adhesion during *Drosophila* heart tube lumen formation. J Cell Biol 182:241–248
Sawamoto K, Wichterle H, Gonzalez-Perez O, Cholfin JA, Yamada M, Spassky N, Murcia NS, Garcia-Verdugo JM, Marín O, Rubenstein JLR et al (2006) New neurons follow the flow of cerebrospinal fluid in the adult brain. Science 311:629–632
Schmid BC, Rezniczek GA, Fabjani G, Yoneda T, Leodolter S, Zeillinger R (2007) The neuronal guidance cue Slit2 induces targeted migration and may play a role in brain metastasis of breast cancer cells. Breast Cancer Res Treat 106:333–342
Schweitzer J, Löhr H, Bonkowsky JL, Hübscher K, Driever W (2013) Sim1a and Arnt2 contribute to hypothalamo-spinal axon guidance by regulating Robo2 activity via a Robo3-dependent mechanism. Development 140:93–106
Seiradake E, von Philipsborn AC, Henry M, Fritz M, Lortat-Jacob H, Jamin M, Hemrika W, Bastmeyer M, Cusack S, McCarthy AA (2009) Structure and functional relevance of the Slit2 homodimerization domain. EMBO Rep 10:736–741
Sheldon H, Andre M, Legg JA, Heal P, Herbert JM, Sainson R, Sharma AS, Kitajewski JK, Heath VL, Bicknell R (2009) Active involvement of Robo1 and Robo4 in filopodia formation and endothelial cell motility mediated via WASP and other actin nucleation-promoting factors. FASEB J 23:513–522
Shiau CE, Bronner-Fraser M (2009) N-cadherin acts in concert with Slit1-Robo2 signaling in regulating aggregation of placode-derived cranial sensory neurons. Development 136: 4155–4164

Shiau CE, Lwigale PY, Das RM, Wilson SA, Bronner-Fraser M (2008) Robo2-Slit1 dependent cell-cell interactions mediate assembly of the trigeminal ganglion. Nat Neurosci 11:269–276
Shu T, Li Y, Keller A, Richards LJ (2003a) The glial sling is a migratory population of developing neurons. Development 130:2929–2937
Shu T, Puche AC, Richards LJ (2003b) Development of midline glial populations at the corticoseptal boundary. J Neurobiol 57:81–94
Shu T, Sundaresan V, McCarthy MM, Richards LJ (2003c) Slit2 guides both precrossing and postcrossing callosal axons at the midline in vivo. J Neurosci 23:8176–8184
Silver J, Ogawa MY (1983) Postnatally induced formation of the corpus callosum in acallosal mice on glia-coated cellulose bridges. Science 220:1067–1069
Silver J, Lorenz SE, Wahlsten D, Coughlin J (1982) Axonal guidance during development of the great cerebral commissures: descriptive and experimental studies, in vivo, on the role of preformed glial pathways. J Comp Neurol 210:10–29
Simpson JH, Kidd T, Bland KS, Goodman CS (2000) Short-range and long-range guidance by slit and its Robo receptors. Robo and Robo2 play distinct roles in midline guidance. Neuron 28: 753–766
Slováková J, Speicher S, Sánchez-Soriano N, Prokop A, Carmena A (2012) The actin-binding protein canoe/AF-6 forms a complex with Robo and is required for slit-Robo signaling during axon pathfinding at the CNS midline. J Neurosci 32:10035–10044
Stein E, Huynh-Do U, Lane AA, Cerretti DP, Daniel TO (1998) Nck recruitment to Eph receptor, EphB1/ELK, couples ligand activation to c-Jun kinase. J Biol Chem 273:1303–1308
Stein E, Zou Y, Poo M, Tessier-Lavigne M (2001) Binding of DCC by netrin-1 to mediate axon guidance independent of adenosine A2B receptor activation. Science 291:1976–1982
Stella MC, Trusolino L, Comoglio PM (2009) The Slit/Robo system suppresses hepatocyte growth factor-dependent invasion and morphogenesis. Mol Biol Cell 20:642–657
Suchting S, Heal P, Tahtis K, Stewart LM, Bicknell R (2005) Soluble Robo4 receptor inhibits in vivo angiogenesis and endothelial cell migration. FASEB J 19:121–123
Suda S, Iwata K, Shimmura C, Kameno Y, Anitha A, Thanseem I, Nakamura K, Matsuzaki H, Tsuchiya KJ, Sugihara G et al (2011) Decreased expression of axon-guidance receptors in the anterior cingulate cortex in autism. Mol Autism 2:14
Sundaresan V, Chung G, Heppell-Parton A, Xiong J, Grundy C, Roberts I, James L, Cahn A, Bench A, Douglas J et al (1998a) Homozygous deletions at 3p12 in breast and lung cancer. Oncogene 17:1723–1729
Sundaresan V, Roberts I, Bateman A, Bankier A, Sheppard M, Hobbs C, Xiong J, Minna J, Latif F, Lerman M et al (1998b) The DUTT1 gene, a novel NCAM family member is expressed in developing murine neural tissues and has an unusually broad pattern of expression. Mol Cell Neurosci 11:29–35
Tie J, Pan Y, Zhao L, Wu K, Liu J, Sun S, Guo X, Wang B, Gang Y, Zhang Y et al (2010) MiR-218 inhibits invasion and metastasis of gastric cancer by targeting the Robo1 receptor. PLoS Genet 6:e1000879
Tiveron M-C, Cremer H (2008) CXCL12/CXCR4 signalling in neuronal cell migration. Curr Opin Neurobiol 18:237–244
Unni DK, Piper M, Moldrich RX, Gobius I, Liu S, Fothergill T, Donahoo A-LS, Baisden JM, Cooper HM, Richards LJ (2012) Multiple Slits regulate the development of midline glial populations and the corpus callosum. Dev Biol 365:36–49
Van Vactor D, Wall DP, Johnson KG (2006) Heparan sulfate proteoglycans and the emergence of neuronal connectivity. Curr Opin Neurobiol 16:40–51
Vanderzalm PJ, Pandey A, Hurwitz ME, Bloom L, Horvitz HR, Garriga G (2009) *C. elegans* CARMIL negatively regulates UNC73/Trio function during neuronal development. Development 136:1201–1210
Vargesson N, Luria V, Messina I, Erskine L, Laufer E (2001) Expression patterns of Slit and Robo family members during vertebrate limb development. Mech Dev 106:175–180
Verissimo AR, Herbert JMJ, Heath VL, Legg JA, Sheldon H, Andre M, Swain RK, Bicknell R (2009) Functionally defining the endothelial transcriptome, from Robo4 to ECSCR. Biochem Soc Trans 37:1214–1217

Waltereit R, Leimer U, von Bohlen und Halbach O, Panke J, Holter SM, Garrett L, Wittig K, Schneider M, Schmitt C, Calzada-Wack J et al (2012) Srgap3–/– mice present a neurodevelopmental disorder with schizophrenia-related intermediate phenotypes. FASEB J 26:4418–4428

Wang KH, Brose K, Arnott D, Kidd T, Goodman CS, Henzel W, Tessier-Lavigne M (1999) Biochemical purification of a mammalian slit protein as a positive regulator of sensory axon elongation and branching. Cell 96:771–784

Wang B, Xiao Y, Ding BB, Zhang N, Yuan XB, Gui L, Qian KX, Duan S, Chen Z, Rao Y et al (2003) Induction of tumor angiogenesis by Slit-Robo signaling and inhibition of cancer growth by blocking Robo activity. Cancer Cell 4:19–29

Ward M, McCann C, DeWulf M, Wu JY, Rao Y (2003) Distinguishing between directional guidance and motility regulation in neuronal migration. J Neurosci 23:5170–5177

Ward ME, Jiang H, Rao Y (2005) Regulated formation and selection of neuronal processes underlie directional guidance of neuronal migration. Mol Cell Neurosci 30:378–387

Watari-Goshima N, Ogura K-I, Wolf FW, Goshima Y, Garriga G (2007) *C. elegans* VAB-8 and UNC-73 regulate the SAX-3 receptor to direct cell and growth-cone migrations. Nat Neurosci 10:169–176

Werbowetski-Ogilvie TE, Seyed Sadr M, Jabado N, Angers-Loustau A, Agar NYR, Wu J, Bjerkvig R, Antel JP, Faury D, Rao Y et al (2006) Inhibition of medulloblastoma cell invasion by Slit. Oncogene 25:5103–5112

Whitford KL, Marillat V, Stein E, Goodman CS, Tessier-Lavigne M, Chedotal A, Ghosh A (2002) Regulation of cortical dendrite development by Slit-Robo interactions. Neuron 33:47–61

Wilson SI, Shafer B, Lee KJ, Dodd J (2008) A molecular program for contralateral trajectory: Rig-1 control by LIM homeodomain transcription factors. Neuron 59:413–424

Wilson NKA, Lee Y, Long R, Hermetz K, Rudd MK, Miller R, Rapoport JL, Addington AM (2011) A novel microduplication in the neurodevelopmental gene SRGAP3 that segregates with psychotic illness in the family of a COS proband. Case Rep Genet 2011:1–5

Wong K, Ren XR, Huang YZ, Xie Y, Liu G, Saito H, Tang H, Wen L, Brady-Kalnay SM, Mei L et al (2001) Signal transduction in neuronal migration: roles of GTPase activating proteins and the small GTPase Cdc42 in the Slit-Robo pathway. Cell 107:209–221

Wright KM, Lyon KA, Leung H, Leahy DJ, Ma L, Ginty DD (2012) Dystroglycan organizes axon guidance cue localization and axonal pathfinding. Neuron 76:931–944

Wu W, Wong K, Chen J, Jiang Z, Dupuis S, Wu JY, Rao Y (1999) Directional guidance of neuronal migration in the olfactory system by the protein Slit. Nature 400:331–336

Xu H-T, Yuan X-B, Guan C-B, Duan S, Wu C-P, Feng L (2004) Calcium signaling in chemorepellant Slit2-dependent regulation of neuronal migration. Proc Natl Acad Sci USA 101: 4296–4301

Yang L, Bashaw GJ (2006) Son of sevenless directly links the Robo receptor to rac activation to control axon repulsion at the midline. Neuron 52:595–607

Yao Q, Jin W-L, Wang Y, Ju G (2008) Regulated shuttling of Slit-Robo-GTPase activating proteins between nucleus and cytoplasm during brain development. Cell Mol Neurobiol 28:205–221

Ypsilanti AR, Zagar Y, Chedotal A (2010) Moving away from the midline: new developments for Slit and Robo. Development 137:1939–1952

Yu TW, Hao JC, Lim W, Tessier-Lavigne M, Bargmann CI (2002) Shared receptors in axon guidance: SAX-3/Robo signals via UNC-34/Enabled and a Netrin-independent UNC-40/DCC function. Nat Neurosci 5:1147–1154

Yuan SS, Cox LA, Dasika GK, Lee EY (1999) Cloning and functional studies of a novel gene aberrantly expressed in RB-deficient embryos. Dev Biol 207:62–75

Yuasa-Kawada J, Kinoshita-Kawada M, Rao Y, Wu JY (2009a) Deubiquitinating enzyme USP33/VDU1 is required for Slit signaling in inhibiting breast cancer cell migration. Proc Natl Acad Sci USA 106:14530–14535

Yuasa-Kawada J, Kinoshita-Kawada M, Wu G, Rao Y, Wu JY (2009b) Midline crossing and Slit responsiveness of commissural axons require USP33. Nat Neurosci 12:1087–1089

Zhang X (2010) Prevalence of diabetic retinopathy in the United States, 2005–2008. JAMA 304:649

Zhang H-Y, Zheng S-J, Zhao J-H, Zhao W, Zheng L-F, Zhao D, Li J-M, Zhang X-F, Chen Z-B, Yi X-N (2011) MicroRNAs 144, 145, and 214 are down-regulated in primary neurons responding to sciatic nerve transection. Brain Res 1383:62–70

Zheng W, Geng A-Q, Li P-F, Wang Y, Yuan X-B (2011) Robo4 regulates the radial migration of newborn neurons in developing neocortex. Cereb Cortex 22:2587–2601

Zhou W, Yu W, Xie W, Huang L, Xu Y, Li X (2011a) The role of SLIT-ROBO signaling in proliferative diabetic retinopathy and retinal pigment epithelial cells. Mol Vis 17:1526–1536

Zhou W-J, Geng ZH, Chi S, Zhang W, Niu X-F, Lan S-J, Ma L, Yang X, Wang L-J, Ding Y-Q et al (2011b) Slit-Robo signaling induces malignant transformation through Hakai-mediated E-cadherin degradation during colorectal epithelial cell carcinogenesis. Cell Res 21:609–626

Chapter 8
New Insights into the Roles of the Contactin Cell Adhesion Molecules in Neural Development

Alma N. Mohebiany, Sheila Harroch, and Samuel Bouyain

Abstract In vertebrates, the contactin (CNTN) family of neural cell recognition molecules includes six related cell adhesion molecules that play non-overlapping roles in the formation and maintenance of the nervous system. CNTN1 and CNTN2 are the prototypical members of the family and have been involved, through *cis*- and *trans*-interactions with distinct cell adhesion molecules, in neural cell migration, axon guidance, and the organization of myelin subdomains. In contrast, the roles of CNTN3–6 are less well characterized although the generation of null mice and the recent identification of a common extracellular binding partner have considerably advanced our grasp of their physiological roles in particular as they relate to the wiring of sensory tissues. In this review, we aim to present a summary of our current understanding of CNTN functions and give an overview of the challenges that lie ahead in understanding the roles these proteins play in nervous system development and maintenance.

8.1 Introduction

The proper development of the nervous system is the culmination of several complex and coordinated processes that include the proliferation and differentiation of neural cells, cellular migration, axon growth and guidance, the formation of synapses, and myelination of axons to permit efficient saltatory conduction.

A.N. Mohebiany, B.Sc., M.Sc. • S. Harroch, Ph.D. (✉)
Department of Neuroscience, School of Biological Sciences, Institut Pasteur de Paris,
25-28 Rue du Docteur Roux, 75724 Paris Cedex 15, France
e-mail: sharroch@pasteur.fr

S. Bouyain, D.Phil. (✉)
Division of Molecular Biology and Biochemistry, School of Biological Sciences,
University of Missouri-Kansas City, 5100 Rockhill Road, Kansas City, MO 64110, USA
e-mail: bouyains@umkc.edu

V. Berezin and P.S. Walmod (eds.), *Cell Adhesion Molecules: Implications in Neurological Diseases*, Advances in Neurobiology 8, DOI 10.1007/978-1-4614-8090-7_8,

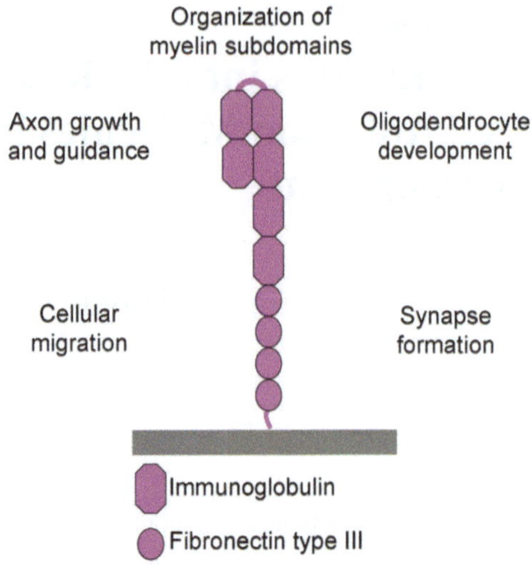

Fig. 8.1 Domain organization of vertebrate CNTNs. The architecture shared by all six vertebrate CNTNs is shown here. They include six N-terminal immunoglobulin domains and four fibronectin type III domains and are attached to the cell surface by a glycophosphatidylinositol anchor. Specific roles for the six CNTNs that have been identified from mouse studies are highlighted here

The information necessary for the appropriate coordination of these events is encoded in the interactions between growth factors, receptor tyrosine kinases, receptor protein tyrosine phosphatases, components of the extracellular matrix, and cell adhesion molecules. It is the precise interplay between these receptors and intracellular signaling machineries that underpins the organization and complexity of neural tissues, and alterations of these binding events are linked to developmental defects as well as the pathologies of human mental disorders.

Cell adhesion molecules (CAMs) represent one of the most diverse groups of cell surface proteins involved in neural development. Among them, members of the immunoglobulin (Ig) superfamily have a prominent role in shaping the nervous system. The extracellular regions of these proteins include almost exclusively Ig and fibronectin type III repeats (FNIII) and yet this seemingly simplistic modular architecture underpins a great variety of distinct neurodevelopmental processes. The contactins (CNTNs) represent one such group of neural CAMs and are involved in all major aspects of neural development. In vertebrates, the family is represented by CNTN1 (F3), CNTN2 (TAG-1/TAX-1/axonin), CNTN3 (PANG/BIG-1), CNTN4 (BIG-2), CNTN5 (NB-2), and CNTN6 (NB-3) (Shimoda and Watanabe 2009; Zuko et al. 2011) whereas a single family member called CONT has been described in *Drosophila* (Faivre-Sarrailh et al. 2004). The vertebrate CNTNs include six Ig and four FNIII modules, are tethered to the cell membrane by a glycophosphatidylinositol (GPI) anchor (Fig. 8.1), and share 40–60 % amino acid sequence identity (Zuko et al. 2011). Recent excellent reviews have focused mostly on subgroups of CNTNs (Shimoda and Watanabe 2009; Zuko et al. 2011) or on one particular developmental aspect (Stoeckli 2010). Here we aim to give an overview of the functions played by members of the CNTN gene family across different species as well as discuss recent results about the structures and functions of these proteins.

8.2 Insights into Contactin Function from *Drosophila* and Zebrafish

8.2.1 The Formation of Septate Junctions in Drosophila *Peripheral Nerves*

In *Drosophila melanogaster*, the CNTN gene family is represented by a single gene named *cont* (Faivre-Sarrailh et al. 2004). The gene product is a GPI-anchored protein (CONT), which shares a similar domain organization as its vertebrate counterparts though it also includes an N-terminal C-type lectin-like domain of unknown function (Fig. 8.2). CONT is found on epithelial cells originating from the ectoderm and importantly on the surface of glial cells called perineurial glial cells. These perineurial cells encase inner glial cells that ensheath the axons of *Drosophila* peripheral nerves. The role of these cell layers is to isolate the nerves from the electrolytes in the hemolymph, thus allowing the propagation of saltatory impulses along the axons (Banerjee et al. 2006b). Specialized cell contacts called septate junctions (SJs) functioning as permeability barriers form between the inner glial cells and the perineurial cells. The expression of CONT on perineurial cells overlaps with that of two other neural CAMs and SJ markers called neuroglian (NRG) and neurexin IV (NRX IV) (Faivre-Sarrailh et al. 2004), which are the *Drosophila* representatives of the L1 and contactin associated-like proteins (CNTNAPs) families, respectively. *Cont* mutant flies do not form SJs, as is the case for *nrg* and *nrx IV* mutants (Banerjee et al. 2006a) and the three proteins only localize correctly at SJs when all three are present. In other words, removal of NRG, NRX IV, or CONT prevents the appropriate localization of the two other cell surface receptors at SJs. The evidence accumulated thus far suggest that NRG, NRX IV, and CONT form a tripartite complex, with NRX IV and CONT associating on the same cell (*cis-interaction*) and binding to NRG expressed on an apposing cell (*trans*-interaction) (Banerjee et al. 2006a; Faivre-Sarrailh et al. 2004) and a clearer picture of how this complex controls the formation of SJs is starting to emerge.

Most of the biochemical insights into the CONT/NRX IV/NRG complex come from domain deletion studies performed on NRX IV (Banerjee et al. 2011). As is the case for the vast majority of CAMs, NRX IV is a modular protein; it includes an N-terminal discoidin domain, a laminin G module, and two laminin G/epidermal growth factor/laminin G repeats followed by a transmembrane region and a cytoplasmic segment (Baumgartner et al. 1996). Deletion of the NRX IV intracellular region has no effect on the localization of the tripartite complex members NRG and CONT, but appears to impair the association of NRX IV with CONT and leads to formation of defective SJs. This deletion phenocopies the effects observed in flies that do not express the cytoplasmic organizer Coracle (Genova and Fehon 2003), consistent with the notion that anchoring NRX IV, and by extension the tripartite complex, to the cytoskeleton is essential for the formation of SJs (Banerjee et al. 2011). Moreover, module deletion in the extracellular region of NRX IV has

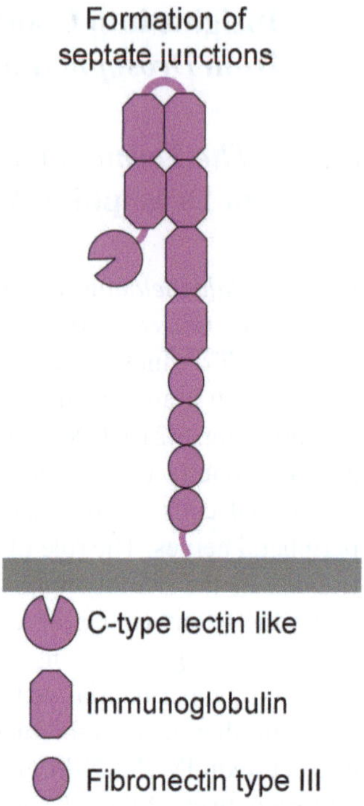

Fig. 8.2 Architecture of the single *Drosophila* CNTN homologue. The domain organization of CONT is identical to the one described for vertebrate CNTNs save for the presence of a C-type lectin-like domain at its N-terminus

identified the first laminin G/epidermal growth factor/laminin G as essential for associating with CONT since a mutant form of NRX IV containing only this repeat in its ectodomain was able to associate with CONT and led to the correct localization of CONT and NRG at SJs (Banerjee et al. 2011). This mutant NRX IV could not support the formation of functional SJs, however, demonstrating that mere association of NRX IV with CONT is not enough for the correct organization of the junctions. Less is known about the biochemical roles of the two other complex members, NRG and CONT. NRG is a homophilic binding protein, capable of inducing aggregation of non-adherent S2 cells when expressed on their surface (Hortsch et al. 1995), but the identities of the domains required for association with CONT and NRX IV remain unclear. The same is true for CONT since domain deletion studies have yet to be performed so that the molecular details of how it forms a complex with both NRX IV and NRG are yet to be uncovered.

The findings accumulated so far in *Drosophila* mirror earlier results obtained in vertebrates. Indeed the wrapping of peripheral axons by glial cells and the formation of SJs in *Drosophila* is reminiscent of the ensheathment of vertebrate axons by myelin and the formation of SJs in paranodal regions (Banerjee et al. 2006b). As is the case in *Drosophila*, mice lacking the *CNTN1* gene have disrupted junctions in paranodal regions (Boyle et al. 2001). The parallel between the roles of CONT and CNTN1 is made stronger when considering that a *cis*-complex of CNTN1 and contactin-associated protein 1 (CNTNAP1) expressed on axons associates specifically with the 155 kDa glial isoform of neurofascin designated NF155, a member of the L1 family and homologue of NRG (Charles et al. 2002). The role of the tripartite complex of CNTN1/CNTNAP1/NF155 in mice neuronal development is considered in more detail below, but it is important to note the evolutionary conserved roles of CONT and CNTN1 as we start considering the more intricate roles of CNTN family members and additional binding partners in the development and maintenance of the vertebrate nervous system.

8.2.2 Axonogenesis and Oligodendrocyte Development in Zebrafish

Zebrafish is an ideal model system for analysis of gene function and is increasingly being used to model human diseases (Lieschke and Currie 2007). Two paralogues of CNTN1 called *Cntn1a* and *Cntn1b* exist in zebrafish. Both are expressed during development although *Cntn1a* transcripts are present earlier than those of *Cntn1b* (Haenisch et al. 2005). Expression of Cntn1a is first detected during axonal growth in the trigeminal neurons that innervate jaw muscles and in Rohon–Beard sensory neurons, which extend into the spinal cord and the epidermis where they function in the detection of information about the environment (Gimnopoulos et al. 2002). Axons from these neurons are the first to grow out indicating that Cntn1a expression occurs concomitantly with axonogenesis. In contrast, Cntn1b expression coincides with later neurodevelopmental events such as pathfinding, synapse formation, and myelination (Haenisch et al. 2005). The role of Cntn2 has also been examined in the development of the zebrafish nervous system and in particular in the nucleus of the medial longitudinal fascicle (nucMLF). Axons from nucMLF grow out and extend into the midbrain where they essentially form a major tract from which axons that develop at a later stage will extend (Wolman et al. 2008). Axons must grow out in the appropriate direction and converge onto an axon tract. Loss of *Cntn2* results in nucMLF axons extending in incorrect directions along with a reduction in the axonal growth rate and a decrease in axon–axon interactions (Wolman et al. 2008). Cntn2 expression is also necessary for the guidance of central axons from Rohon–Beard sensory neurons into the spinal cord to the epidermis as knockdown of Cntn2 using morpholinos induces extensive branching of central Rohon–Beard neurons while their forward advance is drastically reduced (Liu and Halloran 2005). Taken together, these results indicate that expression of Cntn1 and Cntn2 is required during axonal growth in zebrafish and as such underline an additional role in axonal growth compared to fly CONT.

In addition to their involvement in axonal growth and pathfinding, the two paralogues of CNTN1 in zebrafish participate in the differentiation of myelinating glial cells during development as well as in optic nerve regeneration. Indeed, both *Cntn1a* and *Cntn1b* mRNAs have been detected in oligodendrocytes in the developing optic pathway of zebrafish (Haenisch et al. 2005; Schweitzer et al. 2007). Furthermore, *Cntn1a* expression is increased following lesion of the optic nerve in dedifferentiating and re-differentiating oligodendrocytes, a process required for the remyelination of regenerating axons (Ankerhold and Stuermer 1999), yet *Cntn1a* mRNA is absent during remyelination. Interestingly, the regulation of the two CNTN1 paralogues in the lesioned optic nerve is inversely correlated since *Cntn1b* is reduced during the de- and re-differentiation phases, but returns to normal levels during remyelination in goldfish (Haenisch et al. 2005). The reasons underlying these differences in expression levels are unclear, but the sum of these findings links CNTN1 to oligodendrocyte differentiation and myelination, which is consistent with recent work on performed in mice (see below).

8.3 Insights into CNTN Function from Mice Studies

8.3.1 The Roles of CNTN1 and CNTN2 in Cerebellum Development

The work performed in *Drosophila* and zebrafish has identified roles for the contactin gene family in the formation of axoglial junctions, myelination, and axonogenesis. However, the most comprehensive view of CNTN function is afforded by almost two decades of work using mice. In particular, the mouse cerebellum is an excellent system to elucidate the role of CNTN1 and CNTN2 in guiding neural cell development as they are expressed in the postnatal cerebellar cortex (Faivre-Sarrailh et al. 1992; Lee et al. 2000; Yoshihara et al. 1994) and because the anatomical organization and the postnatal development of this structure are already well characterized. Briefly, the cerebellar cortex is made up of four types of neurons: the granule cells, Purkinje cells, and two types of inhibitory neurons: the Golgi cells and the stellate/basket cells (Voogd and Glickstein 1998). The cerebellum is involved in motor control and learning and the coordination of body position and limb movement; more recently it has also been described to contribute to cognitive processes, with impaired function in humans linked to autism, schizophrenia, and mental retardation (Schmahmann 2004). Therefore, studies of the roles played by CNTNs in cerebellar tissues may shed light on their involvement in human mental disorders.

Overall, the presence of CNTN1 and 2 is critical for normal development of the cerebellum. Because these two proteins are closely related (they share 48 % amino acid sequence identity overall), one could speculate that they play redundant roles in cerebellar development. However, expression of CNTN1 under the gene promoter of CNTN2 results in reduced cerebellum size, due to a reduction in granule cell number and a reduction in the growth and fasciculation of their axons (Bizzoca et al. 2003), indicating the importance of proper spatial and temporal regulation for the expression of these two CNTNs. During development of the cerebellum, CNTN2 is expressed by granule cell progenitors as they differentiate; it is present on the cell bodies and elongating parallel fibers of the granule cells (Pickford et al. 1989; Yamamoto et al. 1990). CNTN2 expression is then replaced by that of CNTN1 with the onset of radial cellular migration (Bizzoca et al. 2003; Faivre-Sarrailh et al. 1992; Wolfer et al. 1994). The distinct roles of CNTN1 and CNTN2 are further highlighted by a recent study by Xenaki and colleagues (2011) in which the authors describe opposing roles for these two receptors in the regulation of cerebellar granule neuron (CGN) progenitor proliferation. After birth, expansion of CGN precursors occurs prior to their migration and differentiation and this proliferation is in particular regulated by the Sonic hedgehog (Shh) signaling pathway. In this context, Xenaki and colleagues demonstrated that Shh-promoted CGN proliferation was inhibited when these cells were cultured in the presence of soluble CNTN1, which in turn favored differentiation. However, culture of CGNs in the presence of CNTN2 antagonized the effect of CNTN1 and restored Shh-induced CGN proliferation. How can CNTN1 and CNTN2 produce such distinct outcomes for the proliferation of CGNs? It appears that the inhibitory effect of CNTN1 on CGNs is mediated through interactions with the L1 family NrCAM expressed on the surface of these cells (Xenaki et al. 2011). However, because CNTN2 has also been shown to bind to NrCAM, it would compete with CNTN1 and antagonize its effect on CGNs (Brümmendorf and Lemmon 2001). In addition to underlining the non-overlapping functions of CNTN1 and CNTN2 in cerebellar development, these results highlight an important link between L1 and CNTN family members, further examples of which are mentioned below.

CNTN2 is considered to play a critical role in neuronal migration (Denaxa et al. 2001, 2005), yet its exact function in this process has remained unclear. Studies using blocking antibodies and *CNTN2*$^{-/-}$ mice have attempted to bring some clarity to this process. Incubation of P5 cerebellar slices with anti-CNTN2 serum resulted in decreased radial migration between cerebellar layers and accumulation of CGNs in the external germinal layer of the cerebellum (Wang et al. 2011). This finding suggests that CNTN2 promotes cellular migration. Strikingly, Wang and colleagues also remarked that treatment of isolated CGNs by an anti-CNTN2 antibody resulted in increased migration in a Boyden chamber assay, suggesting that CNTN2, in this context, is an inhibitor of cellular migration. These contradictory results may stem from differences in the maturation of the CGNs used in the experiments, yet they may also highlight the complex function of CNTN2 in neuronal migration and that CNTN2-mediated migration might depend on additional proteins in the microenvironment of CGNs. On the other hand, although CNTN2-deficient mice present alterations in subsets of neural cell types (Denaxa et al. 2005; Fukamauchi et al. 2001; Traka et al. 2003), studies of the cerebella of these mice at P2 and P10 revealed no obvious phenotype for granule neurons and no significant differences in parallel fibers or in the thicknesses of various regions of the cerebella (Fukamauchi et al. 2001). Taken together with the antibody-blocking experiments, these observations likely suggest the presence of compensatory mechanisms that cannot be identified in vitro in the blocking studies (Fukamauchi et al. 2001). In addition, although the gross morphology of the brain and cerebellum seems unaffected by the loss of CNTN2, a more detailed analysis indicates differences at the cellular level. For instance, Xenaki et al. (2011) remarked the presence of ectopic clusters of small granule-like cells, possibly due to impaired migration. Further experiments demonstrated that the presence of these clusters is not due to a delay in radial migration, but rather that the loss of CNTN2 results in a reduction in CGN production so that production continues longer in compensation (Xenaki et al. 2011). Thus, the sum of these investigations indicates that migration defects observed using anti-CNTN2 antibodies or in $CNTN2^{-/-}$ mice may stem from defects in CGN proliferation and maturation.

Axon guidance and pathfinding depend on the interaction between neuronal cell surface molecules and cues from the environment. It was realized soon after its identification that CNTN1 was involved in neurite extension (Gennarini et al. 1991), repulsion (Pesheva et al. 1993), and later in fasciculation (Buttiglione et al. 1996). Its importance in guiding axonal growth was confirmed in mice lacking CNTN1. At P10, these mice display marked ataxia, which increases in severity until P18, at which point the mutation is lethal (Berglund et al. 1999). A similar phenotype was observed in mice from a distinct genetic background (129SVJ × C57BL/6 × Black Swiss vs. BALB/c) with a spontaneous mutation in the *CNTN1* gene (Cui et al. 2004; Davisson et al. 2011). Analyses of the brains dissected from *CNTN1* mutants show that the parallel fibers extending from the granule cells project parallel rather than perpendicular to the dendritic branches of Purkinje cells, which is indicative of guidance defects; the compaction of axons within fascicles (in vitro) and the tightness of parallel fiber fascicles (in vivo) are also disrupted (Berglund et al. 1999), suggesting that CNTN1 mediates neurite–neurite interactions. Axonal growth is not the sole purview of CNTN1 and CNTN2 also mediates this process. As a substrate, CNTN2 was shown to promote neurite extension from CGNs and, accordingly, blocking CNTN2 with an antibody impairs axon growth (Wang et al. 2011). Furthermore, during additional anti-CNTN2-blocking experiments, Wang et al. (2011) observed short processes emerging from immature CGNs instead of the long axon fibers that typically extend from them. In line with a possible role in axonal growth, experiments using rat dorsal root ganglion explants expressing chicken CNTN2 showed that newly synthesized CNTN2 molecules are delivered to the tip of the axon where they are placed into the growth cone membrane (Vogt et al. 1996). After fulfilling their functions, CNTN2 molecules are removed from the growth cone and held at the axon shaft where they may mediate axon–axon contacts and thus fasciculation (Vogt et al. 1996). Interestingly, although the localization of CNTN2 at growth cones indicates a role in axon guidance, *CNTN2* knockout mice appear to have normal axonal pathways (Fukamauchi et al. 2001), which might indicate that the absence of CNTN2 is compensated for by an as of yet unknown mechanism.

8.3.2 The Organization of Axonal Subdomains: An Evolutionary Conserved Role for CNTN1 and CNTN2

Reminiscent of the role of fly, CONT, CNTN1, and CNTN2 are involved in the organization of axonal subdomains in the central and peripheral nervous systems. Myelinated axon fibers are divided into three distinct domains (1) the node of Ranvier, where sodium channels are segregated and are involved in the propagation of action potentials along the axons, (2) the paranodes, which ensure electrochemical insulation from adjacent domains, and (3) the juxtaparanodes, characterized by the clustering of Shaker-type voltage-gated potassium channels (Fig. 8.3) (Salzer 1997; Wang et al. 1993). Efficient saltatory conduction is crucially dependent on the presence of the myelin sheath produced by oligodendrocytes in the central nervous system (CNS) and Schwann cells in the peripheral nervous system (PNS), but also on the segregation of ion channels and CAMs expressed at the paranodes and juxtaparanodes as well as the subsequent interactions among them.

In this context, it is particularly noteworthy that CNTN1 is expressed on the axolemma at the paranodes in peripheral nerves and that paranodal junctions are disrupted in mice lacking the *CNTN1* gene (Boyle et al. 2001). The cell surface receptors present at paranodal axon–glia contacts have been identified and involve CNTN1 and contactin-associated protein (CNTNAP1/CASPR/Paranodin), a vertebrate orthologue of *Drosophila* NRX IV (Bellen et al. 1998) expressed on the axolemma and NF155, an isoform of neurofascin expressed on glial cells (Menegoz et al. 1997; Rios et al. 2000; Tait et al. 2000). These three receptors form a tripartite complex, with CNTN1 and CNTNAP1 interacting in *cis* on the surface of axons and binding in *trans* to NF155 (Charles et al. 2002) (Fig. 8.3). Interestingly, CNTN1 and CNTNAP1 do not interact in solution and interactions between these two proteins are only detected in vitro when they are coexpressed, indicating that they likely associate intracellularly and are transported to the cell membrane together (Peles et al. 1997). Consistent with this notion, CNTNAP1 is not transported to the cell membrane in CNTN1-deficient mice (Boyle et al. 2001). The presence of the CNTN1/CNTNAP1 complex on axons does not appear necessary for the appropriate localization of NF155 at paranodal loops (Boyle et al. 2001), but in the absence of NF155 neither CNTN1 nor CNTNAP1 could be identified at the paranodes (Sherman et al. 2005). As could be expected from these findings, paranodal junctions and the distribution of Shaker-type potassium channels are disrupted in CNTN1-, CNTNAP1-, or NF155-deficient mice, and nerve conduction velocity is greatly reduced in mice lacking either *CNTN1* or *CNTNAP1* (Bhat et al. 2001; Boyle et al. 2001).

Interestingly, the role of CNTN2 at juxtaparanodes mirrors the one played by CNTN1 at the paranodes. Indeed, CNTN2 is expressed both on axons (Karagogeos et al. 1991) and on glial cells at the juxtaparanodal junctions (Traka et al. 2002). As described above in the case of CNTN1, CNTN2 is engaged in a ternary complex at the paranodes involving a homologue of CNTNAP1 called CNTNAP2. This protein is expressed on the axolemma and found in a *cis*-complex with axonal CNTN2 and

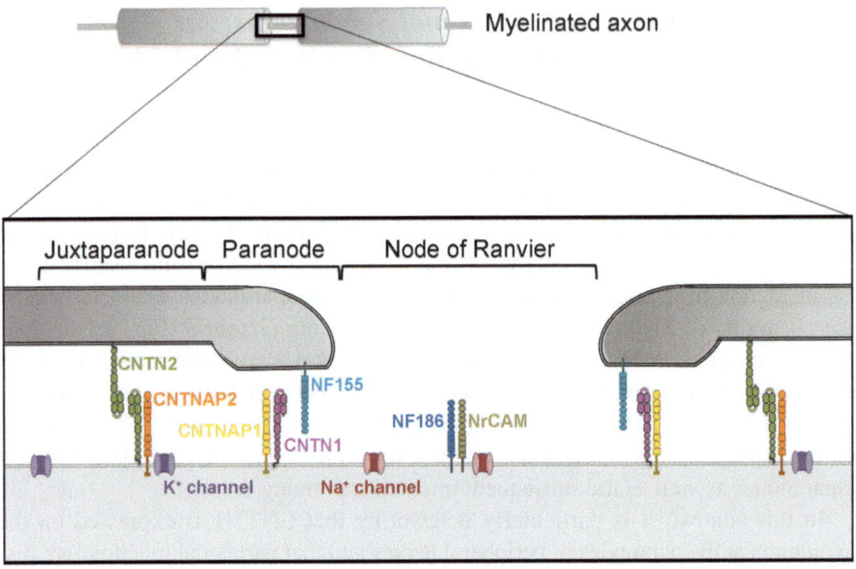

Fig. 8.3 CNTN1 and CNTN2 are involved in the organization of myelin subdomains. CNTN1, CNTNAP1, and NF155 form a ternary complex that is essential for the axon–glia contacts at the paranode. Similarly, glial- and neuron-expressed CNTN2 associates with CNTNAP2 and Shaker-type potassium channels at the juxtaparanode

an apposing CNTN2 expressed on oligodendrocytes and Schwann cells (Fig. 8.3) (Traka et al. 2002, 2003). In essence, the glial-expressed CNTN2 replaces NF155 in the ternary complex of CNTN1, CNTNAP1, and NF155 necessary for formation of paranodal contacts. In contrast to *CNTN1*-null mice, there were no observable differences in nerve conduction in the CNS or the PNS between wild-type and CNTN2-deficient mice, but shaker-type potassium channels and CNTNAP2 were aberrantly localized in the absence of CNTN2 (Traka et al. 2003). This disrupted localization is explained by the *cis*-interactions between CNTN2, CNTNAP2, and the K^+ channels that occur as early as P8. Satisfyingly, the analysis of *CNTNAP2*-null mice published concomitantly as that of *CNTN2*-null mice indicates that there is no alteration in nerve conduction in the absence of CNTNAP2 whereas the distribution of CNTN2 and K^+ channels is also disrupted (Poliak et al. 2003). The mechanism that accounts for the assembly of the tripartite complex at juxtaparanodal junctions also resembles that of CNTN1. The two groups that generated the *CNTN2*- and *CNTNAP2*-null mice independently reported that CNTN2 does not interact with CNTNAP2 in *trans* and association between these proteins is observed only when they are coexpressed in cells (Poliak et al. 2003; Traka et al. 2003). An unexpected twist came from recent work by Savvaki and colleagues who were able to rescue the localization of CNTNAP2 and potassium channels by expressing CNTN2 only in oligodendrocytes (Savvaki et al. 2010). The authors suggest that interaction between CNTN2 and CNTNAP2 in *trans* is sufficient to drive the correct distribution of juxtaparanodal components at least in the CNS. This *trans*-interaction would be unexpected in light of previous studies showing that CNTN2 cannot bind in *trans* with CNTNAP2 (Traka et al. 2003), but the authors surmise that such an interaction could involve release of CNTN2 from the surfaces of oligodendrocytes and formation of a complex between membrane-anchored CNTN2 and released CNTN2. This released CNTN2 would then serve as an adaptor to mediate the interaction with CNTNAP2, thus allowing for proper distribution of juxtaparanode components (Savvaki et al. 2010). The authors also suggest that juxtaparanodal complex formation may differ in the CNS and PNS and that the strict requirement for expression of CNTN2 on the axolemma might remain in the PNS, but whether this is the case or not remains unclear.

8.3.3 The Role of CNTN1 in Oligodendrocyte Development and Myelination

In addition to its role in the organization of axonal regions, CNTN1 is an important player in the development of oligodendrocytes. A key finding was the demonstration that Notch expressed by oligodendrocyte precursor cells (OPCs) is a functional ligand for CNTN1. The *trans*-interaction between these two cell surface molecules triggers cleavage of the Notch intracellular region and its translocation to the nucleus, which results in activation of genes involved in myelination such as myelin-associated glycoprotein (Hu et al. 2003). On the other hand, CNTN1 is also expressed on the surface of OPCs where it associates with integrin α6β1 to form a complex that functions in oligodendrocyte survival and in myelination (Laursen et al. 2009). According to the authors of the study, this CNTN1–integrin complex is activated by the extracellular matrix (ECM) component laminin and the presence of a soluble form of L1, which binds to CNTN1 expressed on OPCs (White et al. 2008). Activation of the CNTN1–integrin complex alters the phosphorylation state of the Src tyrosine kinase Fyn, a key regulator in oligodendroglial development (Czopka et al. 2010; Krämer et al. 1999; Laursen et al. 2009; White et al. 2008).

More recently, a layer of complexity was added when it was demonstrated that a complex between CNTN1 and receptor protein tyrosine phosphatase zeta (PTPRZ) controls critical aspects of oligodendrocyte maturation. In particular, binding of a soluble form of PTPRZ to CNTN1 expressed on OPCs is sufficient to arrest OPC proliferation and initiate their differentiation into mature, myelinating oligodendrocytes (Lamprianou et al. 2011). These results are consistent with the findings that PTPRZ is involved in remyelination (Harroch et al. 2002) and that PTPRZ-null mice present altered oligodendroglial populations with a significant increase in the number of OPCs concomitant with reduction of mature cell numbers (Lamprianou et al. 2011). One of the most intriguing results from this work is that the binding of a soluble form of PTPRZ to CNTN1 could repress OPC proliferation. Indeed, since CNTN1 is a GPI-anchored molecule, signals to stop cellular proliferation would be likely transduced by a CNTN1 co-receptor. A good candidate for this role would be receptor protein tyrosine phosphatase alpha (PTPRA). Indeed, this receptor binds to CNTN1 expressed on the same cell and increased proliferation of OPCs is observed in mice lacking *PTPRA* (Wang et al. 2012), thus resembling the phenotype of *PTPRZ*-null mice. Although PTPRA could be the CNTN1 co-receptor needed to stop OPC proliferation, its role in promoting differentiation of OPCs into mature, myelinating oligodendrocytes remains in doubt. Instead, it has been proposed that the binding of PTPRZ to CNTN1 is necessary to recruit additional CAMs and/or ECM components to mediate glial cell maturation (Lamprianou et al. 2011). The identities of these proteins are not known, but tenascin-R is believed to be one such partner as it binds both PTRPZ and CNTN1 (Lamprianou et al. 2011). The sum of these findings indicates that CNTN1 mediates critical aspects of oligodendrocyte proliferation and maturation, yet specific defects in oligodendrocyte populations have yet to be reported in *CNTN1*-null mice. A second interesting aspect is that the results highlighted above implicate CNTN1 and several distinct *cis-* or *trans-binding* partners in the development of oligodendroglial cells. Thus, the question remains whether these findings are linked and represent snapshots along the path of oligodendrocyte development and myelination or if distinct processes involving Notch, integrins, and PTPRZ are all necessary for the correct development of these cells.

8.3.4 Initial Insights into the Functions of CNTN3, 4, 5, and 6

Although insights on the functions of CNTN3, 4, 5, and 6 are still limited, the results accumulated so far indicate that they are also implicated in the proper development of the brain. Because these molecules have been the topic of a recent comprehensive review (Zuko et al. 2011), we will only briefly consider here their sites of expression and their roles. Northern blot analyses demonstrated that CNTN3 is not present in significant amounts during embryogenesis, but is expressed as early as postnatal day 2 and is found in the adult brain. Using in situ hybridization in adult rat brains, CNTN3 expression was detected in the granule cell layers of the olfactory bulb and in neurons in superficial layers of the cerebral cortex. *CNTN3* mRNA was also detected moderately in the granule cells of the dentate gyrus and much more strongly in the Purkinje cells of the cerebellum. However, no *CNTN3*-null mouse has been reported to date so it remains difficult to discern what its possible function in neural development could be. More is known about the closely related CNTN4 (CNTN3 and 4 share 64 % amino acid sequence identity overall). As is the case for CNTN3, expression of CNTN4 increases from birth and is maximal during adulthood while its expression is also limited to specific subsets of neurons in the hippocampus, the hypothalamus, and the cerebellum (Yoshihara et al. 1995). In particular, CNTN4 is expressed strongly in subsets of olfactory sensory neurons and in subsets of glomeruli in the olfactory bulb. Perhaps not surprisingly, CNTN4-null mice exhibit guidance defect in the olfactory system, suggesting that CNTN4 mediates neuronal wiring in this sensory region (Kaneko-Goto et al. 2008).

The expression of CNTN5 in brains of mice is restricted in time. It can be detected by Western blotting soon after birth in the cerebellum and the cerebrum. CNTN5 expression stops at P3 in the cerebellum, but is still found in the cerebrum during adulthood though maximal levels are reached at P14 (Ogawa et al. 2001). In situ hybridization and immunohistochemistry analyses were used to detect strong CNTN5 expression in regions involved in the auditory pathway and mice lacking CNTN5 exhibit altered responses to acoustic stimuli (Li et al. 2003). This receptor is expressed at glutamatergic synapses in regions of the auditory pathway and this sensory defect was later attributed to a decrease in the formation of synapses as well as an increase in apoptosis in the ventral cochlear nucleus and medial nucleus of the trapezoid body (Toyoshima et al. 2009a, b). An interesting aspect of CNTN5 biology is that it forms a *cis*-complex with amyloid precursor-like protein 1 at presynaptic membranes in the auditory system (Osterfield et al. 2008; Shimoda et al. 2012) though the physiological significance of these interactions remains unclear. Finally, the last member of the CNTN family, CNTN6, is detected in the cerebellum and cerebrum after birth. Expression in the cerebrum reaches its peak at P7 after which levels decline (Lee et al. 2000). In contrast, expression in the cerebellum increases until adulthood. As with CNTN3, 4, and 5, expression of CNTN6 is restricted to neurons in clearly defined regions of the brain (Lee et al. 2000). The brains of *CNTN6*-null mice look normal, but these animal exhibit defects in motor coordination (Takeda et al. 2003). This phenotype is consistent with expression of CNTN6 in a specific region of the cerebellum called vestibulo-cerebellum, which in involved in the control of balance and eye movements. Furthermore, CNTN6 is expressed in granules cells in the cerebellum during its development and is involved in the formation of glutamatergic synapses between parallel fibers and Purkinje cells (Sakurai et al. 2009). The involvement of CNTN6 in the formation of glutamatergic synapses is not restricted to the cerebellum, however, and extends to the hippocampus (Sakurai et al. 2010). Interestingly, and similarly to CNTN1, CNTN6 was also shown to bind Notch during the maturation of oligodendrocytes (Cui et al. 2004), but no abnormality in myelination or oligodendrocyte function was reported in *CNTN6*-null mice, though it may be revealed in future phenotypic analyses.

Overall, we still have limited knowledge about the functions of the "other" CNTNs, but the picture that emerges is that, contrary to CNTN1 and 2, CNTN3, 4, 5, and 6 are expressed mostly at postnatal stages as well as during adulthood and in very specific subsets of neurons. The involvement of CNTN4 and CNTN5 in the olfactory and auditory systems, respectively, is an interesting twist in the functions of the CNTN gene family and it will be informative to analyze the phenotype of *CNTN3*-null mice to determine to which extent this molecule participates in the wiring of sensory systems. Interestingly, receptor protein tyrosine phosphatase gamma (PTPRG), a homologue of PTPRZ, is a potential physiological ligand for CNTN3–6 and is also expressed in sensory neurons, indicating that PTPRG/CNTN complexes may participate in the development and maintenance of these tissues (Bouyain and Watkins 2010; Lamprianou et al. 2006).

8.4 The CNTN Family in Human Pathologies

8.4.1 A Form of Congenital Myopathy Is Linked to CNTN1

There is, thus far, only a single report of a mutation in *CNTN1* in humans. It results in the introduction of a premature stop codon in the third Ig domain and is believed to cause a lethal form of congenital myopathy (Compton et al. 2008) in four patients from a consanguineous family. These infants were born prematurely and three out of the four died shortly after birth. The surviving patient died after 1 month and had low birth weight, reduced muscle mass, and an absence of spontaneous movement. Although the morphology of all organs was normal, detailed examination of the skeletal muscle revealed several abnormalities including the absence of integrin α7 (Compton et al. 2008). Because *CNTN1* is expressed at the neuromuscular junction (Compton et al. 2008), it was suggested that the muscle defects stemmed from aberrant neuromuscular transmission and in particular that the absence of functional CNTN1 affects the dystrophin-associated protein complex. Interestingly, this myopathic phenotype is related to the severe ataxia observed in *CNTN1*-null mice, although there are no observable defects in muscle tissues in the two existing CNTN1-deficient mice (Berglund et al. 1999; Davisson et al. 2011). This lack of similarity between the mouse and human myopathic phenotypes may be due to differences in the expression pattern of CNTN1 between the two species. Nevertheless, it remains that this work has linked a mutation in *CNTN1* to a human disease and suggests that further investigation of the role of CNTN1 in the formation of neuromuscular junctions is warranted.

8.4.2 Autoimmune Disorders and CNTN2

Multiple sclerosis (MS) is a complex autoimmune disorder characterized by a degradation of the myelin sheath in the CNS. In the past, analyses of brains from deceased patients have identified lesions to the white matter. However, in recent years, lesions to the cortical gray matter have also been observed (Rudick and Trapp 2009), and it was demonstrated that axonal injury correlates with the disability of those afflicted with the disease (Bjartmar et al. 2003). Importantly, a proteomic screening of potential autoantigens identified CNTN2 as an autoimmune target in MS (Derfuss et al. 2009). In particular, Derfuss and colleagues (2009) established that CNTN2 was recognized by autoantibodies and that T cells from MS patients had a significantly higher proliferative response to CNTN2, which was also associated with secretion of interferon-gamma and interleukin-17. Furthermore, transfer of CNTN2-specific T cells into rats induced experimental autoimmune encephalomyelitis characterized by inflammation of cortex and spinal cord gray matter. No demyelination or axonal lesions were identified in these animals, however, but they could be induced if antibodies to myelin oligodendrocyte glycoprotein were also injected in the rats. The importance of CNTN2 autoimmunity to the etiology of MS was dampened by a recent study in which antibodies against CNTN2 were only detected in a small number of patients and the authors could find little correlation between the presence of these antibodies and the clinical presentation of the disease (Boronat et al. 2012). Nonetheless, these recent investigations indicate that an autoimmune response against CNTN2 might be a contributing factor to the gray matter lesions observed in MS.

In addition to MS, the presence in serum of autoantibodies against juxtaparanodal components, including CNTN2 and in particular CNTNAP2 or potassium channels, has been linked to disorders such as limbic encephalitis (characterized by cognitive impairment and seizures), neuromyotonia (characterized by muscle hyperactivity or stiffness), or the extremely rare Morvan's syndrome (characterized by CNS hyperexcitability) (Vincent et al. 2006; Irani et al. 2010, 2011; Quek et al. 2012). Importantly, improvement of symptoms is often observed in patients following immunosuppressive therapy, thus emphasizing the immune causes of these conditions. Although these recent studies fall far from providing a clear mechanistic link between the presence of antibodies to CNTN2 or CNTNAP2 in neurological disorders, they indicate that additional work is warranted to determine how antibodies to juxtaparanodal components contribute to the diverse pathologies mentioned here.

8.4.3 A Link Between CNTNs and Autism Spectrum Disorders?

Perhaps one of the most interesting findings on the roles of CNTNs in human pathologies has been the association between several CNTN family members and autism spectrum disorders (ASDs) (Burbach and van der Zwaag 2009). One of the first implications of a CNTN family member in ASD came in 2009 when Roohi and colleagues identified copy number variations [CNVs, a structural variation characterized by an aberrant number of copies of a DNA region (Stankiewicz and Lupski 2010)] that resulted in an interruption of in the gene encoding CNTN4 in three ASD patients (Roohi et al. 2009). These findings were particularly interesting because they came on the heels of a previous report in which a homozygous deletion was identified in the 5′ untranslated region of CNTN3 in an autistic patient (Morrow et al. 2008). A disruption of *CNTN4* in a child suffering from 3p deletion syndrome and exhibiting verbal and nonverbal developmental delays consistent with an ASD diagnosis has also been identified (Fernandez et al. 2004, 2008). CNVs in *CNTN5* and *CNTN6* have since been identified in ASD patients (Burbach and van der Zwaag 2009) although the exact impact that these variations have on disease etiology remains to be determined (van Daalen et al. 2011).

8.5 Structural Insights into CNTN Function

8.5.1 The Horseshoe Motif: A Hallmark of Vertebrate CNTNs and in Other Neural CAMs

The six vertebrate CNTNs share a common domain organization. Each of their extracellular regions includes six Ig domains and four FNIII repeats followed by a GPI anchor that tethers them to the cell membrane. Structural information on CNTNs is limited to their N-terminal Ig repeats, which include a structural arrangement called a horseshoe-like motif that mediates protein–protein interactions. This motif was first identified in the crystal structure of domains Ig1 to Ig4 of axonin, the chicken orthologue of CNTN2 (Freigang et al. 2000). In this structure, the domain pairs Ig1–Ig2 and Ig3–Ig4 are arranged in antiparallel fashion and extensive contacts between Ig1 and Ig4 on the one hand and between Ig2 and Ig3 on the other hand occlude in excess of 2,100 Å^2, which in essence locks this protein fragment in a compact U-shaped structure (Fig. 8.4). Residues participating in the interdomain interface are conserved in vertebrate CNTNs, indicating that this motif could be found in additional family members, which was confirmed by recent structural work (Bouyain and Watkins 2010; Lamprianou et al. 2011). Importantly, the arrangement of Ig domains found in the horseshoe-like motif found in CNTN2 is not unique to this family. It was first reported in the insect protein hemolin (Su et al. 1998) and similar structures were found in the extracellular regions of the neural CAMs Dscam (Meijers et al. 2007; Sawaya et al. 2008) and the L1 family member neurofascin (Liu et al. 2010). What is particularly striking about the horseshoe motif is that it accommodates several distinct binding modes using the same domain arrangement, from the homophilic interactions mediated by CNTN2, neurofascin, and Dscam to the heterophilic interactions mediated by CNTN1, 3, 4, 5, and 6.

8.5.2 Homophilic Interactions Mediated by CNTN2

The expression of CNTN2 in non-adherent myeloma cells leads to the formation of cell aggregates consistent with its role promoting cell–cell contacts via homophilic interactions between two CNTN2 molecules expressed on apposing cells (Freigang et al. 2000). As indicated above, neuronal CNTN2 binds in *cis* to CNTNAP2 and interacts in *trans* with glial CNTN2 to drive the formation of CNTN2–CNTNAP2 heterotrimers that are necessary for the correct localization of Shaker-type K^+ channels in myelinated axons (Poliak et al. 2003). A first glimpse into the homophilic interactions mediated by CNTN2 came more than a decade ago when the crystal structure of Ig1–Ig4 from chicken CNTN2 was determined (Freigang et al. 2000). Perhaps more important than the identification of a horseshoe motif in the N-terminal region of CNTN2 was the realization that contacts between symmetry-related molecules that bury 1,260 Å^2 could represent the arrangement of homophilic dimers of

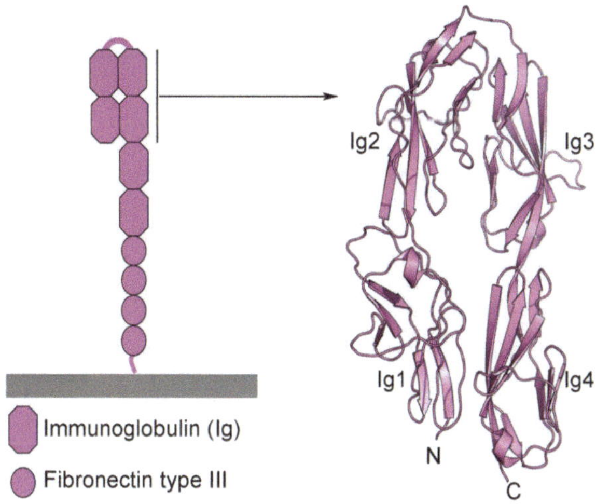

Fig. 8.4 The horseshoe-like motif in chicken CNTN2. The structure of the first four Ig domains of chicken CNTN2 is shown here in a ribbon diagram (Freigang et al. 2000). The four domains are arranged in a *U-shape* form reminiscent of a horseshoe. The letters *N* and *C* denote the N- and C-termini, respectively

CNTN2 (Fig. 8.5a). In this structure, a loop in the third Ig domain of a protomer nestles against the second Ig domain of a second protomer. In particular, the interface includes a second loop linking two strands in the second Ig domain. Mutation of two residues to alanine in this loop prevented the aggregation of chicken CNTN2-expressing myeloma cells, indicating that this region was involved in homophilic cell interactions (Fig. 8.5c) (Freigang et al. 2000). Furthermore, this loop corresponds to a two-residue insertion in the CNTN2 sequence compared those of CNTN1, 3, 4, 5, and 6, which have not been reported to form *trans* homodimers and are instead involved in heterophilic interactions with the ectodomains of the tyrosine phosphatases PTPRG and PTPRZ (Bouyain and Watkins 2010; Lamprianou et al. 2011) as will be detailed below (Fig. 8.5c).

However, these findings were put in question when the same group determined the crystal structure of human CNTN2 (Fig. 8.5b) (Mörtl et al. 2007). Although the Ig1–Ig4 fragment of human CNTN2 adopted the same horseshoe-like arrangement, contacts between two horseshoes were different than the ones found in chicken. Analysis of this structure suggested that the homophilic contacts involved interactions between an Ig1–Ig2 pair in one protomer and a symmetry-related Ig1–Ig2 pair from another protomer. Contacts between these two molecules occur at and near the Ig1–Ig2 interface, bury a much larger surface area of 2,240 Å^2, and, interestingly, include the same residues that prevented the dimerization of chicken CNTN2 when they were mutated (Fig. 8.5b, c) (Freigang et al. 2000). These structural discrepancies are puzzling because one would expect chicken and human CNTN2 to share a similar dimerization mode based on the high amino acid conservation between these two proteins (77 % sequence amino acid sequence identity in the

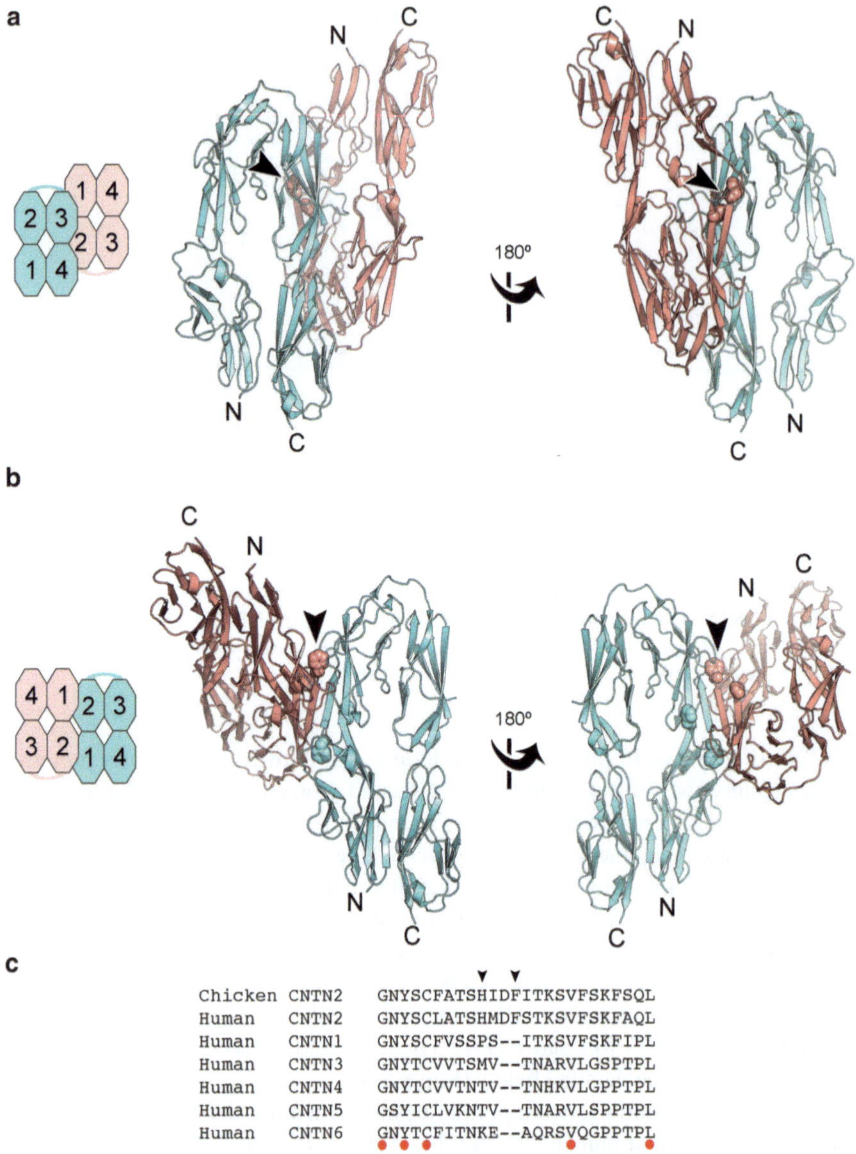

Fig. 8.5 Homophilic interactions mediated by the horseshoe-like motif in chicken CNTN2. (**a**) The CNTN2 dimer identified from the structure of chicken CNTN2 is shown here in a *ribbon diagram* (Freigang et al. 2000) in two different views along with a *cartoon* that aims to provide a simpler view of the dimer. Residues mutated to alanine that disrupt dimer formation are shown as *spheres* and are indicated by the *black arrows*. This CNTN2 dimer is not symmetrical, unlike the human CNTN2 dimer shown in panel (**b**). The two subunits are colored *salmon* and *light blue*, respectively. The letters *N* and *C* denote the N- and C-termini, respectively. (**b**) The CNTN2 dimer identified from the structure of human CNTN2 is shown here in a *ribbon diagram* (Mörtl et al. 2007) in two different views along with a cartoon that aims to provide a simpler view of the dimer. Residues mutated to alanine in chicken CNTN2 that disrupt dimer formation are shown as spheres and are indicated by the two *black arrows*. (**c**) Sequence alignment of vertebrate CNTNs at the potential dimerization interface in Ig2. The *black arrows* denote the two residues that were mutated to alanine in chicken CNTN2, which disrupted dimer formation (Freigang et al. 2000). These residues are not conserved in CNTN1, 3, 4, 5, or 6. *Red discs* denote conserved residues

horseshoe-like region). One possible reason would be that both proteins were purified from bacterial lysates, which may have led to incorrect association in the case of the chicken molecule because of the lack of protein glycosylation (He et al. 2009) though it would be unclear why such a problem would have been avoided in the case of the human protein. More importantly, questions remain as to whether these interactions are physiologically relevant given the differences in the chicken and human dimer arrangement. Recent structural results obtained for the L1 family member neurofascin tip the balance towards the arrangement observed in human CNTN2. As is the case for CNTN2, neurofascin forms homodimers and the horseshoe-like regions of two neurofascin molecules associate similarly to human CNTN2 (Liu et al. 2010), suggesting that the dimerization mode observed for human CNTN2 may represent the physiologically relevant homodimers that drive the homophilic interactions between apposing cells and that this dimerization mode is shared with L1 family members.

8.5.3 Contactins and Their Receptor Protein Tyrosine Phosphatase-Binding Partners

In contrast to CNTN2, the other five members of the CNTN family do not participate in homophilic interactions and instead engage in heterophilic interactions with two homologous receptor protein tyrosine phosphatases called PTPRZ and PTPRG. Binding between PTPRZ and CNTN1 on the one hand and between PTPRG and CNTN3, 4, 5, and 6 on the other hand are also mediated by the horseshoe motif (Fig. 8.6a). However, the interfaces involve a flat surface formed by the juxtaposition of domains Ig2 and Ig3, thus illustrating the versatility of the horseshoe motif as it supports both homophilic and heterophilic interactions. These interactions also involve the inactive N-terminal carbonic anhydrase (CA) of PTPRG and PTPRZ, a unique feature among the family of receptor protein tyrosine phosphatases (Bouyain and Watkins 2010; Peles et al. 1995). As mentioned above, the in vivo role of the PTPRZ/CNTN1 complex is to control the development of oligodendrocyte precursor cells, but it is suspected that this complex may also participate in neurite outgrowth (Peles et al. 1995; Sakurai et al. 1997). In contrast, roles for complexes involving PTPRG with CNTN3, 4, 5, and 6 have yet to be uncovered, but, given the expression of these proteins in sensory systems, a plausible hypothesis is that these receptor complexes are involved in the development of regions of the nervous system associated with sensory functions.

The crystal structures of the PTPRZ/CNTN1 (Lamprianou et al. 2011) and PTPRG/CNTN4 (Bouyain and Watkins 2010) complexes have provided satisfying insights into the heterophilic interactions mediated by these CNTNs. These complexes are, overall, arranged similarly (Fig. 8.6b–d). In both CNTN1 and CNTN4, the binding sites for PTPRG and PTPRZ consist essentially of two discrete 9–14 amino acid segments, one in Ig2 and the other in Ig3 (Fig. 8.6e). These segments

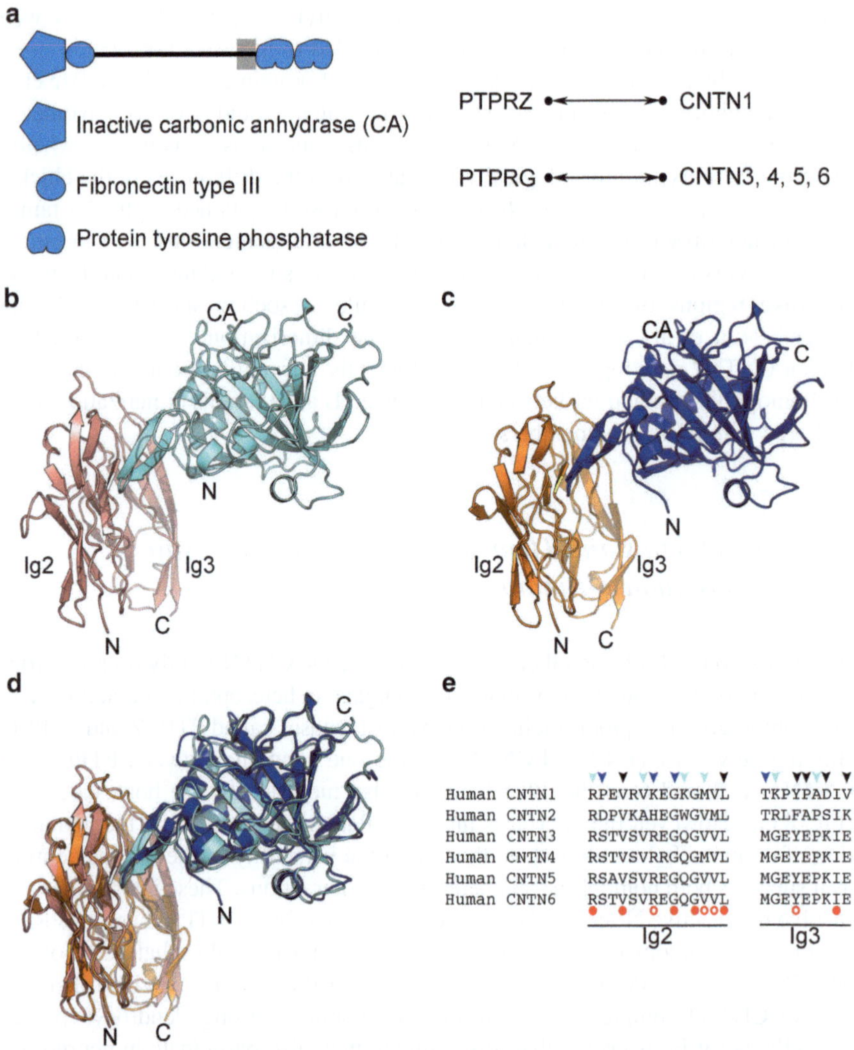

Fig. 8.6 Heterophilic interactions mediated by the horseshoe-like regions of CNTN1 and CNTN4. (**a**) Domain organization of the receptor protein tyrosine phosphatases PTPRG and PTPRZ and overview of their distinct CNTN-binding specificities. (**b**) *Ribbon diagram* depicting the complex between Ig2 and Ig3 of human CNTN1 (*salmon*) and the carbonic anhydrase (CA) domain of human PTPRZ (*light blue*). The letters *N* and *C* denote the N- and C-termini, respectively. (**c**) *Ribbon diagram* depicting the complex between Ig2 and Ig3 of mouse CNTN4 (*orange*) and CA domain of mouse PTPRG (*dark blue*). (**d**) Overlay of the structures shown in panels (**b**) and (**c**) illustrating the closely related arrangements of the two complexes. (**e**) Sequence alignment of human CNTNs in the regions of Ig2 and Ig3 found at the heterophilic interfaces. *Red discs* and *red circles* denote conserved and similar residues, respectively. *Light blue* and *dark blue arrows* indicate contact residues that are specific for the CNTN1/PTPRZ and CNTN4/PTPRG interfaces, respectively. *Black arrows* indicate residues that are found at both interfaces

contact the N-terminal CA domain and in particular a β-hairpin loop that includes residues that mediates about 75 % of the contacts between the CA domain and its associated CNTN. The presence of this loop is critical since a form of PTPRZ lacking it is not able to prevent the binding of a fluorescently labeled form of PTPRZ to OPCs expressing CNTN1 (Lamprianou et al. 2011). Analysis of the non-covalent interactions in the PTPRZ/CNTN1 and PTPRG/CNTN4 complexes has highlighted the residues that were important in providing specificity to the interactions as contact residues in CNTN4 are not conserved in CNTN1, but are strictly conserved in CNTN3, 5, and 6 (Fig. 8.6e). Finally, sequence alignments show that, as could be expected from its different binding properties, CNTN2 shares little resemblance with either CNTN1, 3, 4, 5, or 6 in the PTPRZ- or PTPRG-binding sites.

8.6 New Perspectives in CNTN Biology

8.6.1 A Family of Extracellular Organizers?

The neurodevelopmental functions fulfilled by CNTNs are, unsurprisingly, governed by their binding to other cell surface receptors. However, because the CNTNs do not span the cell membrane and are instead tethered to it by a GPI anchor, they cannot transduce signals across the membrane without associating with a transmembrane protein expressed on the same cell. A good illustration of that is provided by the formation of SJs in *Drosophila* where CONT associates with NRX IV, but it is NRX IV that is linked to the intracellular machinery because its intracellular region binds to the synaptic organizer Coracle. For example, neither CONT nor NRX IV can localize to SJs if the cytoplasmic domain of NRX IV is deleted (Banerjee et al. 2011). This example serves as a reminder that CNTNs must be associated with other cell surface receptors in *cis* and in *trans* for bidirectional signaling events to occur. In essence, the dependency of NRG and NRX IV on CONT for their localization at SJs indicates that CONT can be seen as an extracellular organizer that is necessary for the assembly of a multi-protein complex.

Seen through this lens, it appears that CNTNs would provide a versatile scaffold onto which complexes could be assembled. Indeed, past investigations of CNTN1 provide at least three different complexes that are involved in distinct aspects of neural development (1) the tripartite complex of CNTN1/CNTNAP1/NF155 at paranodal junctions (Boyle et al. 2001; Charles et al. 2002), (2) a complex of CNTN1/NrCAM/PTPRZ that supports neurite outgrowth (Sakurai et al. 1997), and (3) a complex of CNTN1/PTPRZ that likely involves PTPRA and/or tenascin-R during the maturation of oligodendrocytes (Lamprianou et al. 2011; Wang et al. 2012). The ability of CNTN1 to recruit such a diverse array of cell surface receptors or ECM components poses an interesting structural question: how can one protein, albeit a large modular one, interact with so many partners? Thus far, the only complex that has been characterized structurally is the one formed by PTPRZ and

CNTN1 (Lamprianou et al. 2011). Although there is no information on how NF155 may associate with CNTN1, *trans*-associations between L1 and CNTN1 and between NrCAM and CNTN1 have been reported for the chicken homologues of these proteins and involve CNTN1 domains Ig1–Ig2 and Ig2–Ig3, respectively (Brümmendorf et al. 1993; Morales et al. 1993). Since these two L1 family members bind to the horseshoe-like region of CNTN1, NF155 may also bind to CNTN in this region. Because PTPRZ also associates with the horseshoe-like region of CNTN1, it will be of interest to determine to which extent the binding sites for PTPRZ and L1 family members overlap. Furthermore, CNTN1 associates with NrCAM in *cis* and in *trans* with PTPRZ to mediate neurite outgrowth (Sakurai et al. 1997), but CNTN1 also binds to NrCAM to mediate axonal growth of embryonic tectal cells (Morales et al. 1993) as well as the proliferation of Shh-induced proliferation of progenitor CGNs (Xenaki et al. 2011). It remains to be determined, however, if CNTN1 and NrCAM use the same sites to associate in *cis* and *trans* and, more generally, a systematic in vitro characterization of CNTN/L1 family interactions would be a tremendous resource to elucidate the functions played by these receptors during neural development. Similar work is also needed to investigate the known interactions between CNTN1 and CNTN2 and CNTNAP1 and CNTNAP2, but also determine to which extent additional members of the CNTN family, namely, CNTN3–6, interact with the five homologous CNTNAPs (Spiegel et al. 2002; Traut et al. 2006) and, if appropriate, what the physiological outcomes of these interactions may be.

8.6.2 A Likely Involvement of CNTNs in Synapse Formation

Although great strides have been made in understanding the specific roles CNTNs play in nervous system formation and maintenance, exploration of their involvement in synaptic function has only begun. Early evidence for a potential CNTN role in synapse biology came from analysis of CNTN1-deficient mice (Murai et al. 2002) and has been confirmed by microarray analysis showing that expression of CNTN1 is upregulated during synapse formation (Brusés 2010). Although synaptic morphology and basal synaptic transmission were unchanged in *CNTN1*$^{-/-}$ animals, specific defects in long-term depression (a weakening of synaptic strength) were identified in the hippocampal CA1 region of mutant mice. Mechanistically, this change in synaptic plasticity was associated with altered distribution of the CNTN1-binding proteins CNTNAP1 and PTPRZ and a proposed role for CNTN1 was to guide PTPRZ to its proper location (Murai et al. 2002). The involvement of a tripartite complex that includes both CNTN1 and CNTNAP1 draws parallel with the formation of paranodal SJs and would suggest that mice lacking either PTPRZ or CNTNAP1 exhibit similar defects in long-term depression. However, although mice lacking PTPRZ exhibit altered synaptic plasticity, the phenotype is different as long-term potentiation (an increase of synaptic strength) was enhanced in these mice (Niisato et al. 2005). In addition, phenotypic analysis of *CNTNAP1*$^{-/-}$ mice did not reveal alterations in synaptic plasticity (Pillai et al. 2007) even though recent work has shown that CNTNAP1 controls the synaptic content of AMPA receptors in cultured hippocampal neurons (Santos et al. 2012). Thus, although the defects in synaptic plasticity in *CNTN1*$^{-/-}$ mice may involve PTPRZ and CNTNAP1, the specific contributions of these receptors are still nebulous and the importance of CNTN1 at synapses may reflect the formation of a hitherto undetermined molecular complex. As explained above, CNTN1 is not the only family member involved in synaptic function. Microarray analyses in chicken have shown that expression of CNTN5 is upregulated during synaptogenesis (Brusés 2010) and participates in synapse formation in the auditory pathway (Toyoshima et al. 2009b). Finally, CNTN6 also participates in the formation of glutamatergic synapses in the cerebellum and hippocampus. Although these findings clearly suggest a role for CNTNs in synaptic function, it is unclear whether all CNTNs take part in this process. This likely awaits additional analyses of mice lacking single CNTN proteins and further mechanistic studies designed to identify their complete array of binding partners at synapses.

8.6.3 A Clearer Link with Neuropsychiatric Disorders?

Aside from the form of lethal myopathy that has been linked to the absence of functional CNTN1 in humans, the few known examples of alterations in the expression of human CNTNs have all been linked to ASDs (Burbach and van der Zwaag 2009). The rise in the number of reported ASD cases, combined with the burden to the patients and those who care for them, has prompted many research groups to focus their attention to the CNTN family. However, there is thus far no clear mechanistic link between aberrant expression of the CNTNs or mutation in their coding sequences and the etiology of ASDs. Perhaps this is not entirely surprising given the heterogeneous nature of ASDs, yet the absence of ASD-related behavioral defects in the various CNTN-deficient mice created so far has undoubtedly frustrated our grasp of the roles these proteins may play in the disease. The search for such a clearer link is not far-fetched as the causes of ASDs are presumably linked to neurodevelopmental processes such as neuronal migration, axon guidance, or synaptic function, which have all been linked to CNTNs. Although current and future genomic analyses of ASD patients will certainly help delineate the functions played by CNTNs, this task will be greatly helped by renewed behavioral analyses of existing CNTN-deficient mice or a change in the genetic background of these animals such as the one that helped confirm the role of CNTNAP2 in certain forms of ASDs (Peñagarikano et al. 2011). Progress will also come from the systematic investigation of interactions between CNTNs and CNTNAPs as CNTNAP2 has provided many clues on the causes of ASD, but also because CNTNAP5 has been recently linked to the disease (Pagnamenta et al. 2010). Finally, schizophrenia is another neuropsychiatric disorder of particular interest because two of the known CNTN1-binding proteins, namely, PTPRA and PTPRZ, have been linked to the disease (Buxbaum et al. 2008; Takahashi et al. 2011a, b). Thus far, there is no patient data to support a role of CNTN1 in this disorder and schizophrenic-like phenotypes have not been reported for mice lacking *CNTN1*. However, a specific defect could be hidden by the dramatic alterations seen in *CNTN1*$^{-/-}$ mice, which could be circumvented by analyzing mice in which the expression of CNTN1 is limited to specific regions of the nervous system.

Compliance with Ethics Requirements The authors declare that they have no conflicts of interest.

References

Ankerhold R, Stuermer CA (1999) Fate of oligodendrocytes during retinal axon degeneration and regeneration in the goldfish visual pathway. J Neurobiol 41:572–584

Banerjee S, Pillai AM, Paik R, Li J, Bhat MA (2006a) Axonal ensheathment and septate junction formation in the peripheral nervous system of *Drosophila*. J Neurosci 26:3319–3329

Banerjee S, Sousa AD, Bhat MA (2006b) Organization and function of septate junctions: an evolutionary perspective. Cell Biochem Biophys 46:65–77

Banerjee S, Paik R, Mino RE, Blauth K, Fisher ES, Madden VJ, Fanning AS, Bhat MA (2011) A Laminin G-EGF-Laminin G module in Neurexin IV is essential for the apico-lateral localization of Contactin and organization of septate junctions. PLoS One 6:e25926

Baumgartner S, Littleton JT, Broadie K, Bhat MA, Harbecke R, Lengyel JA, Chiquet-Ehrismann R, Prokop A, Bellen HJ (1996) A *Drosophila* neurexin is required for septate junction and blood-nerve barrier formation and function. Cell 87:1059–1068

Bellen HJ, Lu Y, Beckstead R, Bhat MA (1998) Neurexin IV, caspr and paranodin – novel members of the neurexin family: encounters of axons and glia. Trends Neurosci 21:444–449

Berglund EO, Murai KK, Fredette B, Sekerková G, Marturano B, Weber L, Mugnaini E, Ranscht B (1999) Ataxia and abnormal cerebellar microorganization in mice with ablated contactin gene expression. Neuron 24:739–750

Bhat MA, Rios JC, Lu Y, Garcia-Fresco GP, Ching W, St. Martin M, Li J, Einheber S, Chesler M, Rosenbluth J, Salzer JL, Bellen HJ (2001) Axon-glia interactions and the domain organization of myelinated axons requires neurexin IV/Caspr/Paranodin. Neuron 30:369–383

Bizzoca A, Virgintino D, Lorusso L, Buttiglione M, Yoshida L, Polizzi A, Tattoli M, Cagiano R, Rossi F, Kozlov S, Furley A, Gennarini G (2003) Transgenic mice expressing F3/contactin from the TAG-1 promoter exhibit developmentally regulated changes in the differentiation of cerebellar neurons. Development 130:29–43

Bjartmar C, Wujek JR, Trapp BD (2003) Axonal loss in the pathology of MS: consequences for understanding the progressive phase of the disease. J Neurol Sci 206:165–171

Boronat A, Sepúlveda M, Llufriu S, Sabater L, Blanco Y, Gabilondo I, Solà N, Villoslada P, Dalmau J, Graus F, Saiz A (2012) Analysis of antibodies to surface epitopes of contactin-2 in multiple sclerosis. J Neuroimmunol 244:103–106

Bouyain S, Watkins DJ (2010) The protein tyrosine phosphatases PTPRZ and PTPRG bind to distinct members of the contactin family of neural recognition molecules. Proc Natl Acad Sci USA 107:2443–2448

Boyle ME, Berglund EO, Murai KK, Weber L, Peles E, Ranscht B (2001) Contactin orchestrates assembly of the septate-like junctions at the paranode in myelinated peripheral nerve. Neuron 30:385–397

Brümmendorf T, Lemmon V (2001) Immunoglobulin superfamily receptors: cis-interactions, intracellular adapters and alternative splicing regulate adhesion. Curr Opin Cell Biol 13:611–618

Brümmendorf T, Hubert M, Treubert U, Leuschner R, Tárnok A, Rathjen FG (1993) The axonal recognition molecule F11 is a multifunctional protein: specific domains mediate interactions with Ng-CAM and restrictin. Neuron 10:711–727

Brusés JL (2010) Identification of gene transcripts expressed by postsynaptic neurons during synapse formation encoding cell surface proteins with presumptive synaptogenic activity. Synapse 64:47–60
Burbach JPH, van der Zwaag B (2009) Contact in the genetics of autism and schizophrenia. Trends Neurosci 32:69–72
Buttiglione M, Revest JM, Rougon G, Faivre-Sarrailh C (1996) F3 neuronal adhesion molecule controls outgrowth and fasciculation of cerebellar granule cell neurites: a cell-type-specific effect mediated by the Ig-like domains. Mol Cell Neurosci 8:53–69
Buxbaum JD, Georgieva L, Young JJ, Plescia C, Kajiwara Y, Jiang Y, Moskvina V, Norton N, Peirce T, Williams H, Craddock NJ, Carroll L, Corfas G, Davis KL, Owen MJ, Harroch S, Sakurai T, O'Donovan MC (2008) Molecular dissection of NRG1-ERBB4 signaling implicates PTPRZ1 as a potential schizophrenia susceptibility gene. Mol Psychiatry 13:162–172
Charles P, Tait S, Faivre-Sarrailh C, Barbin G, Gunn-Moore F, Denisenko-Nehrbass N, Guennoc A-M, Girault J-A, Brophy PJ, Lubetzki C (2002) Neurofascin is a glial receptor for the paranodin/Caspr-contactin axonal complex at the axoglial junction. Curr Biol 12:217–220
Compton AG, Albrecht DE, Seto JT, Cooper ST, Ilkovski B, Jones KJ, Challis D, Mowat D, Ranscht B, Bahlo M, Froehner SC, North KN (2008) Mutations in contactin-1, a neural adhesion and neuromuscular junction protein, cause a familial form of lethal congenital myopathy. Am J Hum Genet 83:714–724
Cui X-Y, Hu Q-D, Tekaya M, Shimoda Y, Ang B-T, Nie D-Y, Sun L, Hu W-P, Karsak M, Duka T, Takeda Y, Ou L-Y, Dawe GS, Yu F-G, Ahmed S, Jin L-H, Schachner M, Watanabe K, Arsenijevic Y, Xiao Z-C (2004) NB-3/Notch1 pathway via Deltex1 promotes neural progenitor cell differentiation into oligodendrocytes. J Biol Chem 279:25858–25865
Czopka T, von Holst A, Ffrench-Constant C, Faissner A (2010) Regulatory mechanisms that mediate tenascin C-dependent inhibition of oligodendrocyte precursor differentiation. J Neurosci 30:12310–12322
Davisson MT, Bronson RT, Tadenev ALD, Motley WW, Krishnaswamy A, Seburn KL, Burgess RW (2011) A spontaneous mutation in contactin 1 in the mouse. PLoS One 6:e29538
Denaxa M, Chan CH, Schachner M, Parnavelas JG, Karagogeos D (2001) The adhesion molecule TAG-1 mediates the migration of cortical interneurons from the ganglionic eminence along the corticofugal fiber system. Development 128:4635–4644
Denaxa M, Kyriakopoulou K, Theodorakis K, Trichas G, Vidaki M, Takeda Y, Watanabe K, Karagogeos D (2005) The adhesion molecule TAG-1 is required for proper migration of the superficial migratory stream in the medulla but not of cortical interneurons. Dev Biol 288:87–99
Derfuss T, Parikh K, Velhin S, Braun M, Mathey E, Krumbholz M, Kümpfel T, Moldenhauer A, Rader C, Sonderegger P, Pöllmann W, Tiefenthaller C, Bauer J, Lassmann H, Wekerle H, Karagogeos D, Hohlfeld R, Linington C, Meinl E (2009) Contactin-2/TAG-1-directed autoimmunity is identified in multiple sclerosis patients and mediates gray matter pathology in animals. Proc Natl Acad Sci USA 106:8302–8307
Faivre-Sarrailh C, Gennarini G, Goridis C, Rougon G (1992) F3/F11 cell surface molecule expression in the developing mouse cerebellum is polarized at synaptic sites and within granule cells. J Neurosci 12:257–267
Faivre-Sarrailh C, Banerjee S, Li J, Hortsch M, Laval M, Bhat MA (2004) *Drosophila* contactin, a homolog of vertebrate contactin, is required for septate junction organization and paracellular barrier function. Development 131:4931–4942
Fernandez T, Morgan T, Davis N, Klin A, Morris A, Farhi A, Lifton RP, State MW (2004) Disruption of contactin 4 (CNTN4) results in developmental delay and other features of 3p deletion syndrome. Am J Hum Genet 74:1286–1293
Fernandez T, Morgan T, Davis N, Klin A, Morris A, Farhi A, Lifton RP, State MW (2008) Disruption of Contactin 4 (CNTN4) results in developmental delay and other features of 3p deletion syndrome. Am J Hum Genet 82:1385
Freigang J, Proba K, Leder L, Diederichs K, Sonderegger P, Welte W (2000) The crystal structure of the ligand binding module of axonin-1/TAG-1 suggests a zipper mechanism for neural cell adhesion. Cell 101:425–433

Fukamauchi F, Aihara O, Wang YJ, Akasaka K, Takeda Y, Horie M, Kawano H, Sudo K, Asano M, Watanabe K, Iwakura Y (2001) TAG-1-deficient mice have marked elevation of adenosine A1 receptors in the hippocampus. Biochem Biophys Res Commun 281:220–226

Gennarini G, Durbec P, Boned A, Rougon G, Goridis C (1991) Transfected F3/F11 neuronal cell surface protein mediates intercellular adhesion and promotes neurite outgrowth. Neuron 6:595–606

Genova JL, Fehon RG (2003) Neuroglian, Gliotactin, and the Na^+/K^+ ATPase are essential for septate junction function in *Drosophila*. J Cell Biol 161:979–989

Gimnopoulos D, Becker CG, Ostendorff HP, Bach I, Schachner M, Becker T (2002) Expression of the zebrafish recognition molecule F3/F11/contactin in a subset of differentiating neurons is regulated by cofactors associated with LIM domains. Mech Dev 119(Suppl 1):S135–S141

Haenisch C, Diekmann H, Klinger M, Gennarini G, Kuwada JY, Stuermer CAO (2005) The neuronal growth and regeneration associated Cntn1 (F3/F11/Contactin) gene is duplicated in fish: expression during development and retinal axon regeneration. Mol Cell Neurosci 28:361–374

Harroch S, Furtado GC, Brueck W, Rosenbluth J, Lafaille J, Chao M, Buxbaum JD, Schlessinger J (2002) A critical role for the protein tyrosine phosphatase receptor type Z in functional recovery from demyelinating lesions. Nat Genet 32:411–414

He Y, Jensen GJ, Bjorkman PJ (2009) Cryo-electron tomography of homophilic adhesion mediated by the neural cell adhesion molecule L1. Structure 17:460–471

Hortsch M, Wang YM, Marikar Y, Bieber AJ (1995) The cytoplasmic domain of the *Drosophila* cell adhesion molecule neuroglian is not essential for its homophilic adhesive properties in S2 cells. J Biol Chem 270:18809–18817

Hu Q-D, Ang B-T, Karsak M, Hu W-P, Cui X-Y, Duka T, Takeda Y, Chia W, Sankar N, Ng Y-K, Ling E-A, Maciag T, Small D, Trifonova R, Kopan R, Okano H, Nakafuku M, Chiba S, Hirai H, Aster JC, Schachner M, Pallen CJ, Watanabe K, Xiao Z-C (2003) F3/contactin acts as a functional ligand for Notch during oligodendrocyte maturation. Cell 115:163–175

Irani SR, Alexander S, Waters P, Kleopa KA, Pettingill P, Zuliani L, Peles E, Buckley C, Lang B, Vincent A (2010) Antibodies to Kv1 potassium channel-complex proteins leucine-rich, glioma inactivated 1 protein and contactin-associated protein-2 in limbic encephalitis, Morvan's syndrome and acquired neuromyotonia. Brain 133:2734–2748

Irani SR, Bien CG, Lang B (2011) Autoimmune epilepsies. Curr Opin Neurol 24:146–153

Kaneko-Goto T, Yoshihara S-I, Miyazaki H, Yoshihara Y (2008) BIG-2 mediates olfactory axon convergence to target glomeruli. Neuron 57:834–846

Karagogeos D, Morton SB, Casano F, Dodd J, Jessell TM (1991) Developmental expression of the axonal glycoprotein TAG-1: differential regulation by central and peripheral neurons in vitro. Development 112:51–67

Krämer EM, Klein C, Koch T, Boytinck M, Trotter J (1999) Compartmentation of Fyn kinase with glycosylphosphatidylinositol-anchored molecules in oligodendrocytes facilitates kinase activation during myelination. J Biol Chem 274:29042–29049

Lamprianou S, Vacaresse N, Suzuki Y, Meziane H, Buxbaum JD, Schlessinger J, Harroch S (2006) Receptor protein tyrosine phosphatase gamma is a marker for pyramidal cells and sensory neurons in the nervous system and is not necessary for normal development. Mol Cell Biol 26:5106–5119

Lamprianou S, Chatzopoulou E, Thomas J-L, Bouyain S, Harroch S (2011) A complex between contactin-1 and the protein tyrosine phosphatase PTPRZ controls the development of oligodendrocyte precursor cells. Proc Natl Acad Sci USA 108:17498–17503

Laursen LS, Chan CW, Ffrench-Constant C (2009) An integrin-contactin complex regulates CNS myelination by differential Fyn phosphorylation. J Neurosci 29:9174–9185

Lee S, Takeda Y, Kawano H, Hosoya H, Nomoto M, Fujimoto D, Takahashi N, Watanabe K (2000) Expression and regulation of a gene encoding neural recognition molecule NB-3 of the contactin/F3 subgroup in mouse brain. Gene 245:253–266

Li H, Takeda Y, Niki H, Ogawa J, Kobayashi S, Kai N, Akasaka K, Asano M, Sudo K, Iwakura Y, Watanabe K (2003) Aberrant responses to acoustic stimuli in mice deficient for neural recognition molecule NB-2. Eur J Neurosci 17:929–936

Lieschke GJ, Currie PD (2007) Animal models of human disease: zebrafish swim into view. Nat Rev Genet 8:353–367

Liu Y, Halloran MC (2005) Central and peripheral axon branches from one neuron are guided differentially by Semaphorin3D and transient axonal glycoprotein-1. J Neurosci 25:10556–10563

Liu H, Focia PJ, He X (2010) Homophilic adhesion mechanism of neurofascin, a member of the L1 family of neural cell adhesion molecules. J Biol Chem 286:797–805

Meijers R, Puettmann-Holgado R, Skiniotis G, Liu J-H, Walz T, Wang J-H, Schmucker D (2007) Structural basis of Dscam isoform specificity. Nature 449:487–491

Menegoz M, Gaspar P, Le Bert M, Galvez T, Burgaya F, Palfrey C, Ezan P, Arnos F, Girault JA (1997) Paranodin, a glycoprotein of neuronal paranodal membranes. Neuron 19:319–331

Morales G, Hubert M, Brümmendorf T, Treubert U, Tárnok A, Schwarz U, Rathjen FG (1993) Induction of axonal growth by heterophilic interactions between the cell surface recognition proteins F11 and Nr-CAM/Bravo. Neuron 11:1113–1122

Morrow EM, Yoo S-Y, Flavell SW, Kim T-K, Lin Y, Hill RS, Mukaddes NM, Balkhy S, Gascon G, Hashmi A, Al-Saad S, Ware J, Joseph RM, Greenblatt R, Gleason D, Ertelt JA, Apse KA, Bodell A, Partlow JN, Barry B, Yao H, Markianos K, Ferland RJ, Greenberg ME, Walsh CA (2008) Identifying autism loci and genes by tracing recent shared ancestry. Science 321:218–223

Mörtl M, Sonderegger P, Diederichs K, Welte W (2007) The crystal structure of the ligand-binding module of human TAG-1 suggests a new mode of homophilic interaction. Protein Sci 16:2174–2183

Murai KK, Misner D, Ranscht B (2002) Contactin supports synaptic plasticity associated with hippocampal long-term depression but not potentiation. Curr Biol 12:181–190

Niisato K, Fujikawa A, Komai S, Shintani T, Watanabe E, Sakaguchi G, Katsuura G, Manabe T, Noda M (2005) Age-dependent enhancement of hippocampal long-term potentiation and impairment of spatial learning through the Rho-associated kinase pathway in protein tyrosine phosphatase receptor type Z-deficient mice. J Neurosci 25:1081–1088

Ogawa J, Lee S, Itoh K, Nagata S, Machida T, Takeda Y, Watanabe K (2001) Neural recognition molecule NB-2 of the contactin/F3 subgroup in rat: specificity in neurite outgrowth-promoting activity and restricted expression in the brain regions. J Neurosci Res 65:100–110

Osterfield M, Egelund R, Young LM, Flanagan JG (2008) Interaction of amyloid precursor protein with contactins and NgCAM in the retinotectal system. Development 135:1189–1199

Pagnamenta AT, Bacchelli E, de Jonge MV, Mirza G, Scerri TS, Minopoli F, Chiocchetti A, Ludwig KU, Hoffmann P, Paracchini S, Lowy E, Harold DH, Chapman JA, Klauck SM, Poustka F, Houben RH, Staal WG, Ophoff RA, O'Donovan MC, Williams J, Nöthen MM, Schulte-Körne G, Deloukas P, Ragoussis J, Bailey AJ, Maestrini E, Monaco AP, International Molecular Genetic Study of Autism Consortium (2010) Characterization of a family with rare deletions in CNTNAP5 and DOCK4 suggests novel risk loci for autism and dyslexia. Biol Psychiatry 68:320–328

Peles E, Nativ M, Campbell PL, Sakurai T, Martinez R, Lev S, Clary DO, Schilling J, Barnea G, Plowman GD, Grumet M, Schlessinger J (1995) The carbonic anhydrase domain of receptor tyrosine phosphatase beta is a functional ligand for the axonal cell recognition molecule contactin. Cell 82:251–260

Peles E, Nativ M, Lustig M, Grumet M, Schilling J, Martinez R, Plowman GD, Schlessinger J (1997) Identification of a novel contactin-associated transmembrane receptor with multiple domains implicated in protein-protein interactions. EMBO J 16:978–988

Peñagarikano O, Abrahams BS, Herman EI, Winden KD, Gdalyahu A, Dong H, Sonnenblick LI, Gruver R, Almajano J, Bragin A, Golshani P, Trachtenberg JT, Peles E, Geschwind DH (2011) Absence of CNTNAP2 leads to epilepsy, neuronal migration abnormalities, and core autism-related deficits. Cell 147:235–246

Pesheva P, Gennarini G, Goridis C, Schachner M (1993) The F3/11 cell adhesion molecule mediates the repulsion of neurons by the extracellular matrix glycoprotein J1-160/180. Neuron 10:69–82

Pickford LB, Mayer DN, Bolin LM, Rouse RV (1989) Transiently expressed, neural-specific molecule associated with premigratory granule cells in postnatal mouse cerebellum. J Neurocytol 18:465–478

Pillai AM, Garcia-Fresco GP, Sousa AD, Dupree JL, Philpot BD, Bhat MA (2007) No effect of genetic deletion of contactin-associated protein (CASPR) on axonal orientation and synaptic plasticity. J Neurosci Res 85:2318–2331

Poliak S, Salomon D, Elhanany H, Sabanay H, Kiernan B, Pevny L, Stewart CL, Xu X, Chiu S-Y, Shrager P, Furley AJW, Peles E (2003) Juxtaparanodal clustering of Shaker-like K^+ channels in myelinated axons depends on Caspr2 and TAG-1. J Cell Biol 162:1149–1160

Quek AML, Britton JW, McKeon A, So E, Lennon VA, Shin C, Klein CJ, Watson RE, Kotsenas AL, Lagerlund TD, Cascino GD, Worrell GA, Wirrell EC, Nickels KC, Aksamit AJ, Noe KH, Pittock SJ (2012) Autoimmune epilepsy: clinical characteristics and response to immunotherapy. Arch Neurol 69:582–593

Rios JC, Melendez-Vasquez CV, Einheber S, Lustig M, Grumet M, Hemperly J, Peles E, Salzer JL (2000) Contactin-associated protein (Caspr) and contactin form a complex that is targeted to the paranodal junctions during myelination. J Neurosci 20:8354–8364

Roohi J, Montagna C, Tegay DH, Palmer LE, DeVincent C, Pomeroy JC, Christian SL, Nowak N, Hatchwell E (2009) Disruption of contactin 4 in three subjects with autism spectrum disorder. J Med Genet 46:176–182

Rudick RA, Trapp BD (2009) Gray-matter injury in multiple sclerosis. N Engl J Med 361:1505–1506

Sakurai T, Lustig M, Nativ M, Hemperly JJ, Schlessinger J, Peles E, Grumet M (1997) Induction of neurite outgrowth through contactin and Nr-CAM by extracellular regions of glial receptor tyrosine phosphatase beta. J Cell Biol 136:907–918

Sakurai K, Toyoshima M, Ueda H, Matsubara K, Takeda Y, Karagogeos D, Shimoda Y, Watanabe K (2009) Contribution of the neural cell recognition molecule NB-3 to synapse formation between parallel fibers and Purkinje cells in mouse. Dev Neurobiol 69:811–824

Sakurai K, Toyoshima M, Takeda Y, Shimoda Y, Watanabe K (2010) Synaptic formation in subsets of glutamatergic terminals in the mouse hippocampal formation is affected by a deficiency in the neural cell recognition molecule NB-3. Neurosci Lett 473:102–106

Salzer JL (1997) Clustering sodium channels at the node of Ranvier: close encounters of the axon-glia kind. Neuron 18:843–846

Santos SD, Iuliano O, Ribeiro L, Veran J, Ferreira JS, Rio P, Mulle C, Duarte CB, Carvalho AL (2012) Contactin-associated protein 1 (Caspr1) regulates the traffic and synaptic content of α-amino-3-hydroxy-5-methyl-4-isoxazolepropionic acid (AMPA)-type glutamate receptors. J Biol Chem 287:6868–6877

Savvaki M, Theodorakis K, Zoupi L, Stamatakis A, Tivodar S, Kyriacou K, Stylianopoulou F, Karagogeos D (2010) The expression of TAG-1 in glial cells is sufficient for the formation of the juxtaparanodal complex and the phenotypic rescue of tag-1 homozygous mutants in the CNS. J Neurosci 30:13943–13954

Sawaya MR, Wojtowicz WM, Andre I, Qian B, Wu W, Baker D, Eisenberg D, Zipursky SL (2008) A double S shape provides the structural basis for the extraordinary binding specificity of Dscam isoforms. Cell 134:1007–1018

Schmahmann JD (2004) Disorders of the cerebellum: ataxia, dysmetria of thought, and the cerebellar cognitive affective syndrome. J Neuropsychiatry Clin Neurosci 16:367–378

Schweitzer J, Gimnopoulos D, Lieberoth BC, Pogoda H-M, Feldner J, Ebert A, Schachner M, Becker T, Becker CG (2007) Contactin1a expression is associated with oligodendrocyte differentiation and axonal regeneration in the central nervous system of zebrafish. Mol Cell Neurosci 35:194–207

Sherman DL, Tait S, Melrose S, Johnson R, Zonta B, Court FA, Macklin WB, Meek S, Smith AJH, Cottrell DF, Brophy PJ (2005) Neurofascins are required to establish axonal domains for saltatory conduction. Neuron 48:737–742

Shimoda Y, Watanabe K (2009) Contactins: emerging key roles in the development and function of the nervous system. Cell Adh Migr 3:64–70

Shimoda Y, Koseki F, Itoh M, Toyoshima M, Watanabe K (2012) A cis-complex of NB-2/contactin-5 with amyloid precursor-like protein 1 is localized on the presynaptic membrane. Neurosci Lett 510:148–153

Spiegel I, Salomon D, Erne B, Schaeren-Wiemers N, Peles E (2002) Caspr3 and caspr4, two novel members of the caspr family are expressed in the nervous system and interact with PDZ domains. Mol Cell Neurosci 20:283–297

Stankiewicz P, Lupski JR (2010) Structural variation in the human genome and its role in disease. Annu Rev Med 61:437–455

Stoeckli ET (2010) Neural circuit formation in the cerebellum is controlled by cell adhesion molecules of the Contactin family. Cell Adh Migr 4:523–526

Su XD, Gastinel LN, Vaughn DE, Faye I, Poon P, Bjorkman PJ (1998) Crystal structure of hemolin: a horseshoe shape with implications for homophilic adhesion. Science 281:991–995

Tait S, Gunn-Moore F, Collinson JM, Huang J, Lubetzki C, Pedraza L, Sherman DL, Colman DR, Brophy PJ (2000) An oligodendrocyte cell adhesion molecule at the site of assembly of the paranodal axo-glial junction. J Cell Biol 150:657–666

Takahashi N, Nielsen KS, Aleksic B, Petersen S, Ikeda M, Kushima I, Vacaresse N, Ujike H, Iwata N, Dubreuil V, Mirza N, Sakurai T, Ozaki N, Buxbaum JD, Sap J (2011a) Loss of function studies in mice and genetic association link receptor protein tyrosine phosphatase α to schizophrenia. Biol Psychiatry 70:626–635

Takahashi N, Sakurai T, Bozdagi-Gunal O, Dorr NP, Moy J, Krug L, Gama-Sosa M, Elder GA, Koch RJ, Walker RH, Hof PR, Davis KL, Buxbaum JD (2011b) Increased expression of receptor phosphotyrosine phosphatase-β/ζ is associated with molecular, cellular, behavioral and cognitive schizophrenia phenotypes. Transl Psychiatry 1:e8

Takeda Y, Akasaka K, Lee S, Kobayashi S, Kawano H, Murayama S, Takahashi N, Hashimoto K, Kano M, Asano M, Sudo K, Iwakura Y, Watanabe K (2003) Impaired motor coordination in mice lacking neural recognition molecule NB-3 of the contactin/F3 subgroup. J Neurobiol 56:252–265

Toyoshima M, Sakurai K, Shimazaki K, Takeda Y, Nakamoto M, Serizawa S, Shimoda Y, Watanabe K (2009a) Preferential localization of neural cell recognition molecule NB-2 in developing glutamatergic neurons in the rat auditory brainstem. J Comp Neurol 513:349–362

Toyoshima M, Sakurai K, Shimazaki K, Takeda Y, Shimoda Y, Watanabe K (2009b) Deficiency of neural recognition molecule NB-2 affects the development of glutamatergic auditory pathways from the ventral cochlear nucleus to the superior olivary complex in mouse. Dev Biol 336:192–200

Traka M, Dupree JL, Popko B, Karagogeos D (2002) The neuronal adhesion protein TAG-1 is expressed by Schwann cells and oligodendrocytes and is localized to the juxtaparanodal region of myelinated fibers. J Neurosci 22:3016–3024

Traka M, Goutebroze L, Denisenko N, Bessa M, Nifli A, Havaki S, Iwakura Y, Fukamauchi F, Watanabe K, Soliven B, Girault J-A, Karagogeos D (2003) Association of TAG-1 with Caspr2 is essential for the molecular organization of juxtaparanodal regions of myelinated fibers. J Cell Biol 162:1161–1172

Traut W, Weichenhan D, Himmelbauer H, Winking H (2006) New members of the neurexin superfamily: multiple rodent homologues of the human CASPR5 gene. Mamm Genome 17:723–731

van Daalen E, Kemner C, Verbeek NE, van der Zwaag B, Dijkhuizen T, Rump P, Houben R, van't Slot R, de Jonge MV, Staal WG, Beemer FA, Vorstman JAS, Burbach JPH, van Amstel HKP, Hochstenbach R, Brilstra EH, Poot M (2011) Social responsiveness scale-aided analysis of the clinical impact of copy number variations in autism. Neurogenetics 12:315–323

Vincent A, Lang B, Kleopa KA (2006) Autoimmune channelopathies and related neurological disorders. Neuron 52:123–138

Vogt L, Giger RJ, Ziegler U, Kunz B, Buchstaller A, Hermens WTJMC, Kaplitt MG, Rosenfeld MR, Pfaff DW, Verhaagen J, Sonderegger P (1996) Continuous renewal of the axonal pathway sensor apparatus by insertion of new sensor molecules into the growth cone membrane. Curr Biol 6:1153–1158

Voogd J, Glickstein M (1998) The anatomy of the cerebellum. Trends Neurosci 21:370–375

Wang H, Kunkel DD, Martin TM, Schwartzkroin PA, Tempel BL (1993) Heteromultimeric K^+ channels in terminal and juxtaparanodal regions of neurons. Nature 365:75–79

Wang W, Karagogeos D, Kilpatrick DL (2011) The effects of Tag-1 on the maturation of mouse cerebellar granule neurons. Cell Mol Neurobiol 31:351–356

Wang P-S, Wang J, Zheng Y, Pallen CJ (2012) Loss of protein-tyrosine phosphatase α (PTPα) increases proliferation and delays maturation of oligodendrocyte progenitor cells. J Biol Chem 287:12529–12540

White R, Gonsior C, Krämer-Albers E-M, Stöhr N, Hüttelmaier S, Trotter J (2008) Activation of oligodendroglial Fyn kinase enhances translation of mRNAs transported in hnRNP A2-dependent RNA granules. J Cell Biol 181:579–586

Wolfer DP, Henehan-Beatty A, Stoeckli ET, Sonderegger P, Lipp HP (1994) Distribution of TAG-1/axonin-1 in fibre tracts and migratory streams of the developing mouse nervous system. J Comp Neurol 345:1–32

Wolman MA, Sittaramane VK, Essner JJ, Yost HJ, Chandrasekhar A, Halloran MC (2008) Transient axonal glycoprotein-1 (TAG-1) and laminin-alpha1 regulate dynamic growth cone behaviors and initial axon direction in vivo. Neural Dev 3:6

Xenaki D, Martin IB, Yoshida L, Ohyama K, Gennarini G, Grumet M, Sakurai T, Furley AJW (2011) F3/contactin and TAG1 play antagonistic roles in the regulation of sonic hedgehog-induced cerebellar granule neuron progenitor proliferation. Development 138:519–529

Yamamoto M, Hassinger L, Crandall JE (1990) Ultrastructural localization of stage-specific neurite-associated proteins in the developing rat cerebral and cerebellar cortices. J Neurocytol 19:619–627

Yoshihara Y, Kawasaki M, Tani A, Tamada A, Nagata S, Kagamiyama H, Mori K (1994) BIG-1: a new TAG-1/F3-related member of the immunoglobulin superfamily with neurite outgrowth-promoting activity. Neuron 13:415–426

Yoshihara Y, Kawasaki M, Tamada A, Nagata S, Kagamiyama H, Mori K (1995) Overlapping and differential expression of BIG-2, BIG-1, TAG-1, and F3: four members of an axon-associated cell adhesion molecule subgroup of the immunoglobulin superfamily. J Neurobiol 28:51–69

Zuko A, Bouyain S, van der Zwaag B, Burbach JPH (2011) Contactins: structural aspects in relation to developmental functions in brain disease. Adv Protein Chem Struct Biol 84:143–180

Chapter 9
The L1 Family of Cell Adhesion Molecules: A Sickening Number of Mutations and Protein Functions

Michael Hortsch, Kakanahalli Nagaraj, and Rula Mualla

Abstract L1-type proteins are transmembrane cell adhesion molecules with an evolutionary well-conserved protein domain structure of usually six immunoglobulin and five fibronectin type III domains. By engaging in many different protein–protein interactions they are involved in a multitude of molecular functions and are important players during the formation and maintenance of metazoan nervous systems. As a result, mutations in L1-type genes cause a great variety of phenotypes, most of which are neurological in nature. In humans, mutations in the *L1CAM* gene are responsible for L1 syndrome and other L1-type genes have been implicated in conditions as varied as mental retardation, autism, schizophrenia, multiple sclerosis, and other disorders. Equally, the overexpression of L1-type proteins appears to have deleterious effects in various types of human tumor cells, where they generally contribute to an increase in cell mobility and metastatic potential.

9.1 Introduction

L1-type proteins are transmembrane cell adhesion molecules (CAMs) and belong to the immunoglobulin superfamily (IgSF) (Moos et al. 1988; Hortsch 1996). Most L1-type proteins contain 13 distinct protein domains, usually six Ig (immunoglobulin) and three to five FN III (fibronectin type III) protein domains (see Fig. 9.1). L1-CAM

M. Hortsch, Ph.D. (✉) • R. Mualla, B.Sc.
Department of Cell and Developmental Biology, University of Michigan, Ann Arbor, MI 48109, USA
e-mail: hortsch@umich.edu

K. Nagaraj, Ph.D.
Department of Applied Zoology, Kuvempu University, Shankaraghatta, Shimoga, India
e-mail: knagarajv@gmail.com; kvnag@kuvempu.ac.in

V. Berezin and P.S. Walmod (eds.), *Cell Adhesion Molecules: Implications in Neurological Diseases*, Advances in Neurobiology 8, DOI 10.1007/978-1-4614-8090-7_9,

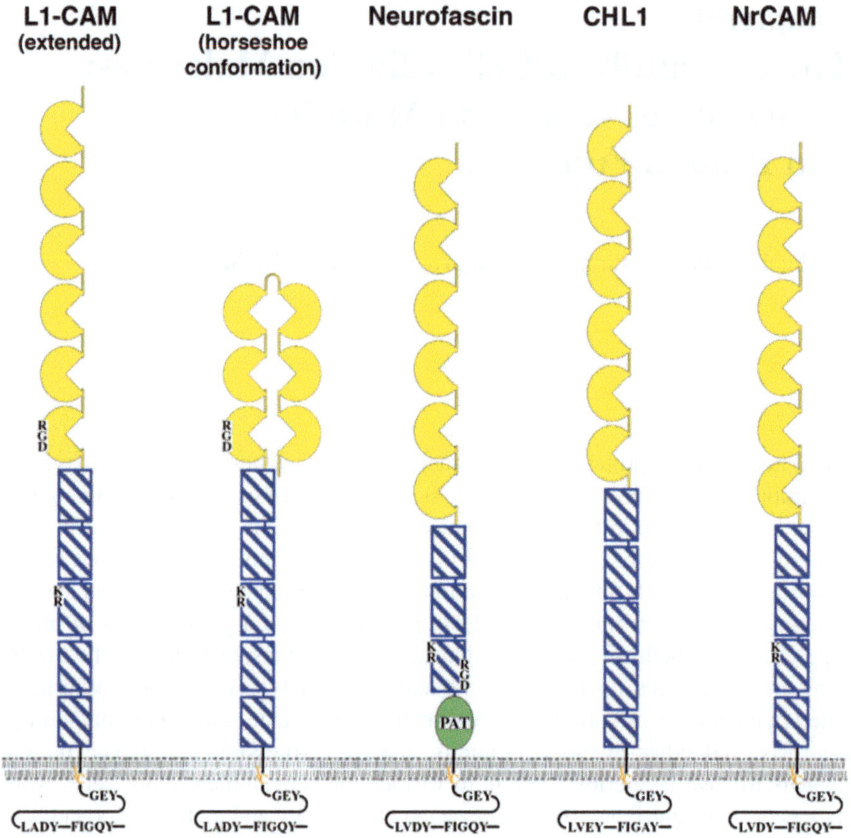

Fig. 9.1 Vertebrate L1-type protein structures. This figure shows the protein domain structure of the four vertebrate L1-type proteins: L1-CAM, Neurofascin, CHL1, and NrCAM. Indicated are several specific protein sequence features, such as conserved integrin-binding RGD motifs, a basic protease target site (KR) in one of the FN III protein domains, the presence of a conserved cysteine residue at the end of the transmembrane segment, which is the target of palmitoylation modification (Ren and Bennett 1998), and the three conserved tyrosine-containing motifs in the cytoplasmic domain. The first two diagrams depict the two putative L1-CAM conformations, extended and horseshoe-shaped, which have been predicted for the L1-CAM ectodomain (Schürmann et al. 2001)

was not only the first L1-type CAM to be identified, characterized, and cloned, but also yielded its name to the entire gene family (Rathjen and Rutishauser 1984; Rathjen and Schachner 1984; Moos et al. 1988).

In this review we will focus on pathological mutations in L1-type genes, which have been described not only in humans, but also in a number of different experimental model systems. The phenotypes caused by L1 mutations reveal an amazingly complex picture reflecting a wide range of biological functions that are associated with L1-type proteins in various species and organs. Although our knowledge about the complex biological functionalities of L1-type proteins is still expanding, we will try to provide a timely overview about our current understanding, how this family of adhesive proteins plays crucial roles in the nervous and other organ systems, and how mutations in and also the overexpression of these proteins cause a variety of phenotypes.

9.2 Structure, Functions, and Genetics of L1-Type CAMs

9.2.1 The Structure of L1-Type Proteins

All L1-type proteins are predominantly, but not exclusively, expressed in the nervous system and belong to the immunoglobulin superfamily. They share a common arrangement of six amino terminal Ig-protein domains, followed by three to five FN III domains and a single transmembrane segment (Fig. 9.1). In humans, the mature L1-CAM protein has 1,256 amino acids with an extracellular part consisting of six Ig-like domains and five FN III-like domains, a single-pass transmembrane domain and a short cytoplasmic C-terminal tail (Wolff et al. 1988; Kobayashi et al. 1991). Genes encoding proteins with this characteristic domain structure form a unique gene family, now referred to as the L1 family of neural cell adhesion molecules (Hortsch 1996, 2000).

Gene duplication events in various metazoan phyla have resulted in multiple L1-type genes per genome (Mualla et al. 2013), and in most chordate species, including humans, four paralogous L1-type genes have been identified. These are now referred to as L1-CAM, CHL1 (Close Homolog of L1), Neurofascin, and NrCAM (neuron–glia-related cell adhesion molecule) (Fig. 9.1) (Hortsch 2000). In the case of Neurofascin and NrCAM proteins, alternative splicing of the initial transcript is responsible for multiple different protein isoforms (Hassel et al. 1997; Wang et al. 1998). The expression of the alternatively spliced Neurofascin protein isoforms is cell and tissue specific and also developmentally regulated (Hassel et al. 1997; Collinson et al. 1998). In some Neurofascin protein isoforms several of the FN III domains are either deleted or substituted by a PAT domain (Fig. 9.1) (Davis et al. 1993; Volkmer et al. 1992). This Neurofascin protein domain is rich in the amino acids proline, alanine, and threonine (thus termed "PAT") and appears to be the target of O-linked glycosylation. These Neurofascin splice variants exhibit significant functional differences, not only in their interactions with various extracellular ligands (Volkmer et al. 1992), but also in their cell-specific expression and subcellular localization in neuronal cells (Davis et al. 1996; Zonta et al. 2008).

The Ig domains found in L1-type molecules were originally assigned to the C2 set of Ig-like domains. However, a comparison with other Ig domain proteins revealed that the domains in L1-type proteins belong to a novel structural subset of the Ig superfamily, now referred to as the I set (Harpaz and Chothia 1994; Bateman et al. 1996). Although the homophilic adhesive function of L1-type proteins involves multiple extracellular protein domains, it appears to be centered around the second Ig domain (Zhao et al. 1998). A number of vertebrate L1-type proteins (specifically L1-CAM and Neurofascin) also contain RGD motifs in their ectodomains, which functionally interact with RGD-specific integrins (Ruppert et al. 1995; Montgomery et al. 1996; Felding-Habermann et al. 1997; Yip et al. 1998; Koticha et al. 2005). Based on a general domain homology to the insect Ig domain protein Hemolin, Su et al. (1998) postulated that the 11 extracellular protein domains of L1-CAM exist in two different conformational states, one being extended and the other in a horseshoe shape (Fig. 9.1). Subsequently, structural analyses of the L1-CAM ectodomain gave some support to this notion (He et al. 2009; Schürmann et al. 2001; Wei and Ryu 2012). However, how these two postulated conformational states of the L1-CAM protein correlate with its functional interactions and activities remains unclear.

The size of the cytoplasmic domain in L1-type proteins ranges from 85 to 148 residues with several segments containing characteristic tyrosine-containing amino acid motifs that exhibit the highest degree of sequence conservation throughout the entire L1 gene family (Fig. 9.1). Two of these tyrosine-containing motifs are part of the cytoplasmic Ankyrin-binding site of L1-type proteins (Hortsch et al. 1998a; Zhang et al. 1998). The phosphorylation of the FIGQY motif is downstream of FGFR signaling and abolishes Ankyrin binding (Garver et al. 1997; Jenkins et al. 2001; Whittard et al. 2006). Interestingly, the entire L1 cytoplasmic domain is not required for homophilic L1–L1 interactions to occur (Hortsch et al. 1995; Wong et al. 1995). Nevertheless, extracellular and intracellular interactions involving L1-type proteins often influence and regulate each other (Hortsch et al. 1998a) and Ankyrin binding is important for a number of different L1 functions (Hortsch et al. 2009; Ooashi and Kamiguchi 2009; Guan and Maness 2010; Nakamura et al. 2010; Chen and Hing 2008; Buhusi et al. 2008; Ango et al. 2004; Nishimura et al. 2003).

Most vertebrate L1-type genes contain a well-conserved 12-nucleotide miniexon, which encodes an RSLE amino acid motif in the cytoplasmic L1 protein domain. Only in CHL1 proteins has the presence of an RSLE miniexon not been demonstrated. Due to differential splicing, these amino acids are not included in non-neuronal L1-type molecules (Reid and Hemperly 1992; Takeda et al. 1996). The insertion of this RSLE motif into the cytoplasmic domain of L1-type proteins generates a tyrosine-based signal (YxxL) that results in the sorting of L1-CAM protein to the growth cone and induces the AP-2-mediated endocytosis of L1-CAM and presumably other L1-type paralogous proteins via Clathrin-coated pits (Kamiguchi et al. 1998; Kamiguchi and Lemmon 1998). Non-vertebrate L1-type genes do not contain an RSLE-encoding miniexon. As reported for the *Drosophila* L1 molecule Neuroglian (Bieber et al. 1989), some of their transcripts undergo rather different, neuron-specific splicing processes (Hortsch et al. 1990), the functional ramifications of which are currently unknown. Therefore, the AP-2-mediated endocytosis of L1-type proteins, which regulates their cell surface expression, appears to be restricted to vertebrate species or non-vertebrate L1-type proteins contain other, yet unidentified, endocytosis-inducing signals.

9.2.2 The Evolutionary Origin of the L1 Gene Family

Figure 9.2 indicates that genes belonging to the L1 family of CAMs can be identified in most metazoan phyla and probably arose together with the appearance of first primitive nervous systems in evolution about 1,200–1,500 million years ago (Mualla et al. 2013). The genomes of several metazoan phyla, such as arthropods, only contain one L1-type gene, whereas other phyla, including most chordate species, appear to harbor multiple different L1-type genes in their genomes. With the exception of tunicates, all chordate genomes contain at least four L1-type paralogs, which are now known as L1-CAM, CHL1, Neurofascin, and NrCAM (Hortsch 2000). As shown in Fig. 9.1, these paralogous proteins all exhibit the typical L1-type protein structure and other L1 typical features. Consequently, these proteins share many similarities, including six Ig domains, three to five FN III domains, a transmembrane region, and a highly conserved cytoplasmic domain (Hortsch 2000). The existence of four paralogous genes in chordate species is now believed to be the result of two sequential genome-wide duplication events, which occurred during early chordate evolution (Kappen et al. 1989; Schughart et al. 1989). An additional genome-wide duplication event has occurred in the teleost lineage (Amores et al. 1998; Christoffels et al. 2004) resulting in two genes for each vertebrate L1 paralog for a total of eight L1-type genes per genome (Mualla et al. 2013).

9.2.3 Biological Functions of L1-Type CAMs

The identified biological functions of the L1-type CAMs cover a wide range and are mostly, but not exclusively, a result of their predominant expression in the nervous system (Hortsch 1996; Wiencken-Barger et al. 2004; Maness and Schachner 2007). Foremost, L1-CAM and other L1-type CAMs have been reported to induce neurite outgrowth (Lemmon et al. 1989; Harper et al. 1994; Hillenbrand et al. 1999; Volkmer et al. 1996; Pruss et al. 2004; Koticha et al. 2005) and to support axon guidance and pathfinding (Hall and Bieber 1997; Wiencken-Barger et al. 2004; Imondi et al. 2007; Demyanenko and Maness 2003); neuronal cell migration (Anderson et al. 2006; Demyanenko et al. 2001; Asou et al. 1992), axonal fasciculation (Wiencken-Barger et al. 2004), and neuronal differentiation (Dihne et al. 2003; Demyanenko et al. 2009; Turner et al. 2009); as well as cell survival (Nishimune et al. 2005; Hulley et al. 1998; Jakovcevski et al. 2009). L1-type proteins also play a prominent role during nervous system regeneration (Becker et al. 2004) and appear to be involved in synapse formation and plasticity (Godenschwege et al. 2006; Saghatelyan et al. 2004; Triana-Baltzer et al. 2006). Several studies also indicate that L1-type CAMs participate in the formation of the myelin sheath that surrounds many axons (Bartsch 2003; Itoh et al. 2005; Wood et al. 1990). These biological functions, which have been attributed to L1-type CAMs, make them promising pharmaceutical agents for aiding regeneration processes after

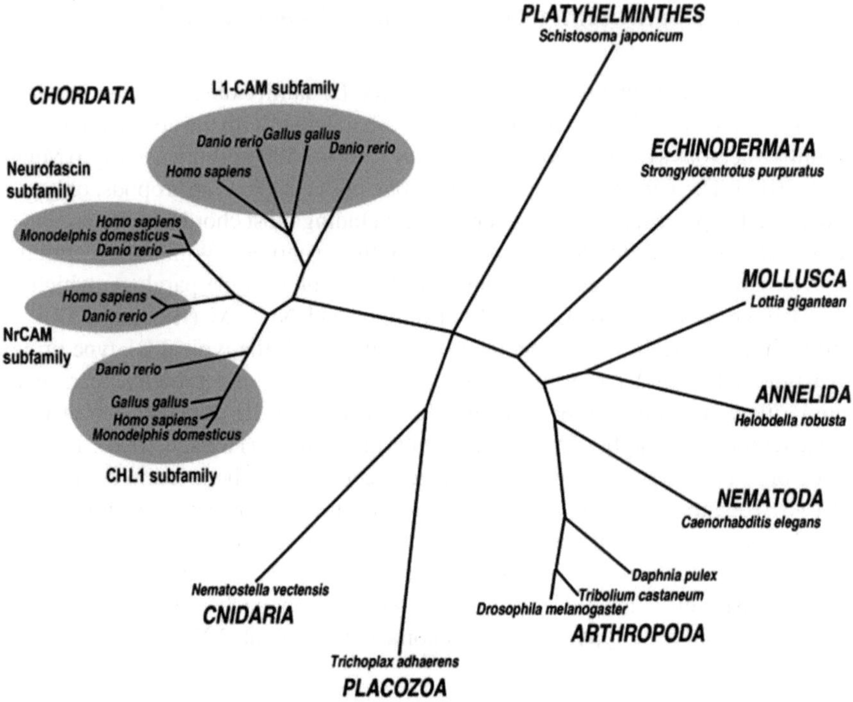

Fig. 9.2 Shown is an unrooted phylogenetic tree of the L1 family in several animal phyla. The species highlighted by gray ovals indicate L1-type paralogs that are only found in chordate species. Whereas many sequences were taken from complete open reading frames, which conform to the six Ig plus five FN III consensus L1 protein domain structure, others were translated from partial cDNA and genomic sequences that were downloaded from GenBank (http://www.ncbi.nlm.nih.gov/genbank/) or the JGI Genome Sequence Database (http://www.jgi.doe.gov). Sequences, which cover the region starting with the basic amino acids following the transmembrane segment to the conserved FIGQY motif, were used for a multiple sequence alignment. This alignment was performed using the online version of the MAFFT program (http://align.bmr.kyushu-u.ac.jp/mafft/online/server/). An unrooted phylogenetic tree was constructed using the Promlk and the Drawtree subroutines of the Phylip v3.65 program package (Felsenstein 1981). Species included in the figure represent 11 different metazoan phyla and include *Schistosoma japonicum*/blood-fluke or bilharzias flatworm (BU799663), *Strongylocentrotus purpuratus*/California purple sea urchin (XP_784933), *Lottia gigantea*/giant owl limpet (JGI Genome Sequence Database scaffold_226:99499–99582), *Helobdella robusta*/Californian leech (JGI Genome Sequence Database scaffold_43: 1086463:1087596, 1086816–1086905 and 1087447–1087596), *Caenorhabditis elegans*/nematode or roundworm (NP_001033395), *Daphnia pulex*/water flea (ACJG01004335), *Tribolium castaneum*/red flour beetle (AAJJ01000894), *Drosophila melanogaster*/fruit fly (NP_727274), *Trichoplax adhaerens*/hairy plate or flat animal (XM_002117949), *Nematostella vectensis*/starlet sea anemone (XM_001637356), *Danio rerio*/zebrafish (CABZ01014841, CABZ01052948, CABZ01022026, NM_001044805, and XM_002662518), *Gallus gallus*/chicken (AADN03013871 and AADN03006373), *Monodelphis domestica*/gray short-tailed opossum (AAFR03023046 and AAFR03020961), and *Homo sapiens*/human (NP_076493, NP_055905, NP_005001, and NP_006605)

spinal tract and other neuronal injuries. Several studies using animal models support this assertion (Chen et al. 2007; Becker et al. 2004; Roonprapunt et al. 2003; Bernreuther et al. 2006).

In arthropod species, the Neuroglian protein represents the product of the single L1-type gene that is present in the genome of species belonging to this phylum. Neuroglian has an additional essential function in stabilizing epithelial integrity as they are part of septate junction complexes (Genova and Fehon 2003; Laval et al. 2008; Wei et al. 2004; Faivre-Sarrailh et al. 2004; Banerjee et al. 2006). Although in vertebrate species the sealing function of epithelial septate junctions has been replaced by tight junctions, two L1-type proteins, Neurofascin and NrCAM, are still components of paranodal septate junctions where they stabilize the molecular architecture of nodes of Ranvier in myelinated nerves (Charles et al. 2002; Jenkins and Bennett 2002; Sherman et al. 2005; Koticha et al. 2006; Davis et al. 1996; Hortsch and Margolis 2003). In addition, L1-CAM is also expressed in many vertebrate epithelia outside the nervous system (Nolte et al. 1999). The physiological role of L1-CAM expression in these epithelia remains unknown, but the addition of anti-L1 antibodies to kidney organ cultures indicates a role in kidney branching morphogenesis (Debiec et al. 1998). Maddaluno et al. (2009) published another interesting finding about the function of L1-type proteins in epithelia. They reported that L1-CAM regulates the transendothelial trafficking of dendritic cells in mice.

Equally, the expression of L1-CAM protein in mammalian leukocytes still remains a mystery (Hubbe et al. 1993; Kowitz et al. 1992; Ebeling et al. 1996) and a topic of speculation (Kadmon et al. 1998). It has been suggested that L1-CAM functions as a co-stimulatory molecule during T-cell activation (Balaian et al. 2000). Another publication reports that a monoclonal antibody specific for L1-CAM disrupts the normal remodeling of lymph node reticular matrix during an immune response in vivo (Di Sciullo et al. 1998). Interestingly, the L1-type protein Neuroglian is also expressed in the moth *Manduca sexta* (tobacco hornworm) plasmatocytes where it contributes to the primitive innate immune functions, which these cells carry out by encapsulating foreign material (Williams 2009; Zhuang et al. 2007; Nardi et al. 2006).

9.2.4 L1 Syndrome: A Wide Spectrum of Phenotypes

The human *L1CAM* gene is located on the X-chromosome at Xq28 (Dietrich et al. 1992). It consists of 29 exons with the first exon of 125 base pairs being part of the 5′ untranslated region (Kallunki et al. 1997). Similar to other L1-type genes, the human *L1CAM* gene is primarily expressed in the nervous system and encodes a protein of 1,257 amino acids, comprising a signal peptide of 19 amino acids and a final processed product of 1,238 amino acids. In non-neural tissue, alternative RNA splicing generates mRNA molecules, which lack exons 2 and 27 of the gene (Jouet et al. 1995; De Angelis et al. 2001) and results in a protein that is 9 amino acids shorter.

Mutations in the human *L1CAM* gene manifest themselves in a wide range of dysfunctions that are usually neurological in origin and appearance. Therefore, the phenotype caused by mutations in the *L1CAM* gene was originally described as four distinct neurological disorders, namely, X-linked hydrocephalus, which is caused by a stenosis of the aqueduct of Sylvius (HSAS) (Rosenthal et al. 1992; Jouet et al. 1994; Bickers and Adams 1949; Finckh et al. 2000; Gu et al. 1996; Fransen et al. 1997), MASA (mental retardation, aphasia, shuffling gait, and adducted thumbs) syndrome (Winter et al. 1989; Schrander-Stumpel et al. 1990; Fryns et al. 1991), X-linked complicated hereditary spastic paraplegia type 1 (SPG1), and X-linked complicated corpus callosum agenesis (X-linked ACC) (Kaplan 1983). The allelic nature of these disorders remained unrecognized until the pathological mutations were identified as affecting the same gene (Fryns et al. 1991; Jouet et al. 1994; Fransen et al. 1994; Vits et al. 1994). Also the term CRASH syndrome (corpus callosum hypoplasia, retardation, adducted thumbs, spastic paraplegia, and hydrocephalus syndrome) has been proposed as a collective name for these disorders (Fransen et al. 1995). Now these terms are usually summarized under the name L1 syndrome (Panicker et al. 2003).

L1 syndrome is an X-linked recessive disorder with an incidence of 1:30,000 newborn males and is caused by mutations in the *L1CAM* gene. Well over 200 different pathogenic *L1CAM* mutations have been identified and reported in the literature. 130 were reviewed by Weller and Gartner (2001). Since then, many additional L1-CAM mutations were reported (Simonati et al. 2006; Senat et al. 2001; Sztriha et al. 2002; Silan et al. 2005; Felsenstein 1981; Okamoto et al. 2004; Hübner et al. 2004; Moya et al. 2002; Panayi et al. 2005; Tegay et al. 2007; Knops et al. 2008; Griseri et al. 2009; Nakakimura et al. 2008; Wilson et al. 2009; Kanemura et al. 2006; Piccione et al. 2010; Rodriguez Criado et al. 2003; Rehnberg et al. 2011) and another 52 new *L1CAM* mutations were recently reported in an online L1-CAM mutations databank (Vos and Hofstra 2010). As mentioned above, the individual phenotype associated with different mutations in the *L1CAM* gene varies considerably, but usually includes various degrees of mental retardation.

The Diversity of Pathogenic L1-CAM Mutations

The mutations affecting the L1-CAM protein have been subdivided into four different classes (Fransen et al. 1998b). Class I mutations lead to a truncation and thereby to a complete absence of L1-CAM protein. These include frameshift mutations (small deletions or insertions) or point mutations resulting in a premature stop codon (nonsense mutations). Class II includes missense mutations resulting in an amino acid substitution in the extracellular part of the L1-CAM protein. Class III includes any mutation in the L1-CAM cytoplasmic domain. Class IV mutations comprise extracellular mutations that result in an aberrant splicing of the L1-CAM pre-mRNA.

The different types of L1-CAM mutations generally correlate with the severity of the observed phenotype. Class 1 mutations in the extracellular part of L1-CAM cause a more severe phenotype; Class 2 extracellular missense mutations, which affect amino acids located on the surface of protein, usually cause a milder phenotype than those affecting amino acids, which are predicted to be buried in the core of an L1-CAM protein domain (Fransen et al. 1998b; Bateman et al. 1996). Mutations affecting the cytoplasmic domain of L1-CAM generally result in a milder phenotype than extracellular mutations (Fransen et al. 1998b; Yamasaki et al. 1997).

L1-CAM Missense Mutations and Functional Defects

In the past, a large number of L1-CAM missense mutations have been published that cause a wide spectrum of neurological abnormalities, including mental retardation, hydrocephalus, shuffling gait, and agenesis of corpus callosum (Finckh et al. 2000; Gu et al. 1996; Rosenthal et al. 1992; Fransen et al. 1995; Jouet et al. 1994; Vits et al. 1994, 1998; Ruiz et al. 1995). Interestingly, almost every family with an identified L1-CAM mutation has its own individual mutation. The pathogenic potential of L1-CAM missense mutations varies considerably and depends on the exact location of the affected residue and the type of amino acid exchange involved. Therefore, these L1-CAM mutations have also been classified as disease-causing, likely disease-causing, likely non-disease-causing polymorphisms, or as unknown (Vos et al. 2010).

L1-CAM missense mutations have been identified in all extracellular protein domains, as well as the cytoplasmic region of the L1-CAM protein (Fig. 9.3). Specifically, disease-causing mutations have been reported in Ig1, Ig2, Ig3, Ig4, and Ig5 and Fn1, Fn2, and Fn5 protein domains of extracellular region of the L1-CAM and also in the cytoplasmic domain. In addition, likely disease-causing mutations have been identified in the L1-CAM Ig6 and Fn1 protein domains. A more complete list of currently known and previously published human L1-CAM mutations can be accessed at the "L1CAM Mutation Database," which is being maintained by the University Medical Center Groningen at http://www.l1cammutationdatabase.info (Vos and Hofstra 2010).

However, the severity of the phenotype and the clinical features vary between different L1-CAM mutations. Certainly the location of the mutated amino acid residue and the affected protein domain (Ig, FN III, or cytoplasmic) do influence and partially determine the severity of the phenotype. This has been analyzed for the severity of L1-CAM-related hydrocephalus and the mortality rate caused by individual L1-CAM mutations (Yamasaki et al. 1997; Bertolin et al. 2010). In general, mutations in the extracellular part of L1-CAM that led to truncation or absence of L1-CAM protein cause a most severe phenotype. In contrast, L1-CAM mutations affecting the cytoplasmic region result in a milder phenotype (Fransen et al. 1998b; Yamasaki et al. 1997). In addition, extracellular mutations affecting amino acids situated on the surface of a domain cause a milder phenotype than those affecting amino acids situated in the core of the protein domains (Bateman et al. 1996; Fransen et al. 1998b). Equally, L1-CAM mutations interfering with homophilic and/or heterophilic protein–protein interactions usually cause more significant neuronal

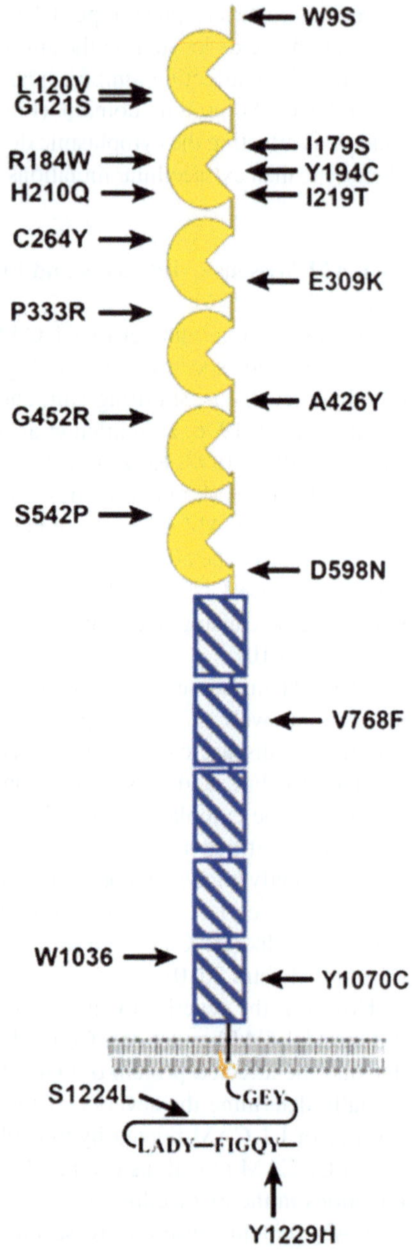

Fig. 9.3 Selection of human pathogenic L1-CAM missense mutations. This figure depicts the position of several pathogenic L1-CAM mutations that have been determined to cause L1 syndrome in humans and that have been experimentally analyzed in various in vitro and in vivo assay systems for their ability to perform specific L1-CAM-associated functions (De Angelis et al. 1999; Nagaraj et al. 2009; Rünker et al. 2003; Michelson et al. 2002; Needham et al. 2001; Moulding et al. 2000; Zhao and Siu 1996; Godenschwege et al. 2006)

dysfunctions (De Angelis et al. 2002). In addition, the variance of the phenotype between affected siblings, who share an identical L1-CAM mutation, suggests a strong epigenetic influence on the expression of specific phenotypic aspects. Using an L1-CAM knockout mouse model, Tapanes-Castillo et al. (2009) identified such a modifier locus for X-linked hydrocephalus on mouse chromosome 5.

9.2.5 Molecular Mechanisms by Which L1-CAM Mutations Cause Neurological Dysfunctions

The phenotypic diversity that is observed for L1-CAM mutations is well documented and it has been speculated that the multitude of L1-CAM interactions with itself and its various binding partners may play an important role in this heterogeneity. Many extracellular L1-CAM mutations have been demonstrated to disrupt L1-CAM's homophilic interaction and have a reduced ability to stimulate neurite outgrowth in vitro (De Angelis et al. 1999; Zhao and Siu 1996) (Fig. 9.3). As L1-CAM exerts many of its physiological functions by its strong homophilic adhesive ability, the relative contribution of homophilic versus heterophilic L1-CAM interactions to the phenotypic expression of mutational defects remains an interesting question. Regardless, it appears reasonable to assume that alterations in ligand-specific L1-CAM binding properties represent one central pathological mechanism, which is at play in L1 syndrome (De Angelis et al. 2002).

As evidenced by two studies that used a transgenic mouse model expressing L1-6D, which lacks the sixth Ig domain of L1-CAM and as a result has no homophilic and RGD-dependent integrin interactions, the homophilic adhesive function of L1-CAM is involved in some phenotypic aspects of L1 syndrome, specifically hydrocephalus. However, its severity is also influenced by other genetic modifiers (Tapanes-Castillo et al. 2009; Itoh et al. 2004). Based on their observations these authors hypothesize that a co-receptor for L1-CAM-mediated neurite outgrowth is involved and some pathogenic mutations affect neurite outgrowth or branching by disrupting the interaction with this co-receptor.

Not surprisingly, some pathogenic L1-CAM mutations also disrupt some of L1-CAM's known heterophilic interactions. For example, many pathological L1-CAM mutations influence the heterophilic interaction with TAX-1/Axonin-1 (De Angelis et al. 1999, 2002) (Fig. 9.3). Similarly, the functional interaction between L1-CAM and EGFR (epidermal growth factor receptor) interaction is impaired by some L1-CAM missense mutations (Nagaraj et al. 2009). In the cytoplasmic domain, two pathogenic L1-CAM mutations (S1124L and Y1229H, Fig. 9.3) affect the binding of the L1-CAM cytoplasmic domain to the Spectrin–Actin cytoskeleton and abolish endocytosis of L1-CAM (Buhusi et al. 2008; Needham et al. 2001). It appears reasonable to hypothesize that other heterophilic L1-CAM interactions, such as the binding of Neuropilin-1 that partially mediates L1-CAM's axon guidance function (Soker et al. 1998; Castellani et al. 2000) and the interaction with RGD-specific integrins that leads to cell migration and myelination (Haney et al. 1999; Mechtersheimer et al. 2001), are equally changed by some pathogenic L1-CAM mutations. Finally, some mutations appear to influence the tertiary structure of the L1-CAM protein, which in turn disrupts protein–protein interactions involving L1-CAM indirectly (Cheng and Lemmon 2004).

Another mechanism by which mutations influence or abrogate L1-CAM function is their effect on the cell surface expression of the L1-CAM protein (De Angelis et al. 1999, 2002; Michelson et al. 2002; Moulding et al. 2000). Particularly, the pathological missense mutations I179S and Y194C affect L1-CAM's neurite outgrowth inducing capability by decreasing the cell surface localization of the mutant L1-CAM protein (Michelson et al. 2002) (Fig. 9.3). The missense mutation C264Y in the extracellular domain of L1-CAM, which is known to cause HSAS (hydrocephalus with stenosis of aqueduct of Sylvius) in affected humans, impairs cellular L1-CAM protein trafficking (Rünker et al. 2003) (Fig. 9.3). Transfection studies in vitro demonstrate that this mutant L1-CAM protein is not expressed at the cell surface, but instead is located intracellularly, most likely within the endoplasmic reticulum. This was further confirmed by an in vivo analysis using a transgenic mouse line that expresses the C264Y mutant L1-CAM protein (Rünker et al. 2003).

There are also multiple lines of evidence that pathogenic L1-CAM mutations alter interactions with the intracellular Actin–Spectrin membrane skeleton and the cytoplasmic machinery that regulates L1-CAM-mediated axonal branching. The L1-CAM cytoplasmic domain (L1CD) is involved in axonal branching and the interaction between L1-CAM and the Actin cytoskeleton is also critical for this activity (Cheng et al. 2005a). Some extracellular mutations, such as I219T and W1036L (Fig. 9.3), alter the interaction between L1-CAM and the Actin cytoskeleton by changing L1-CAM's conformation. This observation might simply be a reflection of extracellular adhesive events regulating the binding of Ankyrin to the cytoplasmic L1 domain (Dubreuil et al. 1996; Hortsch et al. 1998b). Ankyrin acts as the linker between the L1 cytoplasmic domain and the Actin–Spectrin network. Alternatively, Cheng et al. speculated that axonal branching is regulated by L1-CAM interacting extracellularly in *cis* with an unknown co-receptor. The pathogenic mutations I219T and W1036L may disrupt this interaction. The S542P mutation (Fig. 9.3), which exhibits a reduction in both L1-CAM protein surface expression and homophilic adhesion, also elicits a decrease in the number of axonal branching (Cheng and Lemmon 2004).

Some pathogenic L1-CAM mutations have also been reported to affect L1-CAM-associated signaling processes. Using in vitro as well as in vivo *Drosophila* assay systems, we demonstrated that two pathogenic human L1-CAM mutations (E309K and Y1070C, Fig. 9.3), which exhibit wild-type levels of homophilic adhesion, have a reduced ability to induce L1-CAM-dependent EGFR signaling in vitro and are unable to rescue L1 loss-of-function conditions in vivo (Nagaraj et al. 2009).

Pathogenic L1-CAM mutations have various effects on the L1-CAM protein. Many missense L1-CAM mutations are predicted to distort the structure of individual domains and as a result to affect the intracellular processing of the L1-CAM protein and potentially reduce its cell surface expression. A recent study by Bertolin et al. (2010) demonstrated that some frameshift mutations and all nonsense mutations result in truncated L1-CAM proteins, which have carboxy termini in different extracellular L1-CAM protein domains. The observed neurological dysfunctions that are associated with these *L1-CAM* mutations give support to the notion that the severity of the L1 syndrome phenotype correlates with the severity of the molecular effect of the individual mutation and is also dependent on the epigenetic context into which a particular mutation is placed (Yamasaki et al. 1997).

9.2.6 Mutational Defects of L1-Type Genes in Various Model Animal Systems

Drosophila melanogaster

L1-type genes and proteins have been identified in many other metazoan species (Mualla et al. 2013). *Drosophila* Neuroglian (Nrg) was the second L1-type protein identified and described (Bieber et al. 1989). Coincidentally, the *neuroglian* (*nrg*) gene is localized on the fly's X-chromosome at cytological location 7F1. Hall and Bieber (Hall and Bieber 1997) have analyzed and described three mutant lines. All are late embryonic lethal mutations that alter or abolish *Drosophila* Neuroglian expression during development. The nrg^1 mutation is a protein null allele that is caused by an inversion with breakpoints at chromosomal locations 6E-7F1 the proximal breakpoint residing in the *nrg* transcription unit (Bieber et al. 1989; Hall and Bieber 1997). The mutation nrg^2 represents a hypomorphic mutation with markedly reduced expression levels of both Neuroglian protein isoforms. The mutation *nrg3* is temperature sensitive and also represents a late embryonic lethal allele when raised at a nonpermissive temperature (Hall and Bieber 1997). At a nonpermissive temperature the Nrg protein in homozygous nrg^3-mutant animals is mislocalized inside Nrg expressing cells and is not transported to the cell surface.

Neuroglian is also expressed in two protein isoforms, one being restricted to neuronal cells (Nrg^{180}) and the other to non-neuronal cells (Nrg^{167}) (Hortsch et al. 1990). However, the differential cell-specific splicing of the *nrg* transcript in *Drosophila* differs considerably in its effect on the resulting L1 protein isoforms from that described for L1-CAM in vertebrates (De Angelis et al. 2001; Miura et al. 1991). Although the Neuroglian protein is predominantly expressed in the nervous system throughout the life cycle of the fly (Bieber et al. 1989), the non-neuronal Nrg^{167} isoform also exhibits a high level of expression in most, if not all, *Drosophila* epithelia. In arthropod epithelia Neuroglian is an essential part of septate junction protein complexes and thereby stabilizes epithelial integrity (Genova and Fehon 2003; Laval et al. 2008; Wei et al. 2004; Faivre-Sarrailh et al. 2004; Banerjee et al. 2006).

The lack of Neuroglian expression in vivo causes abnormalities in embryonic motor neuron projections of both the intersegmental and the segmental nerves (Hall and Bieber 1997). Temperature shift experiments using homozygous nrg^3-mutant animals also revealed axonal pathfinding defects in the larval ocellar sensory system, which are mediated by EGFR and FGFR (fibroblast growth factor receptor) signaling (Garcia-Alonso et al. 2000; Kristiansen et al. 2005). Similarly, Neuroglian appears to mediate sensory axon advances in the *Drosophila* embryonic nervous system (Martin et al. 2008). Neuroglian is also an important player during the development of the adult mushroom body where it controls axonogenesis, axon bundling, axon branching, and guidance through signaling mechanisms that are different from the ocellar sensory system (Goossens et al. 2011). However, the physiological role of the fly's sole L1-type protein is not restricted to neuronal cells and to axonal pathfinding. It is also involved in dendritic arborization (Yamamoto et al. 2006) and in the proper differentiation of certain glial cells (Chen and Hing 2008). In addition, by using a S213L Neuroglian missense mutation, Godenschwege et al. uncovered an essential role of this *Drosophila* L1-type protein at the pupal giant synapse, which is independent of Neuroglian's role in axonal pathfinding (Godenschwege et al. 2006). A different Neuroglian missense mutation (G92R, designated as *ibx* for *icebox*) not only causes a nonlethal central brain morphology phenotype, but when homozygous in female flies results in a specific deficiency in female mating behavior (Carhan et al. 2005). Male flies show no behavioral defects, nor are other female behaviors visibly affected.

Although the *Drosophila* Neuroglian protein only exhibits a moderate level of amino acid identity when compared with mouse or human L1-CAM (Bieber et al. 1989; Zhao and Hortsch 1998), all major L1 features, including the characteristic L1 protein domain structure and a highly conserved intracellular Ankyrin binding domain, are preserved in the fly ortholog. In contrast to vertebrate species, arthropod genomes have only one L1-type gene in their genome. However, the similarities between human L1-CAM and *Drosophila* Neuroglian extend to several functional aspects. The expression of human L1-CAM in transgenic flies rescues some of *nrg* loss-of-function axonal pathfinding defects (Kristiansen et al. 2005). Similarly, human L1-CAM expression rescues a central nervous system synaptic phenotype in the fly that is caused by the lack of Neuroglian protein (Godenschwege et al. 2006). This surprising functional conservation between two members of the L1 gene family that are separated by more than 600 million years of evolution has made it possible to analyze pathogenic human L1-CAM proteins under in vivo conditions for specific functional aspects (Godenschwege et al. 2006; Nagaraj et al. 2009).

Caenorhabditis elegans

The nematode *Caenorhabditis elegans* genome encodes two L1-type genes, *sax-7/lad-1* and *lad-2* (Chen et al. 2001; Wang et al. 2008), which appear to serve different, nonoverlapping functions (Chen and Zhou 2010). The SAX-7/LAD-1 protein has all the hallmarks of L1-type proteins (Chen et al. 2001) and when mutated or deleted causes a pleiotropic phenotype, which includes embryonic and gonadal malformations (Chen et al. 2001), the misorganization of ganglia and abnormal positioning of neuronal cells in the adult (Sasakura et al. 2005), as well as embryonic lethality, inappropriate axon trajectories, and uncoordinated movements (Wang et al. 2005). In contrast, the LAD-2 protein, which is expressed in only a subset of nematode neurons, has a traditional L1-type ectodomain, but is truncated at its cytoplasmic tail and is missing the Ankyrin-binding domain (Wang et al. 2008). Animals that are mutant for the *lad-2* gene primarily exhibit axonal pathfinding defects. These defects are caused when the LAD-2 protein is unable to fulfill its function as a co-receptor together with PLX-2/Plexin to bind MAB-20/Sema 2 protein (Wang et al. 2008).

Mus musculus

In mice, knockouts of the *L1CAM* gene (L1-KO mice) exhibit a phenotype similar to that observed in humans with L1 syndrome. These phenotypic aspects include a reduced corticospinal tract, abnormal pyramidal decussation, decreased axonal association with non-myelinating Schwann cells, ventricular dilation, and hypoplasia of the cerebellar vermis (Itoh et al. 2004; Dahme et al. 1997; Fransen et al. 1998a; Demyanenko et al. 1999, 2001). Using a co-culture in vitro assay with cells isolated from an L1-CAM-deficient mouse line, Castellani et al. revealed a role for L1-CAM in the Sema3A signaling pathway of axonal guidance (Castellani et al. 2000). Their finding suggests that some L1-CAM mutations may also disrupt Sema3A's chemorepulsive signaling activity in the growth cone (Castellani et al. 2000). This situation is reminiscent of the phenotype that has been observed for the L1-type protein LAD-2 in nematodes (Wang et al. 2008).

9.2.7 L1-CAM Expression in Cancer Cells: The Multidimensional Nature of L1-CAM Function

Expression of L1-CAM in Various Human Tumors

As described above, L1-CAM and other vertebrate L1-type proteins are primarily expressed in the nervous system (Rathjen and Schachner 1984; Sanes et al. 1986). Not surprisingly, L1-CAM expression has been reported in a number of human tumors of neuroectodermal origin, specifically gliomas and neuroblastomas (Figarella-Branger et al. 1990; Izumoto et al. 1996; Tsuzuki et al. 1998). L1-CAM expression in gliomas correlates with increased tumor invasion (Izumoto et al. 1996) and the downregulation of L1-CAM expression decreases glioma growth in a mouse model (Bao et al. 2008). Whereas L1-CAM expression in adult gliomas is associated with reduced survival, the opposite was reported for pediatric gliomas (Wachowiak et al. 2007). This suggests that L1-CAM expression has different effects in different types of neuroectodermal tumors.

Some recent publications have reported an aberrant expression of L1-CAM in several non-neuronal types of human cancer (Gavert et al. 2008). For example, L1-CAM protein is often expressed during advanced stages of colon cancer development (Raveh et al. 2009; Gavert et al. 2005). In addition, L1-CAM was found in certain ovarian cancers (Zecchini et al. 2008), renal cell carcinoma (Allory et al. 2005), human cutaneous malignant melanoma (Thies et al. 2002; Fogel et al. 2003; Linnemann et al. 1989), pancreatic adenocarcinomas (Sebens Müerköster et al. 2007), breast cancers (Valladares et al. 2006; Shtutman et al. 2006; Gutwein et al. 2000), and various lung cancer cell lines (Katayama et al. 1997). This suggests that the expression of L1-type proteins might play an important role in the development and progression of different types of cancers. The study by Gavert et al. also demonstrates that when transfected into LS174T human colon carcinoma cells,

L1-CAM expression results in an increased growth rate and cell motility and in an enhancement of the cells' tumorigenic capacity (Gavert et al. 2008). When injected into the spleen of nude mice, L1-CAM-transfected LS174T cells gained the ability to form liver metastases (Gavert et al. 2007). In colon cancer tissue samples, L1-CAM expression is correlated with higher levels of nuclear β-catenin and these cells are exclusively localized at the invasive front of the tumor tissue (Gavert et al. 2005). The authors provide ample evidence that in human colon carcinoma cells, L1-CAM is a target of aberrantly activated β-catenin–TCF signaling and increases their metastatic potential (Gavert et al. 2008).

Similar to colon carcinoma, L1-CAM has been detected at the invasive front of epithelial ovarian carcinoma tissue cells and its expression is associated with a poor clinical prognosis and increased levels of metastasis (Zecchini et al. 2008). Also in renal cell carcinoma (RCC), the presence of L1-CAM is linked to a propensity to develop metastasis, and when coupled with the loss of Cyclin D1 expression, L1-CAM was defined as an independent prognostic factor for metastasis occurrence in a multivariate analysis (Allory et al. 2005).

Other studies show that L1-CAM is also expressed in a significant portion of various histological subtypes of human cutaneous malignant melanomas. Again it is a good predictor for metastatic ability and indicates a poor prognosis in melanoma patients (Thies et al. 2002; Fogel et al. 2003; Linnemann et al. 1989). Furthermore, the downregulation of L1-CAM expression reduces the migration and invasiveness of metastatic B16 cells in vitro (Meier et al. 2006). A gene array analysis of malignant melanoma tissues demonstrated that L1-CAM RNA is expressed at more than tenfold higher levels when compared with benign lesions or normal skin samples (Talantov et al. 2005). L1-CAM expression has also been reported in neuroendocrine carcinomas of the skin (Deichmann et al. 2003).

For pancreatic adenocarcinomas, one study reported that L1-CAM was expressed in 80 % (16 of 20 samples) of all tissue sections analyzed (Sebens Müerköster et al. 2007). However, other studies reported L1-CAM expression to be less common in pancreatic adenocarcinoma (2 of 111 samples) (Kaifi et al. 2006a) and 5 out of 63 pancreatic neuroendocrine tumors (Kaifi et al. 2006b). Again L1-CAM expression in pancreatic adenocarcinomas is a valid indicator for a poor clinical prognosis (Tsutsumi et al. 2011; Chen et al. 2011). L1-CAM is also detectable in some breast cancers (Valladares et al. 2006; Shtutman et al. 2006; Gutwein et al. 2000). In MCF7 breast cancer cells, L1-CAM expression disrupts E-cadherin-containing adherens junctions and increases cell scattering and motility (Shtutman et al. 2006). This correlates with the observation that L1-CAM expression is more abundant in metastatic breast cancer cells and lower in non-metastatic breast cancer cells (Valladares et al. 2006).

L1-CAM and Neurofascin proteins both contain an RGD motif in their sixth Ig or third FN III domain, respectively, both of which have been shown to interact with RGD-specific integrins (Ruppert et al. 1995; Montgomery et al. 1996; Felding-Habermann et al. 1997; Blaess et al. 1998). The RGD motif in the sixth L1-CAM Ig domain appears to increase the incident of metastasis formation in some tumor cells expressing L1-CAM. An investigation by Duczmal et al. showed that the L1-CAM-mediated migration of human MED-B1 tumor cells is RGD dependent and can be blocked by αγβ3 integrin-specific antibodies (Duczmal et al. 1997). The evidence that RGD motifs in Neurofascins and L1-CAMs can serve as ligands for integrins has led researchers to argue for a functional significance of this interaction in L1-mediated tumor progression (Duczmal et al. 1997; Montgomery et al. 1996).

L1-Type Protein Expression Affects Signaling Pathways in Cancer Cells

L1-CAM and other vertebrate L1-type proteins play an active role in cancer development by affecting different signaling pathways and thereby contributing directly to tumor progression. Several groups have demonstrated that in ovarian cancer cells L1-CAM-dependent ERK activation is associated with increased FGFR, EGFR, and hepatocyte growth factor receptor (HGFR) activity (Zecchini et al. 2008; Stoeck et al. 2007; Novak-Hofer et al. 2008). These and a number of other studies suggest that L1-CAM plays a role in carcinogenesis by its ability to activate the ERK signaling pathway (Gast ct al. 2008; Schaefer et al. 1999; Silletti et al. 2004; Schmid et al. 2000). L1-CAM itself is also phosphorylated by Erk2 and interacts with different components of the ERK pathway (Silletti et al. 2004; Schmid et al. 2000). This ultimately results in the expression of proteins that contribute to cell motility and cell invasion. In addition, L1-CAM-mediated ERK activation was shown to involve Src (Gast et al. 2008; Silletti et al. 2004). The reported association of the L1-CAM cytoplasmic domain with RamBPM also indicates a direct link between L1-CAM expression and the MAPK/ERK signaling pathway (Cheng et al. 2005b).

The co-expression of ADAM10 (A Disintegrin and Metalloproteinase domain-containing protein 10) and L1-CAM in invasive colon cancer tumor cell indicates that the proteolytic processing of L1-CAM may have a role in invasive tumor development (Gavert et al. 2007). The L1-CAM protein is often cleaved by MMPs and the shedded L1-CAM ectodomain can interact with integrins, RTKs, or L1-CAM on the surface of the same or of neighboring cells (Gavert et al. 2007). Similarly, the ectodomain of the Nr-CAM protein is cleaved by matrix metalloproteinases, which results in an enhancement of cell motility, proliferation, ERK and AKT activation, and ultimately oncogenesis (Conacci-Sorrell et al. 2005).

9.3 Mutant Phenotypes Associated with Vertebrate L1-CAM Paralogs

9.3.1 Neurofascin

Neurofascin is one of three vertebrate L1-CAM paralogs and is also primarily expressed in the nervous system (Rathjen et al. 1987). The multiple protein isoforms that are expressed from the *neurofascin* gene vary in the number of their FN III extracellular domains (Hassel et al. 1997). Some Neurofascin protein isoforms also contain an unusual PAT domain, which is usually positioned close to the transmembrane segment (see Fig. 9.1) (Volkmer et al. 1992).

Two major Neurofascin isoforms, Nfasc155 and Nfasc186, are expressed at nodes of Ranvier in myelinated axons (Tait et al. 2000). The neuronal isoform Nfasc186 is required for the clustering of voltage-gated Na^+ channels at nodes of Ranvier (Howell et al. 2006). It thereby controls rapid impulse conduction in these axons (Zonta et al. 2008). In contrast, Nfasc155 is a glia cell-specific protein isoform and is required for the correct assembly of paranodal junctions (Sherman et al. 2005). In *neurofascin* null mice, neither paranodal adhesion junctions nor nodal complexes are formed (Sherman et al. 2005). This demonstrates the essential function of these two major Neurofascin protein isoforms for the formation of these structures. Nfasc155 null mutant mice exhibit severe ataxia, motor paralysis, and death before the third postnatal week (Pillai et al. 2009). In the absence of glia cell-specific Nfasc155, paranodal axonal junctions fail to form, axonal domains do not segregate, and myelinated axons undergo degeneration. Furthermore, in vivo deletion of Neurofascin Ig domains 5 and 6 reveals a requirement for specific Neurofascin protein domains in myelinated axons (Thaxton et al. 2010).

In cases of multiple sclerosis, a disruption of Neurofascin localization at nodes of Ranvier often appears to precede subsequent demyelination (Howell et al. 2006; Lonigro and Devaux 2009). Both Nfasc186 and Nfasc155 proteins have been found in areas of inflammation, demyelination, and remyelination in postmortem brains of multiple sclerosis patients (Howell et al. 2006). Mathey et al. identified Neurofascin as an autoimmune target in patients with multiple sclerosis (Mathey et al. 2007). This suggests a direct involvement of Neurofascin in immune-mediated axonal injury. The alteration of oligodendrocyte Nfasc155 expression that accompanies inflammation and demyelination processes indicates a chronic disruption of the axon–glia cell interaction. This will eventually result in the destruction of the Nfasc186/Na^+_v nodal complexes (Howell et al. 2006). Recently, the involvement in the progression of multiple sclerosis of two newly described protein isoforms of Nfasc155 has been analyzed in more detail (Pomicter et al. 2010). Using conditional knockout mice the authors show that Nfasc155 high and Nfasc155 low are exclusively expressed by oligodendrocytes within the CNS. The timing and expression levels of these two Nfasc155 isoforms are distinctly regulated. Nfasc155 low is incapable of preserving paranodal structures, thus indicating that Nfasc155 high is required for paranodal stability. Comparisons between Nfasc155 high and Nfasc155 low in human samples revealed significant alterations in multiple sclerosis plaques (Pomicter et al. 2010).

9.3.2 NrCAM

NrCAM (neuron–glia-related cell adhesion molecule) was the third vertebrate L1-type gene/protein to be identified and like its paralogs is primarily expressed in the nervous system (Grumet et al. 1991; Grumet 1997). In mice, lack of NrCAM has been implicated in the formation of lens cataracts (More et al. 2001). It is also involved in the formation of nodes of Ranvier (Custer et al. 2003) and in addiction vulnerability (Ishiguro et al. 2006). In humans, NrCAM has been associated with autism (Sakurai et al. 2006; Bonora et al. 2005) and is overexpressed in papillary thyroid carcinomas (Gorka et al. 2007).

The absence of NrCAM causes the formation of cataracts in murine lenses (More et al. 2001). In NrCAM-deficient mice, the authors observed a general disorganization of lens fibers with ensuing cellular disintegration and an accumulation of cellular debris. This mirrors the phenotype found in Ankyrin-B-deficient mice and points to an important interaction between NrCAM and Ankyrin-B in lens fiber cells (More et al. 2001). Similar to Neurofascin, NrCAM is also expressed at nodes of Ranvier and is implicated in node formation and maintenance (Custer et al. 2003). Na^+ channels and Ankyrin G sequestration at developing nodes is delayed in NrCAM null mutant mice.

The genetic mapping of a locus involved in substance abuse vulnerabilities to mouse chromosome 7 identified a positive linkage with several NrCAM haplotypes (Ishiguro et al. 2006). Differential gene display identified the *NrCAM* gene as a drug-regulated gene that is expressed in neurons linked to reward and memory (Ishiguro et al. 2006). NrCAM knockout mice exhibit reduced opiate- and stimulant-conditional preferences. These observations suggest that in humans NrCAM may also be a drug-regulated gene, whose variants are likely linked to vulnerabilities in drug addiction and reward. In another study, wild-type, heterozygous, and NrCAM null mice were tested for a cognitive and behavioral phenotype (Matzel et al. 2008). These different genotypes were assessed using five different learning tasks (such as Lashley maze, odor discrimination, passive avoidance, spatial water maze, and fear conditioning). NrCAM null mutant mice are viable, have normal body weight, and exhibit normal levels of general activity. However, they display an increased propensity to enter stressful areas of novel environments, exhibit higher sensitivity to pain (hot), and are more sensitive to the aversive effects of foot shock. This behavioral phenotype suggests that NrCAM might play a central role in the regulation of general cognitive abilities and might serve a critical function in regulating impulsivity as well as susceptibly to drug abuse and addiction. In humans, two genetic linkage studies have identified the *NrCAM* gene as a potential candidate to be associated with autism susceptibility and with substance abuse (Sakurai et al. 2006; Bonora et al. 2005). Together with the results obtained from the mouse models, which are cited above, this indicates that NrCAM has an important function in the formation and/or maintenance of the brain's reward circuitry.

Like L1-CAM and Neurofascin, NrCAM is also overexpressed in at least one type of human cancer, specifically papillary thyroid carcinomas (PTCs) (Gorka et al. 2007). The level of NrCAM mRNA and protein overexpression in tumor tissues appears to be independent of the primary tumor stage (PT) or its size. How NrCAM induction and upregulation might potentially influence the pathogenesis and the behavior of papillary thyroid cancer cells still remains to be evaluated.

9.3.3 CHL1

The Close Homologue of L1, CHL1 (or *CALL* for Cell Adhesion L1-Like), is the fourth vertebrate L1-type paralog (Holm et al. 1996) and is located on human chromosome 3p26.1 (Wei et al. 1998). CHL1 protein promotes neurite outgrowth (Holm et al. 1996; Hillenbrand et al. 1999), and the gene expression of *CHL1* and *L1CAM* in the mouse and rat nervous system shows overlapping but distinct patterns in neuronal and glia cell populations (Hillenbrand et al. 1999). In contrast to the other three vertebrate L1 paralogs, relatively little is known about CHL1's physiological and molecular functions. Nevertheless, the *CHL1* gene appears to be associated with several interesting genetic conditions and phenotypes.

In cortical slices from CHL1 knockout mice the migration of cortical neurons proceeds at a slower rate of radial migration and migratory cells accumulate in the intermediate and ventricular/subventricular zones (Demyanenko et al. 2004). In neocortical areas, especially in the visual and somatosensory cortex, CHL1 appears to regulate neuronal connectivity (Demyanenko et al. 2004). CHL1 also has a role in regulating the uncoating of Clathrin-coated synaptic vesicles (Leshchyns'ka et al. 2006). CHL1 deficiency or disruption of the CHL1/Hsc70 complex results in an accumulation of abnormally high levels of Clathrin-coated synaptic vesicles. These observed abnormalities of Clathrin-dependent synaptic vesicle recycling have the potential to cause or to contribute to brain malfunctions in humans and mice that carry mutations in their *CHL1* gene.

Nikonenko et al. (2006) reported an enhanced perisomatic inhibition and impaired long-term potentiation in the CA1 region of juvenile CHL1-deficient mice. These authors analyzed the functional role of CHL1 in the synaptic transmission in the CA1 region hippocampus comparing juvenile CHL1-deficient and wild-type mice. The inhibitory postsynaptic currents evoked in pyramidal cells by minimal stimulation of perisomatically projecting interneurons were increased in mice lacking CHL1 when compared with wild-type littermates. Also, the long-term potentiation (LTP) at CA3–CA1 excitatory synapses was reduced under physiological conditions in CHL1-deficient mice. A quantitative immunohistochemical analysis revealed that CA1 interneurons usually express CHL1 protein. This suggests that CHL1 is important for the regulation of inhibitory synaptic transmission in this and potentially other interneuron populations. This observed enhancement of inhibitory transmission in CHL1-deficient mice contrasts with the previous finding of a reduced inhibition in L1-CAM-deficient mice and illustrates a functional difference between these two paralogous L1-type adhesion molecules (Nikonenko et al. 2006).

Mutations in the murine *CHL1* gene have been found to alter the connectivity and morphology of several brain regions (Heyden et al. 2008). In addition, CHL1 acts in a gene dosage-dependent manner to control murine brain development and to influence behavior and cognitive abilities. Mice deficient for CHL1 display alterations in emotional reactivity and motor coordination (Pratte et al. 2003). These mice also display signs of decreased stress and a modification of exploratory behavior and show impairments in a Rotarod test. However, they were able to move as fast as control mice in a T-maze test. The observed changes have been attributed to an attention deficit. CHL1-deficient mice have normal learning abilities, but exhibit a widespread impairment in working memory duration (Kolata et al. 2008). Montag-Saliaz et al. demonstrated that the absence of CHL1 in mice results in aberrant hippocampal mossy fiber and olfactory nerve projections, which might explain the reduced reactivity towards novel environments that is exhibited by these mice (Montag-Sallaz et al. 2003). Together, these observations that have been made using CHL1-deficient mouse models suggest an important role for CHL1 in short-term memory retention in the adult brain.

Similar to the other three vertebrate L1-type CAMs, mutations in the *CHL1* gene and protein have been implicated in several human disease conditions. *CHL1* has been identified as a prime candidate gene for an autosomal form of mental retardation and a translocation breakpoint in intron five of the *CHL1* gene at 46,Y, t(X;3)(p22.1;p26.3) was described in a man with nonspecific mental retardation (Frints et al. 2003). In addition, a haploinsufficiency for the *CHL1* gene has been reported in a mentally retarded patient with 3p-syndrome (Angeloni et al. 1999). A missense mutation (Leu17Phe) in the signal peptide of CHL1 exhibits a positive association with the occurrence of schizophrenia in a group of 282 Japanese patients (Sakurai et al. 2002). This association of CHL1 with schizophrenia was later confirmed by a second study of 560 schizophrenia cases and 576 controls in a Han Chinese population (Chen et al. 2005). These two reports indicate that CHL1 might somehow be involved in the etiology of schizophrenia.

9.4 Conclusions

L1-type genes and their protein products appeared rather early during metazoan evolution when the first primitive neuronal nets became part of animal body plans. From the beginning, the overall L1-type protein domain structure, a predominant expression of L1-type genes in neural cells, and many characteristic protein–protein interactions (both homo- as well as heterophilic) and molecular functions have been remarkably well conserved. However, gene duplication events have allowed for some variations in the L1 protein structure and for the development of novel molecular interactions and physiological functions (Mualla et al. 2013). Vertebrate species usually contain four L1-type genes in their genome, designated as *L1-CAM*, *Neurofascin*, *NrCAM*, and *CHL1*. Loss-of-function conditions have been studied in several animal model systems and usually include a broad range of neurological dysfunctions. In humans, mutations in the *L1CAM* gene have been analyzed in great detail as they cause an X-linked recessive disorder, now known as L1 syndrome. Well over 200 different *L1-CAM* mutations have been identified in individual families. The severity of the phenotype usually correlates with the impact of the specific mutation on the structure of the L1-CAM protein and its interactions with other proteins. However, epigenetic effects also contribute to a significant variability of specific phenotypic aspects.

Though L1-type genes are predominantly expressed in the nervous system, their protein products have also been identified in a number of other tissues, specifically leukocytes and epithelia. Their physiological functions in these non-neural tissues still remain poorly understood. Overexpression of L1-type proteins has been reported in a wide range of different human tumors, where they affect different signaling pathways and contribute to tumor progression and metastasis. As our understanding of the multifaceted normal physiological role of these important proteins is still incomplete and as both loss-of-function and gain-of-function conditions for L1-type genes/proteins cause clinically relevant disorders, the published findings paint a complex and sometimes confusing picture. Certainly, more research is needed for a more complete insight into the physiological role of L1-type CAMs and how they are implicated into various pathological processes.

Compliance with Ethics Requirements The authors declare that they have no conflicts of interest.

References

Allory Y, Matsuoka Y, Bazille C, Christensen EI, Ronco P, Debiec H (2005) The L1 cell adhesion molecule is induced in renal cancer cells and correlates with metastasis in clear cell carcinomas. Clin Cancer Res 11(3):1190–1197

Amores A, Force A, Yan YL, Joly L, Amemiya C, Fritz A, Ho RK, Langeland J, Prince V, Wang YL, Westerfield M, Ekker M, Postlethwait JH (1998) Zebrafish hox clusters and vertebrate genome evolution. Science 282(5394):1711–1714

Anderson RB, Turner KN, Nikonenko AG, Hemperly J, Schachner M, Young HM (2006) The cell adhesion molecule l1 is required for chain migration of neural crest cells in the developing mouse gut. Gastroenterology 130(4):1221–1232. doi:10.1053/j.gastro.2006.01.002, S0016-5085(06)00003-5 [pii]

Angeloni D, Lindor NM, Pack S, Latif F, Wei MH, Lerman MI (1999) CALL gene is haploinsufficient in a 3p-syndrome patient. Am J Med Genet 86(5):482–485

Ango F, di Cristo G, Higashiyama H, Bennett V, Wu P, Huang ZJ (2004) Ankyrin-based subcellular gradient of neurofascin, an immunoglobulin family protein, directs GABAergic innervation at purkinje axon initial segment. Cell 119(2):257–272

Asou H, Miura M, Kobayashi M, Uyemura K, Itoh K (1992) Cell adhesion molecule L1 guides cell migration in primary reaggregation cultures of mouse cerebellar cells. Neurosci Lett 144(1–2):221–224

Balaian LB, Moehler T, Montgomery AM (2000) The human neural cell adhesion molecule L1 functions as a costimulatory molecule in T cell activation. Eur J Immunol 30(3):938–943

Banerjee S, Pillai AM, Paik R, Li J, Bhat MA (2006) Axonal ensheathment and septate junction formation in the peripheral nervous system of *Drosophila*. J Neurosci 26(12):3319–3329

Bao S, Wu Q, Li Z, Sathornsumetee S, Wang H, McLendon RE, Hjelmeland AB, Rich JN (2008) Targeting cancer stem cells through L1CAM suppresses glioma growth. Cancer Res 68(15):6043–6048. doi:10.1158/0008-5472.CAN-08-1079, 68/15/6043 [pii]

Bartsch U (2003) Neural CAMs and their role in the development and organization of myelin sheaths. Front Biosci 8:D477–D490

Bateman A, Jouet M, MacFarlane J, Du JS, Kenwrick S, Chothia C (1996) Outline structure of the human L1 cell adhesion molecule and the sites where mutations cause neurological disorders. EMBO J 15(22):6050–6059

Becker CG, Lieberoth BC, Morellini F, Feldner J, Becker T, Schachner M (2004) L1.1 is involved in spinal cord regeneration in adult zebrafish. J Neurosci 24(36):7837–7842

Bernreuther C, Dihne M, Johann V, Schiefer J, Cui Y, Hargus G, Schmid JS, Xu J, Kosinski CM, Schachner M (2006) Neural cell adhesion molecule L1-transfected embryonic stem cells promote functional recovery after excitotoxic lesion of the mouse striatum. J Neurosci 26(45):11532–11539

Bertolin C, Boaretto F, Barbon G, Salviati L, Lapi E, Divizia MT, Garavelli L, Occhi G, Vazza G, Mostacciuolo ML (2010) Novel mutations in the L1CAM gene support the complexity of L1 syndrome. J Neurol Sci 294(1–2):124–126. doi:10.1016/j.jns.2010.03.030, S0022-510X(10)00156-5 [pii]

Bickers DS, Adams RD (1949) Hereditary stenosis of the aqueduct of Sylvius as a casue of congenital hydrocephalus. Brain 72:246–262

Bieber AJ, Snow PM, Hortsch M, Patel NH, Jacobs JR, Traquina ZR, Schilling J, Goodman CS (1989) *Drosophila* neuroglian: a member of the immunoglobulin superfamily with extensive homology to the vertebrate neural adhesion molecule L1. Cell 59(3):447–460

Blaess S, Kammerer RA, Hall H (1998) Structural analysis of the sixth immunoglobulin-like domain of mouse neural cell adhesion molecule L1 and its interactions with alpha(v)beta3, alpha(IIb)beta3, and alpha5beta1 integrins. J Neurochem 71(6):2615–2625

Bonora E, Lamb JA, Barnby G, Sykes N, Moberly T, Beyer KS, Klauck SM, Poustka F, Bacchelli E, Blasi F, Maestrini E, Battaglia A, Haracopos D, Pedersen L, Isager T, Eriksen G, Viskum B, Sorensen EU, Brondum-Nielsen K, Cotterill R, Engeland H, Jonge M, Kemner C, Steggehuis K, Scherpenisse M, Rutter M, Bolton PF, Parr JR, Poustka A, Bailey AJ, Monaco AP (2005) Mutation screening and association analysis of six candidate genes for autism on chromosome 7q. Eur J Hum Genet 13(2):198–207. doi:10.1038/sj.ejhg.5201315, 5201315 [pii]

Buhusi M, Schlatter MC, Demyanenko GP, Thresher R, Maness PF (2008) L1 interaction with ankyrin regulates mediolateral topography in the retinocollicular projection. J Neurosci 28(1):177–188

Carhan A, Allen F, Armstrong JD, Hortsch M, Goodwin SF, O'Dell KM (2005) Female receptivity phenotype of icebox mutants caused by a mutation in the L1-type cell adhesion molecule neuroglian. Genes Brain Behav 4(8):449–465

Castellani V, Chedotal A, Schachner M, Faivre-Sarrailh C, Rougon G (2000) Analysis of the L1-deficient mouse phenotype reveals cross-talk between Sema3A and L1 signaling pathways in axonal guidance. Neuron 27(2):237–249

Charles P, Tait S, Faivre-Sarrailh C, Barbin G, Gunn-Moore F, Denisenko-Nehrbass N, Guennoc AM, Girault JA, Brophy PJ, Lubetzki C (2002) Neurofascin is a glial receptor for the paranodin/Caspr-contactin axonal complex at the axoglial junction. Curr Biol 12(3):217–220

Chen W, Hing H (2008) The L1-CAM, neuroglian, functions in glial cells for Drosophila antennal lobe development. Dev Neurobiol 68(8):1029–1045. doi:10.1002/dneu.20644

Chen L, Zhou S (2010) "CRASH"ing with the worm: insights into L1CAM functions and mechanisms. Dev Dyn 239(5):1490–1501. doi:10.1002/dvdy.22269

Chen L, Ong B, Bennett V (2001) LAD-1, the *Caenorhabditis elegans* L1CAM homologue, participates in embryonic and gonadal morphogenesis and is a substrate for fibroblast growth factor receptor pathway-dependent phosphotyrosine-based signaling. J Cell Biol 154(4): 841–856

Chen QY, Chen Q, Feng GY, Lindpaintner K, Chen Y, Sun X, Chen Z, Gao Z, Tang J, He L (2005) Case-control association study of the close homologue of L1 (CHL1) gene and schizophrenia in the Chinese population. Schizophr Res 73(2–3):269–274

Chen J, Wu J, Apostolova I, Skup M, Irintchev A, Kugler S, Schachner M (2007) Adeno-associated virus-mediated L1 expression promotes functional recovery after spinal cord injury. Brain 130(Pt 4):954–969. doi:10.1093/brain/awm049, 130/4/954 [pii]

Chen MM, Lee CY, Leland HA, Silletti S (2011) Modification of the L1-CAM carboxy-terminus in pancreatic adenocarcinoma cells. Tumour Biol 32(2):347–357. doi:10.1007/s13277-010-0127-4

Cheng L, Lemmon V (2004) Pathological missense mutations of neural cell adhesion molecule L1 affect neurite outgrowth and branching on an L1 substrate. Mol Cell Neurosci 27(4):522–530

Cheng L, Itoh K, Lemmon V (2005a) L1-mediated branching is regulated by two ezrin-radixin-moesin (ERM)-binding sites, the RSLE region and a novel juxtamembrane ERM-binding region. J Neurosci 25(2):395–403

Cheng L, Lemmon S, Lemmon V (2005b) RanBPM is an L1-interacting protein that regulates L1-mediated mitogen-activated protein kinase activation. J Neurochem 94(4):1102–1110. doi:10.1111/j.1471-4159.2005.03254.x, JNC3254 [pii]

Christoffels A, Koh EG, Chia JM, Brenner S, Aparicio S, Venkatesh B (2004) Fugu genome analysis provides evidence for a whole-genome duplication early during the evolution of ray-finned fishes. Mol Biol Evol 21(6):1146–1151. doi:10.1093/molbev/msh114msh114 [pii]

Collinson JM, Marshall D, Gillespie CS, Brophy PJ (1998) Transient expression of neurofascin by oligodendrocytes at the onset of myelinogenesis: implications for mechanisms of axon-glial interaction. Glia 23(1):11–23

Conacci-Sorrell M, Kaplan A, Raveh S, Gavert N, Sakurai T, Ben-Ze'ev A (2005) The shed ectodomain of Nr-CAM stimulates cell proliferation and motility, and confers cell transformation. Cancer Res 65(24):11605–11612

Custer AW, Kazarinova-Noyes K, Sakurai T, Xu X, Simon W, Grumet M, Shrager P (2003) The role of the ankyrin-binding protein NrCAM in node of Ranvier formation. J Neurosci 23(31):10032–10039

Dahme M, Bartsch U, Martini R, Anliker B, Schachner M, Mantei N (1997) Disruption of the mouse L1 gene leads to malformations of the nervous system. Nat Genet 17(3):346–349

Davis JQ, McLaughlin T, Bennett V (1993) Ankyrin-binding proteins related to nervous system cell adhesion molecules: candidates to provide transmembrane and intercellular connections in adult brain. J Cell Biol 121(1):121–133

Davis JQ, Lambert S, Bennett V (1996) Molecular composition of the node of Ranvier: Identification of ankyrin-binding cell adhesion molecules neurofascin (mucin+/third FNIII domain-) and NrCAM at nodal axon segments. J Cell Biol 135(5):1355–1367

De Angelis E, MacFarlane J, Du JS, Yeo G, Hicks R, Rathjen FG, Kenwrick S, Brümmendorf T (1999) Pathological missense mutations of neural cell adhesion molecule L1 affect homophilic and heterophilic binding activities. EMBO J 18(17):4744–4753

De Angelis E, Brümmendorf T, Cheng L, Lemmon V, Kenwrick S (2001) Alternative use of a mini exon of the L1 gene affects L1 binding to neural ligands. J Biol Chem 276(35):32738–32742

De Angelis E, Watkins A, Schäfer M, Brümmendorf T, Kenwrick S (2002) Disease-associated mutations in L1 CAM interfere with ligand interactions and cell-surface expression. Hum Mol Genet 11(1):1–12

Debiec H, Christensen EI, Ronco PM (1998) The cell adhesion molecule L1 is developmentally regulated in the renal epithelium and is involved in kidney branching morphogenesis. J Cell Biol 143(7):2067–2079

Deichmann M, Kurzen H, Egner U, Altevogt P, Hartschuh W (2003) Adhesion molecules CD171 (L1CAM) and CD24 are expressed by primary neuroendocrine carcinomas of the skin (Merkel cell carcinomas). J Cutan Pathol 30(6):363–368

Demyanenko GP, Maness PF (2003) The L1 cell adhesion molecule is essential for topographic mapping of retinal axons. J Neurosci 23(2):530–538

Demyanenko GP, Tsai AY, Maness PF (1999) Abnormalities in neuronal process extension, hippocampal development, and the ventricular system of L1 knockout mice. J Neurosci 19(12):4907–4920

Demyanenko GP, Shibata Y, Maness PF (2001) Altered distribution of dopaminergic neurons in the brain of L1 null mice. Brain Res Dev Brain Res 126(1):21–30

Demyanenko GP, Schachner M, Anton E, Schmid R, Feng G, Sanes J, Maness PF (2004) Close homolog of L1 modulates area-specific neuronal positioning and dendrite orientation in the cerebral cortex. Neuron 44(3):423–437

Demyanenko GP, Halberstadt AI, Rao RS, Maness PF (2009) CHL1 cooperates with PAK1-3 to regulate morphological differentiation of embryonic cortical neurons. Neuroscience 165(1): 107–115

Di Sciullo G, Donahue T, Schachner M, Bogen SA (1998) L1 antibodies block lymph node fibroblastic reticular matrix remodeling *in vivo*. J Exp Med 187(12):1953–1963

Dietrich A, Korn B, Poustka A (1992) Completion of the physical map of Xq28: the location of the gene for L1CAM on the human X chromosome. Mamm Genome 3(3):168–172
Dihne M, Bernreuther C, Sibbe M, Paulus W, Schachner M (2003) A new role for the cell adhesion molecule L1 in neural precursor cell proliferation, differentiation, and transmitter-specific subtype generation. J Neurosci 23(16):6638–6650
Dubreuil RR, MacVicar G, Dissanayake S, Liu C, Homer D, Hortsch M (1996) Neuroglian-mediated cell adhesion induces assembly of the membrane skeleton at cell contact sites. J Cell Biol 133(3):647–655
Duczmal A, Schollhammer S, Katich S, Ebeling O, Schwartz-Albiez R, Altevogt P (1997) The L1 adhesion molecule supports alpha v beta 3-mediated migration of human tumor cells and activated T lymphocytes. Biochem Biophys Res Commun 232(1):236–239
Ebeling O, Duczmal A, Aigner S, Geiger C, Schöllhammer S, Kemshead JT, Möller P, Schwartz-Albiez R, Altevogt P (1996) L1 adhesion molecule on human lymphocytes and monocytes: expression and involvement in binding to alpha v beta 3 integrin. Eur J Immunol 26(10): 2508–2516
Faivre-Sarrailh C, Banerjee S, Li J, Hortsch M, Laval M, Bhat MA (2004) *Drosophila* contactin, a homolog of vertebrate contactin, is required for septate junction organization and paracellular barrier function. Development 131(20):4931–4942
Felding-Habermann B, Silletti S, Mei F, Siu CH, Yip PM, Brooks PC, Cheresh DA, O'Toole TE, Ginsberg MH, Montgomery AM (1997) A single immunoglobulin-like domain of the human neural cell adhesion molecule L1 supports adhesion by multiple vascular and platelet integrins. J Cell Biol 139(6):1567–1581
Felsenstein J (1981) Evolutionary trees from DNA sequences: a maximum likelihood approach. J Mol Evol 17(6):368–376
Figarella-Branger DF, Durbec PL, Rougon GN (1990) Differential spectrum of expression of neural cell adhesion molecule isoforms and L1 adhesion molecules on human neuroectodermal tumors. Cancer Res 50(19):6364–6370
Finckh U, Schroder J, Ressler B, Veske A, Gal A (2000) Spectrum and detection rate of L1CAM mutations in isolated and familial cases with clinically suspected L1-disease. Am J Med Genet 92(1):40–46
Fogel M, Mechtersheimer S, Huszar M, Smirnov A, Abu-Dahi A, Tilgen W, Reichrath J, Georg T, Altevogt P, Gutwein P (2003) L1 adhesion molecule (CD 171) in development and progression of human malignant melanoma. Cancer Lett 189(2):237–247
Fransen E, Schrander-Stumpel C, Vits L, Coucke P, Van Camp G, Willems PJ (1994) X-linked hydrocephalus and MASA syndrome present in one family are due to a single missense mutation in exon 28 of the L1CAM gene. Hum Mol Genet 3(12):2255–2256
Fransen E, Lemmon V, Van Camp G, Vits L, Coucke P, Willems PJ (1995) CRASH syndrome: clinical spectrum of corpus callosum hypoplasia, retardation, adducted thumbs, spastic paraparesis and hydrocephalus due to mutations in one single gene, L1. Eur J Hum Genet 3(5):273–284
Fransen E, Van Camp G, Vits L, Willems PJ (1997) L1-associated diseases: clinical geneticists divide, molecular geneticists unite. Hum Mol Genet 6(10):1625–1632
Fransen E, D'Hooge R, Van Camp G, Verhoye M, Sijbers J, Reyniers E, Soriano P, Kamiguchi H, Willemsen R, Koekkoek SK, De Zeeuw CI, De Deyn PP, Van der Linden A, Lemmon V, Kooy RF, Willems PJ (1998a) L1 knockout mice show dilated ventricles, vermis hypoplasia and impaired exploration patterns. Hum Mol Genet 7(6):999–1009
Fransen E, Van Camp G, D'Hooge R, Vits L, Willems PJ (1998b) Genotype-phenotype correlation in L1 associated diseases. J Med Genet 35(5):399–404
Frints SG, Marynen P, Hartmann D, Fryns JP, Steyaert J, Schachner M, Rolf B, Craessaerts K, Snellinx A, Hollanders K, D'Hooge R, De Deyn PP, Froyen G (2003) CALL interrupted in a patient with non-specific mental retardation: gene dosage-dependent alteration of murine brain development and behavior. Hum Mol Genet 12(13):1463–1474
Fryns JP, Spaepen A, Cassiman JJ, van den Berghe H (1991) X linked complicated spastic paraplegia, MASA syndrome, and X linked hydrocephalus owing to congenital stenosis of the

aqueduct of Sylvius: variable expression of the same mutation at Xq28. J Med Genet 28(6):429–431

Garcia-Alonso L, Romani S, Jimenez F (2000) The EGF and FGF receptors mediate neuroglian function to control growth cone decisions during sensory axon guidance in *Drosophila*. Neuron 28(3):741–752

Garver TD, Ren Q, Tuvia S, Bennett V (1997) Tyrosine phosphorylation at a site highly conserved in the L1 family of cell adhesion molecules abolishes ankyrin binding and increases lateral mobility of neurofascin. J Cell Biol 137(3):703–714

Gast D, Riedle S, Issa Y, Pfeifer M, Beckhove P, Sanderson MP, Arlt M, Moldenhauer G, Fogel M, Kruger A, Altevogt P (2008) The cytoplasmic part of L1-CAM controls growth and gene expression in human tumors that is reversed by therapeutic antibodies. Oncogene 27(9): 1281–1289. doi:10.1038/sj.onc.1210747, 1210747 [pii]

Gavert N, Conacci-Sorrell M, Gast D, Schneider A, Altevogt P, Brabletz T, Ben-Ze'ev A (2005) L1, a novel target of beta-catenin signaling, transforms cells and is expressed at the invasive front of colon cancers. J Cell Biol 168(4):633–642

Gavert N, Sheffer M, Raveh S, Spaderna S, Shtutman M, Brabletz T, Barany F, Paty P, Notterman D, Domany E, Ben-Ze'ev A (2007) Expression of L1-CAM and ADAM10 in human colon cancer cells induces metastasis. Cancer Res 67(16):7703–7712

Gavert N, Ben-Shmuel A, Raveh S, Ben-Ze'ev A (2008) L1-CAM in cancerous tissues. Expert Opin Biol Ther 8(11):1749–1757. doi:10.1517/14712598.8.11.1749

Genova JL, Fehon RG (2003) Neuroglian, gliotactin, and the Na^+/K^+ ATPase are essential for septate junction function in *Drosophila*. J Cell Biol 161(5):979–989

Godenschwege TA, Kristiansen LV, Uthaman SB, Hortsch M, Murphey RK (2006) A conserved role for Drosophila Neuroglian and human L1-CAM in central-synapse formation. Curr Biol 16(1):12–23

Goossens T, Kang YY, Wuytens G, Zimmermann P, Callaerts-Vegh Z, Pollarolo G, Islam R, Hortsch M, Callaerts P (2011) The Drosophila L1CAM homolog Neuroglian signals through distinct pathways to control different aspects of mushroom body axon development. Development 138(8):1595–1605. doi:10.1242/dev.052787, dev.052787 [pii]

Gorka B, Skubis-Zegadlo J, Mikula M, Bardadin K, Paliczka E, Czarnocka B (2007) NrCAM, a neuronal system cell-adhesion molecule, is induced in papillary thyroid carcinomas. Br J Cancer 97(4):531–538

Griseri P, Vos Y, Giorda R, Gimelli S, Beri S, Santamaria G, Mognato G, Hofstra RM, Gimelli G, Ceccherini I (2009) Complex pathogenesis of Hirschsprung's disease in a patient with hydrocephalus, vesico-ureteral reflux and a balanced translocation t(3;17)(p12;q11). Eur J Hum Genet 17(4):483–490. doi:10.1038/ejhg.2008.191, ejhg2008191 [pii]

Grumet M (1997) Nr-CAM: a cell adhesion molecule with ligand and receptor functions. Cell Tissue Res 290(2):423–428

Grumet M, Mauro V, Burgoon MP, Edelman GM, Cunningham BA (1991) Structure of a new nervous system glycoprotein, Nr-CAM, and relationship to subgroups of neural cell adhesion molecules. J Cell Biol 113(6):1399–1412

Gu SM, Orth U, Veske A, Enders H, Klünder K, Schlösser M, Engel W, Schwinger E, Gal A (1996) Five novel mutations in the L1CAM gene in families with X linked hydrocephalus. J Med Genet 33(2):103–106

Guan H, Maness PF (2010) Perisomatic GABAergic innervation in prefrontal cortex is regulated by ankyrin interaction with the L1 cell adhesion molecule. Cereb Cortex 20(11):2684–2693

Gutwein P, Oleszewski M, Mechtersheimer S, Agmon-Levin N, Krauss K, Altevogt P (2000) Role of Src kinases in the ADAM-mediated release of L1 adhesion molecule from human tumor cells. J Biol Chem 275(20):15490–15497

Hall SG, Bieber AJ (1997) Mutations in the *Drosophila* neuroglian cell adhesion molecule affect motor neuron pathfinding and peripheral nervous system patterning. J Neurobiol 32(3): 325–340

Haney CA, Sahenk Z, Li C, Lemmon VP, Roder J, Trapp BD (1999) Heterophilic binding of L1 on unmyelinated sensory axons mediates Schwann cell adhesion and is required for axonal survival. J Cell Biol 146(5):1173–1184

Harpaz Y, Chothia C (1994) Many of the immunoglobulin superfamily domains in cell adhesion molecules and surface receptors belong to a new structural set which is close to that containing variable domains. J Mol Biol 238(4):528–539

Harper SJ, Bolsover SR, Walsh FS, Doherty P (1994) Neurite outgrowth stimulated by L1 requires calcium influx into neurons but is not associated with changes in steady state levels of calcium in growth cones. Cell Adhes Commun 2(5):441–453

Hassel B, Rathjen FG, Volkmer H (1997) Organization of the neurofascin gene and analysis of developmentally regulated alternative splicing. J Biol Chem 272(45):28742–28749

He Y, Jensen GJ, Bjorkman PJ (2009) Cryo-electron tomography of homophilic adhesion mediated by the neural cell adhesion molecule L1. Structure 17(3):460–471. doi:10.1016/j.str.2009.01.009, S0969-2126(09)00073-2 [pii]

Heyden A, Angenstein F, Sallaz M, Seidenbecher C, Montag D (2008) Abnormal axonal guidance and brain anatomy in mouse mutants for the cell recognition molecules close homolog of L1 and NgCAM-related cell adhesion molecule. Neuroscience 155(1):221–233. doi:10.1016/j.neuroscience.2008.04.080, S0306-4522(08)00645-3 [pii]

Hillenbrand R, Molthagen M, Montag D, Schachner M (1999) The close homologue of the neural adhesion molecule L1 (CHL1): patterns of expression and promotion of neurite outgrowth by heterophilic interactions. Eur J Neurosci 11(3):813–826

Holm J, Hillenbrand R, Steuber V, Bartsch U, Moos M, Lubbert H, Montag D, Schachner M (1996) Structural features of a close homologue of L1 (CHL1) in the mouse: a new member of the L1 family of neural recognition molecules. Eur J Neurosci 8(8):1613–1629

Hortsch M (1996) The L1 family of neural cell adhesion molecules: old proteins performing new tricks. Neuron 17(4):587–593

Hortsch M (2000) Structural and functional evolution of the L1-family: are four adhesion molecules better than one? Mol Cell Neurosci 15(1):1–10

Hortsch M, Margolis B (2003) Septate and paranodal junctions: kissing cousins. Trends Cell Biol 13(11):557–561

Hortsch M, Bieber AJ, Patel NH, Goodman CS (1990) Differential splicing generates a nervous system-specific form of *Drosophila* neuroglian. Neuron 4(5):697–709

Hortsch M, Wang YM, Marikar Y, Bieber AJ (1995) The cytoplasmic domain of the *Drosophila* cell adhesion molecule Neuroglian is not essential for its homophilic adhesive properties in S2 cells. J Biol Chem 270(32):18809–18817

Hortsch M, Homer D, Malhotra JD, Chang S, Frankel J, Jefford G, Dubreuil RR (1998a) Structural requirements for "outside-in" and "inside-out" signaling by *Drosophila* neuroglian, a member of the L1 family of cell adhesion molecules. J Cell Biol 142(1):251–261

Hortsch M, O'Shea KS, Zhao G, Kim F, Vallejo Y, Dubreuil RR (1998b) A conserved role for L1 as a transmembrane link between neuronal adhesion and membrane cytoskeleton assembly. Cell Adhes Commun 5(1):61–73

Hortsch M, Nagaraj K, Godenschwege TA (2009) The interaction between L1-type proteins and ankyrins – a master switch for L1-type CAM function. Cell Mol Biol Lett 14:57–69

Howell OW, Palser A, Polito A, Melrose S, Zonta B, Scheiermann C, Vora AJ, Brophy PJ, Reynolds R (2006) Disruption of neurofascin localization reveals early changes preceding demyelination and remyelination in multiple sclerosis. Brain 129(Pt 12):3173–3185. doi:10.1093/brain/awl290, awl290 [pii]

Hubbe M, Kowitz A, Schirrmacher V, Schachner M, Altevogt P (1993) L1 adhesion molecule on mouse leukocytes: regulation and involvement in endothelial cell binding. Eur J Immunol 23(11):2927–2931

Hübner CA, Utermann B, Tinschert S, Kruger G, Ressler B, Steglich C, Schinzel A, Gal A (2004) Intronic mutations in the L1CAM gene may cause X-linked hydrocephalus by aberrant splicing. Hum Mutat 23(5):526. doi:10.1002/humu.9242

Hulley P, Schachner M, Lubbert H (1998) L1 neural cell adhesion molecule is a survival factor for fetal dopaminergic neurons. J Neurosci Res 53(2):129–134

Imondi R, Jevince AR, Helms AW, Johnson JE, Kaprielian Z (2007) Mis-expression of L1 on pre-crossing spinal commissural axons disrupts pathfinding at the ventral midline. Mol Cell Neurosci 36(4):462–471. doi:10.1016/j.mcn.2007.08.003, S1044-7431(07)00179-0 [pii]

Ishiguro H, Liu QR, Gong JP, Hall FS, Ujike H, Morales M, Sakurai T, Grumet M, Uhl GR (2006) NrCAM in addiction vulnerability: positional cloning, drug-regulation, haplotype-specific expression, and altered drug reward in knockout mice. Neuropsychopharmacology 31(3): 572–584. doi:10.1038/sj.npp.1300855, 1300855 [pii]

Itoh K, Cheng L, Kamei Y, Fushiki S, Kamiguchi H, Gutwein P, Stoeck A, Arnold B, Altevogt P, Lemmon V (2004) Brain development in mice lacking L1-L1 homophilic adhesion. J Cell Biol 165(1):145–154

Itoh K, Fushiki S, Kamiguchi H, Arnold B, Altevogt P, Lemmon V (2005) Disrupted Schwann cell-axon interactions in peripheral nerves of mice with altered L1-integrin interactions. Mol Cell Neurosci 30(1):131–136

Izumoto S, Ohnishi T, Arita N, Hiraga S, Taki T, Hayakawa T (1996) Gene expression of neural cell adhesion molecule L1 in malignant gliomas and biological significance of L1 in glioma invasion. Cancer Res 56(6):1440–1444

Jakovcevski I, Siering J, Hargus G, Karl N, Hoelters L, Djogo N, Yin S, Zecevic N, Schachner M, Irintchev A (2009) Close homologue of adhesion molecule L1 promotes survival of Purkinje and granule cells and granule cell migration during murine cerebellar development. J Comp Neurol 513(5):496–510. doi:10.1002/cne.21981

Jenkins SM, Bennett V (2002) Developing nodes of Ranvier are defined by ankyrin-G clustering and are independent of paranodal axoglial adhesion. Proc Natl Acad Sci USA 99(4): 2303–2308

Jenkins SM, Kizhatil K, Kramarcy NR, Sen A, Sealock R, Bennett V (2001) FIGQY phosphorylation defines discrete populations of L1 cell adhesion molecules at sites of cell-cell contact and in migrating neurons. J Cell Sci 114(Pt 21):3823–3835

Jouet M, Rosenthal A, Armstrong G, MacFarlane J, Stevenson R, Paterson J, Metzenberg A, Ionasescu V, Temple K, Kenwrick S (1994) X-linked spastic paraplegia (SPG1), MASA syndrome and X-linked hydrocephalus result from mutations in the L1 gene. Nat Genet 7(3):402–407

Jouet M, Rosenthal A, Kenwrick S (1995) Exon 2 of the gene for neural cell adhesion molecule L1 is alternatively spliced in B cells. Mol Brain Res 30(2):378–380

Kadmon G, Montgomery AM, Altevogt P (1998) L1 makes immunological progress by expanding its relations. Dev Immunol 6(3–4):205–213

Kaifi JT, Heidtmann S, Schurr PG, Reichelt U, Mann O, Yekebas EF, Wachowiak R, Strate T, Schachner M, Izbicki JR (2006a) Absence of L1 in pancreatic masses distinguishes adenocarcinomas from poorly differentiated neuroendocrine carcinomas. Anticancer Res 26(2A):1167–1170

Kaifi JT, Zinnkann U, Yekebas EF, Schurr PG, Reichelt U, Wachowiak R, Fiegel HC, Petri S, Schachner M, Izbicki JR (2006b) L1 is a potential marker for poorly-differentiated pancreatic neuroendocrine carcinoma. World J Gastroenterol 12(1):94–98

Kallunki P, Edelman GM, Jones FS (1997) Tissue-specific expression of the L1 cell adhesion molecule is modulated by the neural restrictive silencer element. J Cell Biol 138(6): 1343–1354

Kamiguchi H, Lemmon V (1998) A neuronal form of the cell adhesion molecule L1 contains a tyrosine-based signal required for sorting to the axonal growth cone. J Neurosci 18(10):3749–3756

Kamiguchi H, Long KE, Pendergast M, Schaefer AW, Rapoport I, Kirchhausen T, Lemmon V (1998) The neural cell adhesion molecule L1 interacts with the AP-2 adaptor and is endocytosed via the clathrin-mediated pathway. J Neurosci 18(14):5311–5321

Kanemura Y, Okamoto N, Sakamoto H, Shofuda T, Kamiguchi H, Yamasaki M (2006) Molecular mechanisms and neuroimaging criteria for severe L1 syndrome with X-linked hydrocephalus. J Neurosurg 105(5 Suppl):403–412. doi:10.3171/ped.2006.105.5.403

Kaplan P (1983) X linked recessive inheritance of agenesis of the corpus callosum. J Med Genet 20(2):122–124
Kappen C, Schughart K, Ruddle FH (1989) Two steps in the evolution of Antennapedia-class vertebrate homeobox genes. Proc Natl Acad Sci USA 86(14):5459–5463
Katayama M, Iwamatsu A, Masutani H, Furuke K, Takeda K, Wada H, Masuda T, Ishii K (1997) Expression of neural cell adhesion molecule L1 in human lung cancer cell lines. Cell Struct Funct 22(5):511–516
Knops NB, Bos KK, Kerstjens M, van Dael K, Vos YJ (2008) Nephrogenic diabetes insipidus in a patient with L1 syndrome: a new report of a contiguous gene deletion syndrome including L1CAM and AVPR2. Am J Med Genet A 146A(14):1853–1858. doi:10.1002/ajmg.a.32386
Kobayashi M, Miura M, Asou H, Uyemura K (1991) Molecular cloning of cell adhesion molecule L1 from human nervous tissue: a comparison of the primary sequences of L1 molecules of different origin. Biochim Biophys Acta 1090(2):238–240
Kolata S, Wu J, Light K, Schachner M, Matzel LD (2008) Impaired working memory duration but normal learning abilities found in mice that are conditionally deficient in the close homolog of L1. J Neurosci 28(50):13505–13510. doi:10.1523/JNEUROSCI.2127-08.2008, 28/50/13505 [pii]
Koticha D, Babiarz J, Kane-Goldsmith N, Jacob J, Raju K, Grumet M (2005) Cell adhesion and neurite outgrowth are promoted by neurofascin NF155 and inhibited by NF186. Mol Cell Neurosci 30(1):137–148
Koticha D, Maurel P, Zanazzi G, Kane-Goldsmith N, Basak S, Babiarz J, Salzer J, Grumet M (2006) Neurofascin interactions play a critical role in clustering sodium channels, ankyrin G and beta IV spectrin at peripheral nodes of Ranvier. Dev Biol 293(1):1–12
Kowitz A, Kadmon G, Eckert M, Schirrmacher V, Schachner M, Altevogt P (1992) Expression and function of the neural cell adhesion molecule L1 in mouse leukocytes. Eur J Immunol 22(5):1199–1205
Kristiansen LV, Velasquez E, Romani S, Baars S, Berezin V, Bock E, Hortsch M, Garcia-Alonso L (2005) Genetic analysis of an overlapping functional requirement for L1- and NCAM-type proteins during sensory axon guidance in *Drosophila*. Mol Cell Neurosci 28(1):141–152
Laval M, Bel C, Faivre-Sarrailh C (2008) The lateral mobility of cell adhesion molecules is highly restricted at septate junctions in Drosophila. BMC Cell Biol 9:38. doi:10.1186/1471-2121-9-38, 1471-2121-9-38 [pii]
Lemmon V, Farr KL, Lagenaur C (1989) L1-mediated axon outgrowth occurs via a homophilic binding mechanism. Neuron 2(6):1597–1603
Leshchyns'ka I, Sytnyk V, Richter M, Andreyeva A, Puchkov D, Schachner M (2006) The adhesion molecule CHL1 regulates uncoating of clathrin-coated synaptic vesicles. Neuron 52(6):1011–1025. doi:10.1016/j.neuron.2006.10.020, S0896-6273(06)00820-8 [pii]
Linnemann D, Raz A, Bock E (1989) Differential expression of cell adhesion molecules in variants of K1735 melanoma cells differing in metastatic capacity. Int J Cancer 43(4):709–712
Lonigro A, Devaux JJ (2009) Disruption of neurofascin and gliomedin at nodes of Ranvier precedes demyelination in experimental allergic neuritis. Brain 132(Pt 1):260–273
Maddaluno L, Verbrugge SE, Martinoli C, Matteoli G, Chiavelli A, Zeng Y, Williams ED, Rescigno M, Cavallaro U (2009) The adhesion molecule L1 regulates transendothelial migration and trafficking of dendritic cells. J Exp Med 206(3):623–635. doi:10.1084/jem.20081211, jem.20081211 [pii]
Maness PF, Schachner M (2007) Neural recognition molecules of the immunoglobulin superfamily: signaling transducers of axon guidance and neuronal migration. Nat Neurosci 10(1):19–26. doi:10.1038/nn1827, nn1827 [pii]
Martin V, Mrkusich E, Steinel MC, Rice J, Merritt DJ, Whitington PM (2008) The L1-type cell adhesion molecule Neuroglian is necessary for maintenance of sensory axon advance in the Drosophila embryo. Neural Dev 3:10. doi:10.1186/1749-8104-3-10, 1749-8104-3-10 [pii]
Mathey EK, Derfuss T, Storch MK, Williams KR, Hales K, Woolley DR, Al-Hayani A, Davies SN, Rasband MN, Olsson T, Moldenhauer A, Velhin S, Hohlfeld R, Meinl E, Linington C (2007) Neurofascin as a novel target for autoantibody-mediated axonal injury. J Exp Med 204(10):2363–2372

Matzel LD, Babiarz J, Townsend DA, Grossman HC, Grumet M (2008) Neuronal cell adhesion molecule deletion induces a cognitive and behavioral phenotype reflective of impulsivity. Genes Brain Behav 7(4):470–480. doi:10.1111/j.1601-183X.2007.00382.x, GBB382 [pii]

Mechtersheimer S, Gutwein P, Agmon-Levin N, Stoeck A, Oleszewski M, Riedle S, Fogel M, Lemmon V, Altevogt P (2001) Ectodomain shedding of L1 adhesion molecule promotes cell migration by autocrine binding to integrins. J Cell Biol 155(4):661–673

Meier F, Busch S, Gast D, Goppert A, Altevogt P, Maczey E, Riedle S, Garbe C, Schittek B (2006) The adhesion molecule L1 (CD171) promotes melanoma progression. Int J Cancer 119(3): 549–555. doi:10.1002/ijc.21880

Michelson P, Hartwig C, Schachner M, Gal A, Veske A, Finckh U (2002) Missense mutations in the extracellular domain of the human neural cell adhesion molecule L1 reduce neurite outgrowth of murine cerebellar neurons. Hum Mutat 20(6):481–482. doi:10.1002/humu.9096

Miura M, Kobayashi M, Asou H, Uyemura K (1991) Molecular cloning of cDNA encoding the rat neural cell adhesion molecule L1. Two L1 isoforms in the cytoplasmic region are produced by differential splicing. FEBS Lett 289(1):91–95

Montag-Sallaz M, Baarke A, Montag D (2003) Aberrant neuronal connectivity in CHL1-deficient mice is associated with altered information processing-related immediate early gene expression. J Neurobiol 57(1):67–80. doi:10.1002/neu.10254

Montgomery AMP, Becker JC, Siu CH, Lemmon VP, Cheresh DA, Pancook JD, Zhao XN, Reisfeld RA (1996) Human neural cell-adhesion molecule L1 and rat homolog NILE are ligands for integrin alpha(V)beta(3). J Cell Biol 132(3):475–485

Moos M, Tacke R, Scherer H, Teplow D, Früh K, Schachner M (1988) Neural adhesion molecule L1 as a member of the immunoglobulin superfamily with binding domains similar to fibronectin. Nature 334(6184):701–703

More MI, Kirsch FP, Rathjen FG (2001) Targeted ablation of NrCAM or ankyrin-B results in disorganized lens fibers leading to cataract formation. J Cell Biol 154(1):187–196

Moulding HD, Martuza RL, Rabkin SD (2000) Clinical mutations in the L1 neural cell adhesion molecule affect cell-surface expression. J Neurosci 20(15):5696–5702

Moya GE, Michaelis RC, Holloway LW, Sanchez JM (2002) Prenatal diagnosis of L1 cell adhesion molecule mutations. Capabilities and limitations. Fetal Diagn Ther 17(2):115–119, fdt17115 [pii]

Mualla R, Nagaraj K, Hortsch M (2013) A phylogenetic analysis of the L1 family of neural cell adhesion molecules. Neurochem Res 38:1196–1207

Nagaraj K, Kristiansen LV, Skrzynski A, Castiella C, Garcia-Alonso L, Hortsch M (2009) Pathogenic human L1-CAM mutations reduce the adhesion-dependent activation of EGFR. Hum Mol Genet 18(20):3822–3831

Nakakimura S, Sasaki F, Okada T, Arisue A, Cho K, Yoshino M, Kanemura Y, Yamasaki M, Todo S (2008) Hirschsprung's disease, acrocallosal syndrome, and congenital hydrocephalus: report of 2 patients and literature review. J Pediatr Surg 43(5):E13–E17. doi:10.1016/j.jpedsurg.2007.12.069, S0022-3468(08)00002-X [pii]

Nakamura Y, Lee S, Haddox CL, Weaver EJ, Lemmon VP (2010) Role of the cytoplasmic domain of the L1 cell adhesion molecule in brain development. J Comp Neurol 518(7):1113–1132. doi:10.1002/cne.22267

Nardi JB, Pilas B, Bee CM, Zhuang S, Garsha K, Kanost MR (2006) Neuroglian-positive plasmatocytes of Manduca sexta and the initiation of hemocyte attachment to foreign surfaces. Dev Comp Immunol 30(5):447–462

Needham LK, Thelen K, Maness PF (2001) Cytoplasmic domain mutations of the L1 cell adhesion molecule reduce L1-ankyrin interactions. J Neurosci 21(5):1490–1500

Nikonenko AG, Sun M, Lepsveridze E, Apostolova I, Petrova I, Irintchev A, Dityatev A, Schachner M (2006) Enhanced perisomatic inhibition and impaired long-term potentiation in the CA1 region of juvenile CHL1-deficient mice. Eur J Neurosci 23(7):1839–1852

Nishimune H, Bernreuther C, Carroll P, Chen S, Schachner M, Henderson CE (2005) Neural adhesion molecules L1 and CHL1 are survival factors for motoneurons. J Neurosci Res 80(5):593–599

Nishimura K, Yoshihara F, Tojima T, Ooashi N, Yoon W, Mikoshiba K, Bennett V, Kamiguchi H (2003) L1-dependent neuritogenesis involves ankyrinB that mediates L1-CAM coupling with retrograde actin flow. J Cell Biol 163(5):1077–1088

Nolte C, Moos M, Schachner M (1999) Immunolocalization of the neural cell adhesion molecule L1 in epithelia of rodents. Cell Tissue Res 298(2):261–273

Novak-Hofer I, Cohrs S, Grunberg J, Friedli A, Schlatter MC, Pfeifer M, Altevogt P, Schubiger PA (2008) Antibodies directed against L1-CAM synergize with Genistein in inhibiting growth and survival pathways in SKOV3ip human ovarian cancer cells. Cancer Lett 261(2):193–204. doi:10.1016/j.canlet.2007.11.012, S0304-3835(07)00558-7 [pii]

Okamoto N, Del Maestro R, Valero R, Monros E, Poo P, Kanemura Y, Yamasaki M (2004) Hydrocephalus and Hirschsprung's disease with a mutation of L1CAM. J Hum Genet 49(6):334–337. doi:10.1007/s10038-004-0153-4

Ooashi N, Kamiguchi H (2009) The cell adhesion molecule L1 controls growth cone navigation via ankyrin(B)-dependent modulation of cyclic AMP. Neurosci Res 63(3):224–226. doi:10.1016/j.neures.2008.11.009, S0168-0102(08)00294-0 [pii]

Panayi M, Gokhale D, Mansour S, Elles R (2005) Prenatal diagnosis in a family with X-linked hydrocephalus. Prenat Diagn 25(10):930–933. doi:10.1002/pd.1228

Panicker AK, Buhusi M, Thelen K, Maness PF (2003) Cellular signalling mechanisms of neural cell adhesion molecules. Front Biosci 8:D900–D911

Piccione M, Matina F, Fichera M, Lo Giudice M, Damiani G, Jakil MC, Corsello G (2010) A novel L1CAM mutation in a fetus detected by prenatal diagnosis. Eur J Pediatr 169(4):415–419. doi:10.1007/s00431-009-1037-6

Pillai AM, Thaxton C, Pribisko AL, Cheng JG, Dupree JL, Bhat MA (2009) Spatiotemporal ablation of myelinating glia-specific neurofascin (Nfasc NF155) in mice reveals gradual loss of paranodal axoglial junctions and concomitant disorganization of axonal domains. J Neurosci Res 87(8):1773–1793

Pomicter AD, Shroff SM, Fuss B, Sato-Bigbee C, Brophy PJ, Rasband MN, Bhat MA, Dupree JL (2010) Novel forms of neurofascin 155 in the central nervous system: alterations in paranodal disruption models and multiple sclerosis. Brain 133(Pt 2):389–405. doi:10.1093/brain/awp341, awp341 [pii]

Pratte M, Rougon G, Schachner M, Jamon M (2003) Mice deficient for the close homologue of the neural adhesion cell L1 (CHL1) display alterations in emotional reactivity and motor coordination. Behav Brain Res 147(1–2):31–39

Pruss T, Niere M, Kranz EU, Volkmer H (2004) Homophilic interactions of chick neurofascin in trans are important for neurite induction. Eur J Neurosci 20(11):3184–3188. doi:10.1111/j.1460-9568.2004.03773.x, EJN3773 [pii]

Rathjen FG, Rutishauser U (1984) Comparison of two cell surface molecules involved in neural cell adhesion. EMBO J 3(2):461–465

Rathjen FG, Schachner M (1984) Immunocytological and biochemical characterization of a new neuronal cell surface component (L1 antigen) which is involved in cell adhesion. EMBO J 3(1):1–10

Rathjen FG, Wolff JM, Chang S, Bonhoeffer F, Raper JA (1987) Neurofascin: a novel chick cell-surface glycoprotein involved in neurite-neurite interactions. Cell 51(5):841–849

Raveh S, Gavert N, Ben-Ze'ev A (2009) L1 cell adhesion molecule (L1CAM) in invasive tumors. Cancer Lett 282(2):137–145. doi:10.1016/j.canlet.2008.12.021, S0304-3835(08)00969-5 [pii]

Rehnberg M, Jonasson J, Gunnarsson C (2011) Novel L1CAM splice site mutation in a young male with L1 syndrome. Am J Med Genet A 155A(2):439–441. doi:10.1002/ajmg.a.33803

Reid RA, Hemperly JJ (1992) Variants of human L1 cell adhesion molecule arise through alternate splicing of RNA. J Mol Neurosci 3(3):127–135

Ren Q, Bennett V (1998) Palmitoylation of neurofascin at a site in the membrane-spanning domain highly conserved among the L1 family of cell adhesion molecules. J Neurochem 70(5):1839–1849

Rodriguez Criado G, Perez Aytes A, Martinez F, Vos YJ, Verlind E, Gonzalez-Meneses Lopez A, de Terreros G, Sanchez I, Schrander-Stumpel C (2003) X-linked hydrocephalus: another two families with an L1 mutation. Genet Couns 14(1):57–65

Roonprapunt C, Huang W, Grill R, Friedlander D, Grumet M, Chen S, Schachner M, Young W (2003) Soluble cell adhesion molecule L1-Fc promotes locomotor recovery in rats after spinal cord injury. J Neurotrauma 20(9):871–882

Rosenthal A, Jouet M, Kenwrick S (1992) Aberrant splicing of neural cell adhesion molecule L1 mRNA in a family with X-linked hydrocephalus. Nat Genet 2(2):107–112

Ruiz JC, Cuppens H, Legius E, Fryns JP, Glover T, Marynen P, Cassiman JJ (1995) Mutations in L1-CAM in two families with X linked complicated spastic paraplegia, MASA syndrome, and HSAS. J Med Genet 32(7):549–552

Rünker AE, Bartsch U, Nave KA, Schachner M (2003) The C264Y missense mutation in the extracellular domain of L1 impairs protein trafficking *in vitro* and *in vivo*. J Neurosci 23(1):277–286

Ruppert M, Aigner S, Hubbe M, Yagita H, Altevogt P (1995) The L1 adhesion molecule is a cellular ligand for VLA-5. J Cell Biol 131:1881–1891

Saghatelyan AK, Nikonenko AG, Sun M, Rolf B, Putthoff P, Kutsche M, Bartsch U, Dityatev A, Schachner M (2004) Reduced GABAergic transmission and number of hippocampal perisomatic inhibitory synapses in juvenile mice deficient in the neural cell adhesion molecule L1. Mol Cell Neurosci 26(1):191–203

Sakurai K, Migita O, Toru M, Arinami T (2002) An association between a missense polymorphism in the close homologue of L1 (CHL1, CALL) gene and schizophrenia. Mol Psychiatry 7(4):412–415

Sakurai T, Ramoz N, Reichert JG, Corwin TE, Kryzak L, Smith CJ, Silverman JM, Hollander E, Buxbaum JD (2006) Association analysis of the NrCAM gene in autism and in subsets of families with severe obsessive-compulsive or self-stimulatory behaviors. Psychiatr Genet 16(6):251–257. doi:10.1097/01.ypg.0000242196.81891.c900041444-200612000-00011 [pii]

Sanes JR, Schachner M, Covault J (1986) Expression of several adhesive macromolecules (N-CAM, L1, J1, NILE, uvomorulin, laminin, fibronectin, and a heparan sulfate proteoglycan) in embryonic, adult, and denervated adult skeletal muscle. J Cell Biol 102(2):420–431

Sasakura H, Inada H, Kuhara A, Fusaoka E, Takemoto D, Takeuchi K, Mori I (2005) Maintenance of neuronal positions in organized ganglia by SAX-7, a *Caenorhabditis elegans* homologue of L1. EMBO J 24(7):1477–1488

Schaefer AW, Kamiguchi H, Wong EV, Beach CM, Landreth G, Lemmon V (1999) Activation of the MAPK signal cascade by the neural cell adhesion molecule L1 requires L1 internalization. J Biol Chem 274(53):37965–37973

Schmid RS, Pruitt WM, Maness PF (2000) A MAP kinase-signaling pathway mediates neurite outgrowth on L1 and requires Src-dependent endocytosis. J Neurosci 20(11):4177–4188

Schrander-Stumpel C, Legius E, Fryns JP, Cassiman JJ (1990) MASA syndrome: new clinical features and linkage analysis using DNA probes. J Med Genet 27(11):688–692

Schughart K, Kappen C, Ruddle FH (1989) Duplication of large genomic regions during the evolution of vertebrate homeobox genes. Proc Natl Acad Sci USA 86(18):7067–7071

Schürmann G, Haspel J, Grumet M, Erickson HP (2001) Cell adhesion molecule L1 in folded (horseshoe) and extended conformations. Mol Biol Cell 12(6):1765–1773

Sebens Müerköster S, Werbing V, Sipos B, Debus MA, Witt M, Grossmann M, Leisner D, Kötteritzsch J, Kappes H, Kloppel G, Altevogt P, Folsch UR, Schäfer H (2007) Drug-induced expression of the cellular adhesion molecule L1CAM confers anti-apoptotic protection and chemoresistance in pancreatic ductal adenocarcinoma cells. Oncogene 26(19):2759–2768. doi:10.1038/sj.onc.1210076, 1210076 [pii]

Senat MV, Bernard JP, Delezoide A, Saugier-Veber P, Hillion Y, Roume J, Ville Y (2001) Prenatal diagnosis of hydrocephalus-stenosis of the aqueduct of Sylvius by ultrasound in the first trimester of pregnancy. Report of two cases. Prenat Diagn 21(13):1129–1132. doi:10.1002/pd.184 [pii]

Sherman DL, Tait S, Melrose S, Johnson R, Zonta B, Court FA, Macklin WB, Meek S, Smith AJ, Cottrell DF, Brophy PJ (2005) Neurofascins are required to establish axonal domains for saltatory conduction. Neuron 48(5):737–742

Shtutman M, Levina E, Ohouo P, Baig M, Roninson IB (2006) Cell adhesion molecule L1 disrupts E-cadherin-containing adherens junctions and increases scattering and motility of MCF7 breast carcinoma cells. Cancer Res 66(23):11370–11380

Silan F, Ozdemir I, Lissens W (2005) A novel L1CAM mutation with L1 spectrum disorders. Prenat Diagn 25(1):57–59. doi:10.1002/pd.978

Silletti S, Yebra M, Perez B, Cirulli V, McMahon M, Montgomery AM (2004) Extracellular signal-regulated kinase (ERK)-dependent gene expression contributes to L1 cell adhesion molecule-dependent motility and invasion. J Biol Chem 279(28):28880–28888

Simonati A, Boaretto F, Vettori A, Dabrilli P, Criscuolo L, Rizzuto N, Mostacciuolo ML (2006) A novel missense mutation in the L1CAM gene in a boy with L1 disease. Neurol Sci 27(2): 114–117. doi:10.1007/s10072-006-0610-2

Soker S, Takashima S, Miao HQ, Neufeld G, Klagsbrun M (1998) Neuropilin-1 is expressed by endothelial and tumor cells as an isoform-specific receptor for vascular endothelial growth factor. Cell 92(6):735–745, S0092-8674(00)81402-6 [pii]

Stoeck A, Gast D, Sanderson MP, Issa Y, Gutwein P, Altevogt P (2007) L1-CAM in a membrane-bound or soluble form augments protection from apoptosis in ovarian carcinoma cells. Gynecol Oncol 104(2):461–469. doi:10.1016/j.ygyno.2006.08.038, S0090-8258(06)00683-4 [pii]

Su XD, Gastinel LN, Vaughn DE, Faye I, Poon P, Bjorkman PJ (1998) Crystal structure of hemolin: a horseshoe shape with implications for homophilic adhesion. Science 281(5379):991–995

Sztriha L, Vos YJ, Verlind E, Johansen J, Berg B (2002) X-linked hydrocephalus: a novel missense mutation in the L1CAM gene. Pediatr Neurol 27(4):293–296, S088789940200440X [pii]

Tait S, Gunn-Moore F, Collinson JM, Huang J, Lubetzki C, Pedraza L, Sherman DL, Colman DR, Brophy PJ (2000) An oligodendrocyte cell adhesion molecule at the site of assembly of the paranodal axo-glial junction. J Cell Biol 150(3):657–666

Takeda Y, Asou H, Murakami Y, Miura M, Kobayashi M, Uyemura K (1996) A nonneuronal isoform of cell adhesion molecule L1: tissue-specific expression and functional analysis. J Neurochem 66(6):2338–2349

Talantov D, Mazumder A, Yu JX, Briggs T, Jiang Y, Backus J, Atkins D, Wang Y (2005) Novel genes associated with malignant melanoma but not benign melanocytic lesions. Clin Cancer Res 11(20):7234–7242. doi:10.1158/1078-0432.CCR-05-0683, 11/20/7234 [pii]

Tapanes-Castillo A, Weaver EJ, Smith RP, Kamei Y, Caspary T, Hamilton-Nelson KL, Slifer SH, Martin ER, Bixby JL, Lemmon VP (2009) A modifier locus on chromosome 5 contributes to L1 cell adhesion molecule X-linked hydrocephalus in mice. Neurogenetics 11(1):53–71. doi:10.1007/s10048-009-0203-3

Tegay DH, Lane AH, Roohi J, Hatchwell E (2007) Contiguous gene deletion involving L1CAM and AVPR2 causes X-linked hydrocephalus with nephrogenic diabetes insipidus. Am J Med Genet A 143(6):594–598. doi:10.1002/ajmg.a.31536

Thaxton C, Pillai AM, Pribisko AL, Labasque M, Dupree JL, Faivre-Sarrailh C, Bhat MA (2010) In vivo deletion of immunoglobulin domains 5 and 6 in neurofascin (nfasc) reveals domain-specific requirements in myelinated axons. J Neurosci 30(14):4868–4876

Thies A, Schachner M, Moll I, Berger J, Schulze HJ, Brunner G, Schumacher U (2002) Overexpression of the cell adhesion molecule L1 is associated with metastasis in cutaneous malignant melanoma. Eur J Cancer 38(13):1708–1716

Triana-Baltzer GB, Liu Z, Berg DK (2006) Pre- and postsynaptic actions of L1-CAM in nicotinic pathways. Mol Cell Neurosci 33(2):214–226

Tsutsumi S, Morohashi S, Kudo Y, Akasaka H, Ogasawara H, Ono M, Takasugi K, Ishido K, Hakamada K, Kijima H (2011) L1 Cell adhesion molecule (L1CAM) expression at the cancer invasive front is a novel prognostic marker of pancreatic ductal adenocarcinoma. J Surg Oncol 103(7):669–673. doi:10.1002/jso.21880

Tsuzuki T, Izumoto S, Ohnishi T, Hiraga S, Arita N, Hayakawa T (1998) Neural cell adhesion molecule L1 in gliomas: correlation with TGF-beta and p53. J Clin Pathol 51(1):13–17

Turner KN, Schachner M, Anderson RB (2009) Cell adhesion molecule L1 affects the rate of differentiation of enteric neurons in the developing gut. Dev Dyn 238(3):708–715. doi:10.1002/dvdy.21861

Valladares A, Hernandez NG, Gomez FS, Curiel-Quezada E, Madrigal-Bujaidar E, Vergara MD, Martinez MS, Arenas Aranda DJ (2006) Genetic expression profiles and chromosomal alterations in sporadic breast cancer in Mexican women. Cancer Genet Cytogenet 170(2):147–151. doi:10.1016/j.cancergencyto.2006.06.002, S0165-4608(06)00368-2 [pii]

Vits L, Van Camp G, Coucke P, Fransen E, De Boulle K, Reyniers E, Korn B, Poustka A, Wilson G, Schrander-Stumpel C, Winter RM, Schwartz C, Willems PJ (1994) MASA syndrome is due to mutations in the neural cell adhesion gene L1CAM. Nat Genet 7(3):408–413

Vits L, Chitayat D, Van Camp G, Holden JJ, Fransen E, Willems PJ (1998) Evidence for somatic and germline mosaicism in CRASH syndrome. Hum Mutat Suppl 1:S284–S287

Volkmer H, Hassel B, Wolff JM, Frank R, Rathjen FG (1992) Structure of the axonal surface recognition molecule neurofascin and its relationship to a neural subgroup of the immunoglobulin superfamily. J Cell Biol 118(1):149–161

Volkmer H, Leuschner R, Zacharias U, Rathjen FG (1996) Neurofascin induces neurites by heterophilic interactions with axonal NrCAM while NrCAM requires F11 on the axonal surface to extend neurites. J Cell Biol 135(4):1059–1069

Vos YJ, Hofstra RM (2010) An updated and upgraded L1CAM mutation database. Hum Mutat 31(1):E1102–E1109. doi:10.1002/humu.21172

Vos YJ, de Walle HE, Bos KK, Stegeman JA, Ten Berge AM, Bruining M, van Maarle MC, Elting MW, den Hollander NS, Hamel B, Fortuna AM, Sunde LE, Stolte-Dijkstra I, Schrander-Stumpel CT, Hofstra RM (2010) Genotype-phenotype correlations in L1 syndrome: a guide for genetic counselling and mutation analysis. J Med Genet 47(3):169–175. doi:10.1136/jmg.2009.071688, jmg.2009.071688 [pii]

Wachowiak R, Fiegel HC, Kaifi JT, Quaas A, Krickhahn A, Schurr PG, Erttmann R, Schachner M, Kluth D, Sauter G, Izbicki JR (2007) L1 is associated with favorable outcome in neuroblastomas in contrast to adult tumors. Ann Surg Oncol 14(12):3575–3580. doi:10.1245/s10434-007-9608-0

Wang B, Williams H, Du JS, Terrett J, Kenwrick S (1998) Alternative splicing of human NrCAM in neural and nonneural tissues. Mol Cell Neurosci 10(5–6):287–295

Wang X, Kweon J, Larson S, Chen L (2005) A role for the *C. elegans* L1CAM homologue *lad-1/sax-7* in maintaining tissue attachment. Dev Biol 284(2):273–291

Wang X, Zhang W, Cheever T, Schwarz V, Opperman K, Hutter H, Koepp D, Chen L (2008) The C. elegans L1CAM homologue LAD-2 functions as a coreceptor in MAB-20/Sema2 mediated axon guidance. J Cell Biol 180(1):233–246

Wei CH, Ryu SE (2012) Homophilic interaction of the L1 family of cell adhesion molecules. Exp Mol Med 44(7):413–423. doi:10.3858/emm.2012.44.7.050, emm.2012.44.050 [pii]

Wei MH, Karavanova I, Ivanov SV, Popescu NC, Keck CL, Pack S, Eisen JA, Lerman MI (1998) *In silico*-initiated cloning and molecular characterization of a novel human member of the L1 gene family of neural cell adhesion molecules. Hum Genet 103(3):355–364

Wei J, Hortsch M, Goode S (2004) Neuroglian stabilizes epithelial structure during *Drosophila* oogenesis. Dev Dyn 230(4):800–808

Weller S, Gartner J (2001) Genetic and clinical aspects of X-linked hydrocephalus (L1 disease): mutations in the L1CAM gene. Hum Mutat 18(1):1–12

Whittard JD, Sakurai T, Cassella MR, Gazdoiu M, Felsenfeld DP (2006) MAP kinase pathway-dependent phosphorylation of the L1-CAM ankyrin-binding site regulates neuronal growth. Mol Biol Cell 17(6):2696–2706

Wiencken-Barger AE, Mavity-Hudson J, Bartsch U, Schachner M, Casagrande VA (2004) The role of L1 in axon pathfinding and fasciculation. Cereb Cortex 14(2):121–131

Williams MJ (2009) The Drosophila cell adhesion molecule Neuroglian regulates Lissencephaly-1 localisation in circulating immunosurveillance cells. BMC Immunol 10:17. doi:10.1186/1471-2172-10-17, 1471-2172-10-17 [pii]

Wilson PL, Kattman BB, Mulvihill JJ, Li S, Wilkins J, Wagner AF, Goodman JR (2009) Prenatal identification of a novel R937P L1CAM missense mutation. Genet Test Mol Biomarkers 13(4):515–519. doi:10.1089/gtmb.2009.0017

Winter RM, Davies KE, Bell MV, Huson SM, Patterson MN (1989) MASA syndrome: further clinical delineation and chromosomal localisation. Hum Genet 82(4):367–370

Wolff JM, Frank R, Mujoo K, Spiro RC, Reisfeld RA, Rathjen FG (1988) A human brain glycoprotein related to the mouse cell adhesion molecule L1. J Biol Chem 263(24):11943–11947

Wong EV, Cheng GH, Payne HR, Lemmon V (1995) The cytoplasmic domain of the cell-adhesion molecule L1 is not required for homophilic adhesion. Neurosci Lett 200(3):155–158

Wood PM, Schachner M, Bunge RP (1990) Inhibition of Schwann cell myelination *in vitro* by antibody to the L1 adhesion molecule. J Neurosci 10(11):3635–3645

Yamamoto M, Ueda R, Takahashi K, Saigo K, Uemura T (2006) Control of axonal sprouting and dendrite branching by the nrg-ank complex at the neuron-glia interface. Curr Biol 16(16):1678–1683

Yamasaki M, Thompson P, Lemmon V (1997) CRASH syndrome: mutations in L1CAM correlate with severity of the disease. Neuropediatrics 28(3):175–178

Yip PM, Zhao X, Montgomery AM, Siu CH (1998) The Arg-Gly-Asp motif in the cell adhesion molecule L1 promotes neurite outgrowth via interaction with the alphavbeta3 integrin. Mol Biol Cell 9(2):277–290

Zecchini S, Bianchi M, Colombo N, Fasani R, Goisis G, Casadio C, Viale G, Liu J, Herlyn M, Godwin AK, Nuciforo PG, Cavallaro U (2008) The differential role of L1 in ovarian carcinoma and normal ovarian surface epithelium. Cancer Res 68(4):1110–1118. doi:10.1158/0008-5472.CAN-07-2897, 68/4/1110 [pii]

Zhang X, Davis JQ, Carpenter S, Bennett V (1998) Structural requirements for association of neurofascin with ankyrin. J Biol Chem 273(46):30785–30794

Zhao G, Hortsch M (1998) The analysis of genomic structures in the L1 family of cell adhesion molecules provides no evidence for exon shuffling events after the separation of arthropod and chordate lineages. Gene 215(1):47–55

Zhao X, Siu CH (1996) Differential effects of two hydrocephalus/MASA syndrome-related mutations on the homophilic binding and neuritogenic activities of the cell adhesion molecule L1. J Biol Chem 271(12):6563–6566

Zhao X, Yip PM, Siu CH (1998) Identification of a homophilic binding site in immunoglobulin-like domain 2 of the cell adhesion molecule L1. J Neurochem 71(3):960–971

Zhuang S, Kelo L, Nardi JB, Kanost MR (2007) Neuroglian on hemocyte surfaces is involved in homophilic and heterophilic interactions of the innate immune system of Manduca sexta. Dev Comp Immunol 31(11):1159–1167

Zonta B, Tait S, Melrose S, Anderson H, Harroch S, Higginson J, Sherman DL, Brophy PJ (2008) Glial and neuronal isoforms of Neurofascin have distinct roles in the assembly of nodes of Ranvier in the central nervous system. J Cell Biol 181(7):1169–1177. doi:10.1083/jcb.200712154, jcb.200712154 [pii]

Chapter 10
Organisation and Control of Neuronal Connectivity and Myelination by Cell Adhesion Molecule Neurofascin

Julia Ebel, Simone Beuter, Jennifer Wuchter, Martin Kriebel, and Hansjürgen Volkmer

Abstract The neuronal cell adhesion molecule neurofascin is expressed in highly complex temporally and spatially regulated patterns. Accordingly, many different functions have been described including control of neurite outgrowth, clustering of protein complexes at the axon initial segments as well as at the nodes of Ranvier and axoglial contact formation at paranodal segments. At the molecular level, neurofascin provides a link between extracellular interactions of many different interaction partners and cytoskeletal components or signal transduction. Such interactions are subject to intimate regulation by alternative splicing and posttranslational modification. The versatile functional aspects of neurofascin interactions pose it at a central position for the shaping and maintenance of neural circuitry and synaptic contacts which are implicated in nervous system disorders.

10.1 Introduction

The cell adhesion protein neurofascin was first discovered in chick brain by Rathjen et al. (1987). This seminal work provided a first hint on a new cell adhesion molecule implicated in the fasciculation of axons. Later on, in V. Bennetts group, neurofascin was identified as an ankyrin-binding glycoprotein together with the related cell adhesion molecule NrCAM (Davis and Bennett 1994). Soon it became clear that neurofascin is an extraordinary molecule due to its highly complex expression patterns. Complexity is achieved by extensive alternative splicing and posttranslational modification. This goes in parallel with temporally and spatially controlled expression.

J. Ebel • S. Beuter • J. Wuchter • M. Kriebel • H. Volkmer (✉)
Department of Molecular Biology, NMI an der Universität Tübingen,
Markwiesenstrasse 55, 72770 Reutlingen, Germany
e-mail: volkmer@nmi.de

V. Berezin and P.S. Walmod (eds.), *Cell Adhesion Molecules: Implications in Neurological Diseases*, Advances in Neurobiology 8, DOI 10.1007/978-1-4614-8090-7_10,

It is therefore not a surprise that neurofascin is implicated in a wide spectrum of different functions including regulation of neurite outgrowth, assembly and stabilisation of protein clustering in postsynapses and nodes of Ranvier as well as the control of myelination. In this regard, neurofascin acts as a classical cell adhesion molecule which links extracellular interaction to the intracellular cytoskeleton. On the other hand, it also serves as a co-receptor that is linked to signal transduction via interaction with the receptor phosphotyrosine kinase fibroblast growth factor receptor 1(FGFR1).

It is therefore conceivable that neurofascin might be involved in disorders of the nervous system. First indications were reported for demyelinating diseases which coincides with the well-documented role of neurofascin in the stabilisation of axoglial contacts. We shall discuss further lines of evidence that may relate neurofascin functions to neurodevelopmental disorders.

10.2 Four Major Neurofascin Isoforms Become Selectively Expressed in the Nervous System

As a part of the L1 subgroup of the immunoglobulin superfamily of cell adhesion molecules (IgCAMs), neurofascin consists of four types of major structural components (listed from N- to C-terminus): Six Ig-like repeats of the C2 subgroup (Ig1–Ig6), five fibronectin type III-like repeats (FN 1–FN 5), a transmembrane domain and a cytoplasmic domain of 113 amino acid residues in length, the latter being highly conserved within the L1 subgroup (Volkmer et al. 1992; Davis et al. 1993). A further structural element is represented by the so-called PAT/mucin-like domain, which is specific for a subset of neurofascin isoforms and mostly consists of proline, alanine and threonine amino acid residues (Volkmer et al. 1992). The threonine residues in the PAT domain may serve as a target of O-linked glycosylation.

The chick *NFASC* gene of about 72 kb gives rise to an mRNA of 8.5 kb in length (Hassel et al. 1997; Volkmer et al. 1992). Exons 2–29 refer to the extracellular domains; the transmembrane domain is encoded by exon 30 while exons 31–33 account for the cytosolic domain. Alternative splicing of ten exons generates a large variety of different neurofascin isoforms. Alternatively spliced exons encode three larger domains including the third as well as the fifth FNIII-like domains and the PAT domain. Four minor alternatively spliced sequences are located at the NH_2-terminus (exon 3), between Ig-like domains 2 and 3 (exon 8) and at the junction between the Ig-like and the FNIII-like domains (exon 17), and within the cytosolic domain (exon 32). In an unbiased screen using a chick brain cDNA library, 50 different possible neurofascin isoforms have been identified. So far, four major isoforms NF155, NF166, NF180 and NF186 have been studied in detail (Kriebel et al. 2012). These isoforms can be categorised according to selective expression of alternatively spliced sequences (Fig. 10.1). NF155 harbours FNIII-like repeats 1–4, NF166 is built up of FNIII-like repeats-1, -2, -3, NF180 is a version of NF166

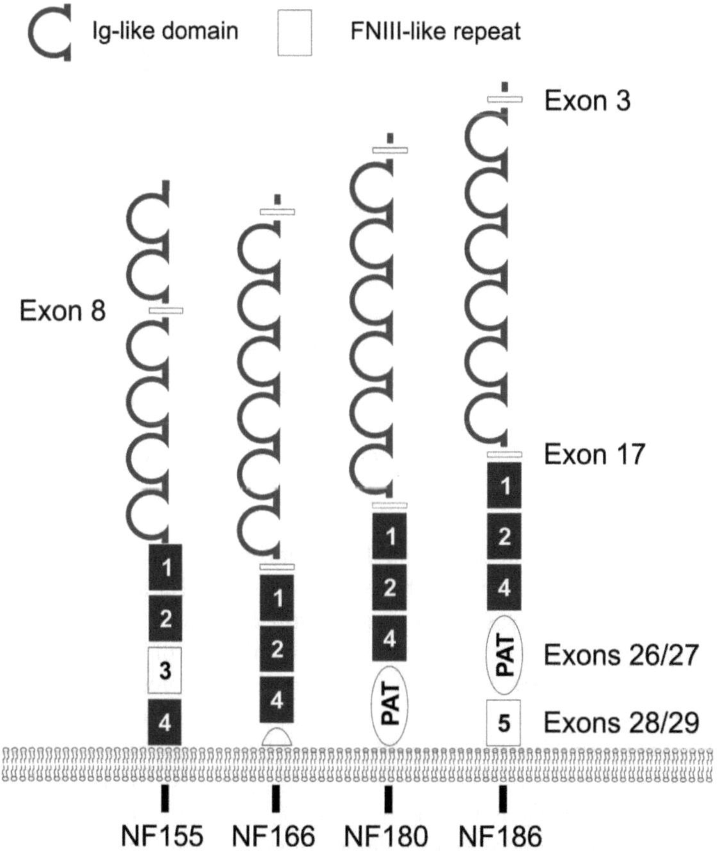

Fig. 10.1 Domain structures of neurofascin isoforms—Immunoglobulin (Ig)-like domains and fibronectin type III (FNIII)-like repeats are represented in the neurofascin (NF) isoforms NF155, NF166, NF180 and NF186 as indicated. Constitutively expressed sequences are represented by *filled symbols* while alternatively spliced sequences are depicted as *open symbols*. The FNIII-like repeats are numbered according to their appearance in the *NFASC* gene. PAT refers to proline–alanine–threonine-rich sequence (also termed mucin-like domain)

extended by the PAT domain and NF186 is further extended by FNIII-like repeat 5. Three small alternatively spliced sequences (exons 3, 8, 17) are expressed in mutually exclusive patterns and can be sorted into two distinct categories: exons 3 and 17 are coexpressed in NF166, 180 and 186 while exon 8 is unique for NF155. Expression of neurofascin isoforms is developmentally regulated and may also be restricted to specific cell types.

Neurofascin NF166 discovered in chicken is the first major isoform to emerge in the embryonic brain. It was found on extending axons in developing fibre tracts in both the central and peripheral nervous systems (Rathjen et al. 1987; Hassel et al. 1997; Pruss et al. 2006). The expression pattern suggests that NF166 is

Table 10.1 Summary of neurofascin interaction partners

	Interaction partner	Function	References
Extracellular interactions	Tenascin-R	Cell adhesion	Volkmer et al. (1998)
	Brevican	Link to perineural net	Hedstrom et al. (2007)
	NrCAM	Neurite outgrowth	Volkmer et al. (1996)
	Contactin-1	Stabilisation of septate-like junctions at paranodes	Volkmer et al. (1998), Gollan et al. (2003)
	Contactin-2	Neurite outgrowth	Pruss et al. (2006)
Signal transduction	FGFR1	Neurite outgrowth Gephyrin clustering	Kirschbaum et al. (2009), Kriebel et al. (2011)
Intracellular interactions	AnkyrinG	Assembly and stabilisation of nodal and AIS protein complexes	Davis et al. (1993)
	Doublecortin	Endocytosis	Kizhatil et al. (2002), Yap et al. (2012)
	Syntenin-1	Unknown	Koroll et al. (2001)

mainly involved in early developmental mechanism like neurite outgrowth. Later in development, NF180 and NF186 become upregulated (Hassel et al. 1997). In the CNS of rodents, NF180 was found on axon initial segments (AIS) of developing hippocampal neurons where it regulates the organisation of postsynaptic components (Burkarth et al. 2007; Kriebel et al. 2011). NF180 expression becomes downregulated in the adult brain when it is replaced by NF186 as the most abundant neurofascin isoform (Kriebel et al. 2011). NF186 is located at the AIS and at the axonal surface of nodes of Ranvier (Davis et al. 1996). NF186 is involved in the maintenance and stabilisation of the AIS and the nodal protein complexes as well as in the stabilisation of axo-axonic input. In contrast to the neuronal isoforms mentioned above, NF155 is mainly expressed by myelinating cells at the axoglial contact site, the paranodal segments (Poliak and Peles 2003). Accordingly, NF155 becomes upregulated in parallel with the onset of myelination (Collinson et al. 1998; Basak et al. 2007). Here, neurofascin expression is crucial for the development of paranodal septate-like junctions. Nevertheless NF155 was also found on ummyelinated axons as well as on granule and Purkinje cell somata in the cerebellum (Davis et al. 1996).

As a cell adhesion molecule, neurofascin is involved in a large variety of extracellular interactions (Table 10.1). Neurofascin undergoes homophilic interactions *in trans* between opposing cellular membranes, which is a common feature of the L1 subgroup of cell adhesion molecules (Grumet 1997; Lemmon et al. 1989; Pruss et al. 2004). Likewise, heterophilic interactions have been shown between neurofascin and NrCAM, contactin-1 or contactin-2 (Volkmer et al. 1996, 1998; Pruss et al. 2006). Further interactions were documented with components of the extracellular matrix including tenascin-R and brevican (Hedstrom et al. 2007).

Neurofascin also links extracellular interactions to intracellular mechanisms. One type of interaction is represented by binding to cytoskeletal components ankyrin$_G$ and doublecortin (Davis and Bennett 1994; Kizhatil et al. 2002). On the other hand, neurofascin interacts with the submembrane scaffold protein syntenin-1 which recognise a PDZ motif at the COOH-terminus of neurofascin (Koroll et al. 2001). Neurofascin is also linked to signal transduction via interaction with FGFR1 (Kirschbaum et al. 2009). Interestingly, neurofascin binds FGFR1 with both the extra- and the intracellular domain independently of each other.

In summary, the four neurofascin isoforms NF155, NF166, NF180 and NF186 are expressed in selected brain areas. The neuronal isoforms are dynamically regulated from NF166 to NF180 to NF186 while so far only one glial isoform, NF155, has been characterised.

10.3 NF166 and NF180 Promote Neurite Outgrowth and Postsynaptic Differentiation

For neurodevelopmental as well as neurodegenerative diseases, axon guidance molecules and synaptic adhesion molecules appeared as candidate disease genes in general (Mitchell 2011; Stoeckli 2012). NF166 and NF180 are implicated in neurite outgrowth and the formation as well as stabilisation of synaptic structures. Neurofascin may therefore represent an interesting candidate gene due to its pivotal role in nervous system development and plasticity. Here, we discuss the function of both NF166 and NF180 which are functionally equivalent (Trinks et al., unpublished results).

In the course of neuronal development, specific circuits are generated by neurite outgrowth and subsequent interaction of pre- and postsynaptic membranes. Neurite outgrowth is controlled by repulsive and attractive guidance cues which direct extending neurites towards appropriate target regions. Neurofascin NF166 is expressed on extending neurites and was shown to be required for axonal fasciculation (Rathjen et al. 1987). Neurite outgrowth is induced by both homophilic and heterophilic interactions (Volkmer et al. 1996; Pruss et al. 2004). However, interactions were found to be specific for different neurons of the developing nervous system. Neurite outgrowth of central tectal neurons is regulated via interaction with NrCAM, while peripheral dorsal root ganglion neurons make use of contactin-2 (Pruss et al. 2006). A further aspect of neurofascin-dependent neurite outgrowth is the temporal regulation of such interactions. NF166 expressed in dorsal root ganglion neurons becomes replaced by NF186 later in development. While NF166 is permissive for neurite outgrowth, NF186 is not permissive or even inhibitory (Pruss et al. 2006; Koticha et al. 2005). NF186 differs from NF166 and NF180 by the additional expression of the FNIII-like repeat 5. Inclusion of this motif shifts the FNIII-like repeats off the membrane plane which may result in a perpendicular dislocation of domains required for the interaction with other receptors *in cis*. One crucial interaction partner is FGFR1. FGFR1 is a receptor tyrosine kinase which interacts with a variety of IgCAMs including neurofascin, NCAM or L1 (Saffell et al. 1997; Kirschbaum et al. 2009). FGFRs are important for many different cellular processes including migration, proliferation and differentiation (Ornitz and Itoh 2001). In addition to canonical FGF stimulation, cell adhesion molecules activate FGFRs by non-canonical interactions (Murakami et al. 2008). Accordingly, FGFR1 activity is required for neurofascin-dependent neurite outgrowth. Unlike NF186, NF166 was shown to interact with FGFR1. Hence, a possible explanation for the functional difference between NF166 and NF186 relies on selective interaction with FGFR1 (Kirschbaum et al. 2009). Neurite outgrowth via neurofascin–FGFR1 interactions involves activation of protein kinase C (PKC) and phosphatidylinositol 3-kinase (PI3K). Protein kinase signalling is therefore a hallmark of neurofascin-dependent signal transduction. In this line, a further requirement for neurofascin-dependent neurite outgrowth is phosphorylation of serine residues 56 and 100 within the cytosolic domain.

In addition to neurite outgrowth, neurofascin is also implicated in the formation and clustering of the postsynaptic scaffold protein gephyrin. Gephyrin is a scaffold protein, located on the postsynaptic site of inhibitory synapses which is involved in the clustering of glycine receptors and $GABA_A$ receptors (Fritschy et al. 2008; Kneussel et al. 1999; Jacob et al. 2005; Yu et al. 2007). The transport of gephyrin to the submembrane compartment depends on its association with collybistin II, a GTP exchange factor (GEF) specific for cdc42 (Kins et al. 2000; Harvey et al. 2004). Gephyrin is a phosphoprotein and phosphorylation of Ser270 by GSK3β has been shown to regulate gephyrin functions (Tyagarajan et al. 2011). Therefore, protein kinase signalling is an important mechanism to control gephyrin clustering.

In immature neurons, NF180 induces the formation of gephyrin clusters which initially become localised to the axon hillock (Burkarth et al. 2007). Transfection of different mutants of neurofascin revealed that neurofascin is required both for the formation of the gephyrin clusters and that it is necessary for the translocation to the axon hillock of hippocampal neurons in vitro. Heterochronic expression of NF186 which is specific for later stages of development interferes with gephyrin clustering, indicating that the switch from neurofascin isoforms expressed early in development to NF186 is accompanied by a shift in functionality.

Later on in development, NF180 redistributes to the axon initial segment (Kriebel et al. 2011). The AIS is a special subcellular compartment with dense clusters of voltage-gated sodium channels required for the initiation of axon potentials (Hedstrom and Rasband 2006). GABAergic innervation of the axon initial segment is conferred by axonal terminals of axo-axonic chandelier cells which represent a subclass of parvalbumin-positive interneurons (Somogyi et al. 1983). At the AIS, NF180 regulates the size of gephyrin clusters opposed to axo-axonic terminals. As for neurite outgrowth, this process relies on activation of the receptor tyrosine kinase function of FGFR1 (Kriebel et al. 2011). Since gephyrin in turn is linked to the assembly of $GABA_A$ receptors, neurofascin may be implicated in the control of GABAergic innervation at the axon initial segment of hippocampal neurons. Accordingly, neurofascin increases gephyrin and $GABA_A$ receptor clustering in vitro.

So far, these findings show a contribution of neurofascin to the organisation of axo-axonic synapses at the axon initial segment. Axo-axonic synapses seem to be involved in schizophrenia which is grouped to the neurodevelopmental disorders (Lewis 2011). In the dorsolateral prefrontal cortex (DLPFC) of patients suffering from schizophrenia, the number of GAT-1-positive terminals at the axon initial segment is reduced while $GABA_A$ receptors are up-regulated. Altered synaptic connectivity in the DLPFC may be related to working memory deficits found in schizophrenia. Interestingly, expression of the neurofascin interaction partner $ankyrin_G$ is also altered in human patients suffering from schizophrenia (Cruz et al. 2009). It is therefore conceivable that neurofascin may also be implicated in schizophrenia although demonstration of a direct contribution of neurofascin still remains elusive.

In conclusion, NF166 and NF180 are involved in neurite outgrowth and the formation of GABAergic differentiation. Therefore, it is conceivable that they might play a role in neurodevelopmental disorders.

10.4 Glial NF155 Stabilises Paranodal Segments

Rapid conduction of axon potentials relies on the process of myelination (Hartline and Colman 2007). Myelin of oligodendrocytes and Schwann cells ensheathes the axons in central nervous system (CNS) and peripheral nervous system (PNS) of vertebrates, thereby establishing nodes of Ranvier as subcellular compartments as defined by distinct molecular markers. Rapid propagation of action potentials relies on the expression and clustering of voltage-gated sodium channels (Na_v) at the nodes of Ranvier (Fig. 10.2). At the flanking paranodal segments, the myelin sheath closely attaches to the axons at both sides of the nodes of Ranvier, thereby restricting the ion channels from diffusion and creating an adhesion zone called paranodal septate-like junction. The juxtaparanodal region is a further important axonal subcompartment which is characterised by shaker-type voltage-gated potassium (K_v 1.1/1.2) (Salzer 2003).

NF155 becomes up-regulated in Schwann cells and in oligodendrocytes at the onset of myelination suggesting neurofascin as a candidate molecule mediating the axoglial interaction (Collinson et al. 1998; Tait et al. 2000; Basak et al. 2007). NF155 on myelinating glia thereby interacts with axonal membrane proteins in order to form axoglial septate-like junctions (Tait et al. 2000). The axonal membrane at the paranodal septate-like junction contains two different cell recognition molecules forming a complex *in cis*: contactin-1, a GPI-anchored protein of the immunoglobulin superfamily and the neurexin-like molecule contactin-associated protein (Caspr), also known as paranodin (Rios et al. 2000). Glial NF155 recruits axonal contactin-1 to the paranodal segment (Gollan et al. 2003; Sherman et al. 2005). The contactin-1–Caspr interaction is required for the transport and intracellular processing of contactin-1 (Gollan et al. 2003). Blocking the interaction between NF155 and the Caspr/contactin-1 complex inhibits myelination (Charles et al. 2002) and genetic ablation of genes encoding Caspr, contactin-1 or NF155 results both in the disruption of the paranodal septate-like junctions and in loss of ion channel segregation and impaired nerve conduction (Boyle et al. 2001; Bhat et al. 2001; Thaxton et al. 2010). In conclusion, interaction of glial NF155 with the axonal Caspr/contactin-1 complex is required for both nodal and paranodal stability.

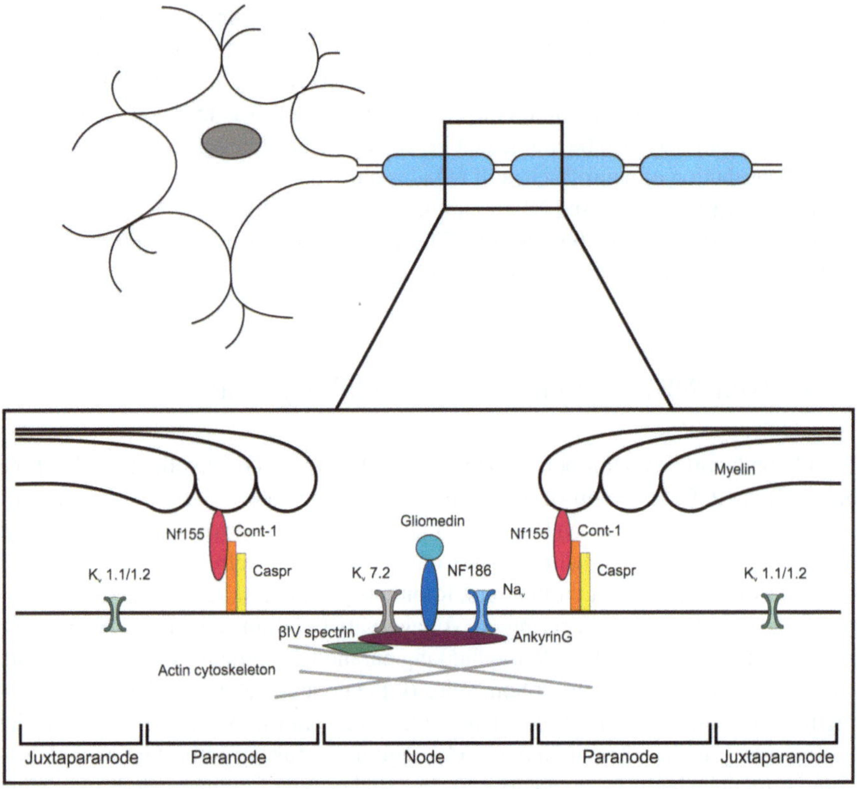

Fig. 10.2 Molecular composition of the node of Ranvier. NF155, NF186, neurofascin isoforms as described; Cont-1, contactin-1; Na_v voltage-gated sodium channels; Kv voltage-gated potassium channels; Caspr, contactin-associated protein

The paranodal complex has an important role not only in the stabilisation of the paranodal region, but also in nodal development. Paranodal septate-like junctions restrict the diffusion of nodal complexes. In the CNS, but not in the PNS, reconstitution of the paranodal septate-like junction in neurofascin null mice by glial expression of NF155 is sufficient to rescue the clustering of Nav channels at the nodes of Ranvier in the CNS (Zonta et al. 2008), further supporting a role for the paranodal septate-like junction in node formation (Feinberg et al. 2010).

10.5 Neurofascin NF186 as an Organiser of Axonal Protein Complexes

While glial NF155 mainly accounts for the stabilisation of paranodal septate-like junctions, axonal NF186 is implicated in the assembly and stabilisation of protein complexes at the AIS and the nodal segments.

Neurofascin has been originally identified as an ankyrin-binding glycoprotein in adult brain (Davis et al. 1993). Immunohistological analysis indicated that NF186 colocalises with $ankyrin_G$ on the AIS of Purkinje neurons and in the nodes of Ranvier of peripheral axons (Davis et al. 1996). $Ankyrin_G$ is a component of the cortical cytoskeleton which links membrane molecules to the actin-based cytoskeleton via spectrins (Bennett and Healy 2009). $Ankyrin_G$ binds to an $ankyrin_G$-binding motif in the cytosolic domain of neurofascin (Davis and Bennett 1994). This motif is shared with other members of the L1 subgroup of neuronal cell adhesion molecules including L1, CHL1 and NrCAM. However, only neurofascin and NrCAM have so far been shown to interact with $ankyrin_G$. NrCAM also colocalises with neurofascin and $ankyrin_G$ at axon initial segments and axonal membranes of the nodes of Ranvier (Davis et al. 1996). In addition to spectrin βIV, $ankyrin_G$ was shown to interact with voltage-gated sodium channels (Na_v) and voltage-gated potassium channels (K_v 7.2, KCNQ2) which form a functional protein complex for the organisation of the AIS and nodal regions together with neurofascin (Hedstrom and Rasband 2006). Therefore, the molecular composition of the AIS and of nodal segments is quite similar.

Neuronal voltage-gated sodium channels are composed of a large pore-forming α subunit and auxiliary β subunits (β1–β4) (Catterall et al. 2005). While Na_v are targeted to the AIS and to nodal segments by $ankyrin_G$ (Lemaillet et al. 2003; Garrido et al. 2003; Gasser et al. 2012), a further interaction has also been reported between Na_v and neurofascin. In particular, the β1 subunit was shown to interact with neurofascin NF186 in nodal segments within the same membrane plane (Ratcliffe et al. 2001). Interestingly, the first Ig-like and second FNIII-like repeat of NF186 may interact with the β1 subunit independently of each other suggesting some kind of a hairpin structure to enable the neurofascin–β1 subunit interaction.

Taken together, neurofascin is part of large protein complex specifically located at the AIS and nodal segments. This complex may link the ankyrin-based cytoskeleton to extracellular interactions. Accordingly, neurofascin was shown to interact with components of the extracellular matrix including brevican and tenascin-R (Volkmer et al. 1998; Hedstrom et al. 2007).

In the PNS, nodal assembly relies on Schwann cells which extend microvilli to the nodal segment (Hedstrom and Rasband 2006). Schwann cells express gliomedin which assemble nodal complexes through interaction with axonal neurofascin (Eshed et al. 2005). Neurofascin becomes first localised to the presumptive nodal region while $ankyrin_G$ becomes clustered only later in development (Lambert et al. 1997). Interference with neurofascin leads to impaired nodal assembly of Na_v, spectrin βIV and $ankyrin_G$, suggesting that neurofascin is downstream of gliomedin clustering and upstream of $ankyrin_G$ clustering (Koticha et al. 2006). Neurofascin isoforms NF186, NF180 and NF166 are expressed in commonly used cultured neurons in vitro and in immature neurons in vivo (Kriebel et al. 2011). It remains therefore elusive which of these axonal isoforms account for nodal assembly. At least NF186 has been shown to rescue the nodal phenotype of a neurofascin knockout mouse (Zonta et al. 2008).

While initial assembly of peripheral nodes of Ranvier relies on neurofascin, this is not the case for the AIS. AIS protein complex assembly is mainly triggered by $ankyrin_G$. Deletion of $ankyrin_G$ leads to dispersal of neurofascin and Na_v (Zhou et al. 1998) while interference with neurofascin expression did not impair the assembly of Na_v, $ankyrin_G$ and spectrin βIV (Hedstrom et al. 2007).

Tyrosine kinase signalling is associated with many mechanisms of development and plasticity. Activation of tyrosine kinase signalling induces the phosphorylation of the $ankyrin_G$-binding motif of neurofascin (Garver et al. 1997). The phosphorylation state of neurofascin selectively regulates the interaction with intracellular components. Phosphorylation abrogates neurofascin–$ankyrin_G$ binding and thereby releases neurofascin from the cytoskeletal tether as implied by enhanced lateral mobility. Phosphorylated neurofascin interacts with doublecortin (Kizhatil et al. 2002) which is a tubulin-binding protein involved in the early organisation of the nervous system. The doublecortin–neurofascin interaction was shown to contribute to the internalisation of neurofascin at somatic and dendritic sites (Yap et al. 2012). While nodal NF186 remains unphosphorylated at the nodal segment as required for $ankyrin_G$ interactions, glial NF155 is phosphorylated suggesting an $ankyrin_G$-independent mechanism of localisation (Jenkins et al. 2001). In addition to alternative splicing, a further degree of complexity is thereby added to neurofascin species by phosphorylation.

It is important to distinguish initial developmental assembly of nodal or AIS protein complexes which may be controlled by NF166, NF180 or NF186, from the maintenance in adult animals which almost exclusively express NF186. Recent experiments using shRNA knockdown technologies in adult rats or inducible conditional knockout mice have greatly expanded our insight into NF186 functions in adult animals (Zonta et al. 2011; Kriebel et al. 2011). At the AIS of cerebellar Purkinje cells, interference with neurofascin expression disintegrates Na_v, $ankyrin_G$ and spectrin βIV (Zonta et al. 2011). As a consequence, spontaneous tonic discharge of axon potentials becomes impaired. Apparently, NF186 is required for long-term stabilisation of the Purkinje cell AIS. By contrast, no impact on the localisation of Na_v has been observed after reduced expression of NF186 at the AIS of dentate gyrus granule cells. Instead, loss of NF186 results in reduced synaptic markers and therefore decreased axo-axonic input (Kriebel et al. 2011). The function of NF186 therefore seems to differ in these two systems. Accordingly, differences in the local organisation of these AIS have been reported. While granule cell AIS are contacted by multiple chandelier cell terminals, the AIS of Purkinje cells is only sparsely innervated. Here, the AIS is mainly contacted by astrocytes (Iwakura et al. 2012). Extracellular adhesion via interaction of neurofascin with extracellular components may be considered as possible mechanisms of how protein complexes at the AIS may trigger stabilisation of presynaptic input. Interaction partners of neurofascin at the AIS are represented by components of the extracellular matrix including tenascin-R and brevican (Hedstrom et al. 2007; Bruckner et al. 2006; Volkmer et al. 1998). Alternatively, neurofascin may homophilically interact with neurofascin as suggested for basket cell terminals (Buttermore et al. 2012; Pruss et al. 2004).

NF186 was also discussed to contribute to the stabilisation of nodal complexes. However, reduction of neurofascin expression in adult animals does not impair long-term nodal clustering of Na_v, spectrin βIV and $ankyrin_G$ which is in contrast to the situation at the AIS (Zonta et al. 2011). Therefore, NF186 appears not to be important for the long-term stabilisation of nodal complexes while it is more implicated in their initial assembly.

10.6 Impaired Neurofascin Functions in Demyelinating Disorders

The localisation of distinct neurofascin isoforms at the nodes of Ranvier as entities of fast saltatory conduction (Davis et al. 1996; Tait et al. 2000) and their corresponding functions in assembling and maintaining this multicellular functional complex (Poliak and Peles 2003) have raised questions about a potential involvement or impairment of neurofascin in the development of associated neurological diseases like multiple sclerosis (MS).

A correlative study in spinal cords of myelin-deficient (md) rats, characterised by severe demyelination of axonal tracts, postnatal death of oligodendrocytes, absence of paranodes and obvious tremor and gait difficulties at postnatal stages, revealed mislocalisation of NF155, normally clustered at septate-like junctions at the paranode. This observation also applied for NF155 interaction partners like Caspr. Moreover juxtaparanodal K_v were not spatially segregated from nodal Na_v underlining the importance of the NF155-containing paranodal protein complex in nodal organisation (Arroyo et al. 2002).

Based on NF155's well-described functions in establishing and maintaining paranodal regions, differences in submembranous localisation of NF155 in chronic relapsing experimental allergic encephalomyelitis (EAE), an animal model displaying characteristics of human MS, pointed towards a need for NF155 in remyelination. Here, lipid raft association of NF155, positively correlating with the differentiation of oligodendrocytes, was found to be perturbed by factors of the extracellular matrix, i.e. fibronectin, infiltrating lesion sites in EAE (Maier et al. 2005).

The importance of NF155 not only for the set-up and maintenance of paranodal subdomains but also for remyelination in the context of demyelinating diseases was further exemplified in a study of post-mortem tissue from MS patients (Howell et al. 2006). Besides loss and alterations of NF155 in actively demyelinating areas of white matter, ectopic expression of NF155 was observed at locations morphologically judged as sites of remyelination, i.e. at contact sites between oligodendrocytes and axons.

First hints towards neurofascin as a target in autoimmune diseases leading to increased inflammation and impairment of saltatory conduction during MS came through the identification of autoantibodies specific for the extracellular domains of both NF155 and NF186 in the plasma of MS patients (Mathey et al. 2007; Hohlfeld et al. 2008). These antibodies when applied to hippocampal tissue slices impaired saltatory conduction, an effect dependent on the complement system as effects only occurred in the presence of fresh sera and could be abolished by heat inactivation of serum components. Testing of corresponding immunglobulins in an EAE mouse model revealed specific detection of the neuronal and nodal isoform NF186 and caused exacerbation of associated clinical symptoms. Pathological signs were accompanied by an increase in ß-amyloid precursor protein, a marker of acute axonal injury, underlining the neuronal character of deterioration.

In this context, the Guillain Barré syndrome, summarising a group of inflammatory neuropathies including acute inflammatory demyelinating polyneuropathy (AIDP), represents another indication characterised by the occurrence of autoantibodies against neurofascin in sera from affected patients (Pruss et al. 2011). By analysing the temporal sequence of nodal and paranodal disorganisation in animal models of AIDP, disorganisation of the nodal factors neurofascin and gliomedin targeted by immunoglobulins and concomitant disruption of sodium channels were identified as an early step in disease progression, preceding obvious morphological alterations of the nodal compartment (Lonigro and Devaux 2009).

The described clinical observations in combination with data derived from animal models for demyelinating diseases further strengthen the notion that an autoimmune response targeting neurofascin as a key player in nodal/paranodal organisation causes initiation of a cascade of pathological alterations ultimately resulting in a failure of fast saltatory conduction and manifestation of clinical symptoms.

10.7 Conclusion

We are at the beginning of an understanding of neurofascin functions for nervous system disorders. While the impact of neurofascin in demyelinating disorders is evident, further studies are required to pinpoint neurofascin functions for neurodevelopmental diseases. Meanwhile it is accepted that a variety of psychiatric disorders like autism spectrum disorders and schizophrenia correlate with altered neuronal circuitry and synaptic connectivity. As a consequence such disorders may be classified as synaptopathies (Brose et al. 2010). This concept may provide a new framework for the understanding of nervous system diseases that is distinct from common theories of chemical imbalance in terms of disturbed neurotransmitter release as a possible cause (Insel 2009). In this line, neurofascin serves as major regulator of neuronal circuitry for both the development of fibre tracts and the stabilisation of synapses. It can therefore be expected that further insight into neurofascin-related functions and signal transduction will increase the understanding of CNS diseases opening new alleys for therapeutical approaches.

Acknowledgements We thank Frank Weise for helpful discussions and critical reading of the manuscript.

Compliance with Ethics Requirements The authors declare that they have no conflicts of interest.

References

Arroyo EJ, Xu T, Grinspan J, Lambert S, Levinson SR, Brophy PJ, Peles E, Scherer SS (2002) Genetic dysmyelination alters the molecular architecture of the nodal region. J Neurosci 22(5):1726–1737

Basak S, Raju K, Babiarz J, Kane-Goldsmith N, Koticha D, Grumet M (2007) Differential expression and functions of neuronal and glial neurofascin isoforms and splice variants during PNS development. Dev Biol 311(2):408–422. doi:10.1016/j.ydbio.2007.08.045, S0012-1606(07)01306-1[pii]

Bennett V, Healy J (2009) Membrane domains based on ankyrin and spectrin associated with cell-cell interactions. Cold Spring Harb Perspect Biol 1(6):a003012. doi:10.1101/cshperspect.a003012, cshperspect.a003012 [pii]

Bhat MA, Rios JC, Lu Y, Garcia-Fresco GP, Ching W, St Martin M, Li J, Einheber S, Chesler M, Rosenbluth J, Salzer JL, Bellen HJ (2001) Axon-glia interactions and the domain organization of myelinated axons requires neurexin IV/Caspr/Paranodin. Neuron 30(2):369–383, S0896-6273(01)00294-X [pii]

Boyle ME, Berglund EO, Murai KK, Weber L, Peles E, Ranscht B (2001) Contactin orchestrates assembly of the septate-like junctions at the paranode in myelinated peripheral nerve. Neuron 30(2):385–397, S0896-6273(01)00296-3 [pii]

Brose N, O'Connor V, Skehel P (2010) Synaptopathy: dysfunction of synaptic function? Biochem Soc Trans 38(2):443–444

Bruckner G, Szeoke S, Pavlica S, Grosche J, Kacza J (2006) Axon initial segment ensheathed by extracellular matrix in perineuronal nets. Neuroscience 138(2):365–375. doi:10.1016/j.neuroscience.2005.11.068, S0306-4522(05)01346-1. [pii]

Burkarth N, Kriebel M, Kranz EU, Volkmer H (2007) Neurofascin regulates the formation of gephyrin clusters and their subsequent translocation to the axon hillock of hippocampal neurons. Mol Cell Neurosci 36(1):59–70. doi:10.1016/j.mcn.2007.06.001, S1044-7431(07)00132-7. [pii]

Buttermore ED, Piochon C, Wallace ML, Philpot BD, Hansel C, Bhat MA (2012) Pinceau organization in the cerebellum requires distinct functions of neurofascin in Purkinje and basket neurons during postnatal development. J Neurosci 32(14):4724–4742. doi:10.1523/JNEUROSCI.5602-11.2012, 32/14/4724 [pii]

Catterall WA, Goldin AL, Waxman SG (2005) International Union of Pharmacology. XLVII. Nomenclature and structure-function relationships of voltage-gated sodium channels. Pharmacol Rev 57(4):397–409. doi:10.1124/pr.57.4.4, 57/4/397 [pii]

Charles P, Tait S, Faivre-Sarrailh C, Barbin G, Gunn-Moore F, Denisenko-Nehrbass N, Guennoc AM, Girault JA, Brophy PJ, Lubetzki C (2002) Neurofascin is a glial receptor for the paranodin/Caspr-contactin axonal complex at the axoglial junction. Curr Biol 12(3):217–220

Collinson JM, Marshall D, Gillespie CS, Brophy PJ (1998) Transient expression of neurofascin by oligodendrocytes at the onset of myelinogenesis: implications for mechanisms of axon-glial interaction. Glia 23(1):11–23

Cruz DA, Weaver CL, Lovallo EM, Melchitzky DS, Lewis DA (2009) Selective alterations in postsynaptic markers of chandelier cell inputs to cortical pyramidal neurons in subjects with schizophrenia. Neuropsychopharmacology 34(9):2112–2124. doi:10.1038/npp.2009.36, npp200936 [pii]

Davis JQ, Bennett V (1994) Ankyrin binding activity shared by the neurofascin/L1/NrCAM family of nervous system cell adhesion molecules. J Biol Chem 269(44):27163–27166

Davis JQ, McLaughlin T, Bennett V (1993) Ankyrin-binding proteins related to nervous system cell adhesion molecules: candidates to provide transmembrane and intercellular connections in adult brain. J Cell Biol 121(1):121–133

Davis JQ, Lambert S, Bennett V (1996) Molecular composition of the node of Ranvier: identification of ankyrin-binding cell adhesion molecules neurofascin (mucin+/third FNIII domain-) and NrCAM at nodal axon segments. J Cell Biol 135(5):1355–1367

Eshed Y, Feinberg K, Poliak S, Sabanay H, Sarig-Nadir O, Spiegel I, Bermingham JR Jr, Peles E (2005) Gliomedin mediates Schwann cell-axon interaction and the molecular assembly of the nodes of Ranvier. Neuron 47(2):215–229. doi:10.1016/j.neuron.2005.06.026, doi:S0896-6273(05)00532-5 [pii]

Feinberg K, Eshed-Eisenbach Y, Frechter S, Amor V, Salomon D, Sabanay H, Dupree JL, Grumet M, Brophy PJ, Shrager P, Peles E (2010) A glial signal consisting of gliomedin and NrCAM clusters axonal Na+channels during the formation of nodes of Ranvier. Neuron 65(4): 490–502. doi:10.1016/j.neuron.2010.02.004, S0896-6273(10)00095-4 [pii]

Fritschy JM, Harvey RJ, Schwarz G (2008) Gephyrin: where do we stand, where do we go? Trends Neurosci 31(5):257–264. doi:10.1016/j.tins.2008.02.006, S0166-2236(08)00093-3 [pii]

Garrido JJ, Giraud P, Carlier E, Fernandes F, Moussif A, Fache MP, Debanne D, Dargent B (2003) A targeting motif involved in sodium channel clustering at the axonal initial segment. Science 300(5628):2091–2094. doi:10.1126/science.1085167300/5628/2091 [pii]

Garver TD, Ren Q, Tuvia S, Bennett V (1997) Tyrosine phosphorylation at a site highly conserved in the L1 family of cell adhesion molecules abolishes ankyrin binding and increases lateral mobility of neurofascin. J Cell Biol 137(3):703–714

Gasser A, Ho TS, Cheng X, Chang KJ, Waxman SG, Rasband MN, Dib-Hajj SD (2012) An AnkyrinG-binding motif is necessary and sufficient for targeting Nav1.6 sodium channels to axon initial segments and nodes of ranvier. J Neurosci 32(21):7232–7243. doi:10.1523/JNEUROSCI.5434-11.2012, 32/21/7232 [pii]

Gollan L, Salomon D, Salzer JL, Peles E (2003) Caspr regulates the processing of contactin and inhibits its binding to neurofascin. J Cell Biol 163(6):1213–1218

Grumet M (1997) Nr-CAM: a cell adhesion molecule with ligand and receptor functions. Cell Tissue Res 290(2):423–428

Hartline DK, Colman DR (2007) Rapid conduction and the evolution of giant axons and myelinated fibers. Curr Biol 17(1):R29–35. doi:10.1016/j.cub.2006.11.042, S0960-9822(06)02523-1 [pii]

Harvey K, Duguid IC, Alldred MJ, Beatty SE, Ward H, Keep NH, Lingenfelter SE, Pearce BR, Lundgren J, Owen MJ, Smart TG, Luscher B, Rees MI, Harvey RJ (2004) The GDP-GTP exchange factor collybistin: an essential determinant of neuronal gephyrin clustering. J Neurosci 24(25):5816–5826

Hassel B, Rathjen FG, Volkmer H (1997) Organization of the neurofascin gene and analysis of developmentally regulated alternative splicing. J Biol Chem 272(45):28742–28749

Hedstrom KL, Rasband MN (2006) Intrinsic and extrinsic determinants of ion channel localization in neurons. J Neurochem 98(5):1345–1352. doi:10.1111/j.1471-4159.2006.04001.x, JNC4001 [pii]

Hedstrom KL, Xu X, Ogawa Y, Frischknecht R, Seidenbecher CI, Shrager P, Rasband MN (2007) Neurofascin assembles a specialized extracellular matrix at the axon initial segment. J Cell Biol 178(5):875–886. doi:10.1083/jcb.200705119, jcb.200705119 [pii]

Hohlfeld R, Meinl E, Dornmair K (2008) B- and T-cell responses in multiple sclerosis: novel approaches offer new insights. J Neurol Sci 274(1–2):5–8. doi:10.1016/j.jns.2008.07.006, S0022-510X(08)00332-8 [pii]

Howell OW, Palser A, Polito A, Melrose S, Zonta B, Scheiermann C, Vora AJ, Brophy PJ, Reynolds R (2006) Disruption of neurofascin localization reveals early changes preceding demyelination and remyelination in multiple sclerosis. Brain 129(Pt 12):3173–3185. doi:10.1093/brain/awl290, awl290 [pii]

Insel TR (2009) Disruptive insights in psychiatry: transforming a clinical discipline. J Clin Invest 119(4):700–705. doi:10.1172/JCI38832, 38832 [pii]

Iwakura A, Uchigashima M, Miyazaki T, Yamasaki M, Watanabe M (2012) Lack of molecular-anatomical evidence for GABAergic influence on axon initial segment of cerebellar Purkinje cells by the pinceau formation. J Neurosci 32(27):9438–9448. doi:10.1523/jneurosci.1651-12.2012

Jacob TC, Bogdanov YD, Magnus C, Saliba RS, Kittler JT, Haydon PG, Moss SJ (2005) Gephyrin regulates the cell surface dynamics of synaptic GABAA receptors. J Neurosci 25(45): 10469–10478

Jenkins SM, Kizhatil K, Kramarcy NR, Sen A, Sealock R, Bennett V (2001) FIGQY phosphorylation defines discrete populations of L1 cell adhesion molecules at sites of cell-cell contact and in migrating neurons. J Cell Sci 114(Pt 21):3823–3835

Kins S, Betz H, Kirsch J (2000) Collybistin, a newly identified brain-specific GEF, induces submembrane clustering of gephyrin. Nat Neurosci 3(1):22–29. doi:10.1038/71096

Kirschbaum K, Kriebel M, Kranz EU, Potz O, Volkmer H (2009) Analysis of non-canonical fibroblast growth factor receptor 1 (FGFR1) interaction reveals regulatory and activating domains of neurofascin. J Biol Chem 284(42):28533–28542. doi:10.1074/jbc.M109.004440, M109.004440 [pii]

Kizhatil K, Wu YX, Sen A, Bennett V (2002) A new activity of doublecortin in recognition of the phospho-FIGQY tyrosine in the cytoplasmic domain of neurofascin. J Neurosci 22(18): 7948–7958

Kneussel M, Brandstatter JH, Laube B, Stahl S, Muller U, Betz H (1999) Loss of postsynaptic GABA(A) receptor clustering in gephyrin-deficient mice. J Neurosci 19(21):9289–9297

Koroll M, Rathjen FG, Volkmer H (2001) The neural cell recognition molecule neurofascin interacts with syntenin-1 but not with syntenin-2, both of which reveal self-associating activity. J Biol Chem 276(14):10646–10654

Koticha D, Babiarz J, Kane-Goldsmith N, Jacob J, Raju K, Grumet M (2005) Cell adhesion and neurite outgrowth are promoted by neurofascin NF155 and inhibited by NF186. Mol Cell Neurosci 30(1):137–148

Koticha D, Maurel P, Zanazzi G, Kane-Goldsmith N, Basak S, Babiarz J, Salzer J, Grumet M (2006) Neurofascin interactions play a critical role in clustering sodium channels, ankyrin G and beta IV spectrin at peripheral nodes of Ranvier. Dev Biol 293(1):1–12. doi:10.1016/j.ydbio.2005.05.028, S0012-1606(05)00363-5 [pii]

Kriebel M, Metzger J, Trinks S, Chugh D, Harvey RJ, Harvey K, Volkmer H (2011) The cell adhesion molecule neurofascin stabilizes Axo-axonic GABAergic terminals at the axon initial segment. J Biol Chem 286(27):24385–24393. doi:10.1074/jbc.M110.212191, M110.212191 [pii]

Kriebel M, Wuchter J, Trinks S, Volkmer H (2012) Neurofascin: a switch between neuronal plasticity and stability. Int J Biochem Cell Biol 44(5):694–697. doi:10.1016/j.biocel.2012.01.012, S1357-2725(12)00016-7 [pii]

Lambert S, Davis JQ, Bennett V (1997) Morphogenesis of the node of Ranvier: co-clusters of ankyrin and ankyrin-binding integral proteins define early developmental intermediates. J Neurosci 17(18):7025–7036

Lemaillet G, Walker B, Lambert S (2003) Identification of a conserved ankyrin-binding motif in the family of sodium channel alpha subunits. J Biol Chem 278(30):27333–27339. doi:10.1074/jbc.M303327200M303327200 [pii]

Lemmon V, Farr KL, Lagenaur C (1989) L1-mediated axon outgrowth occurs via a homophilic binding mechanism. Neuron 2(6):1597–1603

Lewis DA (2011) The chandelier neuron in schizophrenia. Dev Neurobiol 71(1):118–126. doi:10.1002/dneu.20825

Lonigro A, Devaux JJ (2009) Disruption of neurofascin and gliomedin at nodes of Ranvier precedes demyelination in experimental allergic neuritis. Brain 132(Pt 1):260–273. doi:10.1093/brain/awn281, awn281 [pii]

Maier O, van der Heide T, van Dam AM, Baron W, de Vries H, Hoekstra D (2005) Alteration of the extracellular matrix interferes with raft association of neurofascin in oligodendrocytes. Potential significance for multiple sclerosis? Mol Cell Neurosci 28(2):390–401. doi:10.1016/j.mcn.2004.09.012, S1044-7431(04)00233-7 [pii]

Mathey EK, Derfuss T, Storch MK, Williams KR, Hales K, Woolley DR, Al-Hayani A, Davies SN, Rasband MN, Olsson T, Moldenhauer A, Velhin S, Hohlfeld R, Meinl E, Linington C (2007) Neurofascin as a novel target for autoantibody-mediated axonal injury. J Exp Med 204(10):2363–2372. doi:10.1084/jem.20071053, jem.20071053 [pii]

Mitchell KJ (2011) The genetics of neurodevelopmental disease. Curr Opin Neurobiol 21(1): 197–203. doi:10.1016/j.conb.2010.08.009, S0959-4388(10)00129-7 [pii]

Murakami M, Elfenbein A, Simons M (2008) Non-canonical fibroblast growth factor signalling in angiogenesis. Cardiovasc Res 78(2):223–231. doi:10.1093/cvr/cvm086, cvm086 [pii]
Ornitz DM, Itoh N (2001) Fibroblast growth factors. Genome Biol 2 (3):REVIEWS3005
Poliak S, Peles E (2003) The local differentiation of myelinated axons at nodes of Ranvier. Nat Rev Neurosci 4(12):968–980. doi:10.1038/nrn1253
Pruss T, Niere M, Kranz EU, Volkmer H (2004) Homophilic interactions of chick neurofascin in trans are important for neurite induction. Eur J Neurosci 20(11):3184–3188
Pruss T, Kranz EU, Niere M, Volkmer H (2006) A regulated switch of chick neurofascin isoforms modulates ligand recognition and neurite extension. Mol Cell Neurosci 31(2):354–365
Pruss H, Schwab JM, Derst C, Gortzen A, Veh RW (2011) Neurofascin as target of autoantibodies in Guillain-Barre syndrome. Brain 134(Pt 5):e173. doi:10.1093/brain/awq372, author reply e174. awq372 [pii]
Ratcliffe CF, Westenbroek RE, Curtis R, Catterall WA (2001) Sodium channel beta1 and beta3 subunits associate with neurofascin through their extracellular immunoglobulin-like domain. J Cell Biol 154(2):427–434
Rathjen FG, Wolff JM, Chang S, Bonhoeffer F, Raper JA (1987) Neurofascin: a novel chick cell-surface glycoprotein involved in neurite-neurite interactions. Cell 51(5):841–849
Rios JC, Melendez-Vasquez CV, Einheber S, Lustig M, Grumet M, Hemperly J, Peles E, Salzer JL (2000) Contactin-associated protein (Caspr) and contactin form a complex that is targeted to the paranodal junctions during myelination. J Neurosci 20(22):8354–8364
Saffell JL, Williams EJ, Mason IJ, Walsh FS, Doherty P (1997) Expression of a dominant negative FGF receptor inhibits axonal growth and FGF receptor phosphorylation stimulated by CAMs. Neuron 18(2):231–242
Salzer JL (2003) Polarized domains of myelinated axons. Neuron 40(2):297–318, S0896627303006287 [pii]
Sherman DL, Tait S, Melrose S, Johnson R, Zonta B, Court FA, Macklin WB, Meek S, Smith AJ, Cottrell DF, Brophy PJ (2005) Neurofascins are required to establish axonal domains for saltatory conduction. Neuron 48(5):737–742
Somogyi P, Smith AD, Nunzi MG, Gorio A, Takagi H, Wu JY (1983) Glutamate decarboxylase immunoreactivity in the hippocampus of the cat: distribution of immunoreactive synaptic terminals with special reference to the axon initial segment of pyramidal neurons. J Neurosci 3(7):1450–1468
Stoeckli ET (2012) What does the developing brain tell us about neural diseases? Eur J Neurosci 35(12):1811–1817. doi:10.1111/j.1460-9568.2012.08171.x
Tait S, Gunn-Moore F, Collinson JM, Huang J, Lubetzki C, Pedraza L, Sherman DL, Colman DR, Brophy PJ (2000) An oligodendrocyte cell adhesion molecule at the site of assembly of the paranodal axo-glial junction. J Cell Biol 150(3):657–666
Thaxton C, Pillai AM, Pribisko AL, Labasque M, Dupree JL, Faivre-Sarrailh C, Bhat MA (2010) In vivo deletion of immunoglobulin domains 5 and 6 in neurofascin (Nfasc) reveals domain-specific requirements in myelinated axons. J Neurosci 30(14):4868–4876. doi:10.1523/JNEUROSCI.5951-09.2010, 30/14/4868 [pii]
Tyagarajan SK, Ghosh H, Yevenes GE, Nikonenko I, Ebeling C, Schwerdel C, Sidler C, Zeilhofer HU, Gerrits B, Muller D, Fritschy JM (2011) Regulation of GABAergic synapse formation and plasticity by GSK3beta-dependent phosphorylation of gephyrin. Proc Natl Acad Sci USA 108(1):379–384. doi:10.1073/pnas.1011824108, 1011824108 [pii]
Volkmer H, Hassel B, Wolff JM, Frank R, Rathjen FG (1992) Structure of the axonal surface recognition molecule neurofascin and its relationship to a neural subgroup of the immunoglobulin superfamily. J Cell Biol 118(1):149–161
Volkmer H, Leuschner R, Zacharias U, Rathjen FG (1996) Neurofascin induces neurites by heterophilic interactions with axonal NrCAM while NrCAM requires F11 on the axonal surface to extend neurites. J Cell Biol 135(4):1059–1069
Volkmer H, Zacharias U, Norenberg U, Rathjen FG (1998) Dissection of complex molecular interactions of neurofascin with axonin-1, F11, and tenascin-R, which promote attachment and neurite formation of tectal cells. J Cell Biol 142(4):1083–1093

Yap CC, Vakulenko M, Kruczek K, Motamedi B, Digilio L, Liu JS, Winckler B (2012) Doublecortin (DCX) mediates endocytosis of neurofascin independently of microtubule binding. J Neurosci 32(22):7439–7453. doi:10.1523/JNEUROSCI.5318-11.2012, 32/22/7439 [pii]

Yu W, Jiang M, Miralles CP, Li RW, Chen G, de Blas AL (2007) Gephyrin clustering is required for the stability of GABAergic synapses. Mol Cell Neurosci 36(4):484–500. doi:10.1016/j.mcn.2007.08.008, S1044-7431(07)00193-5 [pii]

Zhou D, Lambert S, Malen PL, Carpenter S, Boland LM, Bennett V (1998) AnkyrinG is required for clustering of voltage-gated Na channels at axon initial segments and for normal action potential firing. J Cell Biol 143(5):1295–1304

Zonta B, Tait S, Melrose S, Anderson H, Harroch S, Higginson J, Sherman DL, Brophy PJ (2008) Glial and neuronal isoforms of Neurofascin have distinct roles in the assembly of nodes of Ranvier in the central nervous system. J Cell Biol 181(7):1169–1177. doi:10.1083/jcb.200712154, jcb.200712154 [pii]

Zonta B, Desmazieres A, Rinaldi A, Tait S, Sherman DL, Nolan MF, Brophy PJ (2011) A critical role for neurofascin in regulating action potential initiation through maintenance of the axon initial segment. Neuron 69(5):945–956. doi:10.1016/j.neuron.2011.02.021, S0896-6273(11)00119-X [pii]

Chapter 11
Roles for DSCAM and DSCAML1 in Central Nervous System Development and Disease

María Luz Montesinos

Abstract DSCAMs (Down syndrome cell adhesion molecules) are a group of immunoglobulin-like transmembrane proteins that contain fibronectin III domains. The founding member of the family was isolated in a positional cloning study that sought to identify genes located on chromosome 21 at the locus 21q22.2–q22.3 that is implicated in the neurological and cardiac phenotypes associated with Down's syndrome. In *Drosophila*, Dscam proteins are involved in neuronal wiring, while in vertebrates, the role of these cell adhesion molecules in neurogenesis, dendritogenesis, axonal outgrowth, synaptogenesis, and synaptic plasticity is only just beginning to be understood. In this chapter, we will review the functions ascribed to the two paralogous proteins found in humans, DSCAM and DSCAML1 (DSCAM-like 1), based on findings in knockout mice. The signaling pathways downstream of DSCAM activation and the role of *DSCAM* miss-expression in disease will be also discussed, particularly with regard to the intellectual disability in Down's syndrome.

11.1 Introduction

DSCAM (Down syndrome cell adhesion molecule) is a neural cell adhesion molecule belonging to the immunoglobulin superfamily (Yamakawa et al. 1998). It is strongly expressed in the central nervous system (CNS), especially during development (Agarwala et al. 2001a; Barlow et al. 2001a), and it mediates neuron-to-neuron recognition events that are important for neural circuit formation. This molecule has

M.L. Montesinos, Ph.D. (✉)
Departamento de Fisiología Médica y Biofísica, Universidad de Sevilla,
Av. Sánchez-Pizjuán 4, 41009 Sevilla, Spain
e-mail: mlmontesinos@us.es

V. Berezin and P.S. Walmod (eds.), *Cell Adhesion Molecules: Implications in Neurological Diseases*, Advances in Neurobiology 8, DOI 10.1007/978-1-4614-8090-7_11,

attracted considerable attention for two main reasons. First, the *DSCAM* gene is located on the human chromosome 21, specifically in the so-called Down syndrome critical region (DSCR), trisomy of which is considered a determinant for the intellectual disability associated with Down's syndrome (Delabar et al. 1993; Korenberg et al. 1994; Olson et al. 2007; Belichenko et al. 2009). Second, in *Drosophila, Dscam* is alternatively spliced to generate up to 38,016 different isoforms (Schmucker et al. 2000). This impressive molecular variability regulates neuronal connectivity, as Dscam isoforms have isoform-specific homophilic binding properties (Wojtowicz et al. 2004) that are responsible for neurite self-avoidance. Intriguingly, although vertebrate *DSCAM* genes are not subjected to extensive splicing, they also participate in neuronal wiring. A *DSCAM* paralogue has been identified in vertebrates, *DSCAML1* (*Down Syndrome Cell Adhesion Molecule-Like 1*), which is also expressed in the CNS, although its function remains poorly characterized.

Many excellent reviews have been published on *Drosophila* Dscams (Schmucker 2007; Hattori et al. 2008; Millard and Zipursky 2008; Schmucker and Chen 2009), and therefore, in this chapter we will only briefly refer to the key aspects of *Drosophila* Dscams that are important to analyze the functions and molecular mechanisms underlying the activity of vertebrate DSCAMs. The potential role of *DSCAM* miss-expression in several human diseases will also be discussed.

11.2 Molecular Features of DSCAMs

As mentioned, two *DSCAM* genes exist in humans that both encode transmembrane proteins of about 200 kDa: the founding member of the family, *DSCAM*, located on chromosome 21, and *DSCAML1*, on chromosome 11. As deduced from the cDNA sequence (Yamakawa et al. 1998; Agarwala et al. 2001a), the N-terminal extracellular region of the human DSCAM protein contains ten immunoglobulin-like C2-type (Ig-C2) domains and six fibronectin type III domains, arranged as shown in Fig. 11.1. The Ig-C2 domains each contain about 100 amino acids and they include a pair of conserved cysteines 49–56 amino acids apart that are predicted to form intrachain disulfide bonds. A single transmembrane domain, followed by the intracellular portion of the protein, is also present. Although the cytoplasmic domain lacks known motifs, homology with the Epsin and Shank proteins is observed in two specific regions, suggesting functional links with clathrin-mediated endocytosis and interactions with postsynaptic density proteins (Barlow et al. 2001a). Interestingly, alternative splicing of the transmembrane domain appears to occur, producing an isoform lacking the transmembrane and cytoplasmic domains that may be targeted to the extracellular matrix (Yamakawa et al. 1998). DSCAML1 has a similar domain organization: the extracellular portion of the protein contains nine tandem repeat Ig-C2 domains, and the tenth Ig-C2 domain is located between domains 4 and 5 of the following set of 6 repeated fibronectin type III domains (Agarwala et al. 2001b). However, the transmembrane region of

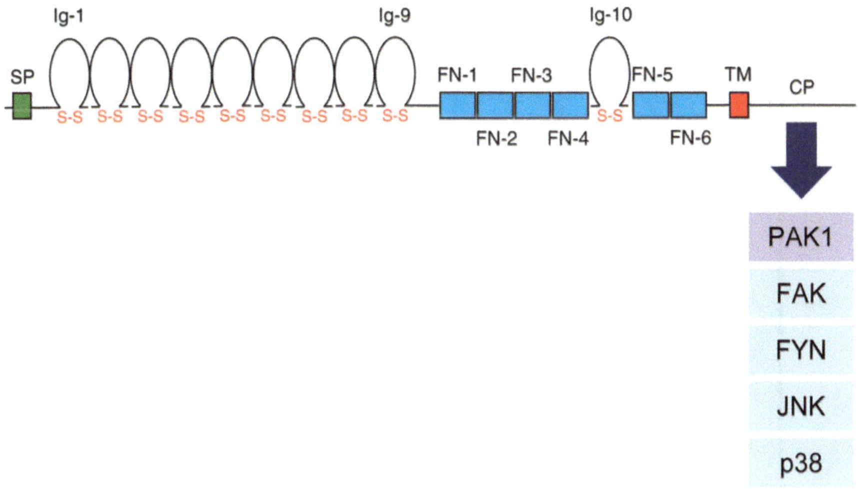

Fig. 11.1 Structure of vertebrate DSCAM protein. Positioned after a signal peptide (SP), the extracellular portion of the protein is formed by several immunoglobulin-like (Ig) and fibronectin type III (FN) domains. The position of the Ig-10 domain between FN domains 4 and 5 is characteristic of DSCAM proteins. The protein also contains a transmembrane (TM) and cytoplasmic (CP) domain. Each Ig domain contains two cysteines (S) that form an intrachain disulfide bond. Kinases known to interact with the cytoplasmic domain are indicated

DSCAML1 is not subjected to alternative splicing, but rather two splice variants are generated involving the region encoding the second Ig-C2 domain (Barlow et al. 2002). In both variants an in-frame deletion of 49 amino acids occurs, affecting either the N-terminal or the C-terminal half of the second Ig-C2 domain. Consequently, the resulting protein lacks this Ig-C2 domain as the appropriate disulfide bond does not form (Barlow et al. 2002).

DSCAMs are highly homologous proteins, with 64 % sequence identity in the extracellular domains and 45 % in the cytoplasmic domains (Agarwala et al. 2001b). Moreover, human and mouse DSCAM proteins share 98 % amino acid identity (Agarwala et al. 2001a; Barlow et al. 2001a) and the domain organization of *Drosophila* Dscam is identical to that of mice and humans (Schmucker et al. 2000). Indeed, the extracellular domain of *Drosophila* Dscam shares 33 % amino acid identity with that of human DSCAM and DSCAML1, although no significant homology has been described for the cytoplasmic domain (Agarwala et al. 2001b; Barlow et al. 2001a).

An important difference between *Drosophila* and vertebrate *DSCAM* genes is the startling number of isoforms produced by alternative splicing in *Drosophila*, as well as in other arthropods. As many as 12, 48, and 33 alternative isoforms can be generated for the second, third, and seventh Ig-C2 domains of *Drosophila* Dscam1 protein, giving rise to 19,008 different ectodomains. Moreover, two transmembrane domains can be produced, resulting in a total of 38,016 potential isoforms (Schmucker et al. 2000). These isoforms bind homophilically via their ectodomains,

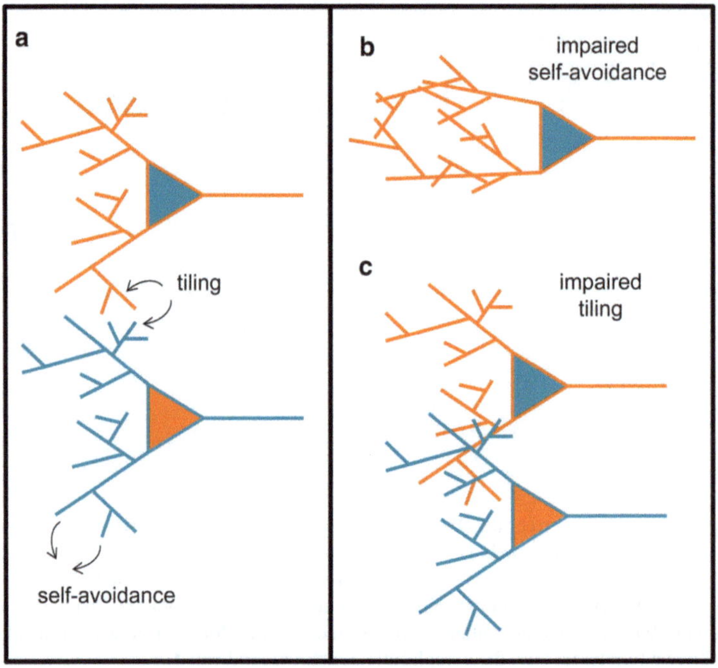

Fig. 11.2 Dendritic self-avoidance and tiling. (**a**) Sister dendrites of a given neuron repel one another (self-avoidance), as do dendrites of neighboring neurons (tiling). (**b**) Impaired self-avoidance causes dendrites to fasciculate and crossover. (**c**) Impaired tiling mechanisms result in the overlapping of the dendritic arbors of neighboring neurons

with little or no heterophilic binding (Wojtowicz et al. 2004, 2007). Interestingly, each neuron expresses a unique set of about 14–50 *Dscam* isoforms (Neves et al. 2004), resulting in a cell surface signature that forms the molecular basis of self-avoidance, the tendency of neurites from the same cell to repel one another in order to ensure correct arborization (Wang et al. 2002; Zhu et al. 2006; Hattori et al. 2007; Hughes et al. 2007; Matthews et al. 2007; Soba et al. 2007. See Fig. 11.2).

Despite exhibiting little (DSCAML1) or no (DSCAM) ectodomain variability, some functionally conservation appears to exist between vertebrate DSCAMs and *Drosophila* Dscam. Using transfected cells in a cell aggregation assay, it was demonstrated that both vertebrate DSCAM (Agarwala et al. 2000) and DSCAML1 (Agarwala et al. 2001b) mediate cell adhesion via homophilic interactions. Importantly, no cell adhesion occurs between cells transfected with *DSCAM* and those expressing *DSCAML*, indicating that heterophilic interactions between these cell types do not occur (Yamagata and Sanes 2008).

High levels of *DSCAM* mRNA expression have been observed in fetal and adult brain, in which at least three transcripts (9.7, 8.5, and 7.6 kb) can be detected (Yamakawa et al. 1998). Pioneering mouse in situ hybridization studies pointed to a role for DSCAM in early development and differentiation, as suggested by the coincidence of *DSCAM* expression with the period of neurite outgrowth (Yamakawa et al. 1998). Although *DSCAM* levels in the brain decrease in adulthood (Agarwala et al. 2001a), relatively high expression is observed in the cortex, the granule cells of the *dentate gyrus*, and the pyramidal cells of CA1, CA2, and CA3, and in cerebellar Purkinje cells, suggesting roles for DSCAM in learning and memory in adulthood (Barlow et al. 2001a). A single 7.6 kb *DSCAML1* transcript is detected in both fetal and adult brain, with expression peaking in adulthood, suggesting distinct developmental roles for DSCAML1 and DSCAM proteins (Agarwala et al. 2001b). In addition, *DSCAM* and *DSCAML1* exhibit complementary expression patterns in some developing CNS regions, including the spinal cord and cortex. For example, *DSCAM* is expressed ventrally in spinal cord, mostly in developing motor neurons, while *DSCAML1* is expressed dorsally in sensory cells (Barlow et al. 2002). Similarly, in the adult cortex and cerebellum, *DSCAM* transcripts predominate in pyramidal cells of cortical layers 3/5 and in cerebellar Purkinje cells, whereas *DSCAML1* expression is stronger in granule cells of cortical layer 2 and in cerebellar granule cells (Barlow et al. 2002).

11.3 Roles of DSCAMs in the Vertebrate Nervous System

11.3.1 Lessons from Drosophila

Four *Drosophila Dscam* genes exist (*Dscam*1–4). Only *Dscam1* is subjected to extensive alternative splicing, while *Dscam2* encodes two splice variants in the seventh Ig-C2 domain, and *Dscam3* and *Dscam4* encode single ectodomains (Millard et al. 2007). Dscam1 participates in axon guidance (Schmucker et al. 2000; Hummel et al. 2003; Zhan et al. 2004) and dendritic self-avoidance (Hughes et al. 2007; Matthews et al. 2007; Soba et al. 2007), thereby fulfilling both axonal and dendritic functions. Interestingly, the subcellular localization of Dscam1 is dependent on its transmembrane domain and the two potential splice variants that can be generated, TM1 and TM2, are targeted to dendrites and axons, respectively (Wang et al. 2004; Shi et al. 2007). Furthermore, Dscam2 controls neuronal tiling (Millard et al. 2007), the tendency of neurites from neighboring neurons not to overlap (Fig. 11.2).

Self-avoidance and tiling are mediated by repulsive forces between neurites from the same or different neurons, respectively. To explain how these processes may be mediated by molecules that bind homophilically, a model has been proposed based on the following observations. Binding is possible only between identical Dscam1 ectodomains (Wojtowicz et al. 2004, 2007), and since sister neurites from a given cell may express the same combination of Dscam1 isoforms, they can interact and bind to one other, activating the corresponding cytoplasmic domains. This initiates a signaling cascade that promotes repulsion by modifying cytoskeletal proteins (Hughes et al. 2007; Matthews et al. 2007). Tiling may be mediated by a similar stepwise process, although in this case neurites of neighboring cells would express the same Dscam isoform. Since Dscam2, 3, and 4 do not suffer extensive molecular diversification, they could fulfill this requirement, and in fact, Dscam2 can regulate axonal tiling of a specific neuronal class (Millard et al. 2007).

Despite the molecular differences discussed, some functional conservation between *Drosophila* and vertebrate DSCAMs appears to exist. This will become more evident as the known functions of vertebrate DSCAMs are discussed in the following sections.

11.3.2 The Role of DSCAMs in Self-Avoidance, Tiling, and Synaptic Connectivity in the Vertebrate Retina

Four different *DSCAM* mutant mice have been characterized in recent years: $Dscam^{del17}$, $Dscam^{tm1.1Kzy}$, $Dscam^{2J}$, and the conditional mutant $Dscam^{F}$ (Table 11.1). Despite exhibiting neurological phenotypes, no gross histological defects were detected in most cases, although retinal disorganization was evident (Fuerst et al. 2008, 2010, 2012). DSCAM is normally expressed in some subtypes of retinal amacrine cells, which exhibit arborization defects in mutant mice. Although the total length of the neurites and the number of branches per unit length are unaffected, the number of self-crossing neurites increases, indicating impaired self-avoidance. The cell bodies that normally form regularly spaced mosaics become randomly distributed or clumped, suggesting a role for DSCAM in tiling (Fuerst et al. 2008). A similar phenotype was also described in other DSCAM-expressing retinal neurons, such as ganglion cells (RGCs; Fuerst et al. 2009; see below).

DSCAM and DSCAML1 exhibit non-overlapping expression patterns in the mouse retina. Retinal rod bipolar cells normally express DSCAML1, and interestingly, these cells exhibit fasciculated dendrites in *DSCAML1* knockout mice. Mosaic spacing of a subset of DSCAML1-expressing amacrine cells is also disrupted in these mice (Fuerst et al. 2009). Together, these observations suggest a role for DSCAML1 in self-avoidance and tiling in specific retinal cell types.

Some light was shed on how DSCAMs might mediate self-avoidance without ectodomain variability by analyzing three types of RGCs in the *DSCAM* mutant retina (Fuerst et al. 2009). Using specific markers to label each RGC type, dendritic fasciculation and cell aggregation were detected in all three subpopulations, although no combination of the different RGC types was observed in these aggregates and fascicles. Based on these findings, the authors proposed that DSCAMs may act as "nonstick coatings," masking other cell-type adhesive tags and passively preventing adhesion between neurons of the same class. In the mutant, this "DSCAM coat" is lost, resulting in the formation of homotypic clumps. Thus, DSCAM does not confer cell-type identity, but rather it prevents cell-type-specific adhesion mediated by other molecules.

Interestingly, protocadherins were recently shown to promote dendritic self-avoidance in mouse retinal and cerebellar Purkinje cells (Lefebvre et al. 2012). Protocadherins are encoded by a complex locus from which 58 proteins are

Table 11.1 DSCAM mutant mice

	Dscam^{del17}	*Dscam*$^{tm1.1Kzy}$	*Dscam*^{2J}	*Dscam*^{F}
Generation	Spontaneous (38-bp deletion in exon 17, resulting in a protein truncated at the second fibronectin domain)	Targeted deletion of the first exon	Spontaneous (4-bp duplication in exon 19, resulting in a protein truncated in the extracellular domain and bearing 84 novel amino acids)	Conditional allele (floxed exon 27, encoding the transmembrane domain)
Viability	Perinatal death on a C57BL/6 background. Viable on a mixed background	Perinatal death on a C57BL/6 background. Viable on a mixed background	Viable (even in inbred C3H background)	–
Retinal self-avoidance and tiling	Impaired	N.D.	Impaired	Impaired
Retinal synaptic lamination	Normal	N.D.	Impaired	Impaired
Retinal cell death	Impaired	N.D.	Impaired	Impaired
Axonal outgrowth and guidance	Altered eye-specific segregation of retinogeniculate projections	Normal spinal commissural axons	N.D.	N.D.
Other relevant phenotypes	Spontaneous seizures, kyphosis, hydrocephalus, decreased motor function, impaired motor learning, malformation of cortex, corpus callosum, and commissural fibers	Defects in respiratory rhythm generator, medullar enlargement	Kyphosis	N.D.
References	Fuerst et al. (2008), Blank et al. (2011), Xu et al. (2011)	Amano et al. (2009), Palmesino et al. (2012)	Fuerst et al. (2010)	Fuerst et al. (2012)

N.D. not determined

produced that differ in their extracellular domains. These genes are expressed stochastically and combinatorially in single neurons, and they can promote isoform-specific homophilic adhesion [Kohmura et al. 1998; Wu and Maniatis 1999; Schreiner and Weiner 2010, reviewed by Zipursky and Sanes (2010)]. Thus, protocadherins may represent the cell-type adhesive tag masked by DSCAMs, at least in some neuronal types. Clearly, the potential functional relationship between DSCAMs and protocadherins (and probably other cell adhesion molecules) requires further investigation.

The role of the DSCAM, DSCAML, and Sidekick proteins (which are structurally related to DSCAMs as they contain immunoglobulin and fibronectin type III domains) has been analyzed in the chick retina (Yamagata and Sanes 2008). Amacrine and bipolar cells make synapses with dendrites of RGCs in the inner plexiform layer of the retina (IPL), a layer that is organized into distinct sublaminae (Fig. 11.3). As in the mouse retina, DSCAM, DSCAML, and Sidekick proteins (collectively referred to as Immunoglobulin Superfamily (IgSF) adhesion molecules) are expressed in non-overlapping sets of cells (Yamagata et al. 2002; Yamagata and Sanes 2008). Interestingly, pre- and postsynaptic cells that synapse in the IPL express the same IgSF (Fig. 11.3). Using interfering RNAs and gain-of-function approaches, IgSF proteins were shown to direct sublamina-specific synaptic connections in the chick retina via homophilic recognition and adhesion (Yamagata and Sanes 2008). Therefore, DSCAMs appear to promote adhesion rather than repulsion in appropriate contexts.

Although initial studies in the retina of $Dscam^{del17}$ mice failed to demonstrate a similar role for DSCAM or DSCAML1 in synaptic connectivity (Fuerst et al. 2008, 2009), recent analyses of new DSCAM knockout mice generated on a different genetic background demonstrated the requirement of DSCAM for synaptic lamination of some amacrine cell types (namely, bNOS-positive and cholinergic amacrine cells; Fuerst et al. 2010). Moreover, specific spatial and/or temporal ablation of *DSCAM* expression in the conditional knockout $Dscam^{F}$ mice (Table 11.1) revealed that while cell spacing and dendritic arborization are mediated by cell-autonomous activity of DSCAM, synaptic connectivity is mediated via a non-cell-autonomous mechanism (Fuerst et al. 2012). Thus, lamination defects in bNOS-positive amacrine cells require the disorganization of non-amacrine cells. Moreover, although *DSCAM* is not expressed by cholinergic amacrine cells, these cells exhibit lamination defects (Fuerst et al. 2012). The precise mechanisms involved in these non-cell-autonomous DSCAM activities remain to be elucidated.

11.3.3 Role of DSCAM in Hippocampal Dendritic Morphogenesis

Given the reported dose-dependent effects of DSCAM (Fuerst et al. 2008, 2012; Amano et al. 2009; Blank et al. 2011) and its overexpression in Down's syndrome brains (Saito et al. 2000), it is essential to better understand not only the effects of DSCAM mutation but also those of DSCAM overexpression. Dendritic arborization

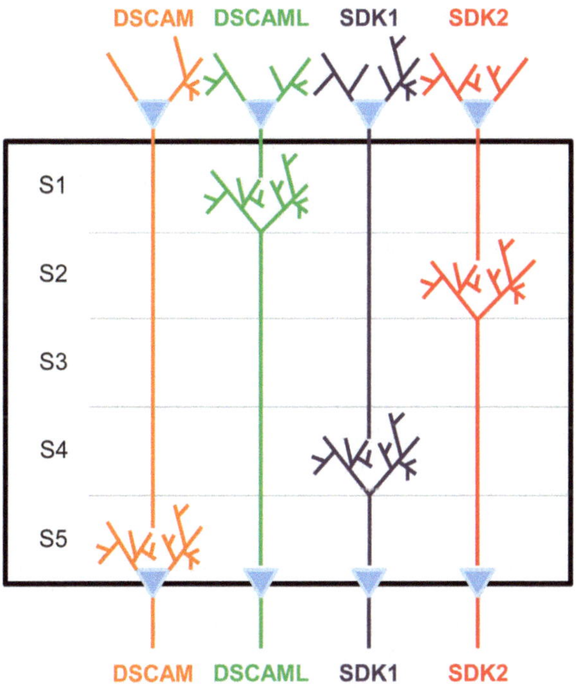

Fig. 11.3 Synaptic lamination in the retinal inner plexiform layer (IPL) is mediated by IgSF molecules. The retinal IPL is subdivided into five sublaminae (S1–S5). Pre- and postsynaptic cells expressing the same IgSF molecule (DSCAM, DSCAML, Sidekick 1, or Sidekick 2) form synapses in specific IPL sublamina [modified from Yamagata and Sanes (2008)]

is dramatically inhibited in transfected hippocampal neurons overexpressing DSCAM, and total dendrite length is reduced (Alves-Sampaio et al. 2010). While the mechanism responsible for this phenotype remains unknown, it is possible that excessive cell surface coating due to DSCAM overexpression impedes other molecular interactions necessary for dendritic branching. Interestingly, dendritic arborization defects are a hallmark of Down's syndrome brains (discussed in Sect. 11.5.1).

11.3.4 Role of DSCAM in Axon Guidance

Dscam is not only known to be involved in axon guidance in *Drosophila* (Schmucker et al. 2000; Hummel et al. 2003; Zhan et al. 2004; Andrews et al. 2008), but in vertebrates it also drives axon guidance in at least some specific neurons. Spinal commissural axons express DCC (Deleted in Colorectal Carcinomas), a receptor involved in axonal guidance in response to Netrin-1, a chemoattractant molecule derived from spinal cord floor-plate cells (Keino-Masu et al. 1996). Commissural axons also express *DSCAM* during development, and pathfinding of these axons is impaired in mouse embryos by *DSCAM* knockdown using siRNAs (Ly et al. 2008).

Similar results were obtained in chick embryos electroporated *in ovo* with *DSCAM*-specific siRNAs (Liu et al. 2009). To fulfill its axon guidance function, DSCAM binds Netrin-1 (Ly et al. 2008; Liu et al. 2009), demonstrating the ability of DSCAM to interact heterophilically with non-related ligands. Interestingly, DSCAM and DCC can form a receptor complex that is disrupted by Netrin-1, although the functional consequences of these interactions remain unclear (Ly et al. 2008). In any case, the role of DSCAM as a Netrin-1 receptor in commissural axon pathfinding appears clear (Ly et al. 2008; Liu et al. 2009), although some conflicting data have also been reported (Palmesino et al. 2012). Strikingly, in *Dscam*$^{tm1.1Kzy}$ mutant mice (Table 11.1), the dorsal root ganglion (DRG) cells that normally express *DSCAM* exhibit reduced neurite extension on DSCAM-coated dishes, suggesting a role of homophilic DSCAM binding in neurite outgrowth (Amano et al. 2009).

To further complicate the scenario, DSCAM also appears to participate in Netrin-1-induced axon repulsion (Purohit et al. 2012), as this bifunctional guidance cue also acts as a chemorepellent in specific cell contexts. In cerebellar neurons, UNC5 protein functions as a Netrin-1 receptor to induce axonal chemorepulsion (Ackerman et al. 1997; Leonardo et al. 1997; Przyborski et al. 1998). Interestingly, attraction promoted by DCC in response to Netrin-1 switches to repulsion when the cytoplasmic domains of UNC5 and DCC interact (Hong et al. 1999). DSCAM and UNC5 were seen to interact via their extracellular domains, an interaction that is enhanced by Netrin-1 (Purohit et al. 2012). In addition, knockdown of *DSCAM* or *UNC5*, or partial simultaneous knockdown of both genes, abolishes the growth cone collapse induced by Netrin-1 in cerebellar granule cells, suggesting that axonal repulsion is mediated by the coordinated action of DSCAM and UNC5 (Purohit et al. 2012). Recently, DSCAM has been shown to physically interact with Draxin, a guidance molecule that mediates the repulsion of spinal commissural axons (Islam et al. 2009) and that inhibits axonal outgrowth of cortical and olfactory bulb neurons (Ahmed et al. 2011).

In summary, vertebrate DSCAM appears to acts as a receptor for heterophilic (and possibly homophilic) axonal guidance cues, mediating either attraction or repulsion.

11.3.5 A Role for DSCAM in Neurogenesis?

Neurogenesis during adulthood is restricted to the subventricular zone (SVZ) and the hippocampal *dentate gyrus* (DG), where progenitor cells are located in the subgranular zone (SGZ). There is evidence that DSCAM may participate in adult hippocampal neurogenesis. Cerebral ischemia is used as an experimental model to induce DG neurogenesis, as this condition increases the proliferation of SGZ progenitor cells (Liu et al. 1998). Global brain ischemia in monkeys (*Macaca fuscata*) induces a decrease in DSCAM labeling in DG granule cells (probably due to the ischemic injury itself), accompanied by an increase in DSCAM immunoreactivity in cells of the SGZ (Yamashima et al. 2006). Using PSA-NCAM as a marker of newly generated SGZ progenitor cells, it was established that about 50 % of DSCAM-positive cells in the SGZ are also positive for PSA-NCAM and that many of these cells are also BrdU positive (indicative of dividing cells), suggesting that they are progenitor cells (Yamashima et al. 2004). These cells were shown to project dendrites later and to express βIII-tubulin, a marker of immature neurons (Yamashima et al. 2006). The remaining 50 % of PSA-NCAM-negative, DSCAM-positive cells in the SGZ were positive for S100β, a marker of immature astrocytes (Boyes et al. 1986).

Although there is no direct proof that DSCAM is involved in neurogenesis (either during development or in adulthood), several observations suggest that DSCAM can interfere with cell cycle progression and/or developmental apoptotic processes. The number of retinal cells expressing DSCAM increases by 250–300 % in *Dscam*^{del17} knockout mice (Table 11.1) due to decreased cell death rather than increased proliferation (Fuerst et al. 2008). Moreover, cell death during retinal developmental is regulated by DSCAM in a dose-dependent manner, and it is independent of other processes regulated by DSCAM in the retina, such as self-avoidance and cell body spacing, ruling out the possibility that cell death is a secondary consequence of dendrite fasciculation or cell clumping due to loss of *DSCAM* function (Fuerst et al. 2012). Interestingly, *Dscam*$^{tm1.1Kzy}$ mutants (Table 11.1) also exhibit more medulla neurons, although this phenotype has not been investigated in detail (Amano et al. 2009). It is tempting to speculate that DSCAM is involved in cell proliferation and/or neuronal cell death during brain development via its interaction with PAK (p21-activated kinase) proteins, which have been implicated in both these processes (see Sect. 11.4).

The putative role of DSCAM in neurogenesis is particularly interesting in the context of Down's syndrome (DS), as defective neurogenesis appears to be involved in this intellectual disability (Contestabile et al. 2007; Guidi et al. 2008, 2011; discussed further in Sect. 11.5.1).

11.3.6 A Possible Role of DSCAM in Synaptic Plasticity

Although the number of vertebrate *DSCAM* isoforms is limited, additional unexpected complexity was recently described by our group (Alves-Sampaio et al. 2010) as we found that at least five different isoforms produced by alternative polyadenylation expressed in the mouse hippocampus. These isoforms bear motifs in the 3′ untranslated regions (3′UTRs) that regulate local dendritic translation of the corresponding transcript. In fact, *DSCAM* mRNA is dendritically localized in the hippocampus and it associates with CPEB1 (Cytoplasmic Polyadenylation Element Binding Protein 1) (Alves-Sampaio et al. 2010). CPEB1 is a key regulator of local dendritic translation and it also facilitates the transport of specific mRNAs into dendrites (Richter 2007). CPEB1 is phosphorylated in response to the activation of NMDA receptors (NMDARs), which results in polyadenylation and concomitant translation of target mRNAs (Huang et al. 2002) known to participate in dendritic

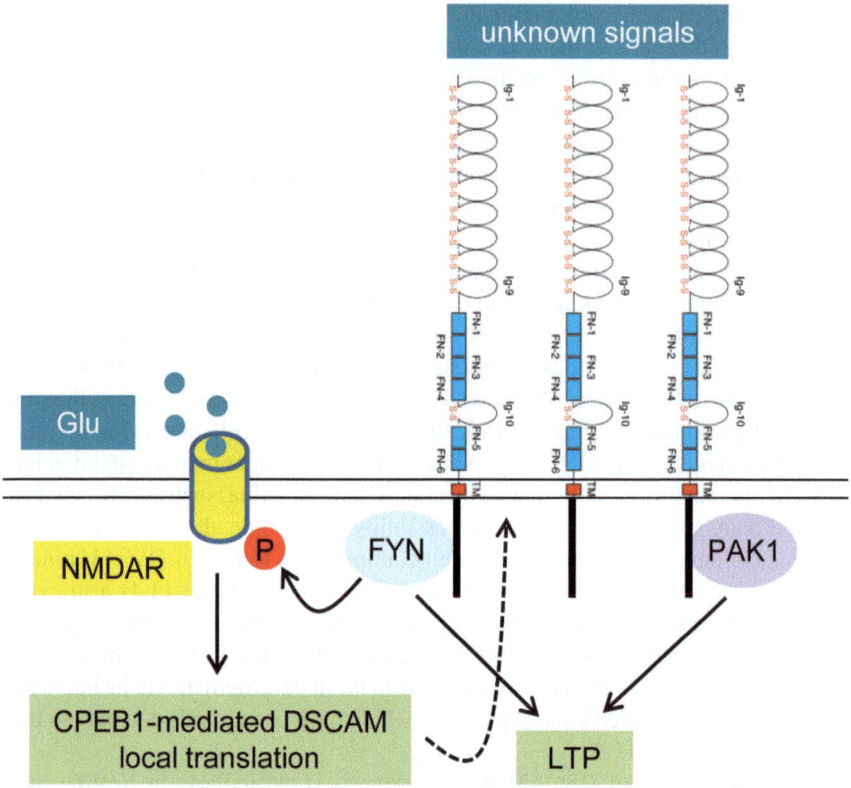

Fig. 11.4 Role of DSCAM in synaptic plasticity: a working model. Postsynaptic DSCAM is activated by unknown mechanisms and it signals through its interaction with PAK1 and FYN kinases, which play important roles in LTP and learning (Grant et al. 1992; Asrar et al. 2009). *DSCAM* mRNA is locally translated in dendrites of hippocampal neurons in response to NMDA receptor (NMDAR) activation by glutamate (Glu), via CPEB1 activation (Alves-Sampaio et al. 2010). This locally translated DSCAM may be inserted into the activated postsynaptic element, thereby enhancing the effects of PAK1 and FYN kinases. Moreover, by positively regulating NMDARs through its phosphorylation (Kojima et al. 1998; Tezuka et al. 1999; Nakazawa et al. 2001), the FYN kinase may enhance this effect via a positive feedback loop

morphogenesis during development and in synaptic plasticity in adulthood (Richter 2007). Using a heterologous reporter system in *Xenopus* oocytes, we demonstrated that the translation of some *DSCAM* 3′UTR isoforms is mediated by CPEB. Moreover, incubation of hippocampal neurons with NMDA increased local dendritic translation of DSCAM (Alves-Sampaio et al. 2010). Although it remains unclear whether locally translated DSCAM can activate PAK proteins, such an interaction may be relevant to the modulation of synaptic plasticity, particularly as PAK1 knockout mice exhibit impaired long-term potentiation (LTP) in the hippocampal CA1 region (Asrar et al. 2009). Together these results suggest that DSCAM could be involved in hippocampal synaptic plasticity (Fig. 11.4) and that some DSCAM developmental functions could be mediated by locally translated protein isoforms.

11.4 Signaling Cascades Downstream of DSCAM

Several proteins that interact with Dscam have been identified in *Drosophila*. While the cytoplasmic domain of Dscam interacts with Pak via the adaptor protein Dock (Schmucker et al. 2000), the sorting nexin DSH3PX1 forms a complex with Dscam and Dock (Worby et al. 2001). Moreover, DSH3PX1 interacts with Wasp, a regulator of the actin cytoskeleton that is also involved in endocytosis, and with AP-50, a clathrin-coat adaptor protein (Worby et al. 2001), suggesting that Dscam can regulate the actin cytoskeleton via Pak and/or Wasp, and that like other cell surface receptors, it may be internalized when activated.

The cytoplasmic domain of *Drosophila* Dscam, which is necessary to transduce homophilic binding into repulsion (Matthews et al. 2007) shows little homology with the cytoplasmic domains of vertebrate DSCAMs, although they appear to engage similar signaling cascades. Human DSCAM interacts directly with PAK1, inducing its activation by phosphorylation (Li and Guan 2004). In mammals, the PAK family is made up of three proteins that share more than 90 % amino acid identity [reviewed by Kreis and Barnier (2009)]. While PAK1 is expressed in the brain and spleen, PAK2 is ubiquitously expressed and PAK3 is predominantly expressed in the brain. Thus, although DSCAM interactions have only been formally described for PAK1, it is possible that other members of the PAK family can mediate DSCAM functions in specific brain regions or subcellular compartments. For example, PAK1 is located in the dendrites and axons of hippocampal and cortical neurons, while PAK3 accumulates in the cell bodies (Hayashi et al. 2002; Ong et al. 2002). Moreover, only PAK1 can shuttle to the nucleus and consequently regulate gene transcription (Li et al. 2002; Singh et al. 2005). PAKs bind to Rho family GTPases (RhoA, Rac1, and Cdc42) and they participate in a number of processes associated with cytoskeletal reorganization, including the regulation of cell proliferation and migration, neurogenesis, developmental apoptosis, neuronal polarity, neurite outgrowth, dendritic branching, spine formation, axonal guidance, and synaptic plasticity [reviewed by Kreis and Barnier (2009)]. The signaling pathways that lie downstream of DSCAM/PAK activation remain largely unknown. Netrin-1, which is involved in axonal pathfinding, enhances the DSCAM-mediated phosphorylation of PAK1, and it is sufficient to stimulate tyrosine phosphorylation of the DSCAM cytodomain (Liu et al. 2009). The FAK and Fyn kinases that mediate Netrin-1-induced phosphorylation of UNC5 (Li et al. 2006) can also interact with DSCAM, and Netrin-1 enhances these interactions (Purohit et al. 2012). Finally, DSCAM binds to and activates JNK and p38 kinases (Li and Guan 2004). In summary, vertebrate DSCAM appears to interact with and activate a number of different kinases, although the functional consequences of these interactions require further investigation.

The cytoplasmic domains of vertebrate DSCAMs and Sidekicks have also been shown to interact with postsynaptic scaffolding proteins, in particular with members of the MAGI (membrane-associated guanylate kinase with inverted orientation) and PSD-95 (postsynaptic density-95) families (Yamagata and Sanes 2010). Interestingly, these interactions appear to be necessary for the synaptic localization and function of the corresponding IgSF adhesion molecule in retinal lamination (Yamagata and Sanes 2010).

11.5 DSCAM in Disease

Initial interest in *DSCAM* was related to its potential role in Down's syndrome (DS). Although the underlying causes of this complex disorder are largely unknown, some evidence implicates DSCAM overexpression in DS neuronal and cardiac defects. As discussed below, in recent years *DSCAM* has also emerged as a candidate gene implicated in other diseases, including neurological and respiratory disorders.

11.5.1 DSCAM in Down's Syndrome: Intellectual Disability and Cardiac Defects

DS is caused by the partial or total trisomy of human chromosome 21 (HSA21). In addition to intellectual disability, DS-affected individuals often exhibit craniofacial dysmorphism, congenital heart defects, and early onset of Alzheimer-like disease (Contestabile et al. 2010). DSCAM dysfunction has been linked with at least two features of DS brains associated with intellectual disability: dendritic abnormalities and defective neurogenesis. As in other intellectual disabilities, morphological defects of dendrites are observed in the cortex and hippocampus of DS-affected individuals, whereby dendrites are shorter and less branched, and the number of dendritic spines is reduced (Marin-Padilla 1972; Suetsugu and Mehraein 1980; Ferrer and Gullotta 1990). As previously discussed, DSCAM overexpression in transfected hippocampal neurons clearly inhibits dendritic branching (Alves-Sampaio et al. 2010). Therefore, overexpression of DSCAM in DS (Saito et al. 2000) may mediate the associated reduction in dendritic complexity.

Neurogenesis during development is also impaired in DS and fewer neurons are found in the cortex, hippocampus, and other regions of DS fetuses and children (Wisniewski 1990; Guidi et al. 2008; Larsen et al. 2008), and there are fewer dividing cells in the DG and neocortical germinal matrix of DS fetuses (Contestabile et al. 2007). These and other deficits in neurogenesis, including defective adult neurogenesis, are also observed in DS models such as Ts65Dn and Ts1Cje mice, and they appear to be due to cell cycle alterations (Clark et al. 2006; Chakrabarti et al. 2007; Contestabile et al. 2007; Bianchi et al. 2009; Ishihara et al. 2009). Impaired proliferation of cerebellar precursors and increased cell death have also been reported in Ts65Dn mice (Contestabile et al. 2009), and this may account for the reduced cerebellar size of DS-affected individuals. Given that DSCAM loss of function results in decreased cell death and an increase in the number of neurons in specific regions (Fuerst et al. 2008, 2012; Amano et al. 2009), overexpression should produce the opposite effect (i.e., reduced neuronal number, as is the case in DS), and this may contribute to the defective neurogenesis in DS.

In addition to impaired neurogenesis and dendrite morphogenesis, the Ts65Dn mouse model of DS exhibits excessive eye-specific segregation of retinal axons in the dorsal lateral geniculate nucleus (LGN) (Blank et al. 2011). This developmental process is normally regulated by both retinal activity and molecular guidance cues [reviewed by Huberman et al. (2008)]. Interestingly, retinogeniculate synaptic refinement is regulated by the DSCR in a dose-dependent fashion, DSCAM playing a critical role in that process (Blank et al. 2011). Accordingly, the axonal guidance capacity of DSCAM may be determinant for this DS phenotype.

Several forms of synaptic plasticity are impaired in DS mouse models, including NMDAR-dependent LTP and LTD (long-term depression) in the CA1 of the hippocampus (Costa and Grybko 2005; Siarey et al. 2005; Scott-McKean and Costa 2011). Interestingly, local dendritic translation of DSCAM occurs in hippocampal neurons after NMDAR activation, suggesting a possible role of DSCAM in synaptic plasticity (Alves-Sampaio et al. 2010). Thus, DSCAM trisomy may contribute to the memory and learning impairments associated with DS. Indeed, we found that dendritic levels of *DSCAM* mRNA and protein were increased in the hippocampus of Ts1Cje mice and that the influence of NMDA on local *DSCAM* translation was lost in Ts1Cje neurons, probably due to saturated signaling via NMDAR (Alves-Sampaio et al. 2010; Troca-Marin et al. 2012).

In addition to the neuronal phenotypes discussed, congenital heart defects (CHD) are frequent in DS-affected individuals. DSCAM is expressed in the heart during cardiac development and has been long proposed as a candidate gene for DS congenital heart disease based on the molecular characterization of partial HSA21 trisomies (Barlow et al. 2001b; Kosaki et al. 2005). Recently, DSCAM and COL6A2 (an extracellular matrix component also encoded by the HSA21) were shown to cooperate in producing CHD in mice (Grossman et al. 2011). Thus, while transgenic mice overexpressing either *DSCAM* or *COL6A2* under the control of a heart-specific promoter were viable, double transgenic mice exhibited cardiac hypertrophy. Interestingly, co-expression of *DSCAM* and *COL6A2* in cardiomyocytes induced a transcriptional response that preferentially affected the expression of genes implicated in cell adhesion and cardiomyopathies, suggesting that excessive adhesion contributes to cardiac defects (Grossman et al. 2011). Finally, it is also noteworthy that *DSCAML1* is located at the 11q23 locus associated with Jacobsen syndrome (Agarwala et al. 2001b), which also causes intellectual disability and cardiac defects (Mattina et al. 2009).

11.5.2 DSCAM in Other Diseases: Epilepsy, Hydrocephalus, Respiratory Disorders, Idiopathic Scoliosis, and Bipolar Disorder

DSCAM has been proposed as a candidate gene that influences the development of other diseases. For example, *DSCAM* overexpression has been reported in the temporal lobe of patients with intractable epilepsy and in an experimental murine model of epilepsy (Shen et al. 2011). Patients with intractable epilepsy do not respond to antiepileptic drugs, and the mossy fiber sprouting and hippocampal synaptic reorganization they exhibit are thought to underlie their recurrent epileptic seizures. Interestingly, upregulation of *DSCAM* expression in pilocarpine-induced epileptic rats is dependent on seizure frequency (Shen et al. 2011). Given the role of DSCAM in axonal outgrowth and guidance, and the observed upregulation of Netrin-1 in the rat hippocampal DG after pilocarpine-induced epilepsy (Lin et al. 2007), a role for DSCAM was proposed in mediating the mossy fiber sprouting and aberrant circuitry linked to chronic epilepsy (Shen et al. 2011).

Detailed phenotypic characterization of *DSCAM* knockout mice also suggests a role for DSCAM in central respiratory disorders, such as congenital central hypoventilation syndrome and sudden infant death syndrome (Amano et al. 2009). Thus, *Dscam*$^{tm1.1Kzy}$ mice (Table 11.1) die shortly after birth due to dysfunction of the medullar neurons responsible for generating the respiratory rhythm. A role for DSCAM in the formation of the neuronal circuits involved in controlling the respiratory pattern was proposed (Amano et al. 2009). Moreover, the dose dependence of the respiratory *Dscam*$^{tm1.1Kzy}$ phenotype suggests that overexpression of *DSCAM* in DS may contribute to the central respiratory defects frequently observed in this syndrome (Ferri et al. 1997, 1998).

Recent characterization of the *Dscam*^{del17} mutant mouse (Table 11.1) revealed severe hydrocephalus due to enlargement of lateral ventricles, although the mechanisms underlying this phenotype remain unknown (Xu et al. 2011). Finally, the findings of association studies have led *DSCAM* to be proposed as a candidate gene that participates in bipolar disorder (Amano et al. 2008) and idiopathic scoliosis (Sharma et al. 2011), although its role in these diseases remain to be confirmed.

11.6 Perspectives

The first *DSCAM* gene was cloned in 1998, and since then, the analysis of *Drosophila* Dscams has produced a number of important discoveries, highlighting the functional importance of these adhesion molecules in neuronal wiring. These advances have formed the basis for vertebrate DSCAM research, which surprisingly suggests that they share similar functions, despite the molecular differences between the vertebrate and invertebrate molecules. Although a number of exciting functions have been proposed for DSCAMs, including dendritic morphogenesis, axon guidance, neurogenesis, and synaptic plasticity, we are now only just beginning to understand the importance of DSCAMs in these processes, and further research effort is required. Perhaps the most intriguing questions relate to the molecular mechanisms engaged by DSCAMs when performing their physiological functions and to the functional relationships between DSCAMs and other cell adhesion molecules. Indubitably, the roles of DSCAMs in a number of neurological, respiratory, and cardiac diseases constitute issues of the utmost importance, which are currently in their infancy.

Acknowledgments Thanks to Francisco J. Tejedor for critical reading of the manuscript. This work was supported by grants from the Ministerio de Economía y Competitividad (Instituto de Salud Carlos III, Spain; grant PI110507), the Junta de Andalucía (grant P09-CTS-4610), and the Fondation Jérôme Lejeune (France).

Compliance with Ethics Requirements The authors declare that they have no conflicts of interest.

References

Ackerman SL, Kozak LP, Przyborski SA, Rund LA, Boyer BB, Knowles BB (1997) The mouse rostral cerebellar malformation gene encodes an UNC-5-like protein. Nature 386:838–842

Agarwala KL, Nakamura S, Tsutsumi Y, Yamakawa K (2000) Down syndrome cell adhesion molecule DSCAM mediates homophilic intercellular adhesion. Brain Res Mol Brain Res 79:118–126

Agarwala KL, Ganesh S, Amano K, Suzuki T, Yamakawa K (2001a) DSCAM, a highly conserved gene in mammals, expressed in differentiating mouse brain. Biochem Biophys Res Commun 281:697–705

Agarwala KL, Ganesh S, Tsutsumi Y, Suzuki T, Amano K, Yamakawa K (2001b) Cloning and functional characterization of DSCAML1, a novel DSCAM-like cell adhesion molecule that mediates homophilic intercellular adhesion. Biochem Biophys Res Commun 285:760–772

Ahmed G, Shinmyo Y, Ohta K, Islam SM, Hossain M, Naser IB, Riyadh MA, Su Y, Zhang S, Tessier-Lavigne M, Tanaka H (2011) Draxin inhibits axonal outgrowth through the netrin receptor DCC. J Neurosci 31:14018–14023

Alves-Sampaio A, Troca-Marin JA, Montesinos ML (2010) NMDA-mediated regulation of DSCAM dendritic local translation is lost in a mouse model of Down's syndrome. J Neurosci 30:13537–13548

Amano K, Yamada K, Iwayama Y, Detera-Wadleigh SD, Hattori E, Toyota T, Tokunaga K, Yoshikawa T, Yamakawa K (2008) Association study between the Down syndrome cell adhesion molecule (DSCAM) gene and bipolar disorder. Psychiatr Genet 18:1–10

Amano K, Fujii M, Arata S, Tojima T, Ogawa M, Morita N, Shimohata A, Furuichi T, Itohara S, Kamiguchi H, Korenberg JR, Arata A, Yamakawa K (2009) DSCAM deficiency causes loss of pre-inspiratory neuron synchroneity and perinatal death. J Neurosci 29:2984–2996

Andrews GL, Tanglao S, Farmer WT, Morin S, Brotman S, Berberoglu MA, Price H, Fernandez GC, Mastick GS, Charron F, Kidd T (2008) Dscam guides embryonic axons by Netrin-dependent and -independent functions. Development 135:3839–3848

Asrar S, Meng Y, Zhou Z, Todorovski Z, Huang WW, Jia Z (2009) Regulation of hippocampal long-term potentiation by p21-activated protein kinase 1 (PAK1). Neuropharmacology 56:73–80

Barlow GM, Micales B, Lyons GE, Korenberg JR (2001a) Down syndrome cell adhesion molecule is conserved in mouse and highly expressed in the adult mouse brain. Cytogenet Cell Genet 94:155–162

Barlow GM, Chen XN, Shi ZY, Lyons GE, Kurnit DM, Celle L, Spinner NB, Zackai E, Pettenati MJ, Van Riper AJ, Vekemans MJ, Mjaatvedt CH, Korenberg JR (2001b) Down syndrome congenital heart disease: a narrowed region and a candidate gene. Genet Med 3:91–101

Barlow GM, Micales B, Chen XN, Lyons GE, Korenberg JR (2002) Mammalian DSCAMs: roles in the development of the spinal cord, cortex, and cerebellum? Biochem Biophys Res Commun 293:881–891

Belichenko NP, Belichenko PV, Kleschevnikov AM, Salehi A, Reeves RH, Mobley WC (2009) The "Down syndrome critical region" is sufficient in the mouse model to confer behavioral, neurophysiological, and synaptic phenotypes characteristic of Down syndrome. J Neurosci 29:5938–5948

Bianchi P, Ciani E, Contestabile A, Guidi S, Bartesaghi R (2009) Lithium restores neurogenesis in the subventricular zone of the Ts65Dn mouse, a model for Down syndrome. Brain Pathol 20:106–118

Blank M, Fuerst PG, Stevens B, Nouri N, Kirkby L, Warrier D, Barres BA, Feller MB, Huberman AD, Burgess RW, Garner CC (2011) The Down syndrome critical region regulates retinogeniculate refinement. J Neurosci 31:5764–5776

Boyes BE, Kim SU, Lee V, Sung SC (1986) Immunohistochemical co-localization of S-100b and the glial fibrillary acidic protein in rat brain. Neuroscience 17:857–865

Chakrabarti L, Galdzicki Z, Haydar TF (2007) Defects in embryonic neurogenesis and initial synapse formation in the forebrain of the Ts65Dn mouse model of Down syndrome. J Neurosci 27:11483–11495

Clark S, Schwalbe J, Stasko MR, Yarowsky PJ, Costa AC (2006) Fluoxetine rescues deficient neurogenesis in hippocampus of the Ts65Dn mouse model for Down syndrome. Exp Neurol 200:256–261

Contestabile A, Fila T, Ceccarelli C, Bonasoni P, Bonapace L, Santini D, Bartesaghi R, Ciani E (2007) Cell cycle alteration and decreased cell proliferation in the hippocampal dentate gyrus and in the neocortical germinal matrix of fetuses with Down syndrome and in Ts65Dn mice. Hippocampus 17:665–678

Contestabile A, Fila T, Cappellini A, Bartesaghi R, Ciani E (2009) Widespread impairment of cell proliferation in the neonate Ts65Dn mouse, a model for Down syndrome. Cell Prolif 42:171–181

Contestabile A, Benfenati F, Gasparini L (2010) Communication breaks-Down: from neurodevelopment defects to cognitive disabilities in Down syndrome. Prog Neurobiol 91:1–22

Costa AC, Grybko MJ (2005) Deficits in hippocampal CA1 LTP induced by TBS but not HFS in the Ts65Dn mouse: a model of Down syndrome. Neurosci Lett 382:317–322

Delabar JM, Theophile D, Rahmani Z, Chettouh Z, Blouin JL, Prieur M, Noel B, Sinet PM (1993) Molecular mapping of twenty-four features of Down syndrome on chromosome 21. Eur J Hum Genet 1:114–124

Ferrer I, Gullotta F (1990) Down's syndrome and Alzheimer's disease: dendritic spine counts in the hippocampus. Acta Neuropathol 79:680–685

Ferri R, Curzi-Dascalova L, Del Gracco S, Elia M, Musumeci SA, Stefanini MC (1997) Respiratory patterns during sleep in Down's syndrome:importance of central apnoeas. J Sleep Res 6:134–141

Ferri R, Curzi-Dascalova L, Del Gracco S, Elia M, Musumeci SA, Pettinato S (1998) Heart rate variability and apnea during sleep in Down's syndrome. J Sleep Res 7:282–287

Fuerst PG, Koizumi A, Masland RH, Burgess RW (2008) Neurite arborization and mosaic spacing in the mouse retina require DSCAM. Nature 451:470–474

Fuerst PG, Bruce F, Tian M, Wei W, Elstrott J, Feller MB, Erskine L, Singer JH, Burgess RW (2009) DSCAM and DSCAML1 function in self-avoidance in multiple cell types in the developing mouse retina. Neuron 64:484–497

Fuerst PG, Harris BS, Johnson KR, Burgess RW (2010) A novel null allele of mouse DSCAM survives to adulthood on an inbred C3H background with reduced phenotypic variability. Genesis 48:578–584

Fuerst PG, Bruce F, Rounds RP, Erskine L, Burgess RW (2012) Cell autonomy of DSCAM function in retinal development. Dev Biol 361:326–337

Grant SG, O'Dell TJ, Karl KA, Stein PL, Soriano P, Kandel ER (1992) Impaired long-term potentiation, spatial learning, and hippocampal development in fyn mutant mice. Science 258:1903–1910

Grossman TR, Gamliel A, Wessells RJ, Taghli-Lamallem O, Jepsen K, Ocorr K, Korenberg JR, Peterson KL, Rosenfeld MG, Bodmer R, Bier E (2011) Over-expression of DSCAM and COL6A2 cooperatively generates congenital heart defects. PLoS Genet 7:e1002344

Guidi S, Bonasoni P, Ceccarelli C, Santini D, Gualtieri F, Ciani E, Bartesaghi R (2008) Neurogenesis impairment and increased cell death reduce total neuron number in the hippocampal region of fetuses with Down syndrome. Brain Pathol 18:180–197

Guidi S, Ciani E, Bonasoni P, Santini D, Bartesaghi R (2011) Widespread proliferation impairment and hypocellularity in the cerebellum of fetuses with down syndrome. Brain Pathol 21:361–373

Hattori D, Demir E, Kim HW, Viragh E, Zipursky SL, Dickson BJ (2007) Dscam diversity is essential for neuronal wiring and self-recognition. Nature 449:223–227

Hattori D, Millard SS, Wojtowicz WM, Zipursky SL (2008) Dscam-mediated cell recognition regulates neural circuit formation. Annu Rev Cell Dev Biol 24:597–620

Hayashi K, Ohshima T, Mikoshiba K (2002) Pak1 is involved in dendrite initiation as a downstream effector of Rac1 in cortical neurons. Mol Cell Neurosci 20:579–594

Hong K, Hinck L, Nishiyama M, Poo MM, Tessier-Lavigne M, Stein E (1999) A ligand-gated association between cytoplasmic domains of UNC5 and DCC family receptors converts netrin-induced growth cone attraction to repulsion. Cell 97:927–941

Huang YS, Jung MY, Sarkissian M, Richter JD (2002) N-methyl-D-aspartate receptor signaling results in Aurora kinase-catalyzed CPEB phosphorylation and alpha CaMKII mRNA polyadenylation at synapses. EMBO J 21:2139–2148

Huberman AD, Feller MB, Chapman B (2008) Mechanisms underlying development of visual maps and receptive fields. Annu Rev Neurosci 31:479–509

Hughes ME, Bortnick R, Tsubouchi A, Baumer P, Kondo M, Uemura T, Schmucker D (2007) Homophilic Dscam interactions control complex dendrite morphogenesis. Neuron 54:417–427

Hummel T, Vasconcelos ML, Clemens JC, Fishilevich Y, Vosshall LB, Zipursky SL (2003) Axonal targeting of olfactory receptor neurons in Drosophila is controlled by Dscam. Neuron 37:221–231

Ishihara K, Amano K, Takaki E, Shimohata A, Sago H, Epstein CJ, Yamakawa K (2009) Enlarged brain ventricles and impaired neurogenesis in the Ts1Cje and Ts2Cje mouse models of Down syndrome. Cereb Cortex 20:1131–1143

Islam SM, Shinmyo Y, Okafuji T, Su Y, Naser IB, Ahmed G, Zhang S, Chen S, Ohta K, Kiyonari H, Abe T, Tanaka S, Nishinakamura R, Terashima T, Kitamura T, Tanaka H (2009) Draxin, a repulsive guidance protein for spinal cord and forebrain commissures. Science 323:388–393

Keino-Masu K, Masu M, Hinck L, Leonardo ED, Chan SS, Culotti JG, Tessier-Lavigne M (1996) Deleted in Colorectal Cancer (DCC) encodes a netrin receptor. Cell 87:175–185

Kohmura N, Senzaki K, Hamada S, Kai N, Yasuda R, Watanabe M, Ishii H, Yasuda M, Mishina M, Yagi T (1998) Diversity revealed by a novel family of cadherins expressed in neurons at a synaptic complex. Neuron 20:1137–1151

Kojima N, Ishibashi H, Obata K, Kandel ER (1998) Higher seizure susceptibility and enhanced tyrosine phosphorylation of N-methyl-D-aspartate receptor subunit 2B in fyn transgenic mice. Learn Mem 5:429–445

Korenberg JR, Chen XN, Schipper R, Sun Z, Gonsky R, Gerwehr S, Carpenter N, Daumer C, Dignan P, Disteche C et al (1994) Down syndrome phenotypes: the consequences of chromosomal imbalance. Proc Natl Acad Sci USA 91:4997–5001

Kosaki R, Kosaki K, Matsushima K, Mitsui N, Matsumoto N, Ohashi H (2005) Refining chromosomal region critical for Down syndrome-related heart defects with a case of cryptic 21q22.2 duplication. Congenit Anom (Kyoto) 45:62–64

Kreis P, Barnier JV (2009) PAK signalling in neuronal physiology. Cell Signal 21:384–393

Larsen KB, Laursen H, Graem N, Samuelsen GB, Bogdanovic N, Pakkenberg B (2008) Reduced cell number in the neocortical part of the human fetal brain in Down syndrome. Ann Anat 190:421–427

Lefebvre JL, Kostadinov D, Chen WV, Maniatis T, Sanes JR (2012) Protocadherins mediate dendritic self-avoidance in the mammalian nervous system. Nature 488:517–521

Leonardo ED, Hinck L, Masu M, Keino-Masu K, Ackerman SL, Tessier-Lavigne M (1997) Vertebrate homologues of C. elegans UNC-5 are candidate netrin receptors. Nature 386:833–838
Li W, Guan KL (2004) The Down syndrome cell adhesion molecule (DSCAM) interacts with and activates Pak. J Biol Chem 279:32824–32831
Li F, Adam L, Vadlamudi RK, Zhou H, Sen S, Chernoff J, Mandal M, Kumar R (2002) p21-activated kinase 1 interacts with and phosphorylates histone H3 in breast cancer cells. EMBO Rep 3:767–773
Li W, Aurandt J, Jurgensen C, Rao Y, Guan KL (2006) FAK and Src kinases are required for netrin-induced tyrosine phosphorylation of UNC5. J Cell Sci 119:47–55
Lin H, Wu LW, Huang YG, Chen YC, Wen XN (2007) Correlation between hippocampal mossy fiber sprouting and synaptic reorganization and mechanisms of temporal lobe epilepsy. Zhonghua Yi Xue Za Zhi 87:341–344
Liu J, Solway K, Messing RO, Sharp FR (1998) Increased neurogenesis in the dentate gyrus after transient global ischemia in gerbils. J Neurosci 18:7768–7778
Liu G, Li W, Wang L, Kar A, Guan KL, Rao Y, Wu JY (2009) DSCAM functions as a netrin receptor in commissural axon pathfinding. Proc Natl Acad Sci USA 106:2951–2956
Ly A, Nikolaev A, Suresh G, Zheng Y, Tessier-Lavigne M, Stein E (2008) DSCAM is a netrin receptor that collaborates with DCC in mediating turning responses to netrin-1. Cell 133:1241–1254
Marin-Padilla M (1972) Structural abnormalities of the cerebral cortex in human chromosomal aberrations: a Golgi study. Brain Res 44:625–629
Matthews BJ, Kim ME, Flanagan JJ, Hattori D, Clemens JC, Zipursky SL, Grueber WB (2007) Dendrite self-avoidance is controlled by Dscam. Cell 129:593–604
Mattina T, Perrotta CS, Grossfeld P (2009) Jacobsen syndrome. Orphanet J Rare Dis 4:9
Millard SS, Zipursky SL (2008) Dscam-mediated repulsion controls tiling and self-avoidance. Curr Opin Neurobiol 18:84–89
Millard SS, Flanagan JJ, Pappu KS, Wu W, Zipursky SL (2007) Dscam2 mediates axonal tiling in the Drosophila visual system. Nature 447:720–724
Nakazawa T, Komai S, Tezuka T, Hisatsune C, Umemori H, Semba K, Mishina M, Manabe T, Yamamoto T (2001) Characterization of Fyn-mediated tyrosine phosphorylation sites on GluR epsilon 2 (NR2B) subunit of the N-methyl-D-aspartate receptor. J Biol Chem 276:693–699
Neves G, Zucker J, Daly M, Chess A (2004) Stochastic yet biased expression of multiple Dscam splice variants by individual cells. Nat Genet 36:240–246
Olson LE, Roper RJ, Sengstaken CL, Peterson EA, Aquino V, Galdzicki Z, Siarey R, Pletnikov M, Moran TH, Reeves RH (2007) Trisomy for the Down syndrome 'critical region' is necessary but not sufficient for brain phenotypes of trisomic mice. Hum Mol Genet 16:774–782
Ong WY, Wang XS, Manser E (2002) Differential distribution of alpha and beta isoforms of p21-activated kinase in the monkey cerebral neocortex and hippocampus. Exp Brain Res 144:189–199
Palmesino E, Haddick PC, Tessier-Lavigne M, Kania A (2012) Genetic analysis of DSCAM's role as a Netrin-1 receptor in vertebrates. J Neurosci 32:411–416
Przyborski SA, Knowles BB, Ackerman SL (1998) Embryonic phenotype of Unc5h3 mutant mice suggests chemorepulsion during the formation of the rostral cerebellar boundary. Development 125:41–50
Purohit AA, Li W, Qu C, Dwyer T, Shao Q, Guan KL, Liu G (2012) Down syndrome cell adhesion molecule (DSCAM) associates with uncoordinated-5C (UNC5C) in netrin-1-mediated growth cone collapse. J Biol Chem 287:27126–27138
Richter JD (2007) CPEB: a life in translation. Trends Biochem Sci 32:279–285
Saito Y, Oka A, Mizuguchi M, Motonaga K, Mori Y, Becker LE, Arima K, Miyauchi J, Takashima S (2000) The developmental and aging changes of Down's syndrome cell adhesion molecule expression in normal and Down's syndrome brains. Acta Neuropathol 100:654–664
Schmucker D (2007) Molecular diversity of Dscam: recognition of molecular identity in neuronal wiring. Nat Rev Neurosci 8:915–920

Schmucker D, Chen B (2009) Dscam and DSCAM: complex genes in simple animals, complex animals yet simple genes. Genes Dev 23:147–156

Schmucker D, Clemens JC, Shu H, Worby CA, Xiao J, Muda M, Dixon JE, Zipursky SL (2000) Drosophila Dscam is an axon guidance receptor exhibiting extraordinary molecular diversity. Cell 101:671–684

Schreiner D, Weiner JA (2010) Combinatorial homophilic interaction between gamma-protocadherin multimers greatly expands the molecular diversity of cell adhesion. Proc Natl Acad Sci USA 107:14893–14898

Scott-McKean JJ, Costa AC (2011) Exaggerated NMDA mediated LTD in a mouse model of Down syndrome and pharmacological rescuing by memantine. Learn Mem 18:774–778

Sharma S, Gao X, Londono D, Devroy SE, Mauldin KN, Frankel JT, Brandon JM, Zhang D, Li QZ, Dobbs MB, Gurnett CA, Grant SF, Hakonarson H, Dormans JP, Herring JA, Gordon D, Wise CA (2011) Genome-wide association studies of adolescent idiopathic scoliosis suggest candidate susceptibility genes. Hum Mol Genet 20:1456–1466

Shen L, Xiao Z, Pan Y, Fang M, Li C, Chen D, Wang L, Xi Z, Xiao F, Wang X (2011) Altered expression of Dscam in temporal lobe tissue from human and experimental animals. Synapse 65:975–982

Shi L, Yu HH, Yang JS, Lee T (2007) Specific Drosophila Dscam juxtamembrane variants control dendritic elaboration and axonal arborization. J Neurosci 27:6723–6728

Siarey RJ, Villar AJ, Epstein CJ, Galdzicki Z (2005) Abnormal synaptic plasticity in the Ts1Cje segmental trisomy 16 mouse model of Down syndrome. Neuropharmacology 49:122–128

Singh RR, Song C, Yang Z, Kumar R (2005) Nuclear localization and chromatin targets of p21-activated kinase 1. J Biol Chem 280:18130–18137

Soba P, Zhu S, Emoto K, Younger S, Yang SJ, Yu HH, Lee T, Jan LY, Jan YN (2007) Drosophila sensory neurons require Dscam for dendritic self-avoidance and proper dendritic field organization. Neuron 54:403–416

Suetsugu M, Mehraein P (1980) Spine distribution along the apical dendrites of the pyramidal neurons in Down's syndrome. A quantitative Golgi study. Acta Neuropathol 50:207–210

Tezuka T, Umemori H, Akiyama T, Nakanishi S, Yamamoto T (1999) PSD-95 promotes Fyn-mediated tyrosine phosphorylation of the N-methyl-D-aspartate receptor subunit NR2A. Proc Natl Acad Sci USA 96:435–440

Troca-Marin JA, Alves-Sampaio A, Montesinos ML (2012) Deregulated mTOR-mediated translation in intellectual disability. Prog Neurobiol 96:268–282

Wang J, Zugates CT, Liang IH, Lee CH, Lee T (2002) Drosophila Dscam is required for divergent segregation of sister branches and suppresses ectopic bifurcation of axons. Neuron 33:559–571

Wang J, Ma X, Yang JS, Zheng X, Zugates CT, Lee CH, Lee T (2004) Transmembrane/juxtamembrane domain-dependent Dscam distribution and function during mushroom body neuronal morphogenesis. Neuron 43:663–672

Wisniewski KE (1990) Down syndrome children often have brain with maturation delay, retardation of growth, and cortical dysgenesis. Am J Med Genet Suppl 7:274–281

Wojtowicz WM, Flanagan JJ, Millard SS, Zipursky SL, Clemens JC (2004) Alternative splicing of Drosophila Dscam generates axon guidance receptors that exhibit isoform-specific homophilic binding. Cell 118:619–633

Wojtowicz WM, Wu W, Andre I, Qian B, Baker D, Zipursky SL (2007) A vast repertoire of Dscam binding specificities arises from modular interactions of variable Ig domains. Cell 130:1134–1145

Worby CA, Simonson-Leff N, Clemens JC, Kruger RP, Muda M, Dixon JE (2001) The sorting nexin, DSH3PX1, connects the axonal guidance receptor, Dscam, to the actin cytoskeleton. J Biol Chem 276:41782–41789

Wu Q, Maniatis T (1999) A striking organization of a large family of human neural cadherin-like cell adhesion genes. Cell 97:779–790

Xu Y, Ye H, Shen Y, Xu Q, Zhu L, Liu J, Wu JY (2011) Dscam mutation leads to hydrocephalus and decreased motor function. Protein Cell 2:647–655

Yamagata M, Sanes JR (2008) Dscam and Sidekick proteins direct lamina-specific synaptic connections in vertebrate retina. Nature 451:465–469

Yamagata M, Sanes JR (2010) Synaptic localization and function of Sidekick recognition molecules require MAGI scaffolding proteins. J Neurosci 30:3579–3588

Yamagata M, Weiner JA, Sanes JR (2002) Sidekicks: synaptic adhesion molecules that promote lamina-specific connectivity in the retina. Cell 110:649–660

Yamakawa K, Huot YK, Haendelt MA, Hubert R, Chen XN, Lyons GE, Korenberg JR (1998) DSCAM: a novel member of the immunoglobulin superfamily maps in a Down syndrome region and is involved in the development of the nervous system. Hum Mol Genet 7:227–237

Yamashima T, Tonchev AB, Vachkov IH, Popivanova BK, Seki T, Sawamoto K, Okano H (2004) Vascular adventitia generates neuronal progenitors in the monkey hippocampus after ischemia. Hippocampus 14:861–875

Yamashima T, Popivanova BK, Guo J, Kotani S, Wakayama T, Iseki S, Sawamoto K, Okano H, Fujii C, Mukaida N, Tonchev AB (2006) Implication of "Down syndrome cell adhesion molecule" in the hippocampal neurogenesis of ischemic monkeys. Hippocampus 16:924–935

Zhan XL, Clemens JC, Neves G, Hattori D, Flanagan JJ, Hummel T, Vasconcelos ML, Chess A, Zipursky SL (2004) Analysis of Dscam diversity in regulating axon guidance in Drosophila mushroom bodies. Neuron 43:673–686

Zhu H, Hummel T, Clemens JC, Berdnik D, Zipursky SL, Luo L (2006) Dendritic patterning by Dscam and synaptic partner matching in the Drosophila antennal lobe. Nat Neurosci 9:349–355

Zipursky SL, Sanes JR (2010) Chemoaffinity revisited: dscams, protocadherins, and neural circuit assembly. Cell 143:343–353

Part II
Cell Adhesion Molecules Not Belonging to the Immunoglobulin Superfamily

Chapter 12
The Adhesion Molecule Anosmin-1 in Neurology: Kallmann Syndrome and Beyond

Fernando de Castro, Pedro F. Esteban, Ana Bribián, Verónica Murcia-Belmonte, Diego García-González, and Diego Clemente

Abstract Anosmin-1 is the glycoprotein encoded by the *KAL1* gene and part of the extracellular matrix, which was first identified as defective in human Kallmann syndrome (KS, characterised by hypogonadotropic hypogonadism and anosmia); biochemically it is a cell adhesion protein. The meticulous biochemical dissection of the anosmin-1 domains has identified which domains are necessary for the protein to bind its different partners to display its biological effects. Research in the last decade has unravelled different roles of anosmin-1 during CNS development (axon pathfinding, axonal collateralisation, cell motility and migration), some of them intimately related with the cited KS but not only with this. More recently, anosmin-1 has been identified in other pathological scenarios both within (multiple sclerosis) and outside (cancer, atopic dermatitis) the CNS.

12.1 Introduction

Anosmin-1 is the protein encoded by the *KAL1* gene. This is a glycoprotein of the extracellular matrix which was first identified as defective in human Kallmann syndrome (KS) characterised by hypogonadotropic hypogonadism and anosmia.

F. de Castro, M.D., Ph.D. (✉) • P.F. Esteban, Ph.D. • V. Murcia-Belmonte, Ph.D.
D. García-González, Ph.D. • D. Clemente, Ph.D.
Grupo de Neurobiología del Desarrollo-GNDe,
Hospital Nacional de Parapléjicos-SESCAM, Toledo, Spain
e-mail: fdec@sescam.jccm.es; pfesteban@externas.sescam.jccm.es; vmurcia@sescam.jccm.es; digago@sescam.jccm.es; dclemente@sescam.jccm.es

A. Bribián, Ph.D.
Grupo de Neurobiología del Desarrollo-GNDe,
Hospital Nacional de Parapléjicos-SESCAM, Toledo, Spain

Institute of Bioengineering of Catalonia, Parc Cientific de Barcelona and Cell Biology Department, Universidad de Barcelona, Barcelona, Spain
e-mail: abribian@ibec.pcb.ub.es

V. Berezin and P.S. Walmod (eds.), *Cell Adhesion Molecules: Implications in Neurological Diseases*, Advances in Neurobiology 8, DOI 10.1007/978-1-4614-8090-7_12,

Its biochemical characteristics allow for its classification as a cell adhesion protein. Beyond all these classical facts about anosmin-1, research in the last decade has involved the protein in different aspects of central nervous system (CNS) development, some of them related with the cited KS but not only with this and, more recently, anosmin-1 has been identified in other pathological scenarios, outside the CNS, too. In parallel, the meticulous biochemical dissection of the anosmin-1 domains has substantially unravelled which are necessary for the protein to bind receptors and other partners and which are indispensable for anosmin-1 to display its biological effects.

12.2 The *KAL1* Gene and the Protein Anosmin-1: Structure, Expression and Regulation

12.2.1 The KAL1 *Gene*

The gene responsible for the X-linked form of KS, *KAL1*, was first identified in 1991 (Franco et al. 1991; Legouis et al. 1991). The *KAL1* gene comprises 14 exons, is located on the X chromosome (Xp22.3), and has an inactive homologous gene on the Y chromosome (del Castillo et al. 1992), which together with the fact that this gene partially escapes X-chromosome inactivation (Franco et al. 1991; Shapiro et al. 1979) would explain the higher prevalence of the disease in males (Dode et al. 2003; Laitinen et al. 2011). Orthologs have been identified in a variety of invertebrate and vertebrate species, from the nematode worm *C. elegans* to rodents and primates, but despite the high degree of sequence identity among species shown by the protein anosmin-1 coded by the gene *KAL1*; no ortholog has been identified in mouse or rat. The *KAL1* locus is adjacent to the pseudoautosomal region 1 (PAR1) from where a 9 Mb block has been removed or translocated from a common ancestor of mouse and rat (Ross et al. 2005) and while the addition of genes to the PAR1 region from autosomes seems to have occurred in eutherians, macropodid marsupials and monotremes; the loss of PAR1 genes from the mouse X-chromosome is evident (Mangs and Morris 2007). Additionally, this genomic region is not stably propagated in bacteria (Perry et al. 2001) and may, in fact, be highly variable (Church et al. 2009). A high proportion of the genes located in the human PAR1 and the proximal Xp22.3 region has not been identified in the mouse genome or is located on autosomal chromosomes (Gläser et al. 1999). The development of the olfactory system is a well-conserved process throughout evolution, from amphibians to primates, requiring a similar molecular mechanism in all of them and, instead of losing the *KAL1* gene during evolution, it has been suggested that the *KAL1* ortholog in mouse is extremely divergent from the human one, or a compensating mechanism has been originated (Lutz et al. 1993).

12.2.2 The Protein Anosmin-1

The protein encoded by the *KAL1* gene is anosmin-1, which is secreted to the extracellular matrix (ECM). Anosmin-1 is a 680 amino acid glycoprotein that presents a modular disposition of different domains. It contains a cysteine rich region (CR domain), a whey acidic protein-like (WAP) domain similar to that found in serin protease inhibitors, four fibronectin type III domains (FnIII) present in several other ECM and cell-adhesion molecules and a C-terminal region with a high content of basic residues histidines and prolines (Legouis et al. 1993). There are five potential heparan sulphate binding motifs along the protein sequence that seem to be important for the proper localisation of the protein within the ECM (Robertson et al. 2001) and 6 putative *N*-glycosylation sites (Rugarli et al. 1996) whose functionality and role in anosmin-1 function remains to be elucidated. The proposed extended domain arrangement and flexible interdomain connections would make anosmin-1 present large basic charged surfaces within the FnIII repeats that could be important in the binding to heparan sulphates and to other interacting proteins (Robertson et al. 2001).

12.2.3 Expression of KAL1 and Anosmin-1. Regulation of the Expression of the KAL1 Gene

In humans, the pattern of expression of the *KAL1* gene and anosmin-1 has been established by in situ hybridisation and immunohistochemistry. *KAL1* transcripts are detected as early as 5 weeks of development, and the expression of the protein has been described as restricted to basement membranes and interstitial matrices of discrete embryonic areas, as should be expected for an ECM protein. Structures that express *KAL1* include the developing olfactory bulb (OB), retina, cerebellum, spinal cord, inner ear and kidney, correlating with the distribution of clinical significant abnormalities in KS patients (Duke et al. 1995; Hardelin et al. 1999; Lutz et al. 1994). Although in GnRH$^+$-neurons *KAL1* transcripts and anosmin-1 have not been detected, both are present in the olfactory nerve fibres and migratory pathways followed by GnRH$^+$-neurons in their way to the hypothalamus, in close association with these cells (Hardelin et al. 1999).

KAL1 expression within the developing olfactory system has been observed in different species. In humans, expression can be seen in the outer olfactory nerve layer of the OB, in olfactory and granule cells as well as in glial cells (Duke et al. 1995; Lutz et al. 1994). In other species such as rat, chicken or zebrafish, expression has been described in the mitral cell layer of the OB, the lateral olfactory tract (LOT) and the olfactory cortex (Ardouin et al. 2000; Ayari and Soussi-Yanicostas 2007; Ayari et al. 2012; Clemente et al. 2008; Dellovade et al. 2003; Lutz et al. 1993; Soussi-Yanicostas et al. 2002; Yanicostas et al. 2009).

Expression has also been described in the cerebellum in human foetus (Lutz et al. 1994) and in several species such as chick, the Asian musk shrew, rat and zebrafish, associated to Purkinje cells, neurons in deep nuclei, the internal granular layer and astrocytes (Ardouin et al. 2000; Ayari et al. 2012; Clemente et al. 2008; Dellovade et al. 2003; Gianola et al. 2009; Legouis et al. 1993; Rugarli et al. 1993; Soussi-Yanicostas et al. 1996, 2002). More recently *KAL1* expression has been shown in muscle, cultured human skeletal muscle cells, basal layer of the skin, dermal cells such as vascular endothelial cells, fibroblasts and cultured human keratonicytes (Raju and Dalakas 2005; Tengara et al. 2010). Anosmin-1 expression has been shown in the germ cells of the testis during spermatogenesis and in the granulosa cells and oocytes in mature ovaries in the marsupial tammar wallaby, which suggests that in addition to the regulatory role in GnRH migration and therefore in the onset of hypogonadotropic hypogonadism, *KAL1* could also work locally in the gonads regulating spermatogenesis and folliculogenesis (Hu et al. 2011).

Apart from the fact that *KAL1* partially escapes X-chromosome inactivation (Franco et al. 1991; Shapiro et al. 1979), little is known regarding the regulation and control of the expression of this gene. In recent years regulation of *KAL1* expression in different tumour tissues and cell lines and inflammatory disorders has been observed, suggesting a possible role of anosmin-1 in the pathogenesis of some of them. Different factors such hypoxia, the methylation of CpG islands in the promoter of the gene and molecules involved in inflammatory processes could be involved in the regulation of the expression of the *KAL1* gene (Arikawa et al. 2011; Jian et al. 2009; Kawamata et al. 2003; Mihara et al. 2006; Raju and Dalakas 2005; Tengara et al. 2010).

12.2.4 Mechanisms of Action of Anosmin-1

Despite recent significant advances, the mode of action of anosmin-1 is not completely comprehended and interactions with different proteins of the ECM, as well as with membrane receptors, have been suggested as major events in the regulation of the activity of this protein. The best known mechanism of action of anosmin-1 is the interaction with the fibroblast growth factor receptor 1 (FGFR1) and the modulation of the activation of this receptor, linking two of the genes responsible for KS (Ayari and Soussi-Yanicostas 2007; Bribián et al. 2006; Dode et al. 2003; García-González et al. 2010; González-Martínez et al. 2004; Hu et al. 2009; Murcia-Belmonte et al. 2010). Interaction between anosmin-1 and FGFR1 has been reported by co-immunoprecipitation (CoIP) (Ayari and Soussi-Yanicostas 2007; Bribián et al. 2006; González-Martínez et al. 2004), and it has been determined that the WAP and FnIII.1 domains together and the FnIII.3 domain by itself interact with FGFR1 (Hu et al. 2009; Murcia-Belmonte et al. 2010). Not surprisingly, some of the missense mutations in the *KAL1* gene found in KS patients lie within the FnIII domains involved in the interaction with FGFR1 (N267K, E514K, F517L), impeding or greatly reducing the interaction with the receptor and rendering non-functional proteins (Cariboni et al. 2004; Hu et al. 2009; Murcia-Belmonte et al. 2010). Mutations within the WAP domain have also been reported in KS patients, mainly missense mutations replacing some of the cysteine residues and presumably disrupting the formation of disulphide bonds and the correct folding or structure of the domain. An intact WAP domain seems to be required for anosmin-1 biological activity (Bülow et al. 2002; González-Martínez et al. 2004; Hu et al. 2004), but in some scenarios the mutations in the WAP domain do not have a negative effect on the activity of the protein (Bülow et al. 2002; Hu et al. 2004).

Approximately two-thirds of the mutations described in the *KAL1* gene are deletions or frameshift or nonsense mutations that could affect the overall length or composition of the protein, sufficient to account for the disease (Hu and Bouloux 2011), what conflicts with the fact that a truncated N-terminal protein comprising the CR-WAP-FnIII.1 domains, behaves as the full-length protein in some scenarios regarding some of the biological functions of anosmin-1 involving FGFR1 (Bülow et al. 2002; González-Martínez et al. 2004; Hu et al. 2004; Murcia-Belmonte et al. 2010). It has been suggested that the function of the different anosmin-1 domains could be determined or conditioned by the extracellular environment and that the protein could be involved in different phenotypic effects, depending on the cell type or the interaction with different receptors or binding proteins (Andrenacci et al. 2006; Bülow et al. 2002). Therefore, it could be speculated that the biological response exerted by the mutated forms of anosmin-1, could be different from that elicited by the full-length protein or the truncated N-terminal protein lacking the last three FnIII domains, since these proteins could present a different binding capacity to FGFR1 and to other receptors or molecules of the ECM.

The interaction of anosmin-1 with heparan sulphates (HS) present in the ECM seems to be important for the localisation and binding of anosmin-1 to the ECM (Hu et al. 2004; Rugarli et al. 1996; Soussi-Yanicostas et al. 1996), and interaction between anosmin-1 and syndecan-1 and glypican-1 regulates the migration of neuroblasts in the *C. elegans* embryo (Hudson et al. 2006). In fact, it has been shown that the action of anosmin-1 in *C. elegans* is dependent on the presence of HS that contain iduronic-acid and 6O-sulphate groups, but not on HS containing 2O-sulphate groups (Bülow and Hobert 2004). Both FGFR1 and anosmin-1 require heparan sulphate proteoglycans (HSPGs) for their biological functions, since HS are essentials for FGF-FGFR complex formation and receptor activation (Guimond and Turnbull 1999; Hu et al. 2004; Rugarli et al. 1996; Soussi-Yanicostas et al. 1996). It has been hypothesised that due to its diffusible nature, anosmin-1 would exert opposing effects depending on the binding dynamics to FGF2–FGFR1–HS complexes. HS-bound anosmin-1 will associate with pre-formed FGF2–FGFR1 pairs and facilitate FGF2–FGFR1–HS signalling enabling FGFR1-mediated cell migration, the role of anosmin-1 being the presentation of the appropriate HS to the complex.

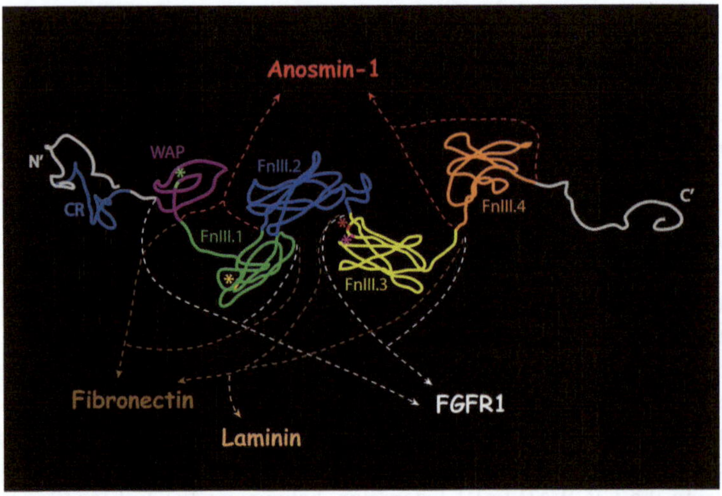

Fig. 12.1 Anosmin-1 interaction map. The WAP-FnIII.1 and FnIII.3 domains allow anosmin-1 to interact with specific sites within FGFR1 and mutations within these domains (N267K in FnIII.1, *orange asterisk*; E514K and F517L in FnIII.3, *red* and *pink asterisk*, respectively) impede this union. The N-terminal truncated protein comprising the CR-WAP-FnIII.1 domains, A1Nt, is still able to bind to the receptor through the WAP and FnIII.1 domains and is biologically functional. On the contrary, in the full-length protein both E514K and F517L substitutions impede the FnIII.3 interaction with FGFR1 and produce a non-functional protein. The N-terminal region would not be able to exert its function correctly, maybe due to the induction by these mutations of conformational changes or to an unstable coupling to the receptor. Mutations in the WAP domain (C172R, *green asterisk*) give rise to functional or non-functional proteins depending on the cellular environment. The FnIII.1 and FnIII.3 domains interact with fibronectin, but only FnIII.3 does interact with laminin. Unlike in the case of FGFR1, the E514K and F517L substitutions do not hinder these interactions and only the F517L mutation reduces the binding to fibronectin. Homophilic anosmin-1/anosmin-1 interactions occur via the FnIII.1 and FnIII.4 domains

In this scenario, the nature of the cell surface HS could be important in determining the anosmin-1-mediated responses. On the other hand, when anosmin-1 levels are high, HS-unbound anosmin-1 could diffuse freely and bind directly to FGFR1 on the cell surface, hampering the formation of the FGF2–FGFR1–HS complex (Hu et al. 2009).

Anosmin-1 has been also shown to interact with other components of the ECM such as uPA (Hu et al. 2004), fibronectin, laminin and anosmin-1 itself (Bribián et al. 2008; Murcia-Belmonte et al. 2010) (Fig. 12.1). Although some of the domains involved in the interaction with FGFR1 participate in the binding to these proteins, the mutations found in FnIII.3 that impede anosmin-1 binding to FGFR1 have little or no effect in the binding to fibronectin and laminin. This, together with the fact that the FnIII.4 domain also participates in the binding to anosmin-1, suggests a mechanism of action independent of FGFR1 (Bribián et al. 2008; Murcia-Belmonte et al. 2010).

12.3 Anosmin-1 Biological Effects

12.3.1 Anosmin-1 in Cell Migration

As a classical ECM protein, anosmin-1 is involved in substrate adhesion and cell migration. This is the case of GnRH-1 neurons, over which anosmin-1 plays a chemoattractive role. Assays performed in GN11 cells (immortalised migrating GnRH-1 human neurons) demonstrated that anosmin-1 with mutations found in KS patients is unable to induce the migration of these cells (Cariboni et al. 2004). This protein is also a chemoattractant cue for neuroblasts migrating out of the subventricular zone (SVZ) on their way to the OB during perinatal development and, together with FGF2, which exerts a motogenic effect; both regulate the migration of these neuronal precursors (García-González et al. 2010). Interestingly, this chemoattraction is mainly mediated through FGFR1 (García-González et al. 2010; Murcia-Belmonte et al. 2010). This fact renders anosmin-1 as one of the first, to date, identified molecules that are involved in the migration of neuroblasts from the SVZ to the OB, before the rostral migratory stream has reached its mature conformation (García-González et al. 2010; Peretto et al. 2005; Petreanu and Álvarez-Buylla 2002). The arrival to the OB of newly generated neuroblasts from the SVZ is crucial for the maintenance of the olfactory function in rodents (Lois and Alvarez-Buylla 1994; Lois et al. 1996). Related to this, anosmin-1 has been shown to play a different role in glial cells: in oligodendrocyte precursors cells (OPC), the relative concentration of anosmin-1 and FGF-2 modulates OPC migration through their interaction with FGFR1 in the optic nerve during development and in the adult brain in mice (Bribián et al. 2006; Clemente et al. 2011). More specifically, anosmin-1 inhibits the motogenic effect of FGF-2 via FGFR1 (Bribián et al. 2006; Clemente et al. 2011). But this effect on migration is tightly related to cell adhesiveness and, in this cell type, anosmin-1 has a stronger adhesive effect than laminin and fibronectin, reducing cell motility in consequence (Bribián et al. 2006, 2008). In OPCs, anosmin-1 mechanism of action is FGFR1 independent and only due to interactions with other ECM proteins (laminin or fibronectin), including anosmin-1 itself in a homophilic way (Bribián et al. 2008). The OPC migration from oligodendrogenic sites to their final emplacement is fundamental for the correct myelination and function of the CNS (de Castro and Bribián 2005).

12.3.2 Axon Outgrowth and Collateral Formation

The first biological effect described for anosmin-1 was its role as a substrate promoting neurite growth in postnatal mouse cerebellar neurons (Soussi-Yanicostas et al. 1998). In that study, cerebellar neurons that co-culture with anosmin-1-expressing CHO cells showed a reduction in neurite growth and an induction in neurite fasciculation. The implication of anosmin-1 has also been found in other

CNS developmental processes in mammals. In rats, anosmin-1 finely attracts OB projection neuron axons (mitral and tufted cells) forming the LOT during a precise developmental time frame (Soussi-Yanicostas et al. 2002). Besides, anosmin-1 enhances collateral branching of LOT axons and exerts a chemoattractive effect on its collateral branches within the piriform cortex (Soussi-Yanicostas et al. 2002). Similar effects are observed in the cerebellum, where anosmin-1 promotes neuritic elongation and strongly increases the budding of collateral branches and the extension of terminal arbours in Pukinje cells from embryonic (E17) and newborn (P0) rats and in axonal regeneration after axotomy (Gianola et al. 2009). In human embryonic GnRH olfactory neuroblasts (FNC-B4 cells), anosmin-1 induces neurite outgrowth and cytoskeletal rearrangements through FGFR1-dependent mechanisms (González-Martínez et al. 2004).

12.4 Anosmin-1 Effects in Other Vertebrates

Two orthologs of the *KAL1* gene have been found in teleosts: *kal1a* and kal1b, encoding anosmin-1a and anosmin-1b, respectively. It has been demonstrated to be an essential requirement for anosmin-1a, but not for anosmin-1b, in GnRH cell migration in zebrafish and medaka (Okubo et al. 2006; Whitlock et al. 2005). In addition, while *kal1a* and *kal1b* display distinct transcription patterns during zebrafish development, both genes are strongly expressed in another migrating cell population from the posterior lateral line primordium (Ardouin et al. 2000), for which migration anosmin-1 is crucial (Yanicostas et al. 2008). More precisely, anosmin-1 seems to play a key role for proper activation of the CXCR4b/SDF1a and/or CXCR7/SDF1a signalling pathways (Yanicostas et al. 2008). Regarding olfactory system development in zebrafish, anosmin-1a depletion impairs the fasciculation of olfactory axons and their terminal targeting within the OB. In this sense, *kal1a* inactivation induces a severe decrease in the number of GABAergic and dopaminergic OB neurons (Yanicostas et al. 2009).

Related to anosmin-1/FGF-2/FGFR1 signalling, a recent study developed in chick embryos has described the effect of anosmin-1 on FGF-8 signalling (Endo et al. 2012). This work highlights the strong influence of anosmin-1 on three morphogen agents, belonging to some well-known families, such as FGF, BMP and WNT, known as key actors in the formation of the neural crest and craniofacial development (Sauka-Spengler and Bronner-Fraser 2008; Trainor et al. 2002) More specifically, Endo et al. (2012) show that anosmin-1 is synthesised locally in the neural crest microenvironment, up-regulating *FGF8* and *BMP5* gene expression. Anosmin-1 also enhances FGF8 activity, while inhibiting both BMP5 and WNT3a activities, being therefore crucial for the formation of cranial neural crest. This study supports the idea of the relevance of this ECM protein at temporally and spatially regulating growth factor activities during embryonic development (Endo et al. 2012) as we and others have suggested (See above).

12.4.1 Anosmin-1 Function in Invertebrates

In *C. elegans*, anosmin-1 ortholog is *kal-1*. In this nematode *kal-1* is involved in two important events in the epithelial morphogenesis: ventral enclosure and male tail formation (Rugarli et al. 2002). Thus, as well as in rodents, *kal-1* affects neurite outgrowth in vivo by modulating neurite branching (Rugarli et al. 2002) and also promotes migration of ventral neuroblasts prior to epidermal enclosure (Hudson et al. 2006). However, in this case, *kal-1* does not modulate FGF signalling in neuroblast migration but it seems that *kal-1* interacts with multiple HSPGs to promote cell migration.

12.5 Anosmin-1 in Kallmann Syndrome

KS is a genetically heterogeneous developmental disease, characterised by hypogonadotropic hypogonadism and anosmia, and its prevalence has roughly been estimated to be from 1:8,000 to 1:10,000 in men (Seminara et al. 1998) and around 1:40,000 in women (Dodé and Hardelin 2009). The presence of a defective sense of smell, whether partial (hyposmia) or complete (anosmia), distinguishes KS from normosmic idiopathic hypogonadotropic hypogonadism with a normal sense of smell (nIHH), which can be associated with mutations in the *GnRHR* and *GPR54* genes. Due to hypothalamic GnRH-1 deficiency, males with KS show cryptorchidism, testicular atrophy and microphallus at birth and then subsequent failure to undergo a normal puberty during adolescence. Females with KS usually present primary amenorrhea or infertility.

First described by the Spanish pathologist Aureliano Maestre de San Juan in 1856, KS is defined by the association of the presence of small testes (hypogonadotropic hypogonadism) with complete (anosmia) or incomplete (hyposmia) olfaction disturbance (Maestre de San Juan 1856). A more detailed description of the syndrome was reported almost a century later using patients from different affected pedigrees with hypogonadism and anosmia (Kallmann et al. 1944). They hint at a broader spectrum of clinical defects and identify the familial nature in the clinical syndrome that was seen in both sexes and accompanied by multiple congenital anomalies. In 1954, de Morsier first noted the link between hypogonadism and neuroanatomical defects, including agenesis of the olfactory bulb and tract and other midline neuroanatomical defects (de Morsier 1954). Since then, it is commonly accepted that hypogonadotropic hypogonadism in KS is caused by the migratory arrest of GnRH-1 neurons, failing to enter the telencephalon and lack of GnRH secretion leads to a complete or partial failure of pubertal development (Wray et al. 1989; Hayes et al. 1998; Schwanzel-Fukuda et al. 1989). However, in the case of hyposmia/anosmia a wide array of anatomical defects could be responsible for the lack of smell ranging from the agenesis/hypoplasia of the OBs to defects in the formation of the olfactory nerve or the lateral olfactory tract and including other causes that may explain this sensory problem.

There are different modes of KS transmission described to date, that include X chromosome-linked recessive, autosomal recessive, autosomal dominant with incomplete penetrance, and most probably digenic/oligogenic inheritance [for more details, see Dodé and Hardelin (2009)]. Multiple genetic causes have been identified so far in the development of this disorder: *KAL1*, *FGFR1*, *PROK2*, *PROKR2*, *NELF*, *KISSR1*, *CHD7*, *SEMA3A* and *FGF8* (Young et al. 2012; Legouis et al. 1991; Dode et al. 2003, 2006; Kim et al. 2008; Pitteloud et al. 2007; Falardeau et al. 2008).

As described above, *KAL1*, the gene causing the X-linked form was the first to be identified in KS (Franco et al. 1991; Legouis et al. 1991). *KAL1* is localised in the Xp22.3 region and encodes anosmin-1, protein that shows a strong homology with adhesion molecules involved in neuronal migration and axonal pathfinding. As previously exposed in this work, *KAL1* has been extensively studied along the last decades, although the absence of an identifiable murine ortholog has denied researchers the opportunity to create and study *Kal1* knock-out mice (see above). Other gene related to KS is *KAL2*, responsible for the autosomal dominant variant of KS, which encodes FGFR1 (Dode et al. 2003). Numerous mutations on *FGFR1* have been described in several functional domains of this receptor. Genotype-phenotype correlations have shown that some clinical features associated with KS satellite symptoms, such as loss of nasal cartilage, hearing impairment, and anomalies of the limbs seem to be mainly associated with *KAL2* mutations. The role of FGFR1 in the normal development of the olfactory bulb explains the association of anosmia with GnRH-1 deficiency in FGFR1-mutated patients. Phenotype analysis indicates that FGFR1 is involved in normal migration of GnRH-1 foetal neurons, but this is clearly not the whole story as a substantial proportion of *KAL2*-mutated individuals have normosmic GnRH-1 deficiency (Martin et al. 2011).

Finally, recent works indicate that FGFR1 signalling is important for the generation of GnRH-1 neurons via the neurotrophic molecule FGF8 (Chung et al. 2010). Interestingly, FGFR1 is expressed by GnRH-1 cells (Gill et al. 2004), and FGFR1 hypomorphic animals show a dramatic reduction in the number of GnRH-1 neurons (Chung and Tsai 2010). Even more, FGF8 is involved in the induction and differentiation of the mouse nasal placode (Kawauchi et al. 2005), and the loss of this morphogen results in the absence of the vomeronasal organ and GnRH-1 neurons (Chung and Tsai 2010). Together with this, the region of the nasal placode from which the GnRH-1 cells emerge is missing in the homozygous FGF8 hypomorphs (Kawauchi et al. 2005). In this sense, recent findings have shown that anosmin-1 promotes the neural crest formation and controls, among other growth factors, FGF8 activity in chick embryo (Endo et al. 2012). However, it is noteworthy that since the entire region in the nasal placode is missing in these mice (homozygous FGF8 hypomorphs), the actual impact of these molecules on the development of GnRH-1 neurons and their precursors is not clear and therefore remains to be elucidated. All these observations together, focused on the description of FGFR1 and FGF8 mutations, shed light on the pathogenesis of GnRH-1 deficiency in general, not just KS (Martin et al. 2011; Villanueva and de Roux 2010).

12.6 Roles of Anosmin-1 in Other Diseases

12.6.1 Anosmin-1 in Multiple Sclerosis

Multiple sclerosis (MS) is the most frequent demyelinating disease in young adults. MS lesions are characterised by demyelination, inflammation, axonal loss and reactive gliosis (Frohman et al. 2006). Accompanying these events, a spontaneous, and sometimes extensive remyelination, is also possible under specific circumstances (Patani et al. 2007; Patrikios et al. 2006). However, in chronic MS lesions remyelination is absent or limited to the plaque border (Barkhof et al. 2003; Bramow et al. 2010; Prineas and Connell 1979). It is plausible that the permissive environment that allows the migration of OPCs during development should be present in this kind of demyelinating diseases, including MS, to produce an effective remyelination (Dubois-Dalcq et al. 2005). Nonetheless, the blockade of oligodendroglial progenitor differentiation is a major determinant of remyelination failure in chronic MS lesions (Kuhlmann et al. 2008). Associated to demyelinated plaques, alterations have been described in the expression pattern of several molecules involved in OPC biology during development such as CXCL1/GRO-α (Omari et al. 2005), semaphorin 3A and 3 F (Williams et al. 2007) and sonic hedgehog (Wang et al. 2008). Regarding the FGF2/anosmin-1 system, the distribution of FGF2 and anosmin-1 varies between the different kinds of demyelinated plaques in MS patients, showing a complementary spatial pattern (Clemente et al. 2011). In areas with active remyelinating activity, i.e. active lesions and the periplaque of chronic-active plaques (Frohman et al. 2006), FGF2 is up-regulated in infiltrating as well as microglia-derived macrophages, whereas anosmin-1 is absent (Clemente et al. 2011). In contrast, where the remyelination process is completely compromised i.e. demyelinated area of chronic-active and chronic-inactive plaques (Frohman et al. 2006), FGF2 is totally absent but Anosmin-1 appears filling the entire extension of both regions (Clemente et al. 2011). Although not ascribed to any cell type, anosmin-1 may be synthesised by astrocytes in these two particular regions, as it occurs in the cerebellum during development (Gianola et al. 2009). In addition, anosmin-1 is present in 13–14 % of the nude axons that pass through the demyelinated area but not in the periplaque or in its adjacent normal appearing white matter (Clemente et al. 2011). Therefore, up-regulated anosmin-1 during remyelination shows a similar profile to that found during human and other mammal development (Dellovade et al. 2003; Duke et al. 1995; Lutz et al. 1994; Soussi-Yanicostas et al. 2002). This is a striking functional histopathological observation and confirms that, in adult MS patients, axons acquire developmental features (Bribián et al. 2008; Soussi-Yanicostas et al. 2002). Similar re-expressions of ECM proteins in MS plaques have been previously shown. PSA-NCAM is re-expressed on 14 % of demyelinated axons in the plaques of chronic lesions but not in the periplaque or in the normal appearing white matter (Charles et al. 2002), and the glycosaminoglycan hyaluronan, as well as its binding transmembrane glycoprotein CD44, also accumulates in the core of chronic MS lesions (Back et al. 2005).

The up-regulation of different adhesion molecules in those regions where remyelination does not occur spontaneously could be interpreted in different ways. It may represent a part of a more general developmental programme reinitiated by neurons in order to protect themselves after the demyelinating injury. But it represents a non-desirable consequence since this neuroprotective activity possibly could render the axon non-permissive to remyelination. However, the mechanism by which adhesion molecules interfere with re-myelination is not well understood. A negative regulation of myelination by this kind of molecules could involve three different and hypothetical mechanisms (1) by triggering negative signals that impede oligodendrocyte maturation; (2) by steric inhibition, preventing a close contact between axons and oligodendrocytes; (3) by strengthening the adhesion of OPCs and inhibiting their migration. In the case of PSA-NCAM, it acts as a negative signal for myelination, probably by preventing adhesion of OPC processes to axons (Charles et al. 2000), but it also should be down-regulated to allow OPC differentiation (Decker et al. 2002). In this sense, PSA-NCAM expression persists in those regions containing unmyelinated fibres, such as the mossy fibres of the dentate gyrus and axons from the supraoptic and paraventricular nuclei, which remain unmyelinated throughout life (Seki and Arai 1991, 1993). During human foetal forebrain development, myelination starts in those areas where PSA-NCAM is down-regulated (Jakovcevski et al. 2007), which reinforces the data observed in mice. Hyaluronan also impedes remyelination in the corpus callosum by inhibiting OPC maturation when injected 5 days after lysolecithin-mediated demyelination (Back et al. 2005), but there are no data about its selective persistence in different tracts of the adult brain. The lack of data about anosmin-1 distribution in adulthood (of either human or other mammals) does not allow the consideration of similar roles of this molecule in adulthood or MS lesions. On the other hand, there are data suggesting that PSA-NCAM is not necessary for OPC motility but for the directional movement of OPCs in response to PDGF (Zhang et al. 2004), while anosmin-1 has been shown to antagonise FGF2 motogenic effect (Clemente et al. 2011; Bribián et al. 2006). Attending to developmental data about anosmin-1 actions on OPCs, two putative roles for axonal anosmin-1 in MS lesions could be considered (1) anosmin-1 may interfere with the FGF2 effects on OPC migration (Clemente et al. 2011; Bribián et al. 2006); (2) since homophilic interactions are important for axon-OPC recognition during development (Bribián et al. 2008), anosmin-1 may facilitate OPC recognition/adhesion and thereby facilitate remyelination.

Therefore, to mimic developmental conditions and induce effective remyelination, several aspects are needed in the FGF2/anosmin-1 system. First, FGF2 should be present and anosmin-1 absent in areas of successful remyelination (Clemente et al. 2011). Second, not only axons, but adult OPCs, should acquire embryonic characteristics (Clemente et al. 2011; Bribián et al. 2006, 2008), as it has been previously shown in MS for other OPC-specific genes (Arnett et al. 2004; Capello et al. 1997). Third, OPCs may express the fundamental receptor involved in anosmin-1 actions, FGFR1 (Hu et al. 2009; Murcia-Belmonte et al. 2010; Clemente et al. 2011). The first circumstance is present in areas with spontaneous remyelination (active lesions and the periplaque of chronic-active lesion). However, in the demyelinated area of MS chronic lesions, FGF2 is totally absent and the lack of anosmin-1 in OPCs could affect their recognition of the nude anosmin-1-expressing axons and thus compromise their ability to repair the damage. Finally, a subpopulation of OPCs within areas where FGF2 is up-regulated expresses FGFR1, which suggests that FGFR1^{+}-OPCs are the cells recruited by this growth factor into these zones or it might up-regulate FGFR1 in OPCs, as observed previously in vitro (Bansal et al. 1996).

In the last years, a new concept about the relationship between the shape of the oligodendrocyte and the ECM content has emerged. Kippert et al. (2009) showed that the cell surface area of this cell type is critically dependent on actomyosin contractility and is regulated by physical properties of the supporting matrix. These authors also demonstrated that the presence of ECM proteins with known non-permissive growth properties within the CNS blocks oligodendrocyte surface spreading, which is accompanied by changes in the rate of endocytosis (Kippert et al. 2009). An implication of these findings would be that changes in the rigidity of the scarred MS lesion due to reactive astrocyte protein secretion may unbalance the intracellular and extracellular forces and inhibit oligodendrocyte differentiation. As has been described along this chapter, the core of MS chronic lesions presents an astrocyte-mediated deposition of different extracellular matrix proteins (anosmin-1, PSA-NCAM and hyaluronan). A first implication of their presence in this specific area could be an increase in the rigidity of the demyelinated area compared to the surrounded periplaque, and therefore, an unbalance in the cellular forces that drive the inhibition of remyelination. Another possibility unexplored to date is that, although these extracellular cues would not alter the physical properties of the demyelinated area, they would change the activity of signalling molecules that regulate intracellular force (specifically those related with RhoA) and could also inhibit remyelination (Baer et al. 2009; Bauer and Ffrench-Constant 2009). As described above, anosmin-1 is a firm candidate for this, since it induces cytoskeletal rearrangements through FGFR1-dependent mechanisms involving Cdc42/Rac1 activation, two members of the Rho family of small GTPases (González-Martínez et al. 2004). However, further experiments are needed to establish whether anosmin-1 also participates in controlling actomyosin contractility and, therefore, oligodendrocyte cell shape and differentiation.

12.6.2 Anosmin-1 in Other Pathologies

Besides their above-mentioned different roles in pathologies affecting the CNS (KS and MS), anosmin-1 has been also involved in totally different diseases. For example, anosmin-1 produced by epidermal keratinocytes in response to calcium concentrations or anti-inflammatory cytokines may modulate epidermal nerve density in atopic dermatitis (Tengara et al. 2010) and it has been also implicated in the response to immunoglobulin therapy in dermatomiositis (Raju and Dalakas 2005).

Finally, a relationship between anosmin-1 and cancer has also been reported. For instance, up-regulation of *KAL1* in breast tumour tissue compared to normal breast tissue has been described (Arikawa et al. 2011), while in a metastasising human oesophageal squamous cell carcinoma cell line, *KAL1* is down-regulated when compared to the non-metastasising parental line (Kawamata et al. 2003). Jian et al. (2009) demonstrated that *KAL1* gene expression plays an important role in cancer metastasis and protection from apoptosis of the tumoural cell (Jian et al. 2009). Screening of colon, lung and ovarian cancer cDNA panels indicated a significant decrease in *KAL1* expression in comparison to corresponding uninvolved tissues. However, *KAL1* expression increased with the progression of cancer from early stages (I and II) to later stages (III and IV) of the cancer, and a direct correlation between the TGF-β and *KAL1* expression in colon cancer cDNA has been reported. In colon cancer cell lines, TGF-β induces *KAL1* gene expression and secretion of anosmin-1 protein. Interestingly, hypoxia induces anosmin-1 expression that, in turn, protects cancer cells from apoptosis activated by hypoxia and increases cancer cell motility (Jian et al. 2009).

12.7 Concluding Remarks

The new roles of anosmin-1 in different biological processes, as well as in pathology, outline a scenario far beyond the olfactory system and Kallmann syndrome. Anosmin-1 plays relevant direct roles in cell adhesion, cell migration, axonal outgrowth and collateralisation, but the indirect effects of the protein, modulating FGF-signalling (via FGFR1), have been recently complemented by other morphogenic pathways like BMP and Wnt. Research has progressed significantly in the last years regarding anosmin-1: we have dissected the different domains of the protein which are relevant for its binding to the different known partners (FGFR1, laminin, fibronectin, anosmin-1 itself, HS, others), and it has been pointed as an important actor in the pathogenesis of MS, cancer metastasis and allergic processes. In this sense, anosmin-1 would be considered as one of the most actively studied ECM proteins, and its binding properties describe a panoply of biological effects which should be increasingly considered in development and in adulthood and in physiological and pathological scenarios, biasing the scientific community by the meaning of its name, undoubtedly linked to olfaction: nowadays, if a scientist smells anomin-1 in the air, she/he should consider this protein as a relevant actor in physiology and pathology, specially, in the nervous system.

Acknowledgements Our research is currently supported by grants from the Spanish Ministerio de Economía y Competitividad-MINECO (ADE10-0010, RD07-0060-2007, SAF2009-07842), and Fundación Eugenio Rodríguez Pascual (Spain) to FdC, Association pour la Recherche en Escleròse en Plaques-ARSEP (France) to DCL and FdC, and Gobierno de Castilla-La Mancha (PI2009/26 and PI2009/29,) to DCL and PFE, respectively. VMB is a PhD student who had a fellowship from

Gobierno de Castilla-La Mancha (MOV2007-JI/19) and is currently hired under RD07-0060-2007. DGG is a PhD student who had a fellowship from Gobierno de Gastilla-La Mancha (PI2007-66) and is currently hired by SESCAM. DCL, FdCS and PFE are hired by SESCAM. ABA is the recipient of a Sara Borrell contract from the Spanish Ministerio de Economía y Competitividad-MINECO.

Compliance with Ethics Requirements The authors declare that they have no conflicts of interest.

References

Andrenacci D, Grimaldi MR, Panetta V, Riano E, Rugarli EI, Graziani F (2006) Functional dissection of the Drosophila Kallmann's syndrome protein DmKal-1. BMC Genet 7:47

Ardouin O, Legouis R, Fasano L, vid-Watine B, Korn H, Hardelin J, Petit C (2000) Characterization of the two zebrafish orthologues of the KAL-1 gene underlying X chromosome-linked Kallmann syndrome. Mech Dev 90:89–94

Arikawa E, Quellhorst G, Han Y, Pan H, Yang J (2011) RT2 profilerTM PCR arrays: pathway-focused gene expression profiling with qRT-PCR. Technical article. SuperArray Bioscience Corporation, Frederick, MD

Arnett HA, Fancy SP, Alberta JA, Zhao C, Plant SR, Kaing S, Raine CS, Rowitch DH, Franklin RJ, Stiles CD (2004) bHLH transcription factor Olig1 is required to repair demyelinated lesions in the CNS. Science 306:2111–2115

Ayari B, Soussi-Yanicostas N (2007) FGFR1 and anosmin-1 underlying genetically distinct forms of Kallmann syndrome are co-expressed and interact in olfactory bulbs. Dev Genes Evol 217:169–175

Ayari B, Landoulsi A, Soussi-Yanicostas N (2012) Localization and characterization of kal 1.a and kal 1.b in the brain of adult zebrafish (Danio rerio). Brain Res Bull 88:345–353

Back SA, Tuohy TM, Chen H, Wallingford N, Craig A, Struve J, Luo NL, Banine F, Liu Y, Chang A, Trapp BD, Bebo BF Jr, Rao MS, Sherman LS (2005) Hyaluronan accumulates in demyelinated lesions and inhibits oligodendrocyte progenitor maturation. Nat Med 11:966–972

Baer AS, Syed YA, Kang SU, Mitteregger D, Vig R, Ffrench-Constant C, Franklin RJ, Altmann F, Lubec G, Kotter MR (2009) Myelin-mediated inhibition of oligodendrocyte precursor differentiation can be overcome by pharmacological modulation of Fyn-RhoA and protein kinase C signalling. Brain 132:465–481

Bansal R, Kumar M, Murray K, Morrison RS, Pfeiffer SE (1996) Regulation of FGF receptors in the oligodendrocyte lineage. Mol Cell Neurosci 7:263–275

Barkhof F, Bruck W, De Groot CJ, Bergers E, Hulshof S, Geurts J, Polman CH, van der Valk P (2003) Remyelinated lesions in multiple sclerosis: magnetic resonance image appearance. Arch Neurol 60:1073–1081

Bauer NG, Ffrench-Constant C (2009) Physical forces in myelination and repair: a question of balance? J Biol 8:78

Bramow S, Frischer JM, Lassmann H, Koch-Henriksen N, Lucchinetti CF, Sorensen PS, Laursen H (2010) Demyelination versus remyelination in progressive multiple sclerosis. Brain 133:2983–2998

Bribián A, Barallobre MJ, Soussi-Yanicostas N, de Castro F (2006) Anosmin-1 modulates the FGF-2-dependent migration of oligodendrocyte precursors in the developing optic nerve. Mol Cell Neurosci 33:2–14

Bribián A, Esteban PF, Clemente D, Soussi-Yanicostas N, Thomas JL, Zalc B, de Castro F (2008) A novel role for anosmin-1 in the adhesion and migration of oligodendrocyte precursors. Dev Neurobiol 68:1503–1516

Bülow HE, Hobert O (2004) Differential sulfations and epimerization define heparan sulfate specificity in nervous system development. Neuron 41:723–736

Bülow HE, Berry KL, Topper LH, Peles E, Hobert O (2002) Heparan sulfate proteoglycan-dependent induction of axon branching and axon misrouting by the Kallmann syndrome gene kal-1. Proc Natl Acad Sci USA 99:6346–6351

Capello E, Voskuhl RR, McFarland HF, Raine CS (1997) Multiple sclerosis: re-expression of a developmental gene in chronic lesions correlates with remyelination. Ann Neurol 41:797–805

Cariboni A, Pimpinelli F, Colamarino S, Zaninetti R, Piccolella M, Rumio C, Piva F, Rugarli EI, Maggi R (2004) The product of X-linked Kallmann's syndrome gene (KAL1) affects the migratory activity of gonadotropin-releasing hormone (GnRH)-producing neurons. Hum Mol Genet 13:2781–2791

Charles P, Hernandez MP, Stankoff B, Aigrot MS, Colin C, Rougon G, Zalc B, Lubetzki C (2000) Negative regulation of central nervous system myelination by polysialylated-neural cell adhesion molecule. Proc Natl Acad Sci USA 97:7585–7590

Charles P, Reynolds R, Seilhean D, Rougon G, Aigrot MS, Niezgoda A, Zalc B, Lubetzki C (2002) Re-expression of PSA-NCAM by demyelinated axons: an inhibitor of remyelination in multiple sclerosis? Brain 125:1972–1979

Chung WC, Tsai PS (2010) Role of fibroblast growth factor signaling in gonadotropin-releasing hormone neuronal system development. Front Horm Res 39:37–50

Chung WC, Matthews TA, Tata BK, Tsai PS (2010) Compound deficiencies in multiple fibroblast growth factor signalling components differentially impact the murine gonadotrophin-releasing hormone system. J Neuroendocrinol 22:944–950

Church DM et al (2009) Lineage-specific biology revealed by a finished genome assembly of the mouse. PLoS Biol 7:e1000112

Clemente D, Esteban PF, Del Valle I, Bribian A, Soussi-Yanicostas N, Silva A, de Castro F (2008) Expression pattern of Anosmin-1 during pre- and postnatal rat brain development. Dev Dyn 237:2518–2528

Clemente D, Ortega MC, Arenzana FJ, de Castro F (2011) FGF-2 and Anosmin-1 are selectively expressed in different types of multiple sclerosis lesions. J Neurosci 31:14899–14909

de Morsier G (1954) Studies in cranio-encephalic dysraphia. I. Agenesia of the olfactory lobe (lateral telencephaloschisis) and of the callous and anterior commissures (median telencephaloschisis); olfacto-genital dysplasia. Schweiz Arch Neurol Psychiatr 74:309–361

Decker L, Durbec P, Rougon G, Evercooren AB (2002) Loss of polysialic residues accelerates CNS neural precursor differentiation in pathological conditions. Mol Cell Neurosci 19:225–238

de Castro F, Bribián A (2005) The molecular orchestra of the migration of oligodendrocyte progenitors during development. Brain Res Rev 49:227–241

del Castillo I, Cohen-Salmon M, Blanchard S, Lutfalla G, Petit C (1992) Structure of the X-linked Kallmann syndrome gene and its homologous pseudogene on the Y chromosome. Nat Genet 2:305–310

Dellovade TL, Hardelin JP, Soussi-Yanicostas N, Pfaff DW, Schwanzel-Fukuda M, Petit C (2003) Anosmin-1 immunoreactivity during embryogenesis in a primitive eutherian mammal. Brain Res Dev Brain Res 140:157–167

Dodé C, Hardelin JP (2009) Kallmann syndrome. Eur J Hum Genet 17:139–146

Dode C et al (2003) Loss-of-function mutations in FGFR1 cause autosomal dominant Kallmann syndrome. Nat Genet 33:463–465

Dode C, Teixeira L, Levilliers J, Fouveaut C, Bouchard P, Kottler ML, Lespinasse J, Lienhardt-Roussie A, Mathieu M, Moerman A, Morgan G, Murat A, Toublanc JE, Wolczynski S, Delpech M, Petit C, Young J, Hardelin JP (2006) Kallmann syndrome: mutations in the genes encoding prokineticin-2 and prokineticin receptor-2. PLoS Genet 2:e175

Dubois-Dalcq M, Ffrench-Constant C, Franklin RJ (2005) Enhancing central nervous system remyelination in multiple sclerosis. Neuron 48:9–12

Duke VM, Winyard PJ, Thorogood P, Soothill P, Bouloux PM, Woolf AS (1995) KAL, a gene mutated in Kallmann's syndrome, is expressed in the first trimester of human development. Mol Cell Endocrinol 110:73–79

Endo Y, Ishiwata-Endo H, Yamada KM (2012) Extracellular Matrix Protein Anosmin Promotes Neural Crest Formation and Regulates FGF, BMP, and WNT Activities. Dev Cell 23:305–316

Falardeau J et al (2008) Decreased FGF8 signaling causes deficiency of gonadotropin-releasing hormone in humans and mice. J Clin Invest 118:2822–2831

Franco B, Guioli S, Pragliola A, Incerti B, Bardoni B, Tonlorenzi R, Carrozzo R, Maestrini E, Pieretti M, Taillon-Miller P, Brown CJ, Willard HF, Lawrence C, Graziella PM, Camerino G, Ballabio A (1991) A gene deleted in Kallmann's syndrome shares homology with neural cell adhesion and axonal path-finding molecules. Nature 353:529–536

Frohman EM, Racke MK, Raine CS (2006) Multiple sclerosis–the plaque and its pathogenesis. N Engl J Med 354:942–955

García-González D, Clemente D, Coelho M, Esteban PF, Soussi-Yanicostas N, de Castro F (2010) Dynamic roles of FGF-2 and Anosmin-1 in the migration of neuronal precursors from the subventricular zone during pre- and postnatal development. Exp Neurol 222:285–295

Gianola S, de Castro F, Rossi F (2009) Anosmin-1 stimulates outgrowth and branching of developing Purkinje axons. Neuroscience 158:570–584

Gill JC, Moenter SM, Tsai PS (2004) Developmental regulation of gonadotropin-releasing hormone neurons by fibroblast growth factor signaling. Endocrinology 145:3830–3839

Gläser B, Myrtek D, Rumpler Y, Schiebel K, Hauwy M, Rappold GA, Schempp W (1999) Transposition of SRY into the ancestral pseudoautosomal region creates a new pseudoautosomal boundary in a progenitor of simian primates. Hum Mol Genet 8:2071–2078

González-Martínez D, Kim SH, Hu Y, Guimond S, Schofield J, Winyard P, Vannelli GB, Turnbull J, Bouloux PM (2004) Anosmin-1 modulates fibroblast growth factor receptor 1 signaling in human gonadotropin-releasing hormone olfactory neuroblasts through a heparan sulfate-dependent mechanism. J Neurosci 24:10384–10392

Guimond SE, Turnbull JE (1999) Fibroblast growth factor receptor signalling is dictated by specific heparan sulphate saccharides. Curr Biol 9:1343–1346

Hardelin JP, Julliard AK, Moniot B, Soussi-Yanicostas N, Verney C, Schwanzel-Fukuda M, Ayer-Le Lievre C, Petit C (1999) Anosmin-1 is a regionally restricted component of basement membranes and interstitial matrices during organogenesis: implications for the developmental anomalies of X chromosome-linked Kallmann syndrome. Dev Dyn 215:26–44

Hayes FJ, Seminara SB, Crowley WF, Jr. (1998) Hypogonadotropic hypogonadism. Endocrinol Metab Clin North Am 27:739-63, vii.

Hu Y, Bouloux PM (2011) X-linked GnRH deficiency: role of KAL-1 mutations in GnRH deficiency. Mol Cell Endocrinol 346:13–20

Hu Y, Gonzalez-Martinez D, Kim SH, Bouloux PM (2004) Cross-talk of anosmin-1, the protein implicated in X-linked Kallmann's syndrome, with heparan sulphate and urokinase-type plasminogen activator. Biochem J 384:495–505

Hu Y, Guimond SE, Travers P, Cadman S, Hohenester E, Turnbull JE, Kim SH, Bouloux PM (2009) Novel mechanisms of fibroblast growth factor receptor 1 regulation by extracellular matrix protein anosmin-1. J Biol Chem 284:29905–29920

Hu Y, Yu H, Shaw G, Pask AJ, Renfree MB (2011) Kallmann syndrome 1 gene is expressed in the marsupial gonad. Biol Reprod 84:595–603

Hudson ML, Kinnunen T, Cinar HN, Chisholm AD (2006) C. elegans Kallmann syndrome protein KAL-1 interacts with syndecan and glypican to regulate neuronal cell migrations. Dev Biol 294:352–365

Jakovcevski I, Mo Z, Zecevic N (2007) Down-regulation of the axonal polysialic acid-neural cell adhesion molecule expression coincides with the onset of myelination in the human fetal forebrain. Neuroscience 149:328–337

Jian B, Nagineni CN, Meleth S, Grizzle W, Bland K, Chaudry I, Raju R (2009) Anosmin-1 involved in neuronal cell migration is hypoxia inducible and cancer regulated. Cell Cycle 8:3770–3776

Kallmann FJ, Schoenfeld WA, Barrera SE (1944) The genetic aspects of primary eunuchoidism. Am J Ment Defic 48:203–236

Kawamata H, Furihata T, Omotehara F, Sakai T, Horiuchi H, Shinagawa Y, Imura J, Ohkura Y, Tachibana M, Kubota K, Terano A, Fujimori T (2003) Identification of genes differentially expressed in a newly isolated human metastasizing esophageal cancer cell line, T.Tn-AT1, by cDNA microarray. Cancer Sci 94:699–706

Kawauchi S, Shou J, Santos R, Hebert JM, McConnell SK, Mason I, Calof AL (2005) Fgf8 expression defines a morphogenetic center required for olfactory neurogenesis and nasal cavity development in the mouse. Development 132:5211–5223

Kim HG, Kurth I, Lan F, Meliciani I, Wenzel W, Eom SH, Kang GB, Rosenberger G, Tekin M, Ozata M, Bick DP, Sherins RJ, Walker SL, Shi Y, Gusella JF, Layman LC (2008) Mutations in CHD7, encoding a chromatin-remodeling protein, cause idiopathic hypogonadotropic hypogonadism and Kallmann syndrome. Am J Hum Genet 83:511–519

Kippert A, Fitzner D, Helenius J, Simons M (2009) Actomyosin contractility controls cell surface area of oligodendrocytes. BMC Cell Biol 10:71

Kuhlmann T, Miron V, Cui Q, Wegner C, Antel J, Bruck W (2008) Differentiation block of oligodendroglial progenitor cells as a cause for remyelination failure in chronic multiple sclerosis. Brain 131:1749–1758

Laitinen EM, Vaaralahti K, Tommiska J, Eklund E, Tervaniemi M, Valanne L, Raivio T (2011) Incidence, phenotypic features and molecular genetics of Kallmann syndrome in Finland. Orphanet J Rare Dis 6:41

Legouis R, Hardelin JP, Levilliers J, Claverie JM, Compain S, Wunderle V, Millasseau P, Le PD, Cohen D, Caterina D (1991) The candidate gene for the X-linked Kallmann syndrome encodes a protein related to adhesion molecules. Cell 67:423–435

Legouis R, Lievre CA, Leibovici M, Lapointe F, Petit C (1993) Expression of the KAL gene in multiple neuronal sites during chicken development. Proc Natl Acad Sci USA 90:2461–2465

Lois C, Alvarez-Buylla A (1994) Long-distance neuronal migration in the adult mammalian brain. Science 264:1145–1148

Lois C, Garcia-Verdugo JM, Alvarez-Buylla A (1996) Chain migration of neuronal precursors. Science 271:978–981

Lutz B, Rugarli EI, Eichele G, Ballabio A (1993) X-linked Kallmann syndrome. A neuronal targeting defect in the olfactory system? FEBS Lett 325:128–134

Lutz B, Kuratani S, Rugarli EI, Wawersik S, Wong C, Bieber FR, Ballabio A, Eichele G (1994) Expression of the Kallmann syndrome gene in human fetal brain and in the manipulated chick embryo. Hum Mol Genet 3:1717–1723

Maestre de San Juan A (1856) Teratología: falta total de los nervios olfatorios con anosmia en un individuo en quien existía una atrofia congénita de los testículos y el miembro viril. El Siglo Médico, Madrid 3:211–221

Mangs AH, Morris BJ (2007) The human pseudoautosomal region (PAR): origin, function and future. Curr Genomics 8:129–136

Martin C, Balasubramanian R, Dwyer AA, Au MG, Sidis Y, Kaiser UB, Seminara SB, Pitteloud N, Zhou QY, Crowley WF Jr (2011) The role of the prokineticin 2 pathway in human reproduction: evidence from the study of human and murine gene mutations. Endocr Rev 32:225–246

Mihara M, Yoshida Y, Tsukamoto T, Inada K, Nakanishi Y, Yagi Y, Imai K, Sugimura T, Tatematsu M, Ushijima T (2006) Methylation of multiple genes in gastric glands with intestinal metaplasia: A disorder with polyclonal origins. Am J Pathol 169:1643–1651

Murcia-Belmonte V, Esteban PF, García-González D, de Castro F (2010) Biochemical dissection of Anosmin-1 interaction with FGFR1 and components of the extracellular matrix. J Neurochem 115:1256–1265

Okubo K, Sakai F, Lau EL, Yoshizaki G, Takeuchi Y, Naruse K, Aida K, Nagahama Y (2006) Forebrain gonadotropin-releasing hormone neuronal development: insights from transgenic medaka and the relevance to X-linked Kallmann syndrome. Endocrinology 147:1076–1084

Omari KM, John GR, Sealfon SC, Raine CS (2005) CXC chemokine receptors on human oligodendrocytes: implications for multiple sclerosis. Brain 128:1003–1015

Patani R, Balaratnam M, Vora A, Reynolds R (2007) Remyelination can be extensive in multiple sclerosis despite a long disease course. Neuropathol Appl Neurobiol 33:277–287
Patrikios P, Stadelmann C, Kutzelnigg A, Rauschka H, Schmidbauer M, Laursen H, Sorensen PS, Bruck W, Lucchinetti C, Lassmann H (2006) Remyelination is extensive in a subset of multiple sclerosis patients. Brain 129:3165–3172
Peretto P, Giachino C, Aimar P, Fasolo A, Bonfanti L (2005) Chain formation and glial tube assembly in the shift from neonatal to adult subventricular zone of the rodent forebrain. J Comp Neurol 487:407–427
Perry J, Palmer S, Gabriel A, Ashworth A (2001) A short pseudoautosomal region in laboratory mice. Genome Res 11:1826–1832
Petreanu L, Álvarez-Buylla A (2002) Maturation and death of adult-born olfactory bulb granule neurons: role of olfaction. J Neurosci 22:6106–6113
Pitteloud N, Zhang C, Pignatelli D, Li JD, Raivio T, Cole LW, Plummer L, Jacobson-Dickman EE, Mellon PL, Zhou QY, Crowley WF Jr (2007) Loss-of-function mutation in the prokineticin 2 gene causes Kallmann syndrome and normosmic idiopathic hypogonadotropic hypogonadism. Proc Natl Acad Sci USA 104:17447–17452
Prineas JW, Connell F (1979) Remyelination in multiple sclerosis. Ann Neurol 5:22–31
Raju R, Dalakas MC (2005) Gene expression profile in the muscles of patients with inflammatory myopathies: effect of therapy with IVIg and biological validation of clinically relevant genes. Brain 128:1887–1896
Robertson A, MacColl GS, Nash JA, Boehm MK, Perkins SJ, Bouloux PM (2001) Molecular modelling and experimental studies of mutation and cell-adhesion sites in the fibronectin type III and whey acidic protein domains of human anosmin-1. Biochem J 357:647–659
Ross MT et al (2005) The DNA sequence of the human X chromosome. Nature 434:325–337
Rugarli EI, Lutz B, Kuratani SC, Wawersik S, Borsani G, Ballabio A, Eichele G (1993) Expression pattern of the Kallmann syndrome gene in the olfactory system suggests a role in neuronal targeting. Nat Genet 4:19–26
Rugarli EI, Ghezzi C, Valsecchi V, Ballabio A (1996) The Kallmann syndrome gene product expressed in COS cells is cleaved on the cell surface to yield a diffusible component. Hum Mol Genet 5:1109–1115
Rugarli EI, Di SE, Hilliard MA, Arbucci S, Ghezzi C, Facciolli A, Coppola G, Ballabio A, Bazzicalupo P (2002) The Kallmann syndrome gene homolog in C. elegans is involved in epidermal morphogenesis and neurite branching. Development 129:1283–1294
Sauka-Spengler T, Bronner-Fraser M (2008) Evolution of the neural crest viewed from a gene regulatory perspective. Genesis 46:673–682
Schwanzel-Fukuda M, Bick D, Pfaff DW (1989) Luteinizing hormone-releasing hormone (LHRH)-expressing cells do not migrate normally in an inherited hypogonadal (Kallmann) syndrome. Brain Res Mol Brain Res 6:311–326
Seki T, Arai Y (1991) The persistent expression of a highly polysialylated NCAM in the dentate gyrus of the adult rat. Neurosci Res 12:503–513
Seki T, Arai Y (1993) Distribution and possible roles of the highly polysialylated neural cell adhesion molecule (NCAM-H) in the developing and adult central nervous system. Neurosci Res 17:265–290
Seminara SB, Hayes FJ, Crowley WF Jr (1998) Gonadotropin-releasing hormone deficiency in the human (idiopathic hypogonadotropic hypogonadism and Kallmann's syndrome): pathophysiological and genetic considerations. Endocr Rev 19:521–539
Shapiro LJ, Mohandas T, Weiss R, Romeo G (1979) Non-inactivation of an x-chromosome locus in man. Science 204:1224–1226
Soussi-Yanicostas N, Hardelin JP, Arroyo-Jiménez MM, Ardouin O, Legouis R, Levilliers J, Traincard F, Betton JM, Cabanie L, Petit C (1996) Initial characterization of anosmin-1, a putative extracellular matrix protein synthesized by definite neuronal cell populations in the central nervous system. J Cell Sci 109(Pt 7):1749–1757

Soussi-Yanicostas N, Faivre-Sarrailh C, Hardelin JP, Levilliers J, Rougon G, Petit C (1998) Anosmin-1 underlying the X chromosome-linked Kallmann syndrome is an adhesion molecule that can modulate neurite growth in a cell-type specific manner. J Cell Sci 111(Pt 19): 2953–2965

Soussi-Yanicostas N, de Castro F, Julliard AK, Perfettini I, Chedotal A, Petit C (2002) Anosmin-1, defective in the X-linked form of Kallmann syndrome, promotes axonal branch formation from olfactory bulb output neurons. Cell 109:217–228

Tengara S, Tominaga M, Kamo A, Taneda K, Negi O, Ogawa H, Takamori K (2010) Keratinocyte-derived anosmin-1, an extracellular glycoprotein encoded by the X-linked Kallmann syndrome gene, is involved in modulation of epidermal nerve density in atopic dermatitis. J Dermatol Sci 58:64–71

Trainor PA, Ariza-McNaughton L, Krumlauf R (2002) Role of the isthmus and FGFs in resolving the paradox of neural crest plasticity and prepatterning. Science 295:1288–1291

Villanueva C, de Roux N (2010) FGFR1 mutations in Kallmann syndrome. Front Horm Res 39:51–61

Wang Y, Imitola J, Rasmussen S, O'Connor KC, Khoury SJ (2008) Paradoxical dysregulation of the neural stem cell pathway sonic hedgehog-Gli1 in autoimmune encephalomyelitis and multiple sclerosis. Ann Neurol 64:417–427

Whitlock KE, Smith KM, Kim H, Harden MV (2005) A role for foxd3 and sox10 in the differentiation of gonadotropin-releasing hormone (GnRH) cells in the zebrafish Danio rerio. Development 132:5491–5502

Williams A, Piaton G, Aigrot MS, Belhadi A, Theaudin M, Petermann F, Thomas JL, Zalc B, Lubetzki C (2007) Semaphorin 3A and 3F: key players in myelin repair in multiple sclerosis? Brain 130:2554–2565

Wray S, Grant P, Gainer H (1989) Evidence that cells expressing luteinizing hormone-releasing hormone mRNA in the mouse are derived from progenitor cells in the olfactory placode. Proc Natl Acad Sci USA 86:8132–8136

Yanicostas C, Ernest S, Dayraud C, Petit C, Soussi-Yanicostas N (2008) Essential requirement for zebrafish anosmin-1a in the migration of the posterior lateral line primordium. Dev Biol 320:469–479

Yanicostas C, Herbomel E, Dipietromaria A, Soussi-Yanicostas N (2009) Anosmin-1a is required for fasciculation and terminal targeting of olfactory sensory neuron axons in the zebrafish olfactory system. Mol Cell Endocrinol 312:53–60

Young J, Metay C, Bouligand J, Tou B, Francou B, Maione L, Tosca L, Sarfati J, Brioude F, Esteva B, Briand-Suleau A, Brisset S, Goossens M, Tachdjian G, Guiochon-Mantel A (2012) SEMA3A deletion in a family with Kallmann syndrome validates the role of semaphorin 3A in human puberty and olfactory system development. Hum Reprod 27:1460–1465

Zhang H, Vutskits L, Calaora V, Durbec P, Kiss JZ (2004) A role for the polysialic acid-neural cell adhesion molecule in PDGF-induced chemotaxis of oligodendrocyte precursor cells. J Cell Sci 117:93–103

Chapter 13
Protocadherins in Neurological Diseases

Takahiro Hirabayashi and Takeshi Yagi

Abstract Cadherins were originally isolated as calcium-dependent cell adhesion molecules and are characterized by their cadherin motifs in the extracellular domain. In vertebrates, including humans, there are more than 100 different cadherin-related genes, which constitute the cadherin superfamily. The protocadherin (Pcdh) family comprises a large subgroup within the cadherin superfamily. The Pcdhs are divided into clustered and non-clustered Pcdhs, based on their genomic structure. Almost all the *Pcdh* genes are expressed widely in the brain and play important roles in brain development and in the regulation of brain function. This chapter presents an overview of Pcdh family members with regard to their functions, knockout mouse phenotypes, and association with neurological diseases and tumors.

13.1 Introduction

Cadherins are calcium-dependent adhesion molecules that constitute a superfamily and play crucial roles in cell signaling, development, and other processes. The cadherin superfamily includes classical cadherins, desmosomal cadherins, and protocadherins. All the cadherin superfamily members are transmembrane proteins and include an approximately 110-amino-acid region containing several repeats, designated the extracellular cadherin (EC) domain (Fig. 13.1a). In vertebrates, over 100 cadherin superfamily members have been identified.

A novel cadherin-like protein family isolated from rat and human brain cDNA by the PCR method was termed the "protocadherin" family by Dr. Shintaro Suzuki and colleagues (Sano et al. 1993). Subsequent studies have identified additional

T. Hirabayashi, Ph.D. (✉) • T. Yagi, Ph.D.
KOKORO-Biology group and JST-CREST, Laboratories for Integrated Biology, Graduate School of Frontier Biosciences, Osaka University, 1-3 Yamadaoka, Osaka, Suita 565-0874, Japan
e-mail: hira@fbs.osaka-u.ac.jp; yagi@fbs.osaka-u.ac.jp

V. Berezin and P.S. Walmod (eds.), *Cell Adhesion Molecules: Implications in Neurological Diseases*, Advances in Neurobiology 8, DOI 10.1007/978-1-4614-8090-7_13,

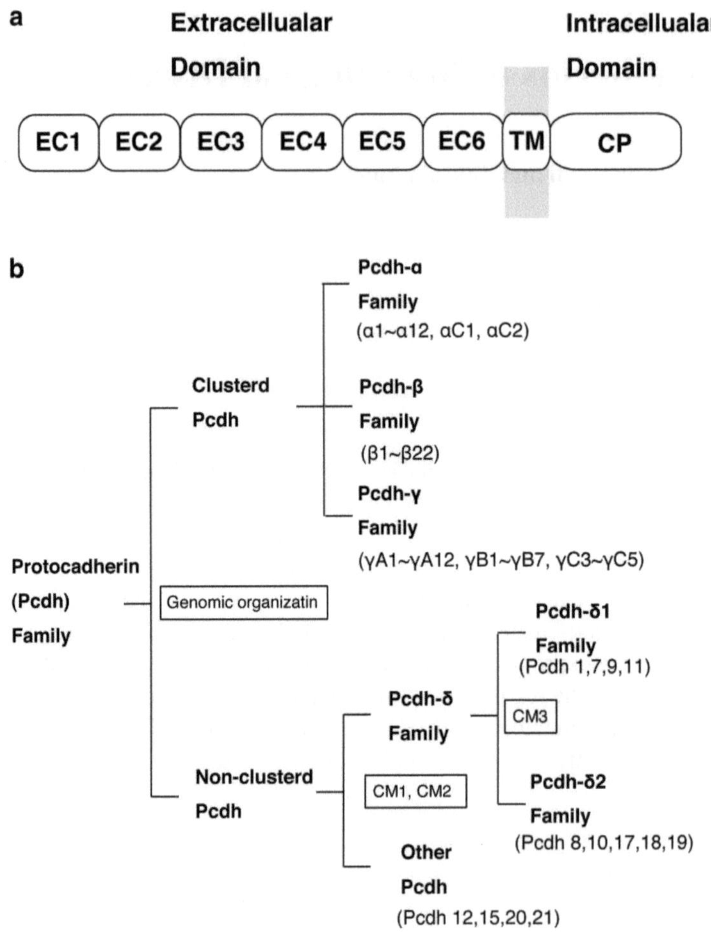

Fig. 13.1 Structure and classification of the Pcdh protein family. (**a**) Pcdh protein structure. Pcdh proteins are type I transmembrane proteins, like classical cadherins, containing an approximately 110-amino-acid region containing several repeats, designated the extracellular cadherin (EC) domain. (**b**) Classification of Pcdhs. The Pcdh family is classified into two groups, clustered Pcdhs and non-clustered Pcdhs, based on genomic organization. Non-clustered Pcdhs are divided into two subgroups, Pcdh-δ and other Pcdhs. Pcdh-δ members contain two highly conserved motifs, designated CM1 and CM2, in their cytoplasmic domain. The Pcdh-δ family can be further subdivided into Pcdh-δ1 and Pcdh-δ2, based on the presence or absence of a CM3 motif

protocadherin (Pcdh) family members. To date, more than 70 Pcdh molecules have been reported, and they constitute the largest subgroup within the cadherin superfamily. All the Pcdh proteins, like classical cadherins, are single-pass type I membrane proteins. The *Pcdh* genes are classified into clustered *Pcdh* and non-clustered *Pcdh* genes based on their genomic structure (Fig. 13.1b).

The clustered *Pcdh* genes consist of three subfamilies, *Pcdh-α*, *Pcdh-β*, and *Pcdh-γ*, which have a clustered genomic organization similar to that of the

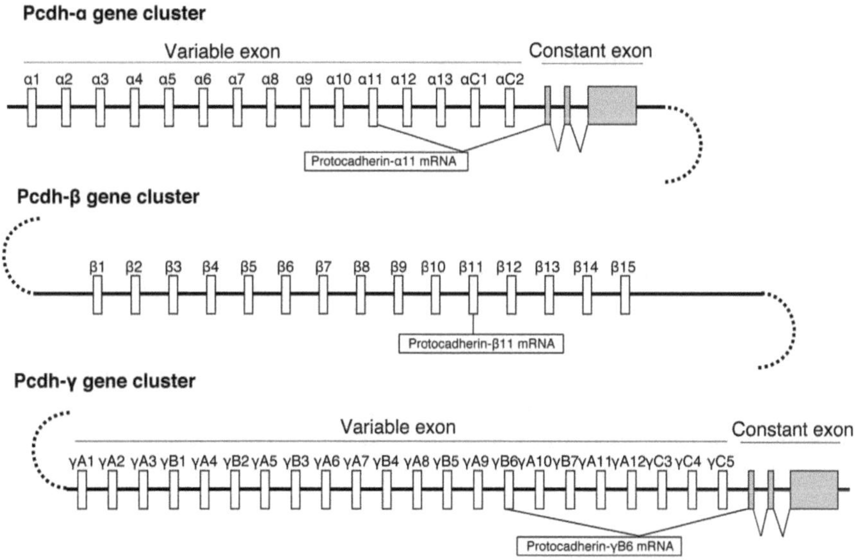

Fig. 13.2 Genomic structure of the clustered *Pcdh*s. The clustered *Pcdhs* consist of three subfamilies, *Pcdh-α*, *Pcdh-β*, and *Pcdh-γ*, which are organized in a unique genomic arrangement on human chromosome 5q31. The human *Pcdh-α* and *Pcdh-γ* clusters consist of variable-region exons and a set of common constant-region exons. Each variable-region exon has a promoter and is *cis*-spliced to the constant-region exons. The *Pcdh-β* genes lack constant exons and consist of unspliced single-exon genes. In humans, a total of 52 genes are encoded by the clustered *Pcdh*s: 15 by *Pcdh-α* (*Pcdhl-αl–α13*, *-αC1*, and *-αC2*), 15 by *Pcdh-β* (*β1–β15*), and 22 by *Pcdh-γ* (*Pcdh-γA1–γA12*, *-γB1–γB7*, and *-γC3–γC5*)

immunoglobulin (*Ig*) and T-cell receptor (*TCR*) genes. The *Pcdh* gene locus has been described in human (chromosome 5q31) (Wu and Maniatis 1999), mouse (chromosome 18c) (Wu et al. 2001), rat (chromosome 18) (Yanase et al. 2004), chicken (Sugino et al. 2004), and zebrafish (Tada et al. 2004). In humans, a total of 52 genes are arranged in the *Pcdh-α*, *Pcdh-β*, and *Pcdh-γ* clusters, which have 15 (*α1–α13*, *αC1*, and *αC2*), 15 (*β1–β15*), and 22 (*γA1–γA12*, *γB1–γB7*, and *γC3–γC5*) genes, respectively (Wu and Maniatis 1999) (Fig. 13.2). In mice, 14 *Pcdh-α*, 22 *Pcdh-β*, and 22 *Pcdh-γ* genes are arrayed in the same direction (Wu et al. 2001).

In contrast, the non-clustered *Pcdh* genes are scattered throughout the genome. Non-clustered *Pcdh* genes can be classified into two subgroups. The *Pcdh*-δ family proteins contain two highly conserved motifs, designated CM1 (27 amino acids) and CM2 (17 amino acids), in the cytoplasmic domain; all other non-clustered *Pcdh*s fall into a separate group. The *Pcdh*-δ family can be further subdivided into the *Pcdh-δ1* and *Pcdh-δ2* groups, based on the presence or absence of a CM3 motif (5 amino acids), which is essential for their interaction with protein phosphatase-1α (PP1α) (Fig. 13.2b).

13.2 Non-clustered Pcdh Functions and Association with Neurological Disease

13.2.1 *Pcdh1*

[Other designations: Axial protocadherin, protocadherin 42]

Pcdh1, formerly named protocadherin 42, was the first *Pcdh* gene to be discovered. The Pcdh1 protein has homophilic adhesive activity, which is calcium dependent, like that of classical cadherins, but weaker (Sano et al. 1993). *Pcdh1* is expressed in various tissues, including the brain, airway, and skin epithelium, and it has been identified as a susceptibility gene for bronchial hyper-responsiveness (BHR), a key pathophysiological feature of asthma (Koppelman et al. 2009). In addition, *Pcdh1* polymorphisms are associated with eczema (Koning et al. 2012). However, no association of *Pcdh1* with neurological diseases has been reported.

13.2.2 *Pcdh7*

[Other designations: BH-protocadherin, NF-protocadherin]

Pcdh7 was first identified in a human gastric adenocarcinoma cell line by a transmembrane domain trapping method for isolating the cDNAs of putative membrane proteins (Sugano et al. 1998) and showed overall significant homology with *Pcdh1* (Yoshida et al. 1998). Its mRNA is predominantly expressed in the brain and heart; thus, it is also called *BH* (brain, heart)-*Pcdh*. There are three isoforms of *Pcdh7*, named *BH-Pcdh-a*, *-b*, and *-c*, which have different cytoplasmic tails. *Pcdh7* is classified into the *Pcdh-δ1* subgroup, although only the BH-Pcdh-c isoform binds to PP1α (Yoshida et al. 1999). Xenopus *NF-Pcdh*, an ortholog of *Pcdh7*, associates with the cellular cofactor TAF1, and both NF-Pcdh and TAF1 are required for formation of the embryonic ectoderm (Bradley et al. 1998) and neural tube (Rashid et al. 2006). MeCP2, the responsible gene for Rett syndrome, binds to an upstream region of the *Pcdh7* gene and downregulates the expressions of *Pcdh7* and *Pcdh-β1* (Miyake et al. 2011); however, the physiological relevance of this relationship is not understood.

13.2.3 *Pcdh8*

[Other designations: Arcadrin, Paraxial protocadherin]

Pcdh8 was identified as a neural activity-regulated cadherin-like molecule by differential cloning from a rat brain cDNA library. The Pcdh8 protein is expressed at synapses and has homophilic binding activity that is calcium dependent, like that of classical protocadherins. An inhibitory anti-Pcdh8 antibody reduces the excitatory postsynaptic potential (EPSP) amplitude and blocks long-term potentiation

(LTP) in rat hippocampal slices, suggesting that *Pcdh8* is involved in synaptic plasticity (Yamagata et al. 1999). Moreover, *Pcdh8* is induced by neural activity and causes the endocytosis of N-cadherin through the TAO2b and p38 MAP kinase pathway, resulting in a reduction in spine number (Yasuda et al. 2007). *Pcdh8-knockout* mice show no visible defects during development and are viable and fertile (Yamamoto et al. 2000), but they have an abnormally high number of dendritic spines (Yasuda et al. 2007).

In humans, the *Pcdh8* gene maps to chromosome 13q (Strehl et al. 1998) where there is a reported linkage to schizophrenia (Shaw et al. 1998). Screening of the *Pcdh8* gene in schizophrenia patients revealed several single-nucleotide polymorphisms (SNPs). However, the genotyping of *Pcdh8* polymorphisms in case–control and proband–parent trio samples revealed no strong association with schizophrenia. These results suggested that any contribution of *Pcdh8* polymorphisms to schizophrenia susceptibility is likely to be weak (Bray et al. 2002).

13.2.4 Pcdh9

[Other designation: cadherin superfamily protein VR4-11]

Pcdh9 was discovered by a database search. The *Pcdh9* gene is predominantly expressed in the brain, and its expression pattern changes during development (Strehl et al. 1998). Structural variations (i.e., copy number variations) of chromosomes have been found in patients with autism spectrum disorders (ASDs), which include autistic disorder, Asperger syndrome, pervasive developmental disorders not otherwise specified, childhood disintegrative disorder, and Rett syndrome. These disorders are characterized by impairments in social interaction and communication and by unusual interests and behaviors. Genome-wide analyses for structural abnormalities by single-nucleotide microarray and karyotyping in autism spectrum patients have shown an increased copy number of the *Pcdh9* intronic gene in some of them (Marshall et al. 2008).

13.2.5 Pcdh10

[Other designation: OL-protocadherin]

Pcdh10 was cloned by screening mouse brain cDNA libraries using a cDNA fragment of human *Pcdh2* as a probe. *Pcdh10* is expressed in various regions of the nervous system, especially in the olfactory and limbic systems; thus, it is also called *OL* (olfactory, limbic)-*Pcdh*. Pcdh10 has homophilic adhesive activity, which is weaker than that of classical cadherins (Hirano et al. 1999). In the chicken, Pcdh10 is strongly expressed along developing axonal fibers and the path of the brachial nerves, suggesting that Pcdh10 may be involved in axon navigation (Nakao et al. 2005, 2008). Pcdh10's cytoplasmic domain interacts with Nck-associated protein 1

(Nap1), a component of the WAVE complex, which regulates actin assembly and cell migration. *Pcdh10* gene-deficient mice show abnormalities in neural projections of the ventral telencephalon, including the thalamocortical, corticothalamic, corticospinal, nigrostriatal, and striatonigral projections. *Pcdh10* gene deficiency also causes striatal axon growth failure in the ventral telencephalon (Uemura et al. 2007). These results suggest that Pcdh10 is essential for neural circuit formation in the ventral telencephalon. In families with autism spectrum disorders, homozygous deletions, including those of the *Pcdh10* gene, have been found (Morrow et al. 2008).

13.2.6 Pcdh11X and Pcdh11Y

[Other designation of protocadherin 11X: protocadherin-X]

[Other designations of protocadherin 11Y: protocadherin-Y, protocadherin22]

Pcdh11X (Yoshida and Sugano 1999) and then *Pcdh11Y* (Blanco et al. 2000) were cloned in a database search. These *Pcdh* genes are located within the hominid-specific Yp11.2/Xq21.3 block of homology between the sex chromosomes. *Pcdh11X* and *11Y* have 98.1 % nucleotide and 98.3 % amino acid identity. Both mRNAs are expressed about equally in all subregions of the brain except for the cerebellum, where *Pcdh11X* expression is predominant. In other organs, both mRNAs are expressed in the heart and other tissues, except for the kidney, liver, muscle, and testis, in which *Pcdh11Y* expression predominates (Blanco et al. 2000).

Late-onset alzheimer's disease (AD) is a neurodegenerative disease and the most common form of dementia, appearing in approximately 50 % of all people over the age of 85. Because late-onset AD is not associated with a genetic factor or family link, it is also called "sporadic AD." In early-onset familial AD, multiple rare mutations are found in responsible genes, such as amyloid beta precursor protein, presenilin 1, and presenilin 2. In late-onset AD, the APOE epsilon 4 allele and a SNP on Xq21.3 in the *Pcdh11X* gene (in patients of European descent living in the USA) have been identified as susceptibility markers (Carrasquillo et al. 2009). However, two subsequent studies demonstrated that the *Pcdh11X* polymorphism is not associated with late-onset AD (Miar et al. 2011; Beecham et al. 2011). Finally, *Pcdh11X* and *Pcdh11Y* have been suggested as possible candidate genes for roles in schizophrenia and schizoaffective disorder, based on genomic analyses of patients (Giouzeli et al. 2004).

13.2.7 Pcdh12

[Other designations: vascular endothelial protocadherin 2 (VE-cadherin-2), protocadherin 14]

Pcdh12 was originally identified in murine endothelial cells by RT-PCR, by using degenerate primers to amplify the highly conserved cytoplasmic and extracellular region of cadherins, as Vascular Endothelial cadherin 2 (VE-cadherin-2), a homolog of VE-cadherin, a classical cadherin (Telo' et al. 1998). *Pcdh12* is strongly expressed in highly vascularized tissues, such as the lung, kidney, liver, spleen, placenta, and heart (Ludwig et al. 2000). *Pcdh12* gene-deficient mice are viable, fertile, and show no obvious histomorphological defects (Rampon et al. 2005). To investigate the possible association between brain morphology and DNA polymorphisms in schizophrenia patients, comparison analyses of brain structural data obtained with MRI and DNA from the peripheral blood cells of patients were performed. These analyses revealed a putative association between a *Pcdh12* polymorphism involving a Ser/Asn substitution at EC6 and cortical folding, suggesting that *Pcdh12* may be important for the development of specific brain areas (Gregório et al. 2009).

13.2.8 Pcdh15

[Other designation: Usher syndrome 1F (USH1F)]

Pcdh15 was cloned as the responsible gene for the phenotype in Ames waltzer mutant mice, which are deaf and have abnormal stereocilia on the outer and inner hair cells of the cochlea (Alagramam et al. 2001). *Pcdh15* is expressed in the developing sensory epithelia of the inner ear, the central nervous system, and retina (Ahmed et al. 2001) and in the epithelia of the kidney, lung, gastrointestinal tract (Murcia and Woychik 2001). This Pcdh protein is exceptionally large, with an extracellular domain containing 27 EC motifs. The functions of Pcdh15 have been extensively studied in relation to Usher syndrome (or Usher's syndrome), a genetic disorder inherited as an autosomal recessive trait that causes hearing and vision loss in school-age children. Usher syndrome is classified clinically into three subtypes: I, II, and III. Usher syndrome type I is the most common and most severe. It is characterized by congenital deafness, disequilibrium owing to problems in the vestibular system, and retinitis pigmentosa. By genetic linkage analysis, seven loci have been mapped for Usher syndrome type 1, termed *USH1A* to *USH1G*. Patients with Usher syndrome type 1F show various mutations in the *Pcdh15* gene, including substitutions causing abnormal splicing or truncation (Ahmed et al. 2001) and a missense mutation in a highly conserved domain (Doucette et al. 2009) (Table 13.1).

Hair cells in the cochlea have bundles of stereocilia (hair bundles) at the apical surface, and these bundles function as a mechanosensor for sound-evoked vibrations. The stereocilia in the hair bundles are connected to one another with tip-link filaments. Pcdh15 and cadherin 23, a classical cadherin, are localized to the lower and upper part of the tip links, respectively, and Pcdh15 homodimers associate with cadherin 23 homodimers to form the tip-link filaments (Kazmierczak et al. 2007). In *Pcdh15*-knockout mice, cochlear and vestibular hair cells show defects in mechanotransduction similar to those of patients with Usher syndrome (Senften et al. 2006).

Table 13.1 Summary of the phenotypes of Pcdh-knockout mice and the diseases related to Pcdh

Molecules		Other designation	Phenotypes in knockout mice		Related neurological disease		Related non-neurological disease	
Clusterd Pcdh	Pcdh-α	CNR (cadherin -like neuronal receptor)	Axon projection defect in olfactory sensory neurons	Hasegawa et al. (2008)	Bipolar disorder	Pedrosa et al. (2008)	Breast cancer	Novak et al. (2008)
			Axon projection defect in serotonergic neurons	Katori et al. (2009)			Wilms' tumor	Dallosso et al. (2009)
			Abnormal behavior (Pcdha A-type)	Fukuda et al. (2008)				
	Pcdh-β				Rett syndrome (Pcdh-β1)	Miyake et al. (2011)	Breast cancer	Novak et al. (2008)
					Autism (Pcdh-β4)	O'Roak et al. (2012)	Wilms' tumor	Dallosso et al. (2009)
	Pcdh-γ		Died within 12 h	Wang et al. (2002b)			Astrocytoma (Pcdh-γA11)	Waha et al. (2005)
			Neuronal cell death and decreased numbers of synapse				Breast cancer	Novak et al. (2008)
							Wilms' tumor	Dallosso et al. (2009)

Non-clustered Pcdh	Pcdh 1	Pcdh 42					Asthma	Koppelman et al. (2009)
							Eczema	Koning et al. (2012)
	Pcdh 7	BH-Pcdh, NF-Pcdh			Rett syndrome	Miyake et al. (2011)	Non-small-cell lung cancer	Huang et al. (2009)
	Pcdh 8	Arcadrin Paraxial Pcdh	Increased dendritic spine number		Schizophrenia	Shaw et al. (1998)	Mantle cell lymphoma	Leshchenko et al. (2010)
							Brest cancer	Yu et al. (2008)
	Pcdh 9				Autism spectrum disorder	Marshall et al. (2008)	Glioblastoma	de Tayrac et al. (2009)
							Oligodendrocytoma	Wang et al. (2012)
	Pcdh 10	OL-Pcdh	Axon projection failure in ventral telencephalon	Uemura et al. (2007)	Autism	Morrow et al. (2008)	Various types of cancer	See text
	Pcdh 11X	Pcdh-X			Late-onset alzheimer's disease	Carrasquillo et al. (2009)		
					Schizophrenia and schizoaffective disorder	Giouzeli et al. (2004)		
	Pcdh 11Y	Pcdh-Y, Pcdh 22			Schizophrenia and schizoaffective disorder	Giouzeli et al. (2004)	Prostate cancer	Terry et al. (2006)
	Pcdh 12	VE-cadherin-2, Pcdh 14	No apparent abnormality	Rampon et al. (2005)	Schizophrenia	Gregório et al. (2009)		
	Pcdh 15	Usher syndrome 1F (USH1F)	Deafness (Ames waltzer mutant mice)	Alagramam et al. (2001)	Usher syndrome type1F	Ahmed et al. (2001)	Hyperlipidemia	Huertas-Vazquez et al. (2010)
					DFNB23	Ahmed et al. (2003)		
	Pcdh 17	Pcdh 68			Schizophrenia	Dean et al. (2007)	Esophageal carcinoma	Haruki et al. (2010)
							Urological cancer	Costa et al. (2011)
	Pcdh 18	Pcdh 68-like			Intellectual disability	Kasnauskiene et al. (2012)	Glioblastoma	Bauer et al. (2011)
	Pcdh 19				EFMR	Dibbens et al. (2008)		
					Dravet-like syndrome	Depienne et al. (2009		
					Autism	Camacho et al. (2012)		
	Pcdh 20				Huntington disease	Becanovic et al. (2010)	Non-small-cell lung cancer	Huang et al. (2009)
	Pcdh 21	Photoreceptor cadherin	Abnormal photoreceptor	Rattner et al. (2001)	Autosomal recessive retinitis pigmentosa	Henderson et al. (2010)		
					Autosomal recessive cone-rod dystrophy	Ostergaard et al. (2010)		

The Kyoto Circling (KCI) rat, which is deaf and shows abnormal, constant circling behavior, was generated by a spontaneous mutation found in the F3 generation of a Sprague–Dawley rat purchased from a breeder in 2003. Histological examination revealed severe defects in the stereocilia in the cochlea, accompanied by a mutation in *Pcdh15* (Naoi et al. 2009).

Cadherin 23 has also been identified as a responsible gene in Usher syndrome type1D (Bolz et al. 2001). In the Ames waltzer mouse, a model for Usher syndrome type1D, various mutations of cadherin 23 have been found, including mutations that create a premature stop codon and an ectopic splice donor site (Di Palma et al. 2001). Mutations in *Pcdh15* have also been reported in patients with nonsyndromic deafness DFNB23 (Ahmed et al. 2003. 2008). Together, these results suggest that Pcdh15 is essential for the development and function of the mechanically sensitive hair bundle through its interactions with cadherin 23.

Besides neurological diseases, Pcdh15 is implicated in familial combined hyperlipidemia: nonsynonymous SNPs are found within the *Pcdh15* gene in families with this disease, and significant decreases in plasma triglyceride and total cholesterol concentration were observed in *Pcdh15*-deficient mice compared with wild-type mice (Huertas-Vazquez et al. 2010). However, the biological role of Pcdh15 in lipid abnormalities has not been identified.

13.2.9 Pcdh17

[Other designation protocadherin 68]

Pcdh17, formerly named protocadherin 68, was cloned as a cadherin-related molecule expressed in rat glomeruli, by RT-PCR with degenerate primers. Postmortem expression analysis of Brodmann's area 46, part of the frontal cortex, in schizophrenia patients revealed that the *Pcdh17* transcript is increased in patients with schizophrenia of short but not long duration. These differences in gene expression may be associated with changes in the symptom profile of schizophrenia (Dean et al. 2007).

13.2.10 Pcdh18

[Other designation: protocadherin 68-like protein]

Pcdh18 was first identified, together with *Pcdh19*, as a novel Pcdh molecule containing a CM-2 motif and is expressed in the brain, heart, kidney, lung, and trachea (Wolverton and Lalande 2001). In a boy with an intellectual disability characterized by severe developmental delay, seizures, microcephaly, hypoplastic corpus callosum, internal hydrocephalus, and dysmorphic features, a single gene deletion was found on chromosome 4q28.3, and the *Pcdh18* gene is considered a possible candidate for the responsible gene (Kasnauskiene et al. 2012).

13.2.11 *Pcdh19*

[Other designation: epilepsy female-restricted with mental retardation (EFMR)]

Pcdh19 was first identified together with *Pcdh18* (Wolverton and Lalande 2001). *Pcdh19* is expressed in the central nervous system and neural retina, and the Pcdh19 protein has a cell adhesive property in chicken (Tai et al. 2010). In zebrafish, Pcdh19 and N-cadherin form a *cis*-complex, which controls cell movements during morphogenesis of the anterior neural tube (Biswas et al. 2010). Although homophilic interactions of Pcdh19 alone are only weakly adhesive, the adhesion mediated by the Pcdh19 and N-cadherin complex appears strong (Emond et al. 2011), suggesting that Pcdh19 may regulate the adhesive property of N-cadherin.

Epilepsy in females with mental retardation (EFMR, also called epilepsy and mental retardation limited to females) is an X-linked disorder characterized by seizure with onset in infancy or early childhood and cognitive impairment. It is unusual in that most X-linked genetic disorders affect males, but in the case of EFMR the carrier males are phenotypically normal, and only females are affected. Various mutations within the *Pcdh19* gene have been identified in families with EFMR, including a mutation resulting in the introduction of premature stop codon, and a missense mutation affecting the adhesive property of *Pcdh19* (Dibbens et al. 2008).

Dravet syndrome (also known as Severe Myoclonic Epilepsy of Infancy, SMEI) is an intractable epilepsy, characterized by severe, uncontrolled seizures, usually triggered by fever, in the first year of life. Dravet syndrome is mainly caused by a heterozygous de novo mutation in the *SCN1A* gene, which encodes the voltage-gated neuronal sodium channel alpha I subunit (Nav1.1). In a large study of a series of patients with Dravet syndrome, 27 % completely lacked or showed rearrangements in one copy of the *SCN1A* gene. Mutations in *Pcdh19* were also found in these patients, suggesting that *Pcdh19* is also a responsible gene for Dravet-like syndrome (Depienne et al. 2009). In addition, because autism is a common symptom in EFMR, *Pcdh19* is considered a possible responsible gene for autism (Camacho et al. 2012).

13.2.12 *Pcdh20*

[Other designation: protocadherin 13]

Pcdh20 was cloned as an olfactory sensory neuron-specific cadherin in a genome-wide microarray screen. Pcdh20 protein is expressed in the brain, especially in the olfactory epithelium and olfactory bulb, and in the pancreas (Lee et al. 2008). Huntington disease is a hereditary neurological disorder that causes early selective neuronal cell death in the striatum and cortex, resulting in a progressive loss of cognitive, physical, and emotional function. The causative mutation is an expansion of the CAG tract in exon 1 of the *huntingtin* gene, which encodes a polyglutamine tract in the huntingtin protein. The CAG tract is normally repeated 10–35 times within the gene, but patients with Huntington disease have from 36 to more than 120

CAG repeats. Polyglutamine-expanded huntingtin protein accumulates in large aggregates that are found in various regions of the brain, and these proteins are considered toxic to neurons.

The YAC 128 transgenic mouse, which expresses the human HTT gene with 128 CAG repeats, is a Huntington disease model. These transgenic mice display a progressive neurological phenotype similar to that of patients with Huntington disease. To analyze transcriptional changes in Huntington disease, a comparative genome-wide expression profile of striatal tissue from YAC128 transgenic mice and wild-type mice was performed. The results revealed 13 genes that showed differential expression between these mice, including that of *Pcdh20*, whose mRNA level is increased in YAC128 transgenic mice as early as 3 months after birth. However, the *Pcdh20* mRNA is downregulated in the caudate nucleus of humans with Huntington disease. These results suggest that a change in Pcdh20 expression level may lead to impaired synaptic connections in the striatum (Becanovic et al. 2010).

13.2.13 Pcdh21

[Other designations: MT-protocadherin, Photoreceptor cadherin]

Pcdh21, originally named photoreceptor cadherin (prCAD), was first identified in bovine retina. Using subtractive hybridization with bovine brain cDNA, cDNAs expressed specifically in the retina were obtained. In situ hybridization of the retina showed *Pcdh21* transcripts only in photoreceptors, and Pcdh21 protein localized to the base of the outer segment of both rods and cones (Rattner et al. 2001). *Pcdh21-knockout* mice are viable, are fertile, and do not display any anatomic abnormalities. However, by electron microscopy, their outer segments appear disorganized and fragmented, and a progressive loss of photoreceptor cells via apoptosis is observed over 5 months. These results indicate that Pcdh21 is necessary for outer segment integrity (Rattner et al. 2001).

Retinitis pigmentosa is an inherited eye disease in which the progressive dysfunction and degeneration of rod photoreceptors are followed by the degeneration of cone photoreceptors. It is characterized by progressive peripheral vision loss, night blindness (nyctalopia), and eventual blindness. In families with an autosomal recessive retinitis pigmentosa, two separate homozygous single-base deletions of the *Pcdh21* gene have been found, leading to a frameshift and a premature stop codon. These deletions are located within the extracellular cadherin domains, which are highly conserved in various species (Henderson et al. 2010).

In addition, a homozygous 1-bp duplication in the *Pcdh21* gene is found in families with autosomal recessive cone-rod dystrophy. The symptoms of this disorder are similar to those of retinitis pigmentosa, although it is characterized by the early loss of cone receptors accompanied by the loss of rod receptors. These mutations in families result in a frameshift and a premature stop codon (Ostergaard et al. 2010). Prominin 1 is a pentaspan transmembrane glycoprotein that associates with Pcdh21 at the base of photoreceptor outer segments. Mutations in prominin 1 have also been reported in patients with autosomal dominant cone-rod dystrophy and with autosomal recessive retinitis pigmentosa (Yang et al. 2008). These results suggest that an association of Pcdh21 with prominin 1 is essential for photoreceptors to function normally.

13.3 Genomic Structure of Clustered *Pcdh* Genes

The clustered *Pcdh* genes were first identified as the cadherin-related neuronal receptor (CNR) family, which weakly interacted with the N-terminus of Fyn tyrosine kinase in a yeast two-hybrid screen with mouse brain cDNA (Kohmura et al. 1998). Subsequently, a large gene cluster was found in the human genome project data by a BLAST search for the *Cnr* gene (Wu and Maniatis 1999). For both the human *Pcdh-α* and *Pcdh-γ* clusters the genomic organization of consists of a set of variable-region exons followed by a set of common constant-region exons. The variable-region exons encode six EC domains, a transmembrane domain, and a short cytoplasmic domain. The constant-region exons encode the rest of the cytoplasmic domain. Each individual variable-region exon has a promoter and is *cis-spliced* to the constant-region exons (Wang et al. 2002a). Moreover, there are alternative splicing forms (types A and B) in the constant-region exons of the *Pcdh-α* genes (Sugino et al. 2000). The *Pcdh-β* cluster lacks constant exons and is instead composed of unspliced single-exon genes that are equivalent to the variable-region exons of the other two clusters. The molecules named protocadherin 2, protocadherin 3, and protocadherin 13 by Sano et al. (1993) are identical to *Pcdh-γC3*, *Pcdh-β12*, and *Pcdh-γA12*, respectively.

13.4 Clustered Pcdh Function and Association with Neurological Disease

The clustered *Pcdh* genes are widely expressed in the central nervous system (Morishita et al. 2004; Frank et al. 2005; Junghans et al. 2008). The most unique feature of the clustered *Pcdh* family is its expression pattern in individual neurons. Single-cell RT-PCR analyses of mouse cerebellar Purkinje cells showed that each Purkinje cell randomly expresses ~2 members of the 12 *Pcdh-α* isoforms (*α1–α12*) (Esumi et al. 2005), ~4 members of the 22 *Pcdh-β* isoforms (Hirano et al. 2012), and ~4 members of the 19 *Pcdh-γ* isoforms (*γA1–γA12*, *γB1*, *γB2*, and *γB4–γB8*) (Kaneko et al. 2006). Similar results were demonstrated by in situ hybridization using specific probes for each isoform. These expression patterns of clustered *Pcdhs* can generate approximately 3×10^{10} variations for each neuron (Yagi 2012). In addition, in any given Purkinje cell, some isoforms of these genes are derived from the maternal allele and others from the paternal one. In contrast, the C-type isoforms in *Pcdh-α* (*αC1* and *αC2*) and *Pcdh-γ* (*γC3–γC5*) are constitutively expressed and are derived from both alleles (Kaneko et al. 2006). Thus, the clustered *Pcdh* members have unique monoallelic and combinatorial expression patterns in individual neurons.

DNaseI hypersensitivity (HS) assays revealed long-range *cis*-regulatory elements in *Pcdh-α* and *Pcdh-γ* that enhance some of the promoter activities (Ribich et al. 2006; Yokota et al. 2011). The *Pcdh-α* enhancer, termed HS5-1, is located downstream of the *Pcdh-α* gene cluster, and its deletion results in strong downregulation of the most 3′ *Pcdh-α* variable-region exons in mice (Ribich et al. 2006). A SNP within HS5-1 is associated with susceptibility to bipolar disorder (Pedrosa et al. 2008).

The CpG methylation state of promoter regions affects gene expression. The *Pcdh* promoters for the expressed *Pcdh* genes are hypo-methylated, and those of unexpressed *Pcdh* genes are hyper-methylated. The promoters for the stochastically expressed isoforms of *Pcdh-α* (*α1–α12*) show mosaic and mixed methylation states, whereas those for the constitutively expressed isoforms (*αC1* and *αC*) are hypomethylated in the mouse brain (Kawaguchi et al. 2008). Methyl-CpG-binding protein 2 (MeCP2) is a DNA-binding protein that binds 5-methylcytosine residues in CpG dinucleotides and represses the transcription from methylated gene promoters. The perturbation of MeCP2 can lead to the misregulation of clustered *Pcdh* transcription (Chahrour et al. 2008). To identify the neuronal MeCP2 target gene, genome microarray and chromatin immunoprecipitation analyses were performed and revealed that MeCP2 binds to the upstream region of *Pcdh-β1*, an isoform of the *Pcdh-β* gene expressed in human neuroblastoma cells. A luciferase assay showed that the transcriptional activity of the *Pcdh-β1* promoter was downregulated by the binding of MeCP2 (Miyake et al. 2011). These results suggest that the expression of *Pcdh* is affected by DNA methylation.

Rett syndrome is a neurological and developmental disorder caused by mutations in MeCP2 (Amir et al. 1999). Its clinical features include seizures, ataxic gait, language dysfunction, autistic behavior, small hands and feet, and slow head growth; it occurs almost exclusively in females (Chahrour and Zoghbi 2007). The *Pcdh-β1* mRNA level is increased in the postmortem brain from Rett syndrome patients (Miyake et al. 2011), suggesting an association between *Pcdh-β1* expression and Rett syndrome. Moreover, a de novo mutation of the *Pcdh-β4* gene is found in patients with sporadic autism (O'Roak et al. 2012); however, little is known about the relationship between these neurological diseases and the *Pcdh-β* genes.

Pcdh-α and Pcdh-γ proteins are localized to both axonal and dendritic synaptic regions (Kohmura et al. 1998; Phillips et al. 2003). Pcdh-α proteins associate with Pcdh-γ proteins in *cis*, and Pcdh-γ enhances the cell surface expression of Pcdh-α (Murata et al. 2004). In addition, Pcdh-α, Pcdh-β, and Pcdh-γ proteins form a trimeric protein complex (Han et al. 2010).

Pcdh-γ proteins have a strictly homophilic adhesive property, and a heteromultimeric *cis*-tetramer functions as a homophilic binding unit (Schreiner and Weiner 2010). The cytoplasmic domain of Pcdh-α and Pcdh-γ associates with two tyrosine kinases, proline-rich tyrosine kinase 2 (PYK2) and focal adhesion kinase (FAK), and the interaction with Pcdh-α or Pcdh-γ inhibits these kinase activities (Chen et al. 2009). The overexpression of PYK2 induces cell death via apoptosis, suggesting that Pcdh-α and Pcdh-γ contribute to neuronal cell survival by inhibiting the

PYK2 pathway. Moreover, Pcdh-γ regulates dendritic arborization in cortical neurons by inhibiting the FAK pathway (Garrett et al. 2012).

Pcdh-α gene-deficient mice are viable, are fertile, and have no apparent abnormality, but their olfactory sensory neurons (Hasegawa et al. 2008) and serotonergic neurons (Katori et al. 2009) display an axon projection defect. Mice in which the Pcdh-αA isoform is downregulated are also viable and fertile, but they show behavioral changes in contextual fear conditioning and spatial working memory (Fukuda et al. 2008). On the other hand, *Pcdh-γ* gene-deficient mice die within 12 h of birth and have a hunched posture, shallow and irregular breathing, and repetitive limb tremors (Wang et al. 2002b). In addition, their spinal cord is smaller than that of wild-type mice, and they have severe neurological defects, including massive cell death via apoptosis and decreased numbers of synapses.

Analyses of knockout mice of the CTCF (CCCTG-binding factor) gene, a key molecule for chromatin conformational change, suggested that CTCF regulates the expression of clustered *Pcdhs*. These mice display neuronal abnormalities, such as decreased dendritic arborization, lack of somatosensory barrel formation, and decreased spine numbers in hippocampal neurons (Hirayama et al. 2012).

The cohesin complex, a key molecule in chromosomal interactions, colocalizes with CTCF. Mice heterozygous for Nipped-B-like (NIPBL), a subunit of the cohesin DNA loading complex, show a significant decrease in *Pcdh-β* gene expression (Kawauchi et al. 2009). Ablation of the cohesin complex subunit SA1 (cohesin-SA1) causes the downregulation of clustered *Pcdh* genes (Remeseiro et al. 2012). Moreover, the cohesin complex subunit Rad21 binds to the promoter and enhancer regions of the clustered *Pcdh*s and may regulate their expression (Monahan et al. 2012). These findings suggest that complex chromatin organization plays an important role in the regulation of *Pcdh* gene expression.

Overall, these results suggest that the diversity of the clustered P*cdh* genes may contribute to selective interactions between neurons and be required for neural circuit formation. However, their detailed functions and disease associations still need to be elucidated.

13.5 *Pcdhs* as Tumor Suppressor Genes

The loss or reduction of classical cadherin expression has been demonstrated in a variety of tumors and may affect cell invasion and metastasis. By large-scale genome analysis, some of the *Pcdhs* have also been reported to act as tumor suppressor genes in various tissues. Microarray-based methylation analysis of astrocytoma showed aberrant methylation of a CpG island within the first exon of a clustered *Pcdh* isoform, *Pcdh-γA11*, compared to normal brain tissue. There is a significant inverse relationship between gene expression and methylation in astrocytoma. After treating astrocytoma cells with 5-aza-2′-deoxycytidine, a demethylating agent, DNA methylation is significantly increased, and the expression of *Pcdh-γA11* gene is reduced. These findings suggest that the adhesive property of

Pcdh-γA11 may play a role in the invasion of astrocytoma cells into the normal brain (Waha et al. 2005). However, the molecular diversity of the *Pcdh*s expressed in astrocytoma is not known, and whether other clustered *Pcdh* isoforms are aberrantly expressed remains to be investigated.

Genome-wide analyses of glioblastoma samples from patients revealed that the non-clustered *Pcdh9* is downregulated in glioblastoma (de Tayrac et al. 2009). Furthermore, patients with *Pcdh9*-expressing glioma have longer survival times than patients with *Pcdh9*-non-expressing glioma. In this study, the *Pcdh9* expression was an independent prognostic factor, and no significant associations were found for gender, chemotherapy, or radiotherapy. Subgroup analysis according to tumor subtype showed that both patients with *Pcdh9*-expressing astrocytoma and *Pcdh9*-expressing oligodendrocytoma have longer survival times than patients with *Pcdh9*-non-expressing tumors. These findings indicate that *Pcdh9* might be a useful biomarker for predicting the prognosis of patients with glioma (Wang et al. 2012).

Collagen XVI, a member of the fibril-associated collagens with interrupted triple helices (FACIT), is upregulated in glioblastoma cells and may affect the migration and invasiveness of glioma cells. Human glioblastoma cells in which the gene for Collagen XVI is knocked down by siRNA show decreased invasiveness and adhesion potential and increased *Pcdh18* mRNA and Pcdh18 protein (Bauer et al. 2011). These results suggest that *Pcdh18* may act as a tumor suppressor in glioma.

In some other tumors, *Pcdh* expressions are decreased, and Pcdh proteins may play a variety of roles in specific tumor types. For example, reduced expression of the clustered *Pcdh*s (*Pcdh-α*, *Pcdh-β*, and *Pcdh-γ*) may be associated with breast cancer (Novak et al. 2008), Wilms' tumor, a pediatric tumor of the kidney (Dallosso et al. 2009), and colorectal tumor (Dallosso et al. 2012); *Pcdh7* and *Pcdh20* with non-small-cell lung cancer (Huang et al. 2009); *Pcdh8* with mantle cell lymphoma (Leshchenko et al. 2010) and breast cancer (Yu et al. 2008). Among the non-clustered *Pcdh*s, *Pcdh10* may be associated with breast cancer (Miyamoto et al. 2005), hematologic cancer (Ying et al. 2007), gastric, colorectal, and pancreatic cancers (Yu et al. 2010), cervical cancer (Narayan et al. 2009), testicular cancer (Cheung et al. 2010), and nasopharyngeal and esophageal carcinomas (Ying et al. 2006); *Pcdh11Y* with prostate cancer (Terry et al. 2006); and *Pcdh17* with esophageal carcinoma (Haruki et al. 2010), bladder cancer, renal cell tumors, and prostate cancer (Costa et al. 2011).

In early-stage non-small-cell lung cancer (NSCLC) patients, two SNPs of the *Pcdh7* gene were identified by genome-wide analysis (Huang et al. 2009). Similarly, in VMRC-LCR cells, an NSCLC cell line, a homozygous loss of the *Pcdh20* gene was identified. *Pcdh20* mRNA is expressed in normal lung tissue but contains a homozygous deletion in other NSCLC cell lines (Imoto et al. 2006).

Pcdh8 is inactivated by mutation or epigenetic silencing, and *Pcdh8* point mutations are found in various breast carcinoma cells. These point mutations cause defects in *Pcdh8*'s ability to inhibit cell growth and migration. In addition, the transfection of a human normal mammary epithelial cell line with *Pcdh8* containing one of these mutations causes the cells' transformation (Yu et al. 2008). Similarly, *Pcdh10* is silenced or downregulated in gastric cancer cell lines. The re-expression of *Pcdh10* in gastric cancer cell lines reduces colony formation in vitro, and less tumor growth is observed in vivo when the transfected cells are injected into mice (Yu et al. 2010). Interestingly, the *Pcdh8*, *Pcdh9*, *Pcdh17*, and *Pcdh20* genes are located on chromosome 13q14–21, which is conserved between humans and mice; thus, their expressions might be regulated by common factors in cellular and brain function (Kim et al. 2011).

13.6 Conclusion

This chapter introduces the *Pcdh* family, which is associated with various neurological diseases, such as bipolar disorder, autism, schizophrenia, and Usher syndrome, as well as a number of tumor types. While the functions of classical cadherins such as E-cadherin and N-cadherin have been extensively studied, the roles of the *Pcdh* family remain poorly understood. In the future, it will be important to determine the functions of the *Pcdh*s to elucidate the mechanisms of neurological diseases and brain functions.

Acknowledgements We thank Dr. Teruyoshi Hirayama for helpful comments.

Compliance with Ethics Requirements The authors declare that they have no conflicts of interest.

References

Ahmed ZM, Riazuddin S, Bernstein SL et al (2001) Mutations of the protocadherin gene PCDH15 cause Usher syndrome type 1F. Am J Hum Genet 69:25–34. doi:10.1086/321277

Ahmed ZM, Riazuddin S, Ahmad J et al (2003) PCDH15 is expressed in the neurosensory epithelium of the eye and ear and mutant alleles are responsible for both USH1F and DFNB23. Hum Mol Genet 12:3215–3223. doi:10.1093/hmg/ddg358

Ahmed ZM, Riazuddin S, Aye S et al (2008) Gene structure and mutant alleles of PCDH15: non-syndromic deafness DFNB23 and type 1 Usher syndrome. Hum Genet 124:215–223. doi:10.1007/s00439-008-0543-3

Alagramam KN, Murcia CL, Kwon HY et al (2001) The mouse Ames waltzer hearing-loss mutant is caused by mutation of Pcdh15, a novel protocadherin gene. Nat Genet 27:99–102. doi:10.1038/83837

Amir RE, Van den Veyver IB, Wan M et al (1999) Rett syndrome is caused by mutations in X-linked MECP2, encoding methyl-CpG-binding protein 2. Nat Genet 23:185–188. doi:10.1038/13810

Bauer R, Ratzinger S, Wales L et al (2011) Inhibition of collagen XVI expression reduces glioma cell invasiveness. Cell Physiol Biochem 27(3–4):217–226 doi:10.1159/000327947

Becanovic K, Pouladi MA, Lim RS et al (2010) Transcriptional changes in Huntington disease identified using genome-wide expression profiling and cross-platform analysis. Hum Mol Genet 19:1438–1452. doi:10.1093/hmg/ddq018

Beecham GW, Naj A, Gilbert JR et al (2011) PCDH11X variation is not associated with late-onset alzheimer disease susceptibility. Psychiatr Genet 20:321–324. doi:10.1097/YPG.0b013e32833b635d.PCDH11X

Biswas S, Emond MR, Jontes JD (2010) Protocadherin-19 and N-cadherin interact to control cell movements during anterior neurulation. J Cell Biol 191:1029–1041. doi:10.1083/jcb.201007008

Blanco P, Sargent CA, Boucher CA et al (2000) Conservation of PCDHX in mammals; expression of human X/Y genes predominantly in brain. Mamm Genome 11(10):906–914. doi:10.1007/s003350010177
Bolz H, von Brederlow B, Ramírez A et al (2001) Mutation of CDH23, encoding a new member of the cadherin gene family, causes Usher syndrome type 1D. Nat Genet 27:108–112. doi:10.1038/83667
Bradley RS, Espeseth A, Kintner C (1998) NF-protocadherin, a novel member of the cadherin superfamily, is required for Xenopus ectodermal differentiation. Curr Biol 8:325–334
Bray NJ, Kirov G, Owen RJ et al (2002) Screening the human protocadherin 8 (PCDH8) gene in schizophrenia. Genes Brain Behav 1(3):187–191
Camacho A, Simón R, Sanz R et al (2012) Cognitive and behavioral profile in females with epilepsy with PDCH19 mutation: two novel mutations and review of the literature. Epilepsy Behav 24:134–137. doi:10.1016/j.yebeh.2012.02.023
Carrasquillo MM, Zou F, Pankratz VS et al (2009) Genetic variation in PCDH11X is associated with susceptibility to late-onset alzheimer's disease. Nat Genet 41:192–198. doi:10.1038/ng.305
Chahrour M, Zoghbi HY (2007) The story of Rett syndrome: from clinic to neurobiology. Neuron 56:422–437. doi:10.1016/j.neuron.2007.10.001
Chahrour M, Jung SY, Shaw C et al (2008) MeCP2, a key contributor to neurological disease, activates and represses transcription. Science 320:1224–1229. doi:10.1126/science.1153252
Chen J, Lu Y, Meng S et al (2009) alpha- and gamma-Protocadherins negatively regulate PYK2. J Biol Chem 284:2880–2890. doi:10.1074/jbc.M807417200
Cheung HH, Lee TL, Davis AJ et al (2010) Genome-wide DNA methylation profiling reveals novel epigenetically regulated genes and non-coding RNAs in human testicular cancer. Br J Cancer 102:419–427. doi:10.1038/sj.bjc.6605505
Costa VL, Henrique R, Danielsen SA, et al (2011) TCF21 and PCDH17 methylation: An innovative panel of biomarkers for a simultaneous detection of urological cancers. Epigenetics. 6:1120-30. doi:10.4161/epi.6.9.16376
Dallosso AR, Hancock AL, Szemes M et al (2009) Frequent long-range epigenetic silencing of protocadherin gene clusters on chromosome 5q31 in Wilms' tumor. PLoS Genet 5:e1000745. doi:10.1371/journal.pgen.1000745
Dallosso AR, Oster B, Greenhough A et al (2012) Long-range epigenetic silencing of chromosome 5q31 protocadherins is involved in early and late stages of colorectal tumorigenesis through modulation of oncogenic pathways. Oncogene 31:4409–4419. doi:10.1038/onc.2011.609
de Tayrac M, Etcheverry A, Aubry M, et al (2009) Integrative genome-wide analysis reveals a robust genomic glioblastoma signature associated with copy number driving changes in gene expression. Genes Chromosomes & Cancer 48:55–68. doi:10.1002/gcc.20618
Dean B, Keriakous D, Scarr E et al (2007) Gene expression profiling in Brodmann's area 46 from subjects with schizophrenia. Aust N Z J Psychiatry 41:308–320
Depienne C, Bouteiller D, Keren B et al (2009) Sporadic infantile epileptic encephalopathy caused by mutations in PCDH19 resembles Dravet syndrome but mainly affects females. PLoS Genet 5:e1000381. doi:10.1371/journal.pgen.1000381
Di Palma F, Holme RH, Bryda EC et al (2001) Mutations in Cdh23, encoding a new type of cadherin, cause stereocilia disorganization in waltzer, the mouse model for Usher syndrome type 1D. Nat Genet 27:103–107. doi:10.1038/83660
Dibbens LM, Tarpey PS, Hynes K et al (2008) X-linked protocadherin 19 mutations cause female-limited epilepsy and cognitive impairment. Nat Genet 40:776–781. doi:10.1038/ng.149
Doucette L, Merner ND, Cooke S et al (2009) Profound, prelingual nonsyndromic deafness maps to chromosome 10q21 and is caused by a novel missense mutation in the Usher syndrome type IF gene PCDH15. Eur J Hum Genet 17:554–564. doi:10.1038/ejhg.2008.231
Emond MR, Biswas S, Blevins CJ, Jontes JD (2011) A complex of Protocadherin-19 and N-cadherin mediates a novel mechanism of cell adhesion. J Cell Biol 195:1115–1121. doi:10.1083/jcb.201108115
Esumi S, Kakazu N, Taguchi Y et al (2005) Monoallelic yet combinatorial expression of variable exons of the protocadherin-alpha gene cluster in single neurons. Nat Genet 37:171–176. doi:10.1038/ng1500

Frank M, Ebert M, Shan W et al (2005) Differential expression of individual gamma-protocadherins during mouse brain development. Mol Cell Neurosci 29:603–616. doi:10.1016/j.mcn.2005.05.001

Fukuda E, Hamada S, Hasegawa S et al (2008) Down-regulation of protocadherin-alpha A isoforms in mice changes contextual fear conditioning and spatial working memory. Eur J Neurosci 28:1362–1376. doi:10.1111/j.1460-9568.2008.06428.x

Garrett AM, Schreiner D, Lobas MA, Weiner JA (2012) γ-protocadherins control cortical dendrite arborization by regulating the activity of a FAK/PKC/MARCKS signaling pathway. Neuron 74:269–276. doi:10.1016/j.neuron.2012.01.028

Giouzeli M, Williams NA, Lonie LJ et al (2004) ProtocadherinX/Y, a candidate gene-pair for schizophrenia and schizoaffective disorder: a DHPLC investigation of genomic sequence. Am J Med Genet B Neuropsychiatr Genet 129B:1–9. doi:10.1002/ajmg.b.30036

Gregório SP, Sallet PC, K-anh D et al (2009) Polymorphisms in genes involved in neurodevelopment may be associated with altered brain morphology in schizophrenia : preliminary evidence. Psychiatry Res 165:1–9. doi:10.1016/j.psychres.2007.08.011

Han M-H, Lin C, Meng S, Wang X (2010) Proteomics analysis reveals overlapping functions of clustered protocadherins. Mol Cell Proteomics 9:71–83. doi:10.1074/mcp.M900343-MCP200

Haruki S, Imoto I, Kozaki K et al (2010) Frequent silencing of protocadherin 17, a candidate tumour suppressor for esophageal squamous cell carcinoma. Carcinogenesis 31:1027–1036. doi:10.1093/carcin/bgq053

Hasegawa S, Hamada S, Kumode Y et al (2008) The protocadherin-alpha family is involved in axonal coalescence of olfactory sensory neurons into glomeruli of the olfactory bulb in mouse. Mol Cell Neurosci 38:66–79. doi:10.1016/j.mcn.2008.01.016

Henderson RH, Li Z, Abd MM et al (2010) Biallelic mutation of protocadherin-21 (PCDH21) causes retinal degeneration in humans. Mol Vis 16:46–52

Hirano S, Yan Q, Suzuki ST (1999) Expression of a novel protocadherin, OL-protocadherin, in a subset of functional systems of the developing mouse brain. J Neurosci 19:995–1005

Hirano S, Kaneko R, Izawa T et al (2012) Single-neuron diversity generated by Protocadherin-β cluster in mouse central and peripheral nervous systems. Front Mol Neurosci 5:90 doi:10.3389/fnmol.2012.00090

Hirayama T, Tarusawa E, Yoshimura Y et al (2012) CTCF is required for neural development and stochastic expression of clustered Pcdh genes in neurons. Cell Rep 2:345–357. doi:10.1016/j.celrep.2012.06.014

Huang Y-T, Heist RS, Chirieac LR et al (2009) Genome-wide analysis of survival in early-stage non-small-cell lung cancer. J Clin Oncol 27:2660–2667. doi:10.1200/JCO.2008.18.7906

Huertas-Vazquez A, Plaisier CL, Geng R et al (2010) A nonsynonymous SNP within PCDH15 is associated with lipid traits in familial combined hyperlipidemia. Hum Genet 127:83–89. doi:10.1007/s00439-009-0749-z

Imoto I, Izumi H, Yokoi S et al (2006) Frequent silencing of the candidate tumor suppressor PCDH20 by epigenetic mechanism in non-small-cell lung cancers. Cancer Res 66:4617–4626. doi:10.1158/0008-5472.CAN-05-4437

Junghans D, Heidenreich M, Hack I et al (2008) Postsynaptic and differential localization to neuronal subtypes of protocadherin beta16 in the mammalian central nervous system. Eur J Neurosci 27:559–571. doi:10.1111/j.1460-9568.2008.06052.x

Kaneko R, Kato H, Kawamura Y et al (2006) Allelic gene regulation of Pcdh-alpha and Pcdh-gamma clusters involving both monoallelic and biallelic expression in single Purkinje cells. J Biol Chem 281:30551–30560. doi:10.1074/jbc.M605677200

Kasnauskiene J, Ciuladaite Z, Preiksaitiene E et al (2012) A single gene deletion on 4q28.3: PCDH18–a new candidate gene for intellectual disability? Eur J Med Genet 55:274–277. doi:10.1016/j.ejmg.2012.02.010

Katori S, Hamada S, Noguchi Y et al (2009) Protocadherin-alpha family is required for serotonergic projections to appropriately innervate target brain areas. J Neurosci 29:9137–9147. doi:10.1523/JNEUROSCI.5478-08.2009

Kawaguchi M, Toyama T, Kaneko R et al (2008) Relationship between DNA methylation states and transcription of individual isoforms encoded by the protocadherin-alpha gene cluster. J Biol Chem 283:12064–12075. doi:10.1074/jbc.M709648200

Kawauchi S, Calof AL, Santos R et al (2009) Multiple organ system defects and transcriptional dysregulation in the Nipbl(+/-) mouse, a model of Cornelia de Lange Syndrome. PLoS Genet 5:e1000650. doi:10.1371/journal.pgen.1000650

Kazmierczak P, Sakaguchi H, Tokita J et al (2007) Cadherin 23 and protocadherin 15 interact to form tip-link filaments in sensory hair cells. Nature 449:87–91. doi:10.1038/nature06091

Kim S-Y, Yasuda S, Tanaka H et al (2011) Non-clustered protocadherin. Cell Adh Migr 5:97–105. doi:10.4161/cam.5.2.14374

Kohmura N, Senzaki K, Hamada S et al (1998) Diversity revealed by a novel family of cadherins expressed in neurons at a synaptic complex. Neuron 20:1137–1151

Koning H, Postma DS, Brunekreef B et al (2012) Protocadherin-1 polymorphisms are associated with eczema in two Dutch birth cohorts. Pediatr Allergy Immunol 23:270–277. doi:10.1111/j.1399-3038.2011.01201.x

Koppelman GH, Meyers DA, Howard TD, Zheng SL, Hawkins GA, Ampleford EJ, Xu J et al (2009) Identification of PCDH1 as a novel susceptibility gene for bronchial hyperresponsiveness. Am J Respir Crit Care Med 180(10):929–935. doi:10.1164/rccm.200810-1621OC

Lee W, Cheng T, Gong Q (2008) Olfactory sensory neuron-specific and sexually dimorphic expression of protocadherin 20. J Comp Neurol 507:1076–1086. doi:10.1002/cne

Leshchenko VV, Kuo P-Y, Shaknovich R et al (2010) Genomewide DNA methylation analysis reveals novel targets for drug development in mantle cell lymphoma. Blood 116:1025–1034. doi:10.1182/blood-2009-12-257485

Ludwig D, Lorenz J, Dejana E et al (2000) cDNA cloning, chromosomal mapping, and expression analysis of human VE-Cadherin-2. Mamm Genome 11:1030–1033. doi:10.1007/s003350010186

Marshall CR, Noor A, Vincent JB et al (2008) Structural variation of chromosomes in autism spectrum disorder. Am J Hum Genet 82:477–488. doi:10.1016/j.ajhg.2007.12.009

Miar A, Alvarez V, Corao AI et al (2011) Lack of association between protocadherin 11-X/Y (PCDH11X and PCDH11Y) polymorphisms and late onset alzheimer's disease. Brain Res 1383:252–256. doi:10.1016/j.brainres.2011.01.054

Miyake K, Hirasawa T, Soutome M et al (2011) The protocadherins, PCDHB1 and PCDH7, are regulated by MeCP2 in neuronal cells and brain tissues: implication for pathogenesis of Rett syndrome. BMC Neurosci 12:81. doi:10.1186/1471-2202-12-81

Miyamoto K, Fukutomi T, Akashi-Tanaka S et al (2005) Identification of 20 genes aberrantly methylated in human breast cancers. Int J Cancer 116:407–414. doi:10.1002/ijc.21054

Monahan K, Rudnick ND, Kehayova PD et al (2012) Role of CCCTC binding factor (CTCF) and cohesin in the generation of single-cell diversity of Protocadherin-α gene expression. Proc Natl Acad Sci USA 109:9125–9130. doi:10.1073/pnas.1205074109

Morishita H, Murata Y, Esumi S et al (2004) CNR/Pcdh alpha family in subplate neurons, and developing cortical connectivity. Neuroreport 15:2595–2599. doi:10.1097/00001756-200412030-00007

Morrow EM, Yoo S-Y, Flavell SW et al (2008) Identifying autism loci and genes by tracing recent shared ancestry. Science 321:218–223. doi:10.1126/science.1157657

Murata Y, Hamada S, Morishita H et al (2004) Interaction with protocadherin-gamma regulates the cell surface expression of protocadherin-alpha. J Biol Chem 279:49508–49516. doi:10.1074/jbc.M408771200

Murcia CL, Woychik RP (2001) Expression of Pcdh15 in the inner ear, nervous system and various epithelia of the developing embryo. Mech Dev 105:163–166

Nakao S, Uemura M, Aoki E et al (2005) Distribution of OL-protocadherin in axon fibers in the developing chick nervous system. Brain Res Mol Brain Res 134:294–308. doi:10.1016/j.molbrainres.2004.11.017

Nakao S, Platek A, Hirano S, Takeichi M (2008) Contact-dependent promotion of cell migration by the OL-protocadherin-Nap1 interaction. J Cell Biol 182:395–410. doi:10.1083/jcb.200802069

Naoi K, Kuramoto T, Kuwamura Y et al (2009) Characterization of the Kyoto circling (KCI) rat carrying a spontaneous nonsense mutation in the protocadherin 15 (Pcdh15) gene. Exp Anim 58:1–10

Narayan G, Scotto L, Neelakantan V et al (2009) Protocadherin PCDH10, involved in tumor progression, is a frequent and early target of promoter hypermethylation in cervical cancer. Genes Chromosomes Cancer 48:983–992. doi:10.1002/gcc

Novak P, Jensen T, Oshiro MM et al (2008) Agglomerative epigenetic aberrations are a common event in human breast cancer. Cancer Res 68:8616–8625. doi:10.1158/0008-5472.CAN-08-1419

O'Roak BJ, Vives L, Girirajan S et al (2012) Sporadic autism exomes reveal a highly interconnected protein network of de novo mutations. Nature 485:246–250. doi:10.1038/nature10989

Ostergaard E, Batbayli M, Duno M et al (2010) Mutations in PCDH21 cause autosomal recessive cone-rod dystrophy. J Med Genet 47:665–669. doi:10.1136/jmg.2009.069120

Pedrosa E, Stefanescu R, Margolis B et al (2008) Analysis of protocadherin alpha gene enhancer polymorphism in bipolar disorder and schizophrenia. Schizophr Res 102:210–219. doi:10.1016/j.schres.2008.04.013

Phillips GR, Tanaka H, Frank M et al (2003) Gamma-protocadherins are targeted to subsets of synapses and intracellular organelles in neurons. J Neurosci 23:5096–5104

Rampon C, Prandini M-H, Bouillot S et al (2005) Protocadherin 12 (VE-cadherin 2) is expressed in endothelial, trophoblast, and mesangial cells. Exp Cell Res 302:48–60. doi:10.1016/j.yexcr.2004.08.024

Rashid D, Newell K, Shama L, Bradley R (2006) A requirement for NF-protocadherin and TAF1/Set in cell adhesion and neural tube formation. Dev Biol 291:170–181. doi:10.1016/j.ydbio.2005.12.027

Rattner A, Smallwood PM, Williams J et al (2001) A photoreceptor-specific cadherin is essential for the structural integrity of the outer segment and for photoreceptor survival. Neuron 32:775–786

Remeseiro S, Cuadrado A, Gómez-López G et al (2012) A unique role of cohesin-SA1 in gene regulation and development. EMBO J 31:2090–2102. doi:10.1038/emboj.2012.60

Ribich S, Tasic B, Maniatis T (2006) Identification of long-range regulatory elements in the protocadherin-alpha gene cluster. Proc Natl Acad Sci USA 103:19719–19724. doi:10.1073/pnas.0609445104

Sano K, Tanihara H, Heimark RL, Obata S, Davidson M, St John T, Taketani S et al (1993) Protocadherins: a large family of cadherin-related molecules in central nervous system. EMBO J 12(6):2249–2256

Schreiner D, Weiner JA (2010) Combinatorial homophilic interaction between gamma-protocadherin multimers greatly expands the molecular diversity of cell adhesion. Proc Natl Acad Sci USA 107:14893–14898. doi:10.1073/pnas.1004526107

Senften M, Schwander M, Kazmierczak P et al (2006) Physical and functional interaction between protocadherin 15 and myosin VIIa in mechanosensory hair cells. J Neurosci 26:2060–2071. doi:10.1523/JNEUROSCI.4251-05.2006

Shaw SH, Kelly M, Smith AB et al (1998) A genome-wide search for schizophrenia susceptibility genes. Am J Med Genet 81:364–376

Strehl S, Glatt K, Liu QM et al (1998) Characterization of two novel protocadherins (PCDH8 and PCDH9) localized on human chromosome 13 and mouse chromosome 14. Genomics 53:81–89. doi:10.1006/geno.1998.5467

Sugano S, Yoshitomo-Nakagawa K, Yu YS et al (1998) Transmembrane-domain trapping: a novel method for isolation of cDNAs encoding putative membrane proteins. DNA Res 5:187–193

Sugino H, Hamada S, Yasuda R et al (2000) Genomic organization of the family of cnr cadherin genes in mice and humans. Genimics 63:75–87. doi:10.1006/geno.1999.6066

Sugino H, Yanase H, Hamada S et al (2004) Distinct genomic sequence of the CNR/Pcdhalpha genes in chicken. Biochem Biophys Res Commun 316:437–445. doi:10.1016/j.bbrc.2004.02.067

Tada MN, Senzaki K, Tai Y et al (2004) Genomic organization and transcripts of the zebrafish Protocadherin genes. Gene 340:197–211. doi:10.1016/j.gene.2004.07.014

Tai K, Kubota M, Shiono K et al (2010) Adhesion properties and retinofugal expression of chicken protocadherin-19. Brain Res 1344:13–24. doi:10.1016/j.brainres.2010.04.065

Telo' P, Breviario F, Huber P et al (1998) Identification of a novel cadherin (vascular endothelial cadherin-2) located at intercellular junctions in endothelial cells. J Biol Chem 273:17565–17572

Terry S, Queires L, Gil-diez-de-medina S et al (2006) Protocadherin-PC promotes androgen-independent prostate cancer cell growth. Prostate 66:1100–1113. doi:10.1002/pros
Uemura M, Nakao S, Suzuki ST et al (2007) OL-Protocadherin is essential for growth of striatal axons and thalamocortical projections. Nat Neurosci 10:1151–1159. doi:10.1038/nn1960
Waha A, Güntner S, Huang TH-M et al (2005) Epigenetic silencing of the protocadherin family member PCDH-gamma-A11 in astrocytomas. Neoplasia 7:193–199. doi:10.1593/neo.04490
Wang X, Su H, Bradley A (2002a) Molecular mechanisms governing Pcdh-gamma gene expression : evidence for a multiple promoter and cis-alternative splicing model. Genes Dev 16:1890–1905. doi:10.1101/gad.1004802.Protocadherins
Wang X, Weiner JA, Levi S et al (2002b) Gamma protocadherins are required for survival of spinal interneurons. Neuron 36:843–854
Wang C, Yu G, Liu J et al (2012) Downregulation of PCDH9 predicts prognosis for patients with glioma. J Clin Neurosci 19:541–545. doi:10.1016/j.jocn.2011.04.047
Wolverton T, Lalande M (2001) Identification and characterization of three members of a novel subclass of protocadherins. Genomics 76:66–72. doi:10.1006/geno.2001.6592
Wu Q, Maniatis T (1999) A striking organization of a large family of human neural cadherin-like cell adhesion genes. Cell 97:779–790
Wu Q, Zhang T, Cheng JF et al (2001) Comparative DNA sequence analysis of mouse and human protocadherin gene clusters. Genome Res 11:389–404. doi:10.1101/gr.167301
Yagi T (2012) Molecular codes for neuronal individuality and cell assembly in the brain. Front Mol Neurosci 5:45. doi:10.3389/fnmol.2012.00045
Yamagata K, Andreasson KI, Sugiura H et al (1999) Arcadlin is a neural activity-regulated cadherin involved in long term potentiation. J Biol Chem 274:19473–19479
Yamamoto A, Kemp C, Bachiller D et al (2000) Mouse paraxial protocadherin is expressed in trunk mesoderm and is not essential for mouse development. Genesis 27:49–57
Yanase H, Sugino H, Yagi T (2004) Genomic sequence and organization of the family of CNR/Pcdhalpha genes in rat. Genomics 83:717–726. doi:10.1016/j.ygeno.2003.09.022
Yang Z, Chen Y, Lillo C et al (2008) Mutant prominin 1 found in patients with macular degeneration disrupts photoreceptor disk morphogenesis in mice. J Clin Invest 118:2908–2916. doi:10.1172/JCI35891DS1
Yasuda S, Tanaka H, Sugiura H et al (2007) Activity-induced protocadherin arcadlin regulates dendritic spine number by triggering N-cadherin endocytosis via TAO2beta and p38 MAP kinases. Neuron 56:456–471. doi:10.1016/j.neuron.2007.08.020
Ying J, Li H, Seng TJ et al (2006) Functional epigenetics identifies a protocadherin PCDH10 as a candidate tumor suppressor for nasopharyngeal, esophageal and multiple other carcinomas with frequent methylation. Oncogene 25:1070–1080. doi:10.1038/sj.onc.1209154
Ying J, Gao Z, Li H et al (2007) Frequent epigenetic silencing of protocadherin 10 by methylation in multiple haematologic malignancies. Br J Haematol 136:829–832. doi:10.1111/j.1365-2141.2007.06512.x
Yokota S, Hirayama T, Hirano K et al (2011) Identification of the cluster control region for the Protocadherin-{beta} genes located beyond the Protocadherin-{gamma} cluster. J Biol Chem 286:31885–31895. doi:10.1074/jbc.M111.245605
Yoshida K, Sugano S (1999) Identification of a novel protocadherin gene (PCDH11) on the human XY homology region in Xq21.3. Genomics 62:540–543. doi:10.1006/geno.1999.6042
Yoshida K, Yoshitomo-Nakagawa K, Seki N et al (1998) Cloning, expression analysis, and chromosomal localization of BH-protocadherin (PCDH7), a novel member of the cadherin superfamily. Genomics 49:458–461. doi:10.1006/geno.1998.5271
Yoshida K, Watanabe M, Kato H et al (1999) BH-protocadherin-c, a member of the cadherin superfamily, interacts with protein phosphatase 1 alpha through its intracellular domain. FEBS Lett 460:93–98
Yu JS, Koujak S, Nagase S et al (2008) PCDH8, the human homolog of PAPC, is a candidate tumor suppressor of breast cancer. Oncogene 27:4657–4665. doi:10.1038/onc.2008.101
Yu B, Yang H, Zhang C et al (2010) High-resolution melting analysis of PCDH10 methylation levels in gastric, colorectal and pancreatic cancers. Neoplasma 57:247–252. doi:10.4149/neo_2010_03_247

Chapter 14
Neural Cell Adhesion Molecules Belonging to the Family of Leucine-Rich Repeat Proteins

Malene Winther and Peter S. Walmod

Abstract Leucine-rich repeats (LRRs) are motifs that form protein–ligand interaction domains. There are approximately 140 human genes encoding proteins with extracellular LRRs. These encode cell adhesion molecules (CAMs), proteoglycans, G-protein-coupled receptors, and other types of receptors. Here we give a brief description of 36 proteins with extracellular LRRs that all can be characterized as CAMs or putative CAMs expressed in the nervous system. The proteins are involved in multiple biological processes in the nervous system including the proliferation and survival of cells, neuritogenesis, axon guidance, fasciculation, myelination, and the formation and maintenance of synapses. Moreover, the proteins are functionally implicated in multiple diseases including cancer, hearing impairment, glaucoma, Alzheimer's disease, multiple sclerosis, Parkinson's disease, autism spectrum disorders, schizophrenia, and obsessive–compulsive disorders. Thus, LRR-containing CAMs constitute a large group of proteins of pivotal importance for the development, maintenance, and regeneration of the nervous system.

M. Winther, M.Sc.
Laboratory of Neural Plasticity, Department of Neuroscience and Pharmacology, Faculty of Health Sciences, University of Copenhagen, Copenhagen, Denmark

P.S. Walmod, M.Sc. (✉)
Laboratory of Neural Plasticity, Department of Neuroscience and Pharmacology, Faculty of Health Sciences, University of Copenhagen, Copenhagen, Denmark

Laboratory of Neural Plasticity, University of Copenhagen, Copenhagen, Symbion, Box 39, Fruebjergvej 3, 2100 Copenhagen Ø, Denmark
e-mail: psw@sund.ku.dk

V. Berezin and P.S. Walmod (eds.), *Cell Adhesion Molecules: Implications in Neurological Diseases*, Advances in Neurobiology 8, DOI 10.1007/978-1-4614-8090-7_14,

14.1 Introduction

Traditionally, cell adhesion molecules (CAMs) have been divided into four groups: cadherins, selectins, integrins, and members of the immunoglobulin (Ig) superfamily. However, within recent years several other protein families with members able to mediate homo- or heterophilic cell adhesion have been identified. These include for instance Neurexins and Neuroligins (Bottos et al. 2011). Moreover, the family of leucine-rich repeat (LRR)-containing proteins includes numerous CAMs (Nam et al. 2011; Woo et al. 2009a; Karaulanov et al. 2006; Hohenester 2008).

The LRR is a protein–ligand interaction motif. It was discovered in 1985 as part of the primary sequence of leucine-rich α2-glycoprotein, which is named thus because 66 of the 312 amino acids constituting the protein are leucine residues (Takahashi et al. 1985). Subsequently, LRRs have been identified in a large group of functionally diverse proteins, and in the human genome alone, 375 genes are known to express LRR-containing proteins (Ng et al. 2011).

LRR-containing proteins are not restricted to mammals, but are found in numerous organisms including bacteria, plants, and yeast (Kobe and Kajava 2001). Moreover, the LRR motif is located in intracellular, transmembrane, as well as extracellular proteins (Dolan et al. 2007) with a large variety of functions. For instance, LRR-containing proteins include homo- and heterophilically interacting CAMs (Nam et al. 2011; Woo et al. 2009a; Karaulanov et al. 2006; Hohenester 2008; Chen et al. 2006), proteoglycans (McEwan et al. 2006; Park et al. 2008; Dellett et al. 2012), virulence factors (Bierne et al. 2007; Niemann et al. 2004), ribonuclease inhibitors (Kobe and Deisenhofer 1996), scaffold proteins (Thalhammer et al. 2009), cytoskeleton-binding proteins (Kopecki and Cowin 2008; Kostyukova 2008), phosphatases (Brognard and Newton 2008), phosphatase inhibitors (Santa-Coloma 2003), cytosolic kinases (Tsika and Moore 2012), receptor kinases and receptor tyrosine kinases (Skaper 2012b; Dievart et al. 2011), G-protein-coupled receptors (Barker and Clevers 2010; Kong et al. 2010), as well as other non-catalytic receptors and modulators of cell signaling (Botos et al. 2011; Matilla and Radrizzani 2005; Wadelin et al. 2010).

LRRs mediate protein–protein interactions, and from the examples of LRR-containing protein types listed above, it is clear that this protein family is involved in a variety of biologically processes including immune responses (Botos et al. 2011), disease resistance and pathogen recognition in plants (Bonardi et al. 2012), platelet aggregation (Wijeyewickrema et al. 2005), extracellular matrix (ECM) assembly (McEwan et al. 2006; Park et al. 2008; Dellett et al. 2012), cell adhesion (Nam et al. 2011; Woo et al. 2009a; Karaulanov et al. 2006; Hohenester 2008; Chen et al. 2006), activation and modulation of intracellular signal transduction (Brognard and Newton 2008; Santa-Coloma 2003; Tsika and Moore 2012; Skaper 2012b; Matilla and Radrizzani 2005), cytoskeletal organization and dynamics (Kopecki and Cowin 2008; Kostyukova 2008; Matilla and Radrizzani 2005; Yamashiro et al. 2012), and RNA processing (Kobe and Deisenhofer 1996).

14.1.1 Structural Organization of Leucine-Rich Repeats

LRRs are composed of 20–29 amino acids long hydrophobic motifs that contain a highly conserved sequence and a variable sequence. The conserved sequence is characterized by 11 residues, LxxLxLxxNxL, or 12 residues, LxxLxLxxCxxL, where x can be any amino acid, N is either an asparagine, threonine, cysteine, or serine, C is a cysteine or a serine, and L is a leucine (sometimes replaced by a valine, isoleucine, or phenylalanine) (Kobe and Kajava 2001; Kajava 1998).

Ribonuclease inhibitor was the first LRR-containing protein for which a crystal structure was obtained (PDB ID: 2BNH; Fig. 14.1). The structure revealed that an LRR consists of an α-helix (formed by the variable sequence) connected by a loop to a β-strand (formed by the conserved sequence) (Kobe and Deisenhofer 1996). Generally, LRRs exist in tandem arrays of 2–52 repeats, which results in an overall horseshoe-like shape, where all the β-strands are on the concave side and the α-helices are on the convex side (Kobe and Kajava 2001; Kajava 1998; Matsushima et al. 2005).

To prevent that the hydrophobic core of an LRR domain is exposed to the surroundings, extracellular LRR domains are typically flanked by cysteine-rich domains. There is at least one type of cysteine-rich domain flanking the N-terminal of extracellular LRRs (N-flanking; NF) and at least four cysteine-rich domains flanking the C-terminal (C-flanking; CF1–4) [reviewed in Kobe and Kajava (2001)].

When LRR domains form homodimers they seem to dimerize through the concave surface, as demonstrated for the proteoglycan decorin (PDB ID: 1XKU), the D4 domain of the ROBO1 and -2 ligand SLIT2 (PDB ID: 2WFH) (Seiradake et al. 2009), and the ectodomains of AMIGO1 (PDB ID: 2XOT; see below). Likewise, the majority of LRR domains bind to heterophilic ligands with the concave surface, as demonstrated for the ribonuclease inhibitor binding to ribonuclease [PDB ID: 1Z7X; Johnson et al. (2007)] or Netrin-G1 and -G2 in complex with the Netrin-G ligands (NGLs) NGL1 and NGL2, respectively (PDB IDs: 3ZYJ and 3ZYI; see below). However, there are also examples of LRR-containing proteins that mediate interactions with the convex surface of the LRR domain, for example, the platelet cell surface protein GPIBA in complex with Thrombin (PDB ID: IP8V) [reviewed in Bella et al. (2008)].

As mentioned above, typical LRR domains consist of β-strands, α-helices, and interconnecting loops. However, variations from this general composition exist, and based on differences in their consensus sequences, length, etc. LRR motifs can be classified into at least seven subfamilies (RI-like, SDS22-like, cysteine-containing, bacterial, typical, plant-specific, and TpLRR). Four of the seven subfamilies include proteins expressed in animals, and of these four families, three (RI-like, SDS22-like, and cysteine-containing) include proteins with intracellular LRRs, whereas the fourth (the "typical" subfamily) exclusively includes proteins with extracellular LRRs [reviewed in Kobe and Kajava (2001)].

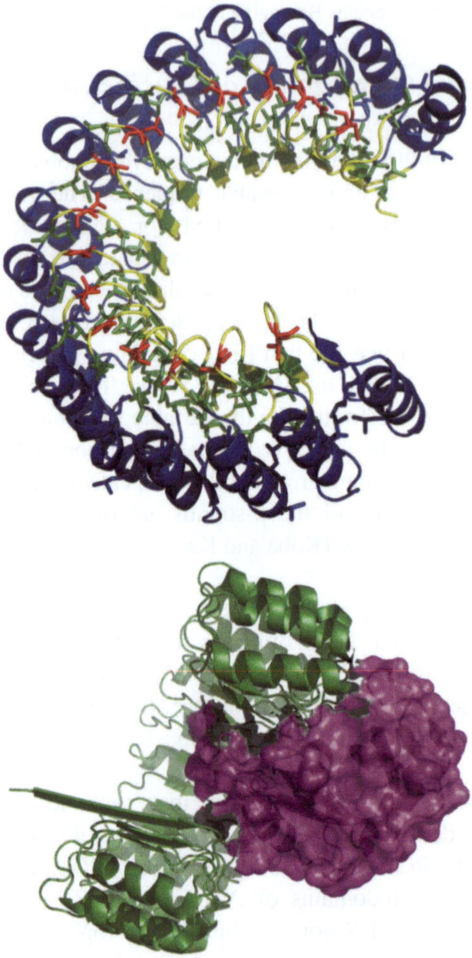

Fig. 14.1 Structural model of Ribonuclease inhibitor. *Top*: Cartoon drawing of the structural model of porcine ribonuclease inhibitor (PDB ID: 2BNH). The model reveals that the protein consists of 16 α-helices (*blue*) connected to β-strands with the overall sequence LxxLxLxxNxL, where x (*yellow*) can be any amino acid, N (*red*, *sticks*) is either an asparagine, threonine, cysteine, or serine, and L (*green*, *sticks*) is a leucine, valine, or an isoleucine. The conserved polar amino acids are located in the loop regions connecting the β-sheets with the α-helices, whereas the conserved nonpolar residues are in the loop regions, and β-sheets, facing toward the α-helices, away from the concave surface. *Bottom*: Structural model of the interaction between human ribonuclease inhibitor (*green*; *cartoon*) and human pancreatic ribonuclease [*purple*; *cartoon* and *surface*; PDB ID: 1Z7X; Johnson et al. (2007)]. The structure demonstrated how the ribonuclease interacts with the concave surface of the LRR domain. The figures were created with PyMOL (DeLano Scientific)

14.1.2 Subfamilies of Proteins with Extracellular Leucine-Rich Repeats

Bioinformatical analysis has shown that ~140 human proteins contain extracellular LRRs. For comparison, *Caenorhabditis elegans, Drosophila melanogaster*, and *Mus musculus*, respectively, express 29, 66, and 135 proteins with extracellular LRRs.

Table 14.1 Human proteins with extracellular LRRs

Class of protein with extracellular LRR	Type I transmembrane	GPI anchored	Secreted	Multi-transmembrane	Total
LRR-Ig/Fn3	35	1	2	0	38
LRR-Tollkin	17	0	2	0	19
LRR-other	1	0	9	16	26
LRR-only	32	6	19	0	57
Total	85	7	32	16	140

Human proteins with extracellular LRRs have been divided into four classes according to their predicted characteristics [modified from Dolan et al. (2007)]

These proteins have been subdivided into four classes: LRR-only (consisting of proteins that contain only LRRs), LRR-Ig/Fn3 (consisting of proteins that in addition to LRRs also contain Ig and/or Fn3 modules), LRR-Tollkin (consisting of proteins that contain a cytoplasmic Toll/interleukin 1 receptor domain or cluster with the Toll proteins), and LRR-other (consisting of proteins that contain other types of domains, e.g., epidermal growth factor (EGF) repeats or a G-protein-coupled receptor domain [see Dolan et al. (2007), for details]. As shown in Table 14.1 proteins with extracellular LRRs can be further subdivided into secreted, lipid-anchored, and various types of transmembrane proteins.

The proteins with extracellular LRRs include several families of CAMs with members that demonstrate a dynamic expression in the developing and adult nervous system (Nam et al. 2011; Woo et al. 2009a; Karaulanov et al. 2006; Hohenester 2008; Chen et al. 2006; Wright and Washbourne 2011).

Of the ten LRR-containing families of CAMs described below, seven families (the AMIGO/Alvin, FLRT, LINGO, LRIT, NGL, NLRR, and SALM families) belong to the LRR-Ig/Fn3 class, two families (the LRRTM and SLITRK families) belong to the LRR-only class, and LRRC15/Lib belongs to the LRR-Tollkin class (Fig. 14.2).

14.1.3 LRR-Containing Non-cell Adhesion Molecules in Diseases

Since LRR-containing proteins constitute a large family of structurally and functionally diverse proteins that are implicated in multiple biological processes, it is not surprising that they are implicated in a number of diseases [reviewed in Matsushima et al. (2005)].

Mutations in the cytoskeleton-regulating SDS22-like LRR-containing protein LRRC50 has been linked to human cystic kidney disease and primary cilia dyskinesia (Zariwala et al. 1993; van Rooijen et al. 2008).

Mutations in the cytoplasmic kinase LRR kinase-2 (LRRK2, also known as Parkinson disease (autosomal dominant) 8 [PARK8], and dardarin) are related to the development of Parkinson's disease (PD). Thus, the LRRK2 mutant G2019S alone has been identified in up to 40 % of the cases of familial PD [reviewed in Tsika and Moore (2012)].

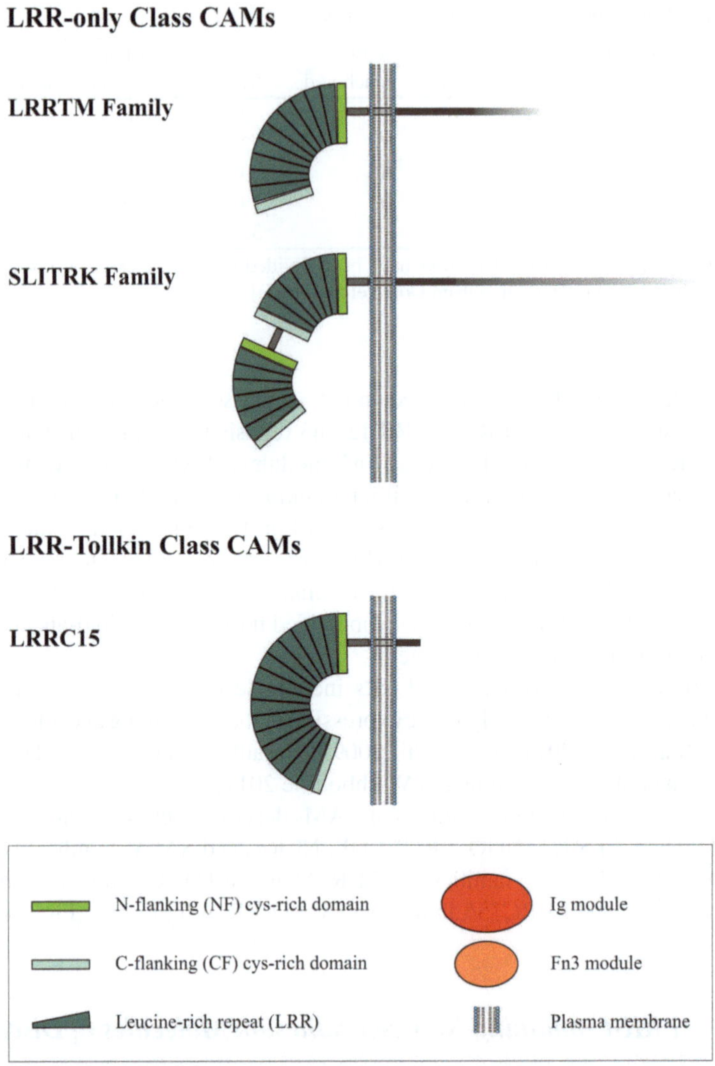

Fig. 14.2 Schematic drawing of the LRR-containing CAMs. The figure shows schematic drawings of the general structural organizations of the LRR-containing CAMs described in the text. The general structure of the proteins within a single family is presented as a single drawing. If there are variations in the general organization of the proteins within the family, this is indicated by a *pale coloring* of the domains in question. Differences in the lengths of the cytoplasmic domains between members within a single family are indicated with an *uneven coloring* of the cytoplasmic domain. The relative lengths of the cytoplasmic domains reflect differences in the number of amino acids constituting the cytoplasmic domains in the respective protein families. The figures were created with CorelDraw 11 (Corel Corporation)

Several LRR-containing proteins are pathogen recognition receptors. Intracellular pathogen recognition receptors include the CATERPILLER (also called NOD-LRR and NACHT-LRR) family of proteins, which has been shown to be important regulators of cell death, cell growth, and immunity. The family also contains proteins

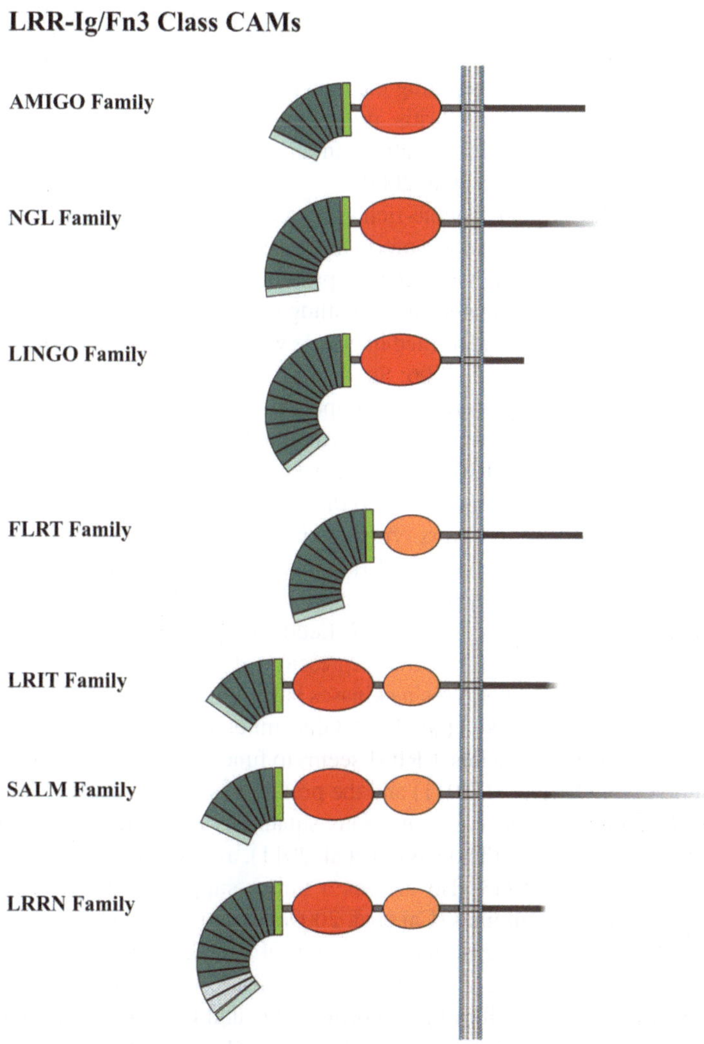

Fig. 14.2 (continued)

associated with several immunological disorders including bare lymphocyte syndrome (MHC class II deficiency), Blau syndrome, Crohn's disease, familial-cold autoinflammatory syndrome, familial Mediterranean fever, multiple sclerosis (MS), myocardial infarction, Muckle–Wells syndrome, neonatal-onset multisystem inflammatory disease, and rheumatoid arthritis [reviewed in Inohara and Nunez (2003) and Ting et al. (2006)]. Pathogen recognition receptors with extracellular LRRs include Toll-like receptors that recognize pathogen-associated molecular patterns and danger-associated molecular patterns. Toll-like receptors are involved in numerous diseases including Alzheimer's disease (AD) and MS [reviewed in Hanamsagar et al. (2012)].

An absence of the small leucine-rich proteoglycans (SLRPs) biglycan, decorin, fibromodulin, and lumican affects collagen fibrillogenesis, and SLRP-deficient mice exhibit several defects including the connective tissue disease Ehlers–Danlos syndrome, osteoporosis, osteoarthritis, muscular dystrophy, and corneal opacification (Ameye and Young 2002). Mutations in another SLRP, nyctalopin, can cause night blindness (Bech-Hansen et al. 2000).

In the brain, the protein leucine-rich, glioma-inactivated 1 (LGI1)/Epitempin exists in two isoforms of which one isoform that is secreted has been shown to regulate synaptic transmission through binding to protein ADAM 22. Data suggest that mutations in *LGI1* are implicated in the pathogenesis of epilepsy, and at least 22 mutations in *LGI1* have been identified in people with autosomal epilepsies (Sirerol-Piquer et al. 2006; Fukata et al. 2006; Staub et al. 2002).

Several LRR-containing proteins are important for the pathogenesis of cancer, and within the group of non-CAMs in particular the LRIG family has received attention. The mammalian LRIG family consists of three members, LRIG1-3 (Guo et al. 2004), that are structurally related to the *Drosophila* protein Kekkon, and like Kekkon, LRIGs modulate receptors tyrosine kinases. Thus, LRIG1 has been found to bind Ret, MET, EGF receptors, and ErbB receptors directly and thereby prevent receptor activation while enhancing receptor downregulation (Gur et al. 2004; Laederich et al. 2004; Shattuck et al. 2007; Ledda et al. 2008). In contrast, reduced expression of LRIG1 can lead to overexpression of ErbB2 (Miller et al. 2008). LRIG1 modulation of receptor tyrosine kinases has implications for the pathogenesis of several types of cancer (Wu et al. 2012; Ghasimi et al. 2012; Powell et al. 2012; Xie et al. 2013). In breast cancers, LRIG1 seems to function as an estrogen-regulated growth suppressor (Krig et al. 2011) and the protein has been suggested as a potential prognostic indicator of, e.g., cutaneous squamous cell carcinoma (Tanemura et al. 2005), prostate cancer (Thomasson et al. 2011), and cervical adenocarcinomas (Muller et al. 2013). LRIG3 also acts as a tumor suppressor by reducing EGF receptor activity in, e.g., gliomas (Cai et al. 2009a, b) and cervical adenocarcinomas (Muller et al. 2013). In contrast, the expression of LRIG2 seems to correlate with EGF receptor activity (Wang et al. 2009a), and LRIG2 is a prognostic marker for poor survival for patients with oligodendrogliomas and early-stage squamous cell carcinoma of the uterine cervix (Holmlund et al. 2009; Hedman et al. 2010).

14.1.4 Cell Adhesion Molecules with Extracellular LRR Domains

In the following sections the members of ten different families of CAMs or potential CAMs with extracellular LRR domains will be described. Although many of the proteins have only been identified recently the combined literature regarding the members of the ten protein families is extensive, and the descriptions are therefore in many cases somewhat sketchy. To compensate for this superficiality we have tried to highlight review and key articles where specific topics are described in more detail.

Some sections will reveal that a given family of proteins has no, or only a minor, known relationship with neurological diseases or that the family members upon a closer inspection may turn out not to be CAMs. However, since the knowledge about the individual proteins in many cases is limited, we have prioritized to include several of the protein families not because they are CAMs related to neurological diseases, but because they are potential CAMs with a potential relationship to neurological diseases.

On the other hand it should be emphasized that the included protein families do not constitute a complete list of LRR-containing CAMs. For instance, Slit proteins and their Robo counter-receptors are excluded from this chapter, but are described in a separate chapter (Ypsilanti and Chedotal 2013). Proteins that might have been included but which we decided to omit include Densin-180 and other members of the LAP family (Thalhammer et al. 2009) and Trk receptors (Zhou et al. 1997).

14.2 Cell Adhesion Molecules Belonging to the LRR-Only Class

14.2.1 The LRRTM Family

The LRR transmembrane neuronal (LRRTM) family consists of four proteins, LRRTM1-4. The family was identified in 2003 (Lauren et al. 2003) and has also been detected in a number of large-scale studies (Ishikawa et al. 1997; Clark et al. 2003; Ota et al. 2004; Gerhard et al. 2004; Bechtel et al. 2007). The proteins have been suggested to be implicated in human handedness and a number of diseases including AD, autism spectrum disorders (ASDs), and schizophrenia (Majercak et al. 2006; Francks et al. 2007; Sousa et al. 2010).

Genes and Proteins

LRRTM1-4 are encoded by separate genes (Table 14.2) that constitute a family with conserved orthologues in vertebrates, but not invertebrates. Interestingly, three of the genes are located within introns of genes encoding different α-Catenin proteins [known to facilitate interactions between cadherins and actin filaments, Pokutta et al. (2008)], and in all cases the *LRRTM* genes are transcribed in the opposite direction as the *CTNN* genes, suggesting coevolution of the of two families; *LRRTM1* is located within *CTNNA2*, *LRRTM2* is located within *CTNNA1*, and *LRRTM3* is located within *CTNNA3* (Lauren et al. 2003); [see also Smith et al. (2011b)]. Recently it was demonstrated that *CTNNA1* and *CTNNA2* contain alternative bidirectional promoters that are used for the transcription of *CTNNA1* and *CTNNA2* as well as *LRRTM2* and *LRRTM1*, respectively. Both promoters predominantly facilitate the transcription of the *LRRTM* genes, but their existence nevertheless points toward a functional

Table 14.2 The human LRRTM family

Gene symbol	Chromosomal location	Synonyms and previous names	Recommended protein name
LRRTM1	2P12	FLJ32082	Leucine-rich repeat transmembrane neuronal protein 1
LRRTM2	5q31	KIAA0416, LRRN2, leucine-rich repeat neuronal 2 protein	Leucine-rich repeat transmembrane neuronal protein 2
LRRTM3	10q22.1		Leucine-rich repeat transmembrane neuronal protein 3
LRRTM4	2p12	FLJ2568	Leucine-rich repeat transmembrane neuronal protein 4

relationship between the expression of LRRTM-family proteins and α-Catenins (Kask et al. 2011).

All four LRRTM proteins have an identical structural organization consisting of an ectodomain composed of 10 LRRs with NF and CF1 domains, followed by a transmembrane domain, and a cytoplasmic domain. Originally, all four LRRTMs were reported to have ~72 amino acids long cytoplasmic domains (Lauren et al. 2003). However, later it has been shown that alternative splicing of LRRTM3 and-4 can also generate isoforms with ~140 amino acids long cytoplasmic domains (Clark et al. 2003; Ota et al. 2004; Gerhard et al. 2004; Bechtel et al. 2007). The human proteins exhibit amino acid identities of 43–63 %, LRRTM2 and -4 demonstrating the lowest degree of identity, and LRRTM3 and -4 the highest degree of identity (Lauren et al. 2003).

Expression

In adult humans, LRRTM1, -2, and -4 are expressed in numerous tissues. However, levels of LRRTM-encoding mRNAs are highest in the brain, and LRRTM3 is almost exclusively located in the central nervous system (CNS). The four proteins are expressed throughout the brain including the amygdala, caudate nucleus, cerebellum, corpus callosum, hippocampus, the olfactory system, and thalamus. See Lauren et al. (2003), Majercak et al. (2006), Francks et al. (2007), Haines and Rigby (2007) and Linhoff et al. (2009) for detailed studies of human and mouse LRRTM expression patterns.

In mouse, the expression of LRRTM1 has been shown to be regulated by the homeobox transcription factor Dlx5, a protein that in humans is associated with the development of, e.g., ASD and hearing loss (Sajan et al. 2011). Human *LRRTM3* has been shown to be transactivated by the DNA-binding protein storkhead box 1 (STOX1) (van Dijk et al. 2010).

At the subcellular level LRRTMs localize to synapses, and several studies demonstrate LRRTMs to be components of the postsynaptic density (Linhoff et al. 2009; Dosemeci et al. 2007; de Wit et al. 2009).

When expressed in HEK293 cells LRRTM2, -3, and -4 localize to the plasma membrane (de Wit et al. 2009), whereas LRRTM1, when overexpressed in several cell types, predominantly localize to intracellular organelles including the ER (Francks et al. 2007; de Wit et al. 2009). These observations suggest that LRRTM1 mainly plays a role for the regulation of intracellular processes, whereas the remaining members of the family serve as receptors located in the plasma membrane. LRRTM1 overexpressed in cerebellar granule neurons localize to cell soma, neurites, and growth cones (Francks et al. 2007), and when overexpressed at low levels in hippocampal neurons LRRTM1 and -2 localize to dendrites, where they co-localize with markers for excitatory, but not inhibitory, synapses (Linhoff et al. 2009).

Interactions

The postsynaptic scaffolding protein PSD-95 binds to the four most C-terminal amino acids of the cytoplasmic domain of LRRTM2 [-ECEV], which resemble a class I PDZ-domain-binding motif. PSD-95 can be immunoprecipitated with LRRTM2 containing the [-ECEV]-sequence, but not with LRRTM2 lacking the motif, suggesting a direct interaction between the C-terminal part of LRRTM2 and one of the three PDZ domains of PSD-95. As a result of the interaction LRRM2 seems to facilitate the recruitment PSD-95 to the cytoplasmic side of the postsynaptic plasma membrane (Linhoff et al. 2009; de Wit et al. 2009). The [-ECEV]-sequence is found in human LRRTM1, -2, and -4, but not LRRTM3 (Uniprot entries Q86UE6, O43300, Q4KMX1, and Q86VH5, respectively), suggesting similar functions for LRRTM1 and -4.

A study of LRR-containing proteins from zebrafish suggests that all four LRRTM proteins can form homophilic interactions as well as heterophilic interactions within the family (Soellner and Wright 2009). However, no mammalian LRRTM-family proteins have been demonstrated to form homophilic interactions. In contrast, LRRTM2 has been shown to mediate cell adhesion by *trans*-interacting with Neurexins (Ko et al. 2009). Thus, if LRRTM-family proteins do form homophilic interactions, these interactions are most likely *cis*-interactions.

In mammals, Neurexins constitute a family of three presynaptic transmembrane proteins, Neurexin-1, -2, and -3, located in the plasma membrane. Alternative promoters permit the generation of α-Neurexins (that extracellularly contain six LNS domains with three interspersed EGF-like domains) and the much shorter β-Neurexins (that extracellularly contains only a single LNS domain). In addition, α-Neurexins contain five sites for alternative splicing in the region encoding the ectodomain, two of which are also present in β-Neurexins. These splice sites enable the generation of multiple splice variants of both α- and β-Neurexins [reviewed in Craig and Kang (2007) and Knight et al. (2011)].

LRRTM2 binds both Neurexin-1α- and -β. Like for another family of postsynaptic Neurexin ligands, the Neuroligins, LRRTM2-Neurexin binding requires the binding of Ca^{2+} to Neurexins. However, in contrast to Neuroligins, which bind to Neurexins irrespective of the presence of an insert at splice site 4, LRRTM2 binds only

Neurexins lacking an insert in splice site 4. Moreover, LRRTM2 and Neuroligin-1 cannot bind Neurexin-1 simultaneously (de Wit et al. 2009; Ko et al. 2009; Siddiqui et al. 2010). In one study, LRRTM2 was found to bind Neurexin-1, but not Neurexin-2α or -3α, whereas another study has demonstrated binding of LRRTM2 to Neurexin-2 and -3 as well as Neurexin-1 (de Wit et al. 2009; Siddiqui et al. 2010). For a recent review, see Wright and Washbourne (2011).

The LRR domain of LRRTM2 mediates *cis*-interactions with the glutamate receptor subunits GluR1/GluA1, GluR2/GluA2 (both AMPA receptor subunits), and NR1 (an NMDA receptor subunit). Knockdown of LRRTM2 causes a 33 % decrease in GluR1-containing glutamate receptors in hippocampal neurons, suggesting that LRRTM2 regulates the amount of AMPA receptors in the plasma membrane, and thereby glutamatergic neurotransmission (de Wit et al. 2009). Similar interactions probably exist between LRRTM1 and glutamate receptor subunits (see below).

Functions

The interactions between LRRTM2 and glutamate receptor subunits seem to facilitate the development and function of excitatory synapses, and also LRRTM1 and -4, and to a lesser degree LRRTM3, are able to stimulate excitatory synaptogenesis (Linhoff et al. 2009).

Knockdown of LRRTM2 in hippocampal neurons leads to a decrease in the density of excitatory but not inhibitory synapses, whereas overexpression of LLRTM2 increases the density of excitatory but not inhibitory synapses (de Wit et al. 2009; Ko et al. 2009). At the molecular level, studies with various chimeric LRRTM2 proteins have demonstrated that the LRR domain of LRRTM2 is both necessary and sufficient for the formation of excitatory synapses (Linhoff et al. 2009).

Knockdown of Neurexin-1, one of the presynaptic binding partners for LLRTM2, prevents LLRTM2-mediated presynaptic differentiation (de Wit et al. 2009), suggesting that LRRTM2–Neurexin *trans*-interactions are required for the development of excitatory synapses. Consistently, it has also been reported that a reduction in excitatory synapses is most pronounced, when both the Neurexin ligands LRRTM1, LRRTM2, and Neuroligin1 and -2 are absent. This further suggests a degree of functional redundancy between LLRTM and Neuroligins in relation to synapse formation. Moreover, the loss of synapses requires synaptic activity. Thus, no synaptic loss is observed if synaptic activity is prevented by administration of neurotransmitter receptor inhibitors or by inhibition of postsynaptic CaM kinase activity (Ko et al. 2011; Soler-Llavina et al. 2011). Interestingly, in vivo studies in mouse suggest that LLRTMs seem to be important for the development of excitatory synapses but not for the maintenance of the synaptic transmission in the mature synapses that in contrast requires the presence of Neuroligins (Soler-Llavina et al. 2011).

Lrrtm1-knockout mice are fertile, they display no obvious phenotype different from that of wild-type mice, and their overall brain morphology is largely unaffected by the LRRTM1 deficiency (Linhoff et al. 2009). However, LRRTM1 deficiency does cause a reduction in the volume of the hippocampus and a reduction in the thickness of the somatosensory cortex (but not the motor cortex, auditory cortex, or prefrontal cortex). In some regions of the hippocampus, the LRRTM1 deficiency results in an increase in spine length, but a decrease in synaptic density, whereas no changes have been observed for the width or density of dendritic spines, the length and thickness postsynaptic densities, or the size of synaptic clefts (Takashima et al. 2011). Synaptic defects in some regions of the hippocampus, defined as changes in the distribution of the vesicular glutamate transporter 1 (VGLUT1]), have also been demonstrated. Interestingly, these effects of LRRTM1 deficiency are most pronounced in the regions of the hippocampus exhibiting the lowest levels of LRRTM2 expression (Linhoff et al. 2009), suggesting a degree of functional redundancy between LRRTM1 and -2.

Two *Lrrtm1*-knockout mice with slightly different genetic backgrounds have been extensively investigated in a number of behavioral tests (Takashima et al. 2011; Voikar et al. 2013). Takashima et al. (2011), but not Voikar et al. (2013), found that *Lrrtm1*-knockout mice demonstrate reduced spontaneous locomotor activity, deficits in spatial memory, and increased freezing (an index of fear memory), whereas only Voikar et al. (2013) observed that *Lrrtm1*-knockout mice avoid small enclosures. However, in both studies it is concluded that *Lrrtm1*-knockout mice demonstrate a reduced ability to adjust or adapt to novel or stressful environments (Takashima et al. 2011; Voikar et al. 2013).

To test whether the behavioral alterations in *Lrrtm1*-knockout mice are related to the observed alterations in the function of excitatory synapses, the effects of the NMDA receptor blocker MK-801 on the behavior of *Lrrtm1*-knockout and wild-type mice have been investigated. Both Takashima et al. and Voikar et al. report that MK-801 administration leads to different locomotor activity in *Lrrtm1*-knockout and wild-type mice, supporting the notion that the abnormal behavior of *Lrrtm1-knockout* mice is related to changes in glutamatergic neurotransmission (Takashima et al. 2011; Voikar et al. 2013).

Lrrtm1-knockout mice demonstrate some of the same characteristics as patients with schizophrenia, including changes in glutamatergic neurotransmission and a reduction in the volume of the hippocampus (Harrison et al. 2003; Adriano et al. 2012). Moreover, as described below, human *LRRTM1* single-nucleotide polymorphism (SNP) alleles have been associated with schizophrenia (Francks et al. 2007). Consequently, the behavior of *Lrrtm1*-knockout and wild-type mice has been observed in response to administration with clozapine, a known drug for the treatment of schizophrenia (Sherwood et al. 2012). However, no difference between the behavior of *Lrrtm1*-knockout and wild-type mice has been found in response to clozapine administration (Takashima et al. 2011). In contrast, the abnormal behavior of *Lrrtm1*-knockout mice is reduced in response to administration of the selective serotonin reuptake inhibitor fluoxetine, a widely prescribed antidepressant (Takashima et al. 2011; Mandrioli et al. 2012).

In summary, *Lrrtm1*-knockout mice demonstrate some characteristics related to schizophrenia, but without exhibiting a pharmacobehavioral phenotype similar to other schizophrenia animal models. Nevertheless, the apparent relationship between LRRTM1 and schizophrenia (see below) suggests that *Lrrtm1*-knockout mice may be useful in future studies of schizophrenia.

Diseases

β-site APP cleaving enzyme 1 (BACE1) is the main protease responsible for cleaving Amyloid precursor protein (APP) to generate Aβ-peptides, and the level of BACE1 is increased in around 50 % of all patients with AD (Tan and Evin 2012). In an attempt to identify proteins responsible for the secretion of Aβ, Majarcak et al. (2006) screened 15,200 genes by siRNA treatment of cells followed by detection of N-terminal APP cleavage products. The screen revealed that expression of LRRTM3 increases BACE1-mediated cleavage of APP, whereas downregulation of LRRTM3 has no effects on the expression of APP, BACE1, or γ-secretase components. However, changes in LRRTM3 expression have no significant effect on BACE1 activity, nor do LRRTM3 and APP demonstrate a high degree of co-localization (Majercak et al. 2006). Another study performed with several different shRNA constructs revealed that knockdown of LRRTM3 results in a reduced processing of APP (Reitz et al. 2012). Recently, the effect of LRRTM3 on APP processing was investigated in in vivo studies employing *Lrrtm3*-knockout mice and AβPPswe/PS1dE9 transgenic mice (mice containing the "Swedish" double point mutations in APP along with a mutation causing mis-splicing of exon 9 of mRNA encoding presenilin 1; all mutations associated with early-onset familial forms of AD). Surprisingly, these studies were unable to detect any effects of LRRTM3 on the processing of APP (Laakso et al. 2012). The potential relationship between LRRTM3 expression and the generation of Aβ-peptides is therefore still unclear.

However, there are other potential factors linking LRRTM3 to AD. A large region of chromosome 10, including the region encoding LRRTM3, has received attention in relation to the development of late-onset AD. Investigations of SNPs potentially related to late-onset AD, as well as other statistical and genetic studies, have observed SNPs in *LRRTM3* as well as *CCTNNA3* (the α-Catenin-encoding gene in which *LRRTM3* is nested in an intron) and identified *LRRTM3* as a potential late-onset AD candidate gene (Reitz et al. 2012; Liang et al. 2007; Thornton-Wells et al. 2008; Edwards et al. 2009).

A family-based association study of four candidate genes, *LRRTM1*, *LRRTM3*, *LRRN1*, and *LRRN3*, in combination with a case–control analysis suggests that common genetic variants of *LRRN3* and *LRRTM3* are associated with a susceptibility to the development of ASDs (Sousa et al. 2010).

Around 10 % of humans are left handed (Sun and Walsh 2006). Interestingly, handedness correlates with left–right asymmetries in relation to the progression of PD (van der Hoorn et al. 2012) and a relationship between non-right handedness and schizophrenia has also been proposed (Dragovic and Hammond 2005).

Francks et al. (2007) have, following experiments including genetic association mapping and gene-functional analysis, proposed that *LRRTM1* is a maternally suppressed gene and that SNP alleles located within *LRRTM1* (and a 80 kb region upstream of *LRRTM1*) are associated with paternally inherited handedness and susceptibility to schizophrenia (Francks et al. 2007). The study received worldwide media attention [as described by McManus et al. (2009)], but also strong criticism by Crow et al. (2009), criticism that was published together with a response by Francks (2009). The notion that there is a relationship between *LRRTM1* SNPs and paternally inherited handedness and schizophrenia as proposed by Francks et al. (2007) has been strengthened by additional data published more recently (Ludwig et al. 2009).

In summary, the LRRTM family consists of four members, of which at least some member can form heterophilic *trans*-interactions with postsynaptic Neurexins, heterophilic *cis*-interactions with glutamate receptor subunits, and intracellular interactions with PSD-95. All members of the family seem to facilitate excitatory synaptogenesis, and the individual members of the family may be related to inherited handedness, susceptibility to schizophrenia ASD, and AD. However, in many respects the individual members of the family are still poorly characterized, both with respect to their basic biochemical properties and their importance for the pathogenesis of various diseases.

14.2.2 The SLITRK Family

The SLIT and NTRK-like (SLITRK) family consists of six members, SLITRK1-6. The family was identified in a search for genes differentially expressed in humans and mice with neural tube defects (Aruga and Mikoshiba 2003). Subsequently, all six proteins constituting the family have been shown to be expressed in the nervous system, and the members of the family have been related to several diseases including cancer, schizophrenia, myotonic dystrophy 1/Steinert disease, Gilles de la Tourette syndrome (TS), and other obsessive–compulsive disorders (OCDs). See also Proenca et al. (2011) and Ko (2012) for recent reviews.

Genes and Proteins

The human and mouse members of the SLITRK gene family (*SLITRK1-6* and *Slitrk1-6*, respectively; Table 14.3) have been identified in several studies between 1998 and 2007 (Clark et al. 2003; Ota et al. 2004; Gerhard et al. 2004; Bechtel et al. 2007; Aruga and Mikoshiba 2003; Nagase et al. 1998, 2001a; Strausberg et al. 2002; Aruga et al. 2003). In both human and mouse the genes are restricted to three different regions: SLITRK1, -5, and -6 are located on chromosome 13 (chromosome 14 in mouse), SLITRK3 on chromosome 3, and SLITRK2 and -4 on the X chromosome (Aruga et al. 2003).

So far, the SLITRK-encoding gene receiving the most attention has been *SLITRK1*. The protein-coding region of this gene is located within a single exon (Aruga et al. 2003). Moreover, this gene has been found to contain tissue-specific differentially methylated regions, a phenomenon that is particularly enriched in relation to genes encoding proteins involved in neurodevelopment and neurodifferentiation (Davies et al. 2012).

Table 14.3 The human SLITRK family

Gene symbol	Chromosomal location	Synonyms and previous names	Recommended protein name
SLITRK1	13q31.1	KIAA1910, leucine-rich repeat-containing 12, LRRC12	SLIT and NTRK-like protein 1
SLITRK2	Xq27.3	KIAA1854, CXorf2, slit-like 1 (Drosophila), SLITL1	SLIT and NTRK-like protein 2
SLITRK3	3q26.1	KIAA0848	SLIT and NTRK-like protein 3
SLITRK4	Xq27.3	DKFZp547M2010	SLIT and NTRK-like protein 4
SLITRK5	13q31.1	bA364G4.2, KIAA0918, leucine-rich repeat-containing 11, LRRC11	SLIT and NTRK-like protein 5
SLITRK6	13q31.1	FLJ22774	SLIT and NTRK-like protein 6

Structurally, all six members of the family are transmembrane proteins containing an ectodomain composed of a tandem of six LRR motifs that are surrounded by both NF and CF domains (Aruga and Mikoshiba 2003). The ectodomain of SLITRK1 has been shown to be *N*-glycosylated and the subcellular localization of SLITRK1 is dependent on the glycosylation of SLITRK1 and/or SLITRK1 transport proteins (Kajiwara et al. 2009). The ectodomain is followed by a single transmembrane domain and a cytoplasmic domain with a length ranging from ~53 to 273 amino acids, depending on the SLITRK (Aruga and Mikoshiba 2003).

SLITRK ectodomains demonstrate high homologies to SLIT proteins, whereas the cytoplasmic domains (with the exception of the cytoplasmic domain of SLITRK1, which is shorter than those of the other SLITRK proteins) contain conserved tyrosines located in sequences homologous to cytoplasmic regions of Trk neurotrophin receptor tyrosine kinases (Aruga and Mikoshiba 2003; Aruga et al. 2003).

Expression

The knowledge of SLITRK protein expression in humans is limited, but studies of mouse, monkey, and human brain tissues suggest a high degree of conservation in the expression of SLITRK1 across mammalian species (Stillman et al. 2009). Moreover, all SLITRKs have been suggested as potential markers of hematopoietic stem cells and progenitor cells, as well as embryonic stem cells (Milde et al. 2007).

In the developing murine nervous system SLITRKs are expressed in several regions in overlapping but distinct expression patterns (Aruga and Mikoshiba 2003; Beaubien and Cloutier 2009). SLIT1-5 are predominantly expressed in the brain (Aruga et al. 2003), whereas SLITRK6 is expressed in numerous tissues including the lung and liver (Aruga 2003). In the brain of 20-week-old human fetuses, SLITRK1 expression has been detected in the developing neocortical plate, subplate zone, striatum, globus pallidus, thalamus, and subthalamus (Abelson et al. 2005). In the adult, all SLITRK transcripts are mainly detected in neural tissues.

Briefly, in the adult mouse, SLITRK1-5 are expressed throughout the brain, but with different expression patterns. Regions expressing SLITRKs include the cerebral cortex, the ventricular layer (strong SLITRK2 expression), the subventricular zone (SLITRK1, -2, -4, and -5), the subplate (SLITRK1 and -4), and the CA3 region of the hippocampus (SLITRK3 and -4). Moreover, SLITRK1-5 are expressed in the thalamus, hypothalamus, the cortical plate, and the pyramidal cell layer of the hippocampus. SLITRK6 expression is restricted to the ventral thalamus, lateral geniculate nucleus, and suprafascicular nucleus (Aruga and Mikoshiba 2003; Aruga et al. 2003) as well as the inner ear (Katayama et al. 2009). For detailed descriptions of the expression of SLITRKs in mouse, see Aruga and Mikoshiba (2003), Stillman et al. (2009), Beaubien and Cloutier (2009), Aruga (2003) and Katayama et al. (2009).

SLITRK1 and -2 overexpressed in PC12 cells localize to the trans-Golgi network in the cell body as well as to the plasma membrane (Aruga and Mikoshiba 2003), and in cortical neurons endogenous SLITRK1 localizes to soma, dendrites, and growth cones (Kajiwara et al. 2009). When expressed at low levels in cultured hippocampal neurons SLITRK3 can be detected on dendrites, but not axons, as a punctate staining that co-localizes with inhibitory, but not excitatory, synapses, suggesting that the protein clusters at presynaptic terminals (Takahashi et al. 2012). SLITRK5 has in striatal neurons been found to localize to PSD-95-positive dendritic spines (Shmelkov et al. 2010).

Interactions

In a cell-based screening assay, protein tyrosine phosphatase δ (PTPδ) was found to interact with the ectodomain of all six SLITRK proteins in a Ca^{2+}-independent manner. In contrast, no interactions were found between SLITRKs and Casprs, CHL1, Leukocyte common antigen related (LAR), NCAM, Neurexins, Neurofascin, or TrkB, and no interactions were found between SLITRK proteins. In the assay, the SLITRK–PTPδ interactions promoted the adhesion of SLITRK-expressing axons to PTPδ-expressing COS cells (Takahashi et al. 2012). This study suggests that all SLITRK proteins mediate cell adhesion through heterophilic *trans*-interactions with PTPδ.

The SLITRK1 intracellular domain (SICD) was used in yeast two-hybrid screenings to scan adult mouse and fetal human brain cDNA libraries for intracellular interaction partners to SLITRK1. Both screens identified 14-3-3 proteins as ligands, and subsequently the SLITRK1-14-3-3 interaction has been confirmed in vivo, and all seven isoforms of 14-3-3 have been found to bind SLITRK1 (Takahashi et al. 2012). The 14-3-3 protein family consists of members that can bind different phospho-motifs [reviewed in Smith et al. (2011a)]. The Ser-695 residue near the C-terminal end of SLITRK1 is situated within one of these 14-3-3-binding motifs, and by performing Ala substitutions in combination with phosphorylation studies, it has been shown that protein kinase A, protein kinase C, and, more specifically, casein kinase 2 can phosphorylate human SLITRK1 Ser-695 (NCBI Ref. Seq. NP_443142.1) and that this phosphorylation strongly promotes SLITRK1–14-3-3 interactions (Kajiwara et al. 2009).

As mentioned above, SLITRK2-6 have longer cytoplasmic domains than SLITRK1, and the C-terminal ends of these SLITRK proteins contain putative Tyr-phosphorylation sites homologous to Tyr-phosphorylation sites in Trk receptors. The phosphorylation sites lie within a phospholipase C (PLC)γ1-binding motif, suggesting that SLITRK2-6 may facilitate PLCγ1-mediated signal transduction (Aruga and Mikoshiba 2003).

Functions

Transient transfection of PC12 and N2a cells with vectors encoding SLITRK1-6, respectively, has demonstrated that the expression of SLITRK2 and -3, and to a lesser extent SLITRK4, -5, and -6, leads to a reduction in NGF-stimulated neurite outgrowth, whereas expression of SLITRK1 leads to a stimulation of neurite outgrowth in a manner that is additive to and qualitatively different from NGF-stimulated neurite outgrowth (Aruga and Mikoshiba 2003). Accordingly, a reduced expression of SLITRK2 and -4 observed in the neural cells of patients with myotonic dystrophy 1 is accompanied by increased neuritogenesis (Marteyn et al. 2011), and whereas SLITRK1 has been found to stimulate neuritogenesis in cortical neurons (Kajiwara et al. 2009; Abelson et al. 2005), a loss-of-function frameshift mutant of SLITRK1 identified in a patient with TS has no effect on neuritogenesis (Abelson et al. 2005).

The different effects of SLITRK proteins on neuritogenesis are related to differences in their cytoplasmic domains. Thus, expression of a SLITRK2 mutant lacking the cytoplasmic domain is able to stimulate neurite outgrowth in a manner comparable to SLITRK1, whereas expression of the cytoplasmic domain of SLITRK2 alone inhibits neurite outgrowth. These observations suggest that sequences in cytoplasmic domain of SLITRK2-6 that are missing in the shorter cytoplasmic domain of SLITRK1 (SICD) somehow can prevent the stimulation of neurite outgrowth (Aruga and Mikoshiba 2003). However, it has also been demonstrated that the phosphorylation of Ser-695 of human SLITRK1 (a residue not conserved between SLITRK1 and -2 [compare, e.g., SLITRK1 NCBI Ref. Seq. NP_443142.1 vs. SLITRK2 GenBank: AAI13013.1]) is necessary for the stimulation of neurite outgrowth (Kajiwara et al. 2009).

The ectodomain of SLITRK1 is secreted in vitro and in vivo in response to cleavage by an ADAM family α-secretase. The process is stimulated by Protein kinase C and inhibited by the ADAM inhibitor TAPI-2. Subsequently, SICD is cleaved by γ-secretase. However, in contrast to APP, which also is cleaved by α- and γ-secretases, causing the cytoplasmic domain of APP to be released as a soluble cytosolic protein, the γ-secretase-cleaved SICD remains attached to the intracellular part of the plasma membrane as peripheral proteins (Kajiwara et al. 2009). The functional relationship between protein kinase C-stimulated SLITRK1 Ser-695 phosphorylation and SLITRK1 α-secretase cleavage remains to be determined, and it is not clear whether there is a functional difference between full-length SLITRK1 and SICD in neurite outgrowth stimulation.

All SLITRK proteins have synaptogenic activity. As mentioned above, all SLITRK proteins are also able to form *trans*-interactions with PTPδ, and these SLITRK–PTPδ interactions seem to be a prerequisite for SLITRK-mediated synaptogenesis. SLITRK2 expression induces the clustering of both GABA vesicular transporter (VGAT), a marker of inhibitory synapses, and VGLUT1, a marker of excitatory synapses, whereas SLITRK3 induces clustering of only VGAT, suggesting that it specifically promotes synaptic differentiation of inhibitory synapses. Moreover, knockdown of SLITRK3 in cultured hippocampal neurons reduces the density of inhibitory synapses, and whereas SLITRK3-knockout mice demonstrate no gross defects in brain morphology, they do demonstrate a reduction in inhibitory synapses in, e.g., the CA1 region of the hippocampus. Consistent with the observed reduction in inhibitory synapses, SLITRK3 knockdown mice demonstrate an increased sensitivity to seizures induced by pentylene–tetrazol (a GABA receptor antagonist) (Takahashi et al. 2012).

To study the effects of SLITRK proteins in vivo, knockout mice have so far been generated for three of the six SLITRKs (Katayama et al. 2009, 2010; Shmelkov et al. 2010; Matsumoto et al. 2011).

SLITRK1-knockout mice demonstrate no external abnormalities or anatomical or histological brain defects. However, they do seem to have a slightly increased postnatal mortality leading to a fraction of homozygous animals at weaning ~20 % lower than the expected Mendelian ratio, and male homozygous SLITRK1-knockout mice have a slightly reduced body weight when compared to wild-type mice. In addition, SLITRK1-knockout mice do exhibit behavioral defects, including an elevated anxiety-like behavior (as determined from, e.g., open-field and elevated plus maze tests) and depression-like behavior (as determined from forces swimming and tail suspension tests). The animals also demonstrate increased levels of norepinephrine and its metabolite 3-methoxy-4-hydroxyphenylglycol (MHPG) in the prefrontal cortex and nucleus accumbens, respectively (Katayama et al. 2010).

SLITRK1 mutations have been related to OCDs like TS and trichotillomania (self-induced hair loss) (see below). Consequently, experiments were performed where SLITRK1-knockout mice were administered with clonidine (Katayama et al. 2010), an α2-adrenergic agonist reported to, e.g., reduce tics, anxiety, hyperactivity/impulsivity, and inattentiveness related to TS and attention-deficit hyperactivity disorder (ADHD) (Sandor 2003; Croxtall 2011). It was found that clonidine reduces the anxiety-like behavior of SLITRK1-knockout mice, suggesting that changes in the noradrenergic system as a result of SLITRK1 deficiency contribute to the anxiety-like behavior of these animals and potentially also are related to the pathogenesis of TS and related neuropsychiatric diseases (Katayama et al. 2010).

Homozygous SLITRK5-knockout mice are born at the expected Mendelian ratio and demonstrate no gross abnormalities. However, a few months after birth both homo- and heterozygous SLITRK5-knockout mice demonstrate facial hair loss followed by skin lesions that turn to chronic wounds; effects believed to be the result of excessive grooming. SLITRK5-knockout mice also demonstrate other signs of OCD (as defined by a marble-burying behavior) as well as anxiety-like behavior (as determined from, e.g., open-field and elevated plus maze tests). Moreover, the volume of the striatum of SLITRK5-knockout mice is smaller than that of wild-type mice, and whereas there is no change in cell soma areas, the dendritic arbors of medium spiny neurons of the striatum are decreased. Finally, the expression of the glutamate receptor subunits NR2A, NR2B, GluR1, and GluR2 is all decreased in the striatum (Shmelkov et al. 2010). The drug Fluoxetine ("Prozac," "Fontex," etc.) is a selective serotonin reuptake inhibitor (Sghendo and Mifsud 2012) used for the treatment of, e.g., OCDs (Choi 2009). Results suggest that Fluoxetine can also prevent the OCD-like behavior of SLITRK5-knockout mice. Thus, treatment of these mice with Fluoxetine leads to a reduction in the duration of grooming to a level similar to wild-type mice (Shmelkov et al. 2010). See also Mah (2010) and Yang and Lu (2011) for reviews.

Homozygous SLITRK6-knockout mice are also born at the expected Mendelian ratio and do not have gross abnormalities (Katayama et al. 2009) or any apparent behavioral abnormalities related to mood, anxiety, learning, or memory, although they do show altered adaptive responses to a novel environment (as determined by open-field and hole-board tests) (Matsumoto et al. 2011). However, as mentioned above, SLITRK6 is expressed in the inner ear, and consistently, SLITRK6-knockout mice demonstrate a loss of neurons in the spiral and vestibular ganglia, a reduction in the cochlear innervation (Katayama et al. 2009), and hearing loss in the mid-frequency range, and SLITRK6-knockout mice can therefore serve as a model for hereditary deafness (Matsumoto et al. 2011). In addition to the decreased neuronal survival of SLITRK6-knockout mice, the sensory epithelia of these mice demonstrate a reduced ability to stimulate neurite outgrowth from spinal ganglion neurons. Consistent with these observations, there is a decreased expression of the neurotrophins Brain-derived neurotrophic factor (BDNF) and Neurotrophin-3 (NT-3) in the inner ear of E14.5 SLITRK6-knockout mice (Katayama et al. 2009).

Diseases

TS is a neurodevelopmental disorder with a strong hereditary element that is characterized by involuntary movements and vocalizations (tics) (Paschou 2013). In a screening of 174 predominantly Caucasian individuals with TS, 3 individuals were found to have mutations in *SLITRK1*, whereas no *SLITRK1* mutations were found in >3,600 control chromosomes. One mutation was a single-base deletion leading to the predicted expression of a mutated and truncated form of SLITRK1 consisting of the 421 N-terminal amino acids (incl. the signal peptide) followed by 27 nonsynonymous amino acid substitutions and a premature stop codon. This mutated protein lacks a part of the ectodomain, as well as the transmembrane and cytoplasmic domains, and as mentioned above, it results in loss of function with respect to stimulation of neuritogenesis. Another mutation (observed in two individuals) was a single-nucleotide substitution (var321) in a part of the 3′ untranslated region of the *SLITRK1* transcript predicted to be a binding site for the miRNA hsa-mir-189. The substitution strengthens the binding of hsa-miR-189 to the *SLITRK1* transcript, hence potentially increasing the degradation of the transcript, reducing the level of SLITRK1, and consequently potentially reducing neuritogenesis in the affected regions (Abelson et al. 2005). The described study by Abelson et al. (2005) has led to several subsequent studies and an intense debate regarding the potential association between *SLITRK1*and TS.

Briefly, in later screens of 82 Caucasian TS patients (Deng et al. 2006), and two families with TS (Fabbrini et al. 2007; Pasquini et al. 2008; Orth et al. 2007), no mutations were observed in *SLITRK1*, and although a study of 92 Austrian TS patients revealed six SNPs in the 3′ untranslated region of *SLITRK1*, five of these were also detected among 192 controls, and only a single variant could potentially be related to the development of TS (Zimprich et al. 2008). Moreover, in a screen that detected six independent occurrences of var321 among a population of Ashkenazi Jews (Keen-Kim et al. 2006), and in a large screening including 989 TS patients (Scharf et al. 2008), it was not possible to correlate the occurrence of var321 mutation to the development of TS.

However, a recent investigation has provided further support for a relationship between the var321 and TS (O'Roak et al. 2010), and a study including 154 nuclear families with TS also suggests an association between *SLITRK1* polymorphisms and the development of TS, although the polymorphisms originally reported by Abelson et al. (2005) were not identified (Miranda et al. 2009; Karagiannidis et al. 2012).

In summary, the potential relationship between *SLITRK1* polymorphisms and the TS pathogenesis is still controversial. The pathogenesis of TS likely includes environmental as well as multiple genetic factors, and although *SLITRK1* polymorphisms seem to be unable to account for many TS cases, it remains one of the candidate genes related to TS pathogenesis.

TS patients often also suffer from OCD or ADHD (see Paschou 2013, for a recent review), and as mentioned above, *SLITRK1* has also been associated with the pathogenesis of trichotillomania. Indeed, two trichotillomania-related *SLITRK1* polymorphisms have been identified. They both lead to amino acid substitutions in the membrane-proximal region of the ectodomain of SLITRK1 (Zuchner et al. 2006). In contrast, the *SLITRK1* var321 has been reported not to be associated with OCD (Wendland et al. 2006).

Results from genome-wide association studies reveal *SLITRK2* to be a candidate gene for bipolar disorder (Smith et al. 2009), and the gene has also been associated with schizophrenia, where two missense mutations (V89M in the second LRR and S549F in the CF domain) have been observed in girls with schizophrenia and in their affected siblings (Piton et al. 2011). Moreover, copy number variations of *SLITRK2* have been associated with X-linked high myopia (near/shortsightedness) (Metlapally et al. 2009), and *SLITRK2* is also a potential candidate gene for development of dyslexia (Huc-Chabrolle et al. 2013). However, all these potential relationships need to be confirmed or deconfirmed with further investigations.

As mentioned above, SLITRK2 and -4 have been found to be downregulated in patients with myotonic dystrophy 1 (Marteyn et al. 2011), a chronic multisystem disease causing, e.g., muscle dystrophy (Romeo 2012). The reduced expression of SLITRK2 and -4 is accompanied by increased neuritogenesis of motor neurons in vitro, leading to a five- to tenfold decrease in the formation of neuromuscular contacts in cocultures of neurons and primary myotubes. Myotonic dystrophy 1 is generally described as a trinucleotide repeat disorder caused by CTG-triplet expansion in the 3′ untranslated leader region of the gene *DMPK* (Romeo 2012) and any relationship between the CTG-triplet expansion and the changes in the expression of SLITRK2 and -4 remains to be determined.

Finally, genome-wide transcriptional profiling has revealed that *SLITRK4* expression is downregulated in the decidua of women with preeclampsia (Loset et al. 2011). However, also in this case, the relationship between changes in gene expression and disease progression remains to be determined.

In summary, the SLITRK family consists of six members, which probably are not able to form homophilic interactions, but which all form heterophilic *trans-interactions* with PTPδ. All members of the family appear to be important for the modulation of neuritogenesis and synaptogenesis, and consequently the respective genes may turn out to be related to a number of diseases. However, biochemically and biologically the proteins are not well characterized and therefore an understanding of the relationships between a given disease and mutations or transcriptional alterations in SLITRK genes is still incomplete.

14.3 Cell Adhesion Molecules Belonging to the LRR-Tollkin Class

14.3.1 LRRC15/Lib

LRR-containing 15 (LRRC15)/LRR protein induced by β-amyloid treatment (Lib) (see Table 14.4) was in 2002 originally described as a protein upregulated in astrocytes in response to treatment of cells with Aβ-peptides or pro-inflammatory cytokines (Satoh et al. 2002). Later, the protein has been identified in several large-scale studies (Gerhard et al. 2004; Muzny et al. 2006). However, subsequent to the first description of the protein, LRRC15 has predominantly been observed to be upregulated in several types of cancer (Reynolds et al. 2003; Satoh et al. 2004; Schuetz et al. 2006; Williams et al. 2008; Reddy and Balk 2006; Stanbrough et al. 2006; Bierkens et al. 2013; Klein et al. 2009), and its relationship to AD and other neuronal diseases remains to be determined.

Table 14.4 Human LRRC15

Gene symbol	Chromosomal location	Synonyms and previous names	Recommended protein name
LRRC15	3q29	Leucine-rich repeat protein induced by β-amyloid homologue, LIB	Leucine-rich repeat-containing protein 15

Gene and Protein

Human LRRC15 was originally identified as a protein upregulated in response to transcriptional activation by the chimeric transcription factor EWS–WT1 (+KTS) characteristic for desmoplastic small round cell tumor (Reynolds et al. 2003). Later, the expression of *LRRC15* has also been shown to be strongly upregulated in the absence of TBP-associated factor 4, a component of the transcription factor TFIID (Fadloun et al. 2007). Loss of TBP-associated factor 4 has also been shown to stimulate Transforming growth factor β (TGFβ) signaling (Davidson et al. 2005), suggesting that *LRRC15* expression in part is regulated by TGFβ.

Human *LRRC15* is located on chromosome 3 (Gerhard et al. 2004; Reynolds et al. 2003; Muzny et al. 2006) and consists of three exons with the entire coding region confined within a single exon. Two other genes, *CPN2* (encoding carboxypeptidase N, subunit 2) and *GP5* (encoding Glycoprotein V/CD42d), are located in the same region of chromosome 3, share structural similarities with *LRRC15*, and encode proteins with a high sequence identity to LRRC15, suggesting that the three genes have originated from genomic duplications (Reynolds et al. 2003).

Mammalian LRRC15 is composed of ~580 amino acids forming an ectodomain of 15 LRR motifs with NF and CF domains, a single transmembrane region, and a short (~20 amino acids) cytoplasmic domain. The protein is both N- and O-glycosylated (Satoh et al. 2002; Reynolds et al. 2003).

Interactions

Results from in vitro studies, where LRRC15 was transiently expressed in COS-7 cells, suggest that the protein is able to form *trans*-homophilic interactions (Satoh et al. 2004). Moreover, the protein has been demonstrated to bind to Fibronectin and to a lesser degree to Matrigel, Collagen IV, and Laminin, but not Aggrecan (Satoh et al. 2005). Thus, LRRC15 may be a CAM mediating cell–cell as well as cell–ECM interactions through *trans*-homo- and heterophilic interactions.

The notion that LRRC15 mediates cell–ECM interactions is supported by the fact that the protein promotes Matrigel invasion of glioma and breast carcinoma cells; a phenomenon that is strongly reduced in response to knockdown of LRRC15 (Reynolds et al. 2003; Satoh et al. 2005).

Expression, Functions, and Diseases

In accordance with the notion that LRRC15 is a CAM, it predominantly localizes to the cell surface upon overexpression in C6 gliomas (Satoh et al. 2002).

Under non-pathological conditions LRRC15 is mainly expressed in the cytotrophoblast cell layer of the placenta and is expressed at very low levels in the brain (Satoh et al. 2002; Reynolds et al. 2003).

In the cerebral cortex of humans without AD, LRRC15 is expressed in neurons, but not in astrocytes. In contrast, the protein is in AD patients expressed in ~50 % of the reactive astrocytes surrounding senile plaques, whereas it is not expressed in the neurons (Satoh et al. 2005). Moreover, in vitro studies have demonstrated that the expression of *LRRC15* can be upregulated 13-fold in astrocytes in response to a 15-h treatment with pre-aggregated Aβ-peptides (Satoh et al. 2002).

LRRC15 expression is also increased in response to treatment with pro-inflammatory cytokines like tumor necrosis factor α, interleukin 1β, and interferon γ (IFNγ) (Satoh et al. 2002). Since, pro-inflammatory cytokines are known to be involved in the pathogenesis of AD (Rubio-Perez and Morillas-Ruiz 2012) these observations further strengthen the potential relationship between LRRC15 expression and AD progression.

In addition to the modulation of LRRC15 expression in AD patients, the proteins have also been found to be upregulated in a number of cancer types including breast cancer (mammary ductal carcinomas), where *LRRC15* transcripts in most tumors demonstrate a >2-fold increase (Satoh et al. 2004; Schuetz et al. 2006; Williams et al. 2008), prostate cancer (Reddy and Balk 2006; Stanbrough et al. 2006), desmoplastic small round cell tumor (Reynolds et al. 2003) (a rare cancer originating in the peritoneum (Hayes-Jordan and Anderson 2011)), and cervix cancer (Bierkens et al. 2013). Moreover LRRC15 is highly expressed in breast-cancer-derived metastases in bone, but not in the brain (Klein et al. 2009).

Interestingly, the expression of LRRC15 seems to be accompanied by a down-regulation of the cell surface-localized fraction of Coxsackievirus-adenovirus receptor (CAR). This can be a problem in relation to adenoviral-based cancer treatment, because the downregulation of CAR obstructs treatment by adenoviral infection (O'Prey et al. 2008). Finally, LRRC15 is highly expressed in carious diseased pulpal tissue, further relating the expression of the protein to pro-inflammatory signals (Cooper et al. 2011).

In summary, LRRC15 is a protein upregulated in cancers and in response to Aβ-exposure and pro-inflammatory signals. However, the effect of LRRC15 expression on the pathogenesis of AD remains to be determined.

14.4 Cell Adhesion Molecules Belonging to the LRR-Ig/Fn3 Class

14.4.1 The AMIGO/Alivin Family

The "Amphoterin-induced gene and ORF" (AMIGO) family also known as "alive" or "activity-dependent LRR and Ig superfamily survival-related protein" (Alivin) family was in 2003 identified independently by two research groups (Kuja-Panula et al. 2003; Ono et al. 2003). Moreover, the different members of the family have been identified in several large-scale studies (Clark et al. 2003; Ota et al. 2004; Gerhard et al. 2004; Bechtel et al. 2007; Hirosawa et al. 1999; Nagase et al. 2001b;

Table 14.5 The human AMIGO family

Gene symbol	Chromosomal location	Synonyms and previous names	Recommended protein name
AMIGO1	1p13.3	Alivin-2, ALI2, amphoterin-induced gene and open reading frame, AMIGO, KIAA1163	Amphoterin-induced protein 1
AMIGO2	12q13.11	Alivin-1, ALI1, amphoterin-induced gene and open reading frame 2, DEGA	Amphoterin-induced protein 2
AMIGO3	3p21	Alivin-3, ALI3, amphoterin-induced gene and open reading frame 3, KIAA1851	Amphoterin-induced protein 3

Gregory et al. 2006). The family consists of three proteins: AMIGO1/Alivin2/Ali2, AMIGO2/Alivin1/Ali1/DEGA, and AMIGO3/Alivin3/Ali3 (Table 14.5).

AMIGO1 was originally identified in a differential display analysis of hippocampal neurons treated with the neuritogenic, heparin-binding protein High mobility group box-1/Amphoterin, and subsequently the two homologues *AMIGO2* and *-3* were cloned (Kuja-Panula et al. 2003). In addition, *AMIGO2* was independently identified in a differential display analysis of genes involved in depolarization- and/or NMDA-dependent survival of cerebellar granule neurons (Ono et al. 2003), and later it was identified as a gene differentially expressed in human gastric adenocarcinomas (DEGA) (Rabenau et al. 2004).

AMIGO2 has been suggested to be implicated in the pathogenesis of AD, PD (Ono et al. 2003; Peltola et al. 2011), and cancer (Rabenau et al. 2004), but the functions of AMIGO-family proteins under normal as well as pathological conditions are still poorly described.

Genes and Proteins

AMIGO1-3 are conserved in mammals, and the genes encode proteins that are structurally related to the *Drosophila* proteins Kekkon 1–3 (Kuja-Panula et al. 2003; Ono et al. 2003).

AMIGO proteins are transmembrane proteins localized in the plasma membrane. They are composed of a ~400 amino acids long ectodomain consisting of six N-terminal LRRs with NF and CF domains, and a membrane-proximal Ig module, a transmembrane domain, and a ~100 amino acids long short cytoplasmic domain (Kuja-Panula et al. 2003; Ono et al. 2003; Chen et al. 2012). The ectodomain of AMIGO1 contains five sites for *N*-linked glycosylation: two in the LRR domain and three in the Ig module. The pronounced glycosylation of the Ig module has been proposed to ensure that the LRR domain is oriented away from (perpendicularly to) the plasma membrane (see below) (Kajander et al. 2011).

Human AMIGO proteins are highly homologous, exhibiting ~50 % similarity; the most conserved regions are the parts of the LRRs encoding the concave parts of the repeats (Kajander et al. 2011), the transmembrane region, and parts of the cytoplasmic domain (Kuja-Panula et al. 2003; Ono et al. 2003). Additionally, the proteins are highly conserved between species (Kuja-Panula et al. 2003).

Expression

During mouse development, *AMIGO1* expression can be detected at E13, and subsequently, the expression becomes pronounced in the brain, especially in the hippocampus. At the protein level, AMIGO1 is detectable from around E14. The level of expression decreases postnatally to the lowest levels around P10, but then increases to higher levels throughout adulthood (Kuja-Panula et al. 2003).

In adult mouse, AMIGO1 is mainly expressed in the nervous system, including the cerebellum, cerebrum, and retina (Kuja-Panula et al. 2003). In the cerebrum the expression of AMIGO1 overlaps with the expression of the potassium channel α-subunit Kv2.1, and the two proteins co-localize in cultured neurons (Peltola et al. 2011). AMIGO2 is mainly expressed in the cerebellum, retina, liver, and lung (Kuja-Panula et al. 2003), but a detailed in situ hybridization study has also revealed a pronounced expression of *AMIGO2* in the CA2 and CA3 regions of the hippocampus (Laeremans et al. 2013). AMIGO3 is ubiquitously expressed and demonstrate no apparent variations in expression between the tissues investigated (Kuja-Panula et al. 2003).

The expression of AMIGO proteins in humans is poorly characterized, but all three proteins are expressed in primary cultures of human microvascular endothelial cells and brain pericytes (Hossain et al. 2012).

In accordance with the way human AMIGO1 was discovered (Kuja-Panula et al. 2003), the expression of *AMIGO1* has in zebrafish been demonstrated to be regulated by Amphoterin (Zhao et al. 2011). Moreover, the expression of AMIGO2 has been shown to be regulated at the transcriptional level by Ca^{2+} influx through voltage-dependent L-type Ca^{2+} channels (Ono et al. 2003).

At the cellular level AMIGO1 was in hippocampal neurons originally found to be located in both the soma and in fasciculated and nonfasciculated axonal but not dendritic processes (Kuja-Panula et al. 2003). However, recently it has been reported that AMIGO1 is also expressed in astroglia and oligodendroglia and that the protein in neurons localizes mainly to dendrites but not axons. Interestingly, an AMIGO1 mutant lacking the LRR domain is mis-allocated to axons rather than dendrites and is transported to the cell surface faster than wild-type AMIGO1 or an AMIGO1 mutant lacking the Ig module (see below)(Chen et al. 2012).

Interactions

All three AMIGO proteins form homophilic as well as heterophilic interactions within the family (Kuja-Panula et al. 2003). The homophilic interactions result in dimerization of the proteins, with the concave side of the LRR domain forming the dimer interface (Kajander et al. 2011) (Fig. 14.3). Dimerization as well as proper

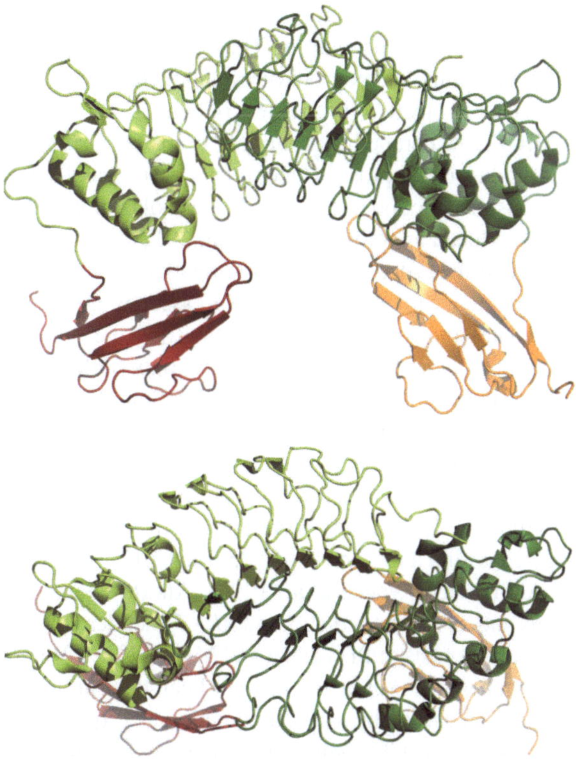

Fig. 14.3 Structural model of an AMIGO1 homodimer. Cartoon drawing of a structural model of two ectodomains of murine AMIGO1 [PDB ID: 2XOT; amino acids 28–372; Kajander et al. (2011)]. The same structure is shown from two different angles (90° rotation). Each ectodomain includes an LRR domain composed of six LRRs with NF and CF domains (*dark/light green*) that is followed by a single Ig module (*red/orange*). The ectodomains form homodimers through interactions between the concave sides of the LRR domains. See text for details. The figures were created with PyMOL (DeLano Scientific)

glycosylation has been shown to be necessary for correct cell surface expression of AMIGO1, suggesting that chaperones in the ER ensure that only dimeric AMIGO molecules are transported to the cell surface (Kajander et al. 2011). Mutant AMIGO molecules lacking the LRR domain may be able to circumvent this quality check, which would explain why such mutants are able to be transported to the plasma membrane faster than wild-type proteins (Chen et al. 2012).

It has been suggested that the dimerization interface used for the direction of AMIGO *cis*-dimers to the cell surface subsequently is used to establish cell–cell interactions via AMIGO *trans*-dimer formation (Kajander et al. 2011).

Functions and Diseases

A downregulation of *AMIGO1* expression in cortical neurons decreases the number, length, and branching of dendrites, but has little effect on axon morphology (Chen et al. 2012), whereas neurite outgrowth is stimulated in hippocampal neurons grown on a coat of recombinant AMIGO1 (Kuja-Panula et al. 2003). Moreover, exposure to the ectodomain of AMIGO1 in solution inhibits the formation of neuritic bundles in cultures of hippocampal neurons, suggesting that AMIGO1 plays a role in fasciculation (Kuja-Panula et al. 2003).

AMIGO2 is, as mentioned above, in mouse expressed in the CA2 and CA3a regions of the hippocampus. These areas are reported to be relatively resistant to neuronal injury and neurotoxicity and therefore it has been suggested that AMIGO2 has neuroprotective properties (Chen et al. 2012). Indeed, overexpression of AMIGO2 in cerebellar granule neurons reduces the fraction of cells undergoing KCl-induced apoptosis (Ono et al. 2003), and an overexpression of AMIGO1 reduces the fraction of SH-SY5Y neuroblastoma cells undergoing Staurosporine or H_2O_2-induced apoptosis (Chen et al. 2012).

The voltage-dependent potassium channel subunit, Kv2.1, that can be immunoprecipitated with AMIGO1 is expressed by most neurons in the CNS. AMIGO1 modulates the function of Kv2.1, shifting the voltage-dependent activation of the channel to more hyperpolarized potentials (Peltola et al. 2011). A biological function of the AMIGO1-regulated modulation of Kv2.1 has not been determined, but since potassium channels have been proposed as potential therapeutic target for the treatment of epilepsy (Wickenden 2002), AMIGO1 is also a potential drug target for diseases related to neuronal excitability.

Human *AMIGO2* has been mapped to 12q13.11 (Ono et al. 2003). This locus is located within a region (12p11.23-12q13.12) related to AD type 5, a late-onset type of AD (Pericak-Vance et al. 1997). Another gene, *LRRK2*, is located at the same locus and as mentioned earlier *LRRK2* encodes a protein that plays a pivotal role in the pathogenesis of PD (Tsika and Moore 2012). Whether AMIGO2 is related to the pathogenesis of AD or PD has not been determined, but a reported case of mental retardation caused by a 1.3 Mb deletion at 12q13.11 has been related to haploinsufficiency of AMIGO2, suggesting that the proper level of AMIGO2 expression is of central importance for cognitive functions (Gimelli et al. 2011).

AMIGO2 is differentially expressed in ~45 % of gastric adenocarcinomas when compared to normal tissue. Knockdown of *AMIGO2* expression in a gastric adenocarcinoma cell line leads to a downregulation of cancer cell migration in vitro and a reduced tumorigenesis in vivo, suggesting that AMIGO2 expression is related to the development or progression of gastric adenocarcinomas (Rabenau et al. 2004).

In summary, the AMIGO family consists of three members, which all can interact with each other, and is believed to form both homophilic *cis*-and *trans-interactions*. The individual members are involved in processes like neuritogenesis, bundling, fasciculation, and neuronal cell survival and may be implicated in the pathogenesis of, e.g., AD or PD. However, all AMIGO proteins are still poorly characterized.

14.4.2 The LRRC4/NGL Family

The family of LRR-containing protein 4 (LRRC4) or Netrin-G ligands (NGLs) includes three members: NGL-1/LRRC4C, NGL-2/LRRC4, and NGL-3/LRRC4B (Table 14.6). The NGLs were cloned in 2003 (Lin et al. 2003) and have also been identified in several large-scale studies and other types of investigations (Clark et al. 2003; Ota et al. 2004; Gerhard et al. 2004; Bechtel et al. 2007; Nagase et al. 2000b;

Table 14.6 The human NGL/LRRC4 family

Gene symbol	Chromosomal location	Synonyms and previous names	Recommended protein name
LRRC4	7q31	Brain tumor-associated protein BAG, nasopharyngeal carcinoma-associated gene 14 protein, NAG14, Netrin-G2 ligand, NGL-2	Leucine-rich repeat-containing protein 4
LRRC4B	19q13.33	DKFZp761A179, HSM, leucine-rich repeats and immunoglobulin-like domains 4, LRIG4, Netrin-G3 ligand, NGL-3	Leucine-rich repeat-containing protein 4B
LRRC4C	11p12	KIAA1580, Netrin-G1 ligand, NGL-1	Leucine-rich repeat-containing protein 4C

Scherer et al. 2003; Grimwood et al. 2004; Kim et al. 2006). As the name implies, some NGLs bind Netrin-G proteins. Netrins are a group of Laminin-related axon guidance molecules including three secreted proteins (Netrin-1, -3, and -4) and two glycosylphosphatidylinositol (GPI)-anchored proteins: Netrin-G1 and –G2 (see Moore et al. 2007 for reviews; Rajasekharan and Kennedy 2009). NGLs have been associated with hearing impairment and the formation or progression of cancer.

Genes and Proteins

The respective NGLs are ~640–709 amino acids long proteins and are composed of an ectodomain containing 9 LRRs with NF and CF domains, followed by one Ig module, a transmembrane domain, and a ~92–116 amino acids long cytoplasmic domain with a conserved C-terminal PDZ-domain-binding motif (Clark et al. 2003; Ota et al. 2004; Gerhard et al. 2004; Bechtel et al. 2007; Lin et al. 2003; Nagase et al. 2000b; Scherer et al. 2003; Grimwood et al. 2004).

The primary sequences of NGL-1, -2, and -3 are highly similar. For instance, the primary sequence of the Ig module of NGL-1 is 72 % identical to that of NGL-2 (Zhang et al. 2005a). Moreover, NGLs are conserved among vertebrates. For instance, the primary sequence of mouse NGL-2 is 97 % and 99 % similar to human and rat NGL-2, respectively (Wu et al. 2007).

The general regulation of NGL expression has not been investigated, but the expression of NGL-2 is reported to be downregulated by the microRNA hsa-miR-381, which in turn is downregulated by NGL-2 (Tang et al. 2011).

Expression

In E10 mouse embryos *NGL1* is expressed in post-mitotic neurons in the telencephalon, in the optic cup, and in motor neurons in the spinal cord (Homma et al. 2009).

At E14, it is highly expressed in the developing cerebral cortex and the striatum (Lin et al. 2003), and at P2 it is expressed in the brain, eye, and inner ear, but also outside the CNS (e.g., heart, lung, kidney, and intestine) (Delprat et al. 2005). In fetal and adult human tissue, *NGL1* is predominantly expressed in neocortical areas, but is also found in the putamen, amygdala, hippocampus, and medulla oblongata (Lin et al. 2003).

NGL2 is mainly expressed in the brain (Kim et al. 2006). In mouse, the expression of *NGL2* increases during development, peaks around P2, and subsequently maintains a relatively stable level (Zhang et al. 2005a). At E12.5, it is expressed throughout the CNS including the cerebrum, cerebellum, spinal cord, pons, and medulla oblongata. The expression is highest in the dorsal root ganglion (DRG), the dentate gyrus, the CA1 and CA2 regions of the hippocampus, and the cerebellum (Wu et al. 2007). In adult rat brain, *NGL2* expression is restricted to the cerebellum, hippocampus, olfactory bulb, and cortex (Kim et al. 2006). In humans, *NGL2* expression has been detected in the brain, but not in 15 other tissues (Zhang et al. 2005a). The expression is highest in the cerebellum, but can also be detected in the cerebral cortex, occipital pole, frontal lobe, temporal lobe, and putamen (Zhang et al. 2005a).

NGL3 is predominantly expressed in the brain (Kim et al. 2006), where it is more ubiquitously expressed than *NGL1* and *-2* (Homma et al. 2009).

At the cellular level, NGL-1 has been found to localize to the cell surface (Lin et al. 2003; Zhang et al. 2005a). In cultured neurons NGLs are located at synapses, where they co-localize with the pre- and postsynaptic markers Synaptophysin and PSD-95 and with the excitatory presynaptic marker VGLUT1, but not the inhibitory synaptic markers VGAT and Gephyrin. Subcellularly, the proteins are located in dendritic spines and at the center of synapses. Moreover, they seem to be more postsynaptically than presynaptically expressed (Kim et al. 2006; Woo et al. 2009b). In Netrin-G1- and -G2-knockout mice, where axon pathfinding is normal, the dendritic distribution of NGL-1 and -2 is more diffuse than in wild-type animals (Nishimura-Akiyoshi et al. 2007).

Interactions

Originally, NGL-1 was found not to form homophilic interactions (Lin et al. 2003), but later the ectodomains of NGL-1 have been reported to interact in a Ca^{2+}-dependent manner (Delprat et al. 2005). However, there are no data suggesting that NGLs form *trans*-homophilic interactions (Woo et al. 2009b). Instead, NGLs form cell–cell interactions through specific *trans*-heterophilic interactions. Postsynaptic NGL-1 binds specifically and Ca^{2+} independently to Netrin-G1, NGL-2 binds specifically to Netrin-G2 (Kim et al. 2006), and NGL-3 binds specifically to Leukocyte common antigen-related (LAR) (Woo et al. 2009b). Accordingly, NGL-1 and -2 are expressed on dendrites interacting with Netrin-G1- and -2-expressing axons, respectively (Nishimura-Akiyoshi et al. 2007). The NGL–Netrin-G complexes have been

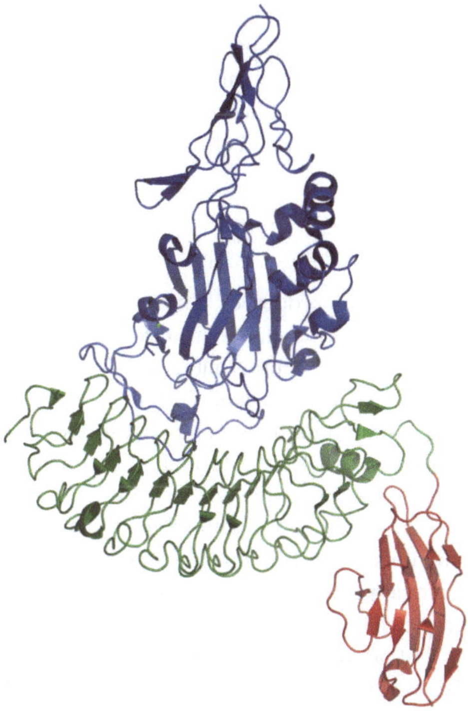

Fig. 14.4 Structural model of an NGL1-Netrin1 dimer. Cartoon drawing of a structural model of the ectodomain of human NGL1 (amino acids 44–444) in interaction with a part of the ectodomain of human Netrin-G1 [amino acids 30–360; PDB ID: 3ZYJ; Seiradake et al. (2011)]. The ectodomain of NGL1 includes an LRR domain composed of nine LRRs with NF and CF domains (*green*) that is followed by a single Ig module (*red*). The modeled part of the ectodomain of Netrin-G1 includes the Laminin N-terminal domain and the first Laminin EGF-like domain (*blue*). The presented interaction represents a heterophilic *trans*-interaction, where the Laminin N-terminal domain of Netrin-G1 interacts with the concave part of the LRR domain of Netrin-G1. See text for details. The figures were created with PyMOL (DeLano Scientific)

crystallized, and the structural models show that the convex part of the NGL LRR domains interacts with the Ig module of the same molecule. Consequently, the concave part of the NGL LRR domains is exposed to the surrounding environment, allowing it to interact with the N-terminal Laminin domains of Netrin-G proteins (Seiradake et al. 2011; Brasch et al. 2011) (Fig. 14.4).

The C-terminal PDZ-domain-binding motifs in the cytoplasmic domain of NGLs bind to the first two PDZ domains of PSD-95, and in the brain, NGL-2 forms a complex with PSD-95 and NMDA receptors, but not AMPA receptors. The NGL-2–PSD-95 interaction facilitates the clustering of both PSD-95 and NGL-2 in dendritic spines (Kim et al. 2006). Moreover, the C-terminal end of NGL-1 has been shown to bind the PDZ domains of the protein Whirlin/USH2D (Delprat et al. 2005).

A membrane-proximal serine residue located in a 5 amino acids long sequence in the cytoplasmic domains of NGLs is phosphorylated by cyclin-dependent kinase-like 5 (CDKL5), and in hippocampal neurons the proteins co-localize at dendritic spines. The CDKL5-mediated phosphorylation of NGLs promotes the interaction between NGLs and PSD-95 (Ricciardi et al. 2012).

Functions and Diseases

NGL proteins have neuritogenic properties. Thus, when used as a coat for thalamic neurons, NGL-1 promotes neuritogenesis in a Netrin-G-dependent manner (Lin et al. 2003), and in a similar way, neurite outgrowth is promoted in neurons grown on top of NGL-2-expressing cells (Wu et al. 2007).

NGL proteins are also important for synaptogenesis (Ricciardi et al. 2012; Berg et al. 2010). NGL-2 is important for the formation and maintenance of excitatory synapses through mechanisms that include extracellular *trans*-interactions with Netrin-G2 and intracellular interactions with PSD-95. Thus, knockdown of NGL-2 in the hippocampus disrupts synaptic interactions in the CA1 region (DeNardo et al. 2012), and in hippocampal neurons, a lack of NGL-2 reduces the number of excitatory, but not inhibitory, synapses and reduces the amplitude of miniature excitatory postsynaptic currents, but not miniature inhibitory postsynaptic currents (Kim et al. 2006). Likewise, NGL-3 expression stimulates postsynaptic differentiation and the formation of excitatory synapses, whereas knockdown of NGL-3 in hippocampal neurons leads to a reduction in the number of excitatory synapses. When expressed in HEK293 cells, NGL-3 stimulates the clustering of the presynaptic markers Synapsin I and VGLUT1 (a marker of excitatory synapses),and to a lesser extent VGAT (a marker of inhibitory synapses) in the axons of cocultured hippocampal neurons. In that setup, NGL-3 is a more potent inducer of synaptogenesis than NGL-1 and -2. Postsynaptically, NGL-3 expression induces the clustering of the scaffolding proteins PSD-95 and Shank, the AMPA receptor subunit GluR2, and the NMDA receptor subunit NR1 (Woo et al. 2009b).

One of the proteins known to bind to the PDZ-domain-binding motif of NGL-1, Whirlin, is expressed in stereocilia of hair cells in the inner ear. The gene encoding Whirlin is one of three genes associated with human Usher syndrome type 2, a severe sensorineural hearing impairment (Pan and Zhang 2012), and defects in Whirlin can lead to abnormally short stereocilia. Whirlin exists in two isoforms: a long isoform containing 3 PDZ domains and a proline-rich domain and a short isoform containing only the proline-rich domain and a single PDZ domain. Interestingly, expression of the short Whirlin isoform is sufficient to rescue both the length of stereocilia and the hair cell degeneration observed in response to lack of Whirlin expression (Mburu et al. 2003). NGL-1 binds to the PDZ domains located in the long isoform of Whirlin (Delprat et al. 2005). Therefore, defects in NGL-1 may play only a minor or no role in Usher syndrome type 2, and the functional significance of Whirlin–NGL-2 interactions remains to be determined.

NGL-2 has also been reported to act as a tumor suppressor. The expression of *LRRC4* is downregulated in brain tumor biopsies when compared to normal brain tissue (Zhang et al. 2005a), whereas a forced expression of NGL-2 in glioblastoma cells reduces tumor growth (Zhang et al. 2005b). Moreover, the expression of NGL-2 inhibits cell migration (Zhang et al. 2005c; Wu et al. 2006) and cell cycle progression, causing an increase in the fraction of cells in the G1 phase of the cell cycle, an increase in the amount of G1-phase proteins cyclin D1 and E, p21, and p27, and a decrease in the amount of cyclin A, proliferating cell nuclear antigen (PCNA), Oncoprotein 18/stathmin 1, and phosphorylated Erk1/2 (Zhang et al. 2005a, b; Wang et al. 2011). The expression of NGL-2 is also reported to inhibit the expression of cytokines, cytokine receptors, and downstream signaling molecules including the CXC chemokine receptor 4 (CXCR4) (Wu et al. 2008).

The small regulatory microRNA, hsa-miR-381, reduces the expression of NGL-2, and consequently hsa-miR-381 increases the proliferation of glioma cells, whereas the expression of NGL-2 inhibits the expression of hsa-miR-381, cell proliferation, and tumor growth (Tang et al. 2011). The expression of another regulatory microRNA, miR-185, is upregulated by NGL-2, and overexpression of miR-185 inhibits glioma cell invasion (Tang et al. 2012).

A study of copy number variations in gastric cancers revealed a case of *LRRC4C* amplification (Vauhkonen et al. 2007), suggesting that NGL-1 might also have functions related to carcinogenesis or cancer progression.

In summary, the NGL family consists of three members, which form *trans-interactions* with Netrin-G1, Netrin-G2, and LAR, respectively. Moreover, the individual members can form *cis*-complexes with, e.g., NMDA receptors through cytoplasmic interactions with PSD-95 and other PDZ-containing proteins. The proteins have neuritogenic and synaptogenic properties and are related to the development of, e.g., Usher syndrome and cancer.

14.4.3 The LINGO Family

The first member of the "LRR and Ig domain-containing Nogo receptor-interacting protein" (LINGO) family was identified in 2003 during an *in silico* gene content analysis of a part of human chromosome 15. The gene was named *LRRN6A* (Carim-Todd et al. 2003), but now the official gene name is *LINGO1* (Table 14.7). The following year the same gene together with two additional members of the family was identified during a database search for homologues to the axon guidance molecule SLIT (Mi et al. 2004), and subsequently a fourth member of the family was cloned (Chen et al. 2006; Haines and Rigby 2008). Moreover, the different members of the family have been detected in several large-scale studies (Clark et al. 2003; Ota et al. 2004; Gerhard et al. 2004; Bechtel et al. 2007; Gregory et al. 2006; Grimwood et al. 2004).

LINGO-1 has been suggested to be implicated in MS, PD, AD, and the recovery after spinal cord injury (Mi et al. 2004; Rudick et al. 2008; McDonald et al. 2011). However, the remaining members of the family are still poorly characterized, and their functions largely remain unknown.

Table 14.7 The human LINGO family

Gene symbol	Chromosomal location	Synonyms and previous names	Recommended protein name
LINGO1	15q24	Leucine-rich repeat and immunoglobulin domain-containing protein 1, leucine-rich repeat neuronal protein 1, leucine-rich repeat neuronal protein 6A, LERN1, LRRN6A	Leucine-rich repeat and immunoglobulin-like domain-containing Nogo receptor-interacting protein 1
LINGO2	9p21.2	Leucine-rich repeat neuronal protein 3, leucine-rich repeat neuronal protein 6C, LERN3, LRRN6C	Leucine-rich repeat and immunoglobulin-like domain-containing Nogo receptor-interacting protein 2
LINGO3	19p13.3	Leucine-rich repeat neuronal protein 2, leucine-rich repeat neuronal protein 6B, LERN2, LRRN6B	Leucine-rich repeat and immunoglobulin-like domain-containing Nogo receptor-interacting protein 3
LINGO4	1q21.3	Leucine-rich repeat neuronal 6D, LRRN6D	Leucine-rich repeat and immunoglobulin-like domain-containing Nogo receptor-interacting protein 4

Genes and Proteins

LINGO proteins are highly homologous, exhibiting 44–61 % pairwise primary sequence identity (Carim-Todd et al. 2003; Mi et al. 2004). Moreover, all the LINGO proteins are highly conserved in mammals, especially LINGO-1 for which the primary sequence for the ectodomain is ~93–100 % identical between human, mouse, rat, monkey, and chicken. However, there are no known invertebrate homologues to *LINGO1* (Carim-Todd et al. 2003; Mi et al. 2004; Mosyak et al. 2006).

LINGO proteins are transmembrane proteins localized in the plasma membrane. The proteins are composed of a ~520 amino acids long ectodomain containing 11-12 LRRs with NF and CF domains, followed by an Ig module, a single transmembrane domain, and a short ~40 amino acids long cytoplasmic domain (Mi et al. 2004; Haines and Rigby 2008). In the ectodomain of LINGO-1, there are eight potential sites for *N*-linked glycosylation; six of these are located in the LRR domain, and the remaining two are located in the Ig module (Mosyak et al. 2006). The cytoplasmic tail of LINGO-1 contains a canonical EGF receptor tyrosine phosphorylation site (Mi et al. 2004; Inoue et al. 2007). The phosphorylation site is conserved between LINGO-1-3, but is not present in LINGO-4 (Llorens et al. 2008).

The regulation of LINGO expression is not well characterized, but nerve growth factor (NGF) has been reported to regulate the expression of LINGO-1 though the activation of the NGF receptor TrkA (Lee et al. 2007).

Expression

In chicken embryos, *cLINGO1* is expressed in the developing spinal cord and brain. In the spinal cord it is expressed in motor neurons and in the dorsal and ventral root ganglia. In the developing brain it is first detected in the prosencephalon and ventral mesencephalon and later in the mesencephalon, telencephalon, and parts of the hindbrain. Moreover, it is expressed in the trigeminal and facial nerves and in the ventral part of the neural retina (Okafuji and Tanaka 2005).

During early development in mouse embryos, *LINGO1* initially exhibits a broad pattern of expression, but gradually the expression becomes restricted to the brain. *LINGO-2* is expressed adjacent to the epithelial lining of the olfactory pit, whereas *LINGO-3* is more widely expressed, especially in mesodermal tissues where it is highly expressed in regions of the brachial arches, head mesoderm, and developing limbs. Lastly, *LINGO-4* is expressed in the cells along the neural tube (Haines and Rigby 2008).

In the adult mouse, the expression of *LINGO1* is especially enriched in the limbic system and the cerebral cortex (Carim-Todd et al. 2003; Haines and Rigby 2008), but generally the expression levels of *LINGO1* and *-2* are low in the brain, and no expression of *LINGO3* and *-4* has been detected (Haines and Rigby 2008). In stem cells from embryonic mouse forebrain, LINGO-1 is co-expressed with Reticulon-4/NogoA, the Reticulon 4 receptor/Nogo-66 receptor (NgR), the Neurotrophin receptor p75, and the tumor necrosis factor receptor TROY (Mathis et al. 2010).

In rat, LINGO-1 is reported to be widely expressed in the brain. In a rostral-to-caudal gradient *LINGO1* was found to be expressed in the CNS, with the highest level of expression in the cortex and the lowest levels of expression in the spinal cord. The expression of *LINGO1* peaks at P1 and subsequently decreases. Immunohistochemically, LINGO-1 has been shown to be expressed in cerebellar granule neurons of P7 rats, but to be more strongly expressed in Purkinje neurons in adult rats (Mi et al. 2004).

In adult human tissues, *LINGO1* is highly expressed in the brain, especially in the cerebral cortex, amygdala, hippocampus, and thalamus, whereas the transcript has not been detected in non-neural tissues (Carim-Todd et al. 2003; Mi et al. 2004).

In rat, the remaining members of the LINGO family have been found to be ubiquitously expressed, but only to be expressed at low levels in the nervous system, when compared to *LINGO1* (Mi et al. 2004).

Interactions

Structural studies of LINGO-1, both when crystallized and in solution, have shown that the protein forms homophilic tetramers, where the individual tetramers create a closed, circular structure around a central hole with a diameter of ~45 Å

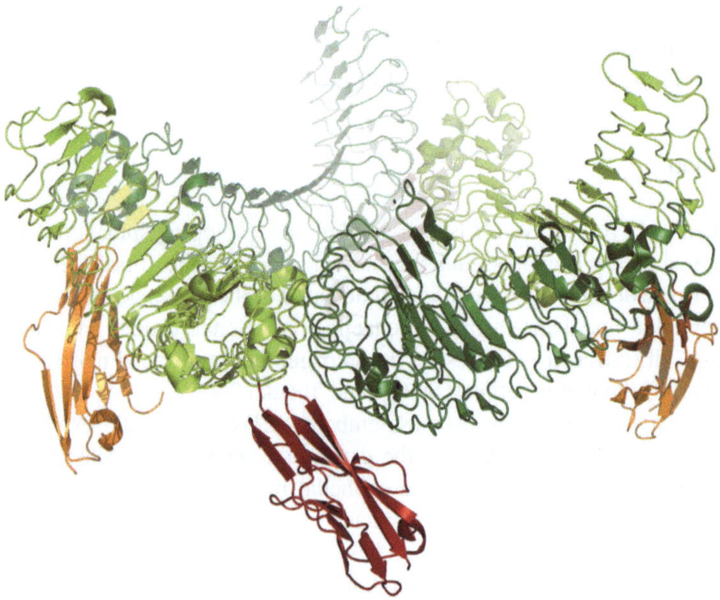

Fig. 14.5 Structural model of a Lingo1 tetramer. Cartoon drawing of a structural model of four ectodomains of human LINGO1 [PDB ID: 2ID5; 477 amino acids; Mosyak et al. (2006)]. Each ectodomain includes an LRR domain composed of 12 LRRs with NF and CF domains (*dark/light green*) that is followed by a single Ig module (*red/orange*). The ectodomains form homodimers through interactions between the concave sides of the LRR domains. In the model there is a near-90° angle between the LRR domain and the Ig module. This creates a cleft where the N-terminal part of the neighboring LINGO1 molecule binds. The structural model is believed to represent homophilic *cis*-interactions where the Ig modules are facing the plasma membrane, and the concave surfaces of the LRR domains (including glycosylations that are not included in the figure) are facing away from the cell surface. However, LINGO molecules are also believed to be able to form homophilic *trans*-interactions [see Mosyak et al. (2006), for details]. The figures were created with PyMOL (DeLano Scientific)

(see Fig. 14.5). According to the structural model, the Ig modules of the four molecules are positioned perpendicularly to the cell surface, whereas the LRR domains are positioned in parallel to the cell surface (Mosyak et al. 2006). Soluble LINGO-1 ectodomains have been shown to bind cell surface-localized LINGO-1 molecules in vitro, hence demonstrating that LINGO-1 molecules can form homophilic interactions under physiologically relevant conditions, and the molecules are also believed to be able to form homophilic *trans*-interactions (Jepson et al. 2012).

In addition to the homophilic *cis*-interactions, LINGO-1 might interact with APP (Bai et al. 2008), and LINGO-1 is one of the components of the Nogo receptor complex, which also includes the GPI-anchored receptor NgR1 and either p75 or TROY. In this heterophilic *cis*-complex NgR1 serves as the ligand-binding part, whereas p75 and TROY serve as signal transducing co-receptor subunits (Mi et al. 2004; Shao et al. 2005; Park et al. 2005; Fournier et al. 2001; Yamashita et al. 2002; Wang et al. 2002).

NgR1 is the receptor for myelin-associated inhibitor (MAI) proteins implicated in axonal regeneration. These include NogoA, myelin-associated glycoprotein (MAG), and oligodendrocyte myelin glycoprotein (OMgP), and binding of NgR1 to either of these MAIs induces intracellular signaling through LINGO-1 and p75/TROY (Rudick et al. 2008; Wang et al. 2002; Bandtlow and Dechant 2004; GrandPre et al. 2000; Domeniconi et al. 2002; Oertle et al. 2003). Recently, olfactomedin 1 (Olfm1), a secreted glycoprotein, has been found also to bind to NgR1 and thereby to interfere with the binding between NgR1, p75, and LINGO-1 (Nakaya et al. 2012).

In retinal ganglion cells LINGO-1 is co-expressed with the BDNF receptor, and TrkB and the two receptors can be co-precipitated, suggesting that they form direct or indirect *cis*-interactions (Fu et al. 2009, 2010).

As mentioned above, LINGO-1 contains an EGF receptor tyrosine phosphorylation site (Mi et al. 2004), and using different methods, LINGO-1 has been shown to interact with the EGF receptor. Interestingly, LINGO-1 expression decreases EGF receptor expression in a dose-dependent manner, whereas the absence of LINGO-1 or the abrogation of extracellular LINGO-1 interactions leads to an increase in the levels of EGF receptor and phosphorylated Akt. It has been proposed that LINGO-1 either directly prevents EGF receptor activation or that it reduces EGF receptor-mediated signal transduction by promoting 0EGF receptor internalization and degradation (Inoue et al. 2007).

In yeast two-hybrid screenings using the intracellular part of LINGO-1 as bait, the serine–threonine kinase WNK lysine-deficient protein kinase 1 (WNK1) and the post-mitotic neuronal-specific zinc finger protein Myelin transcription factor 1-like (Myt1l) were identified as possible interaction partners (Llorens et al. 2008; Zhang et al. 2009). Subsequent studies have shown that the expression of LINGO-1 and WNK1 co-localizes in cortical neurons (Zhang et al. 2009) and that Myt1l can be co-precipitated with LINGO-1, suggesting that LINGO-1 regulates Myt1l transcription factor activity by modulating its subcellular localization (Llorens et al. 2008).

Functions

Neuronal expression of LINGO-1 promotes growth cone collapse and inhibition of neuritogenesis. The effect requires the indirect *trans*-interactions with MAI proteins (NogoA, MAG, or OMgP), the presence of the LINGO-1 EGF receptor tyrosine phosphorylation site, and the expression of WNK1. Thus, the expression of dominant negative LINGO-1 (lacking the phosphorylation site), or the down-regulation of WNK1, leads to an increase in neurite outgrowth (Mi et al. 2004; Zhang et al. 2009).

Both in vitro and in vivo, LINGO-1 is expressed in oligodendrocytes, and this expression inhibits oligodendrocyte differentiation and axon myelination, whereas LINGO-1 knockdown in oligodendrocyte–neuron cocultures in vitro enhances oligodendrocyte differentiation and increases myelination (Lee et al. 2007; Jepson et al. 2012; Mi et al. 2005). Moreover, the spinal cords of *LINGO1*-knockout mice exhibit an increased amount of myelinated axon fibers and an increased fraction of mature oligodendrocytes when compared to the spinal cords of wild-type mice (Mi et al. 2005).

The effects of LINGO-1 expression on oligodendrocyte differentiation have been suggested to be facilitated by *trans*-homophilic LINGO-1 interactions, which subsequently lead to the stimulation of intracellular signal transduction involving the inhibition of the Src-family kinase Fyn and the activation of the small GTPase RhoA (Jepson et al. 2012; Mi et al. 2005; Liang et al. 2004).

Recently, LINGO-1 expression was reported to regulate the differentiation of neuronal stem and progenitor cells, but not of astrocytes (Loov et al. 2012). Cortical neural stem cells from E14 mouse embryos express LINGO-1, but if the protein is neutralized with anti-LINGO-1 antibodies in vitro it leads to an increased fraction of immature neurons as a result of a decrease in neuronal maturation. The effect is probably caused by an increased proliferation of neural progenitors and is accompanied by a decrease in cell death (Loov et al. 2012). Abrogating LINGO-1 interactions with addition of recombinant LINGO-1-Fc protein to in vitro cultures of cerebellar granule neurons also leads to a rescue of apoptosis induced by low potassium concentrations in a manner involving the inhibition of glycogen synthase kinase-3β (GSK3β) (Zhao et al. 2008).

The long-term and in vivo effects of abrogating the function of LINGO-1 are not known, but the phenomenon may have clinical implications in relation to the effect of LINGO-1 on neurodegenerative diseases (Inoue et al. 2007; Bai et al. 2008; Mi et al. 2007).

Diseases

Glaucoma is an eye disease that can lead to blindness. It is characterized by the slow, progressive degeneration of retinal ganglion cells and their axons (Zhang et al. 2012). The presence of LINGO-1 in the adult retina has led to the notion that LINGO-1 antagonists might have protective effects on neurons following retinal injury (Fu et al. 2008; Bessero and Clarke 2010). To test this hypothesis, LINGO-1 antagonists were applied to retinal ganglion cells following ocular hypertension and optic nerve transection, and it was found that LINGO-1 antagonists significantly reduce the loss of retinal ganglion cells after ocular hypertension and promote the survival of retinal ganglion cells after optic nerve transections.

The neurotrophins BDNF, NGF, and neurotrophin 4 (NT-4) have neuroprotective effects on retinal ganglion cells compromised by experimentally induced glaucoma or intraorbital optic nerve crush (Parrilla-Reverter et al. 2009; Colafrancesco et al. 2011); reviewed in Johnson et al. (2009) and Weber et al. (2008). BDNF and NT-4 preferentially bind TrkB, whereas NGF preferentially binds the related receptor, TrkA, and as mentioned above, LINGO-1 is co-expressed with TrkB, whereas TrkA-mediated signaling stimulates the expression of LINGO-1 (Lee et al. 2007; Fu et al. 2009; Skaper 2012a). Interestingly, LINGO-1 antagonists promote BDNF-induced activations of TrkB and the neuroprotective effect of BDNF in the retina, suggesting that LINGO-1 antagonists can be used for the treatment of glaucoma.

Intracellularly, LINGO-1 antagonists reduce RhoA and JNK activities and increase the activity of the anti-apoptotic kinase Akt (Fu et al. 2008).

In mouse brain, APP and LINGO-1 demonstrate comparable expression patterns with respectively high and low expression levels in the CA1-3 regions and dentate gyrus of the hippocampus. Furthermore, a reduction in LINGO-1 expression results in a decrease in the amount of the β-cleavage C-terminal fragment of APP (βCTF)/C99 and the 40–42 Aβ-peptide fragments, in combination with an increase in the secretion of the α-cleavage C-terminal fragment of APP (αCTF)/C83. These findings suggest that LINGO-1 might affect the cleavage of APP as a result of a direct interaction (Bai et al. 2008). However, p75 and NgR1 can also affect APP processing (Costantini et al. 2005; Park et al. 2006), so the effect of LINGO-1 could also be the result of an indirect interaction. Surprisingly, NgR1 expression demonstrates an inverse relationship with Aβ-peptide levels, whereas LINGO-1 appears to promote Aβ-peptide secretion (Bai et al. 2008; Park et al. 2006).

In mice, LINGO-1 expression is elevated in response to damages of midbrain dopaminergic nerve terminals, and in some PD patients increased expression levels of LINGO-1 have been found in the substantia nigra. Moreover, results obtained with primary cell cultures in vitro and with PD in vivo animal models show that the survival, growth, and function of midbrain dopaminergic neurons improve when LINGO-1 expression is reduced, suggesting that LINGO-1 is a potential target for the design of new drug against PD (Inoue et al. 2007).

Surprisingly, genome-wide association studies of LINGO-1 and LINGO2 variations in PD have generally produced negative results, arguing against a role of *LINGO1* variations as a contributing factor in the development of PD [for a recent review, see Deng et al. (2012)]. However, genome-wide association studies of *LINGO1* and *-2* variations have revealed an association with essential tremor, a progressive neurological disorder characterized by symmetric postural or kinetic tremor of the hands and forearms [see Deng et al. (2012), Stefansson et al. (2009), Zeuner and Deuschl (2012) and Vilarino-Guell et al. (2010)].

In rat spinal cord, the axonal expression of LINGO-1 is increased after spinal cord injury (Mi et al. 2004, 2007; Inoue et al. 2007; Bandtlow and Dechant 2004; Ji et al. 2006). Furthermore, LINGO-1 antagonists that block the interaction between LINGO-1 and NgR1 cause a decrease in RhoA activation, promote axonal sprouting in the spinal cord, and improve forelimb and hindlimb function following spinal cord injury (Ji et al. 2006), suggesting that a modulation of *LINGO1* expression following spinal cord injury may be beneficial for the recovery [see Mi et al. (2008) and Gerin et al. (2011), for recent reviews].

Oligodendrocyte precursor cells can give rise to mature oligodendrocyte, Schwann cells, and neurons. However, in MS patients oligodendrocyte precursor cells seem to be prevented from differentiating into remyelination oligodendrocytes (Rivers et al. 2008; Fancy et al. 2009; Zawadzka et al. 2010; Huang and Franklin 2011). LINGO-1 expression has a negative effect on oligodendrocyte differentiation and myelination, and since LINGO-1 is upregulated in humans with MS, the protein has been proposed as a possible target for antagonistic drugs intended to promote remyelination in MS patients (Rudick et al. 2008; McDonald et al. 2011; Inoue et al. 2007).

Indeed, results obtained with experimental autoimmune encephalomyelitis (EAE; the standard animal model for MS) show that the disease course is less severe in *LINGO1*-knockout mice, when compared to wild-type mice (Mi et al. 2007; Wekerle 2008). Likewise, EAE is less severe in animals treated with anti-LINGO-1 antibodies, when compared to control animals (Mi et al. 2007), and the pharmaceutical company Biogen Idec is investigating the safety of anti-LINGO-1 treatment in humans (McDonald et al. 2011). LINGO-1 antagonist seems to promote CNS remyelination by stimulating oligodendrocyte precursor cells directly (Mi et al. 2009). For more information about LINGO-1 in relation to treatment of MS, see Rudick et al. (2008), Huang and Franklin (2011), Pepinsky et al. (2010, 2011a, b).

In addition to the above-mentioned diseases and pathological conditions, the expression of NgR, TROY, LINGO-1, and RhoA, but not p75, is upregulated in the cortices of pediatric patients with tuberous sclerosis and focal cortical dysplasia type llb (Yu et al. 2012). These pathological conditions are characterized by developmental abnormalities in the cytoarchitecture of the cortex and are often associated with intractable focal epilepsies (Blumcke et al. 2009). Consequently, it has been suggested that NgR/TROY/LINGO-1 complexes play a more important role than NgR/p75/LINGO-1 complexes in regulating glial–neuronal and neuronal–neuronal interactions during these pathological conditions (Yu et al. 2012).

Finally, LINGO-1 is, through the interactions with NgR1 and MAIs, potentially implicated in, e.g., schizophrenia, temporal lobe epilepsy, and amyotrophic lateral sclerosis (Bandtlow and Dechant 2004; Llorens et al. 2011). For more information on MAIs and NgR1, see McDonald et al. (2011), Llorens et al. (2011) and Lee and Petratos (2013).

In summary, the LINGO family consists of four members, of which LINGO-1 has been the focus of most investigations. All members of the family are believed to form both homophilic *cis*- and *trans*-interactions. Moreover, LINGO-1 has been found to form direct or indirect heterophilic *cis*-interactions with numerous other proteins including APP, NgR1, p75, TROY, TrkB, and EGFR. LINGO-1 promotes growth cone collapse, inhibits neuritogenesis, oligodendrocyte differentiation, axon myelination, and neuronal cell survival, and has been related to several diseases including glaucoma, AD, PD, and MS.

14.4.4 The FLRT Family

Fibronectin leucine-rich transmembrane protein 1 (FLRT1) was discovered in 1999 in a screen for novel ECM proteins in human adult skeletal muscles. A subsequent search through the human EST database revealed two additional members of the family, FLRT2 and -3 (Lacy et al. 1999) (Table 14.8). In addition, the three members of the family have been detected in several large-scale studies (Ishikawa et al. 1997; Clark et al. 2003; Ota et al. 2004; Gerhard et al. 2004; Nagase et al. 2000a; Otsuki et al. 2005; Taylor et al. 2006; Deloukas et al. 2001). The proteins have been suggested to be implicated in cirrhosis, lupus, schizophrenia, and ASDs (Lesch et al. 2012; Anney et al. 2010; Ahmad et al. 2012; Shirai et al. 2012, 2013).

Table 14.8 The human FLRT family

Gene symbol	Chromosomal location	Synonyms and previous names	Recommended protein name
FLRT1	11q12-q13	Fibronectin-like domain-containing leucine-rich transmembrane protein 1, MGC21624	Leucine-rich repeat transmembrane protein FLRT1
FLRT2	14q24-q32	Fibronectin-like domain-containing leucine-rich transmembrane protein 2, KIAA0405	Leucine-rich repeat transmembrane protein FLRT2
FLRT3	20p11	Fibronectin-like domain-containing leucine-rich transmembrane protein 3, KIAA1469	Leucine-rich repeat transmembrane protein FLRT3

Genes and Proteins

FLRT1, -2, and *-3* are conserved in vertebrates, but have no orthologues in *Drosophila melanogaster* or *Caenorhabditis elegans* (Haines et al. 2006; Bottcher et al. 2004; Smith and Tickle 2006). In vertebrates, *FLRT3* expression is activated by the TGFβ-protein Activin (Ogata et al. 2007) and by Fibroblast growth factor (FGF) (Bottcher et al. 2004).

FLRT proteins are composed of an extracellular domain consisting of 10 LRRs with NF and CF domains, followed by one Fn3 module, a transmembrane domain, and an ~100 amino acids long cytoplasmic domain (Lacy et al. 1999; Tsuji et al. 2004; Robinson et al. 2004). All three proteins contain sites for potential *N*-linked glycosylation: two in FLRT1, five in FLRT2, and four in FLRT3 (Lacy et al. 1999), and FLRT2 has been demonstrated to be glycosylated on at least one of these sites (Chen et al. 2009).

All three proteins are highly conserved among vertebrates. Thus, mouse and human FLRTs demonstrate similarities of 96–97 % (Haines et al. 2006). In contrast, the primary sequences of human FLRT1, -2, and -3 exhibit similarities of only 16–20 % (Lacy et al. 1999). Despite these moderate homologies FLRT2 and -3 are reported to be functionally interchangeable (Muller et al. 2011).

Expression

During mouse development *FLRT1* expression is initially detected in the midbrain and later in the eye, DRG, trigeminal ganglia, near the urogenital ridge, and the facio-acoustic ganglion. *FLRT2* is highly expressed around the ventral midbrain, the developing heart, and the stomach and later in the eye, branchial arches, and limbs. *FLRT3* is expressed in the developing brain, somites, and around the developing heart (Haines et al. 2006). In mouse brain, FLRT3 is around P14 highly expressed in the dentate gyrus (DG) and the CA3 region, but not the CA1 region, of the hippocampus (O'Sullivan et al. 2012). For detailed studies of FLRT3 expression during chicken limb development and FLRT2 during mouse heart and craniofacial development, see Smith and Tickle (2006), Tomas et al. (2011) and Muller et al. (2011), Maretto et al. (2008) and Gong et al. (2009), respectively.

In adult rat brain *FLRT3* is ubiquitously expressed. Highest levels of expression are found in basal ganglia, in pyramidal and dentate granular neurons in the CA3 and CA4 regions of the hippocampus, in the granular layer and Purkinje cells of the cerebellum, and in the granular layer and mitral cells of the olfactory bulb (Tsuji et al. 2004).

Northern blot analyses of adult human tissues have identified low levels of *FLRT1* expression in kidney, skeletal muscle, and brain; *FLRT2* is expressed at high levels in the pancreas and at lower levels in skeletal muscle, brain, and heart, and *FLRT3* is highly expressed in kidney, skeletal muscle, lung, and brain and at lower levels in pancreas, liver, placenta, and heart (Lacy et al. 1999).

At the cellular level FLRT cells localize to the plasma membrane (Haines et al. 2006; Tsuji et al. 2004). In mouse hippocampal neurons FLRT3 enriched in postsynaptic membranes localizes to puncta that partially co-localizes with glutamatergic, but not GABAergic synapses (O'Sullivan et al. 2012).

Interactions

Immunoprecipitation data demonstrate that FLRT proteins form direct or indirect homophilic and heterophilic interactions within the family (Karaulanov et al. 2006). Moreover, FLRT2 and -3 have been shown to mediate cell–cell interactions (Haines et al. 2006) and to promote homotypic cell sorting in a Ca^{2+}-dependent manner (Karaulanov et al. 2006). However, other studies suggest that FLRT3 does not mediate homophilic cell adhesion (Robinson et al. 2004; Yamagishi et al. 2011). Instead, as described below, FLRT proteins are believed to be able to form several heterophilic interactions that may help to explain their effects on cell adhesion and cell sorting even in the absence of *trans*-homophilic interactions.

The ectodomain of *Danio rerio* FLRT3 is reported to bind the IgSF members, MAG and Brother of CDO (BOC) (Soellner and Wright 2009). MAG is known to enhance the myelin stability (Schnaar and Lopez 2009) and BOC is an axon guidance molecule (Connor et al. 2005), but their potential interplay with FLRT3 function is not known. Moreover, it was in a recent study not possible to demonstrate an interaction between mouse FLRT3 and MAG (Yamagishi et al. 2011).

In *Xenopus* and mouse, the ectodomain of FLRT2 binds with high affinity to the Netrin receptor Unc5D in a Netrin-1-independent manner. Likewise, FLRT3 binds with high affinity to Unc5B, and FLRT1 binds with low affinity to Unc5B (Yamagishi et al. 2011; Karaulanov et al. 2009). There are four Netrin receptors, Unc5A-D, but no FLRT proteins have been found to bind with a high affinity to Unc5A, or -C. The Netrin receptors have ectodomains composed of two Ig modules and two thrombospondin type 1 repeats and are known to be involved in Netrin-1-induced axon chemorepulsion (Larrieu-Lahargue et al. 2012). A splice variant of Unsc5D that lacks one of the thrombospondin type 1 repeats does not interact with FLRT2 (Yamagishi et al. 2011), suggesting that the binding site includes this part of the molecule.

Recently, it was shown that mammalian FLRT3 binds a small family of synaptic G-protein-coupled receptors called latrophilins (LPHNs). LPHNs are presynaptic proteins, and they are believed to form *trans*-interaction with postsynaptic FLRT3 molecules (O'Sullivan et al. 2012).

Finally, FLRTs are reported to interact with FGF receptors. In one study, the interaction was found to be extracellular, involving the FLRT Fn3 module (Bottcher et al. 2004), but another study has reported interactions also between the cytoplasmic domains of FLRT2 and FGF receptor 2 (Wei et al. 2011). Surprisingly, it was in a recent study not possible to demonstrate an interaction between FLRT2 and FGF receptor 2 (Yamagishi et al. 2011).

Functions

FLRT3 seems to play important roles in relation to neuronal development. Overexpression of the protein promotes neuritogenesis in, e.g., cerebellar granule and DRG neurons (Tsuji et al. 2004; Robinson et al. 2004), and during chicken development the protein plays a role in the regulation of cell adhesion and establishment of the dorsal–ventral position of the apical ectodermal ridge (Smith and Tickle 2006). In mouse, knockout of *FLRT2* or *-3* during development leads to a disorganized basement membrane in the anterior visceral endoderm region and defects in, e.g., headfold fusion and heart development (Muller et al. 2011; Maretto et al. 2008; Egea et al. 2008). Moreover, studies in mouse and rat have shown that *FLRT3* expression is increased considerably in response to nerve injury, suggesting that FLRT3 plays a part in neuronal recovery and regeneration (Tsuji et al. 2004; Tanabe et al. 2003).

The interaction between FLRT3 and Unc5B regulates cell adhesion in a manner involving the small GTPase Rnd1, which can bind to the cytoplasmic domain of both FLRT3 and Unc5B (Yamagishi et al. 2011; Karaulanov et al. 2009). In *Xenopus* an increased expression of FLRT1, -2, or -3 leads to a reduction in cell–cell interactions, and for FLRT3 this has been demonstrated to be a result of an endocytosis-mediated downregulation of the amount of C-cadherin at the cell surface. This de-adhesion function of FLRT3 is induced by Activin stimulation, but not by FGF (Ogata et al. 2007).

Soluble ectodomains of FLRT1, -2, and -3, produced by extracellular membrane-proximal cleavage, have been detected in conditioned media from embryonic cortical and hippocampal neurons and from HEK cells overexpressing the proteins. The process seems to require metalloprotease, but not γ-secretase, activity (Yamagishi et al. 2011), but FLRT1 has also been proposed as a novel substrate for the extracellular serine protease Thrombin (Gallwitz et al. 2012). The soluble ectodomains of FLRT2 and -3 serve as repulsive axon guidance molecules for neurons expressing Unc5, and for instance the proper migration of neurons from the subventricular zone to the cortical plate is regulated by the expression levels of both FLRT2 and Unc5D (Yamagishi et al. 2011).

LPHNs are the main receptors for the black widow spider venom α-latrotoxin, and activation of the receptors leads to an increase in the intracellular concentration of Ca^{2+} and the subsequent exocytosis of neurotransmitters (Silva and Ushkaryov 2010).

Experiments with hippocampal neurons have demonstrated that abrogation of the *trans*-interaction between postsynaptic FLRT3 molecules and presynaptic LPHNs (either by incubation in the presence of excess soluble chimeric ecto-LPHN proteins or by knockdown of FLRT3 or LPHNs) leads to a reduction in glutamatergic synapse formation and glutamatergic excitatory postsynaptic currents. Moreover, in utero knockdown of FLRT3 leads to a reduction in the density of dendritic spines in dentate gyrus granule neurons (which express FLRT3), but not in pyramidal neurons (which do not express FLRT3) (O'Sullivan et al. 2012).

There seems to be a strong interplay between FGF and FLRT function. FLRTs are reported to bind to FGF receptors, and the expression of FLRTs overlaps with that of some FGFs. Moreover, FGF receptor–FLRT interactions stimulate intracellular FGF receptor-mediated signal transduction pathways, including the Ras-MAPK pathway, which in turn leads to an increased expression of FLRTs (Haines et al. 2006; Bottcher et al. 2004). FGF receptor activity leads to the phosphorylation of FLRT1 in an indirect manner involving Src-family kinase activity. The FLRT1 phosphorylation in turn enhances FGF receptor-mediated Ras-MAPK-dependent stimulation of neurite outgrowth (Wheldon et al. 2010). Despite this apparent interrelationship between FGF receptor and FLRT signaling, FLRT-dependent homotypic cell sorting is independent of FGF receptor-mediated signaling (Karaulanov et al. 2006), and in mouse, FLRT3 knockout does not seem to affect FGF receptor-mediated signaling (Maretto et al. 2008).

Diseases

Despite the many apparent functions of FLRT proteins there are still few known relationships between FLRTs and diseases. *FLRT1* was originally identified in a search for possible gene candidates for laminin 2-positive congenital muscular dystrophies (CMD), a heterogeneous group of muscle disorders (Lacy et al. 1999), but it is still unclear whether FLRTs are implicated in the pathogenesis of this disease.

In a recent microarray analysis, *FLRT1* was shown to be upregulate during cirrhosis, a chronic degenerative disease, where normal liver cells are damaged and replaced by scar tissue (Ahmad et al. 2012).

FLRT2 has been identified as a target for anti-endothelial cell antibodies (AECAs) in patients with the autoimmune disease systemic lupus erythematosus, suggesting that the proteins are potential therapeutic drug targets for the treatment of this disease and potentially other autoimmune diseases (Shirai et al. 2012, 2013; Renaudineau et al. 2002).

FLRT3 is embedded within the *MACROD2* gene, and genome-wide studies have identified *MACROD2* copy number variation in individuals with schizophrenia, brain infarct, ASDs, and MS [see Anney et al. (2010) for references]. Furthermore, both LPHN3 and FLRT3 have been identified as risk genes for attention-deficit/

hyperactivity disorder (ADHD) [reviewed by Lesch et al. (2012)]. However, a direct relationship between changes in the function or expression of *FLRT3* and these diseases remains to be determined.

In summary, the FLRT family consists of 3 members. All members of the family are believed to form *cis*-interactions, whereas FLRT-mediated cell adhesion is facilitated by direct or indirect heterophilic interactions with, e.g., FGF receptors, Netrin receptors, or LPHNs. The different parts of FLRT proteins have different roles that are suggested to be as follows: the extracellular LRRs play a role in homotypic cell sorting and in binding to Netrin receptors and LPHNs; the Fn3 modules are involved in binding to FGF receptors, and the intracellular domains are responsible for modulation of FGF receptor- and Rnd1-mediated signaling.

FLRTs stimulate neuritogenesis and may be involved in neuronal regeneration, but they can also act as soluble repulsive axon guidance molecules. Moreover, FLRT–LPHN interactions seem to affect the formation and function of glutamatergic synapses. However, any relationship between FLRT proteins and neuronal diseases remains to be determined.

14.4.5 The LRIT Family

The LRR, Ig-like, and transmembrane domains (LRIT) family has three members, LRIT1-3. The genes have been cloned by several groups independently, and consequently the proteins are known by several names (Ota et al. 2004; Gerhard et al. 2004; Bechtel et al. 2007; Homma et al. 2009; Gomi et al. 2000; Deloukas et al. 2004; Hillier et al. 2005; Munfus et al. 2007) (Table 14.9).

Biochemical studies suggest that LRIT1 can form homodimers (Gomi et al. 2000), but the interaction has not been characterized, and LRIT proteins have not been demonstrated to mediate cell adhesion. Thus, they may fall outside the scope of this book and will therefore only be briefly described.

All LRIT proteins are composed of an N-terminal domain containing an LRR domain with five LRRs with NF and CF domains, one Ig module, and one Fn3 module, followed by a transmembrane domain and a ~64–76 amino acids long cytoplasmic domain (Homma et al. 2009; Gomi et al. 2000; Munfus et al. 2007).

Apart from being expressed in the retina, LRIT proteins are generally reported not to be expressed in the nervous system (Homma et al. 2009), although LRIT1 recently was reported to be expressed in the pineal gland (Bustos et al. 2011). In the retina, LRIT1 localizes to the outer segments of the photoreceptor cells, where the N-terminal part of the protein is presumed to face the intradiscal space (Gomi et al. 2000), whereas LRIT3 localizes to the dendritic tips of bipolar cells (Zeitz et al. 2013). Surprisingly, a large fraction of LRIT1 proteins are localized not in the plasma membrane, but in the ER with the N-terminal domain facing the ER lumen (Gomi et al. 2000), and based on the function of LRIT3 described below, a similar localization can be expected for LRIT1.

Table 14.9 The human LRIT family

Gene symbol	Chromosomal location	Synonyms and previous names	Recommended protein name
LRIT1	10q23	DKFZP434K091, fibronectin type III, immunoglobulin, and leucine-rich repeat domains 9, FIGLER9, leucine-rich repeat-containing 21, LRRC21, photoreceptor-associated LRR superfamily protein, PAL, retina-specific protein PAL	Leucine-rich repeat, immunoglobulin-like domain, and transmembrane domain-containing protein 1
LRIT2	10q23.2	AC022389.4, leucine-rich repeat-containing 22, LRRC22	Leucine-rich repeat, immunoglobulin-like domain, and transmembrane domain-containing protein 2
LRIT3	4q25	Fibronectin type III, immunoglobulin, and leucine-rich repeat domains 4, FIGLER4, FLJ44691	Leucine-rich repeat, immunoglobulin-like domain, and transmembrane domain-containing protein 3

Because of its prominent expression in the retina, LRIT1 has been proposed as a potential candidate disease gene for inherited retinal degenerations. Moreover, the gene is located in a region of chromosome 10 containing genes implicated in Usher syndrome, a rare disease combining hearing loss and visual impairment (Gomi et al. 2000), but a relationship between LRIT1 and Usher syndrome has not been established.

LRIT3 has been shown to bind FGF receptor 1 (Kim et al. 2012), and the protein is believed to facilitate exit of FGF receptor 1 from the ER in a regulated manner. Co-expression of LRIT3 and FGF receptor 1 in HEK293 cells leads to a relative increase in the amount of FGF receptor 1 and facilitates the activation of PLCγ-mediated signaling in an FGF-independent manner (Kim et al. 2012). Moreover, an analysis of patients with the developmental disease non-syndromic craniosynostosis has led to the identification of two LRIT3 mutants, LRIT3-S494T (a mutation located in the Fn3 module) and LRIT3-C592Y (a mutation located in the transmembrane domain), which when expressed in HEK293 cells also facilitate a relative increase in FGF receptor 1 expression and lead to an increase in FGF receptor-mediated and PLCγ-mediated signaling in an FGF-independent manner (Kim et al. 2012). FGF receptor gain-of-function mutations and PLCγ-mediated signaling are known to be related to craniosynostosis (Miraoui and Marie 2010), and therefore the identified LRIT3 mutants suggest a relationship between LRIT3-mediated changes in FGF receptor signaling and craniosynostosis. In addition to the craniosynostosis-related mutations, four mutations in the region encoding the Ig and Fn3 modules of LRIT3 have been identified in patients with congenital stationary night blindness (Zeitz et al. 2013). These observations suggest that both LRIT1 and -3 have important functions in the retina.

14.4.6 The SALM/LRFN Family

Synaptic adhesion-like molecules (SALMs) also known as LRR and Fn3 domain containing (LRFN) constitute a protein family that originally was identified in 1999 in a screening of human adult and fetal brain cDNA libraries (Kikuno et al. 1999). Subsequently, the different members of the family have been detected in various large-scale screenings (Clark et al. 2003; Ota et al. 2004; Gerhard et al. 2004; Nagase et al. 1999, 2000a; Grimwood et al. 2004; Mungall et al. 2003), but a paper elaborating on the function of the proteins was not published until 2006 (Wang et al. 2006).

The family consists of five members: SALM1/LRFN2, SALMs2/LRFN1, SALMs3/LRFN4, SALMs4/LRFN3, and SALMs5/LRFN5 (Nam et al. 2011), which all are expressed in the nervous system. Pathologically, SALMs have so far been suggested to be implicated in ASD and schizophrenia (Wang et al. 2009b; Xu et al. 2009; de Bruijn et al. 2010; Mikhail et al. 2011).

Genes and Proteins

In both mouse and human, the five genes encoding the respective members of the SALM family are located on four different chromosomes [SALM2 and -4 being located on the same chromosome; Table 14.10; Nam et al. (2011)].

All five members of the SALM family are conserved between mammalian species, and fish have homologues of at least some mammalian SALMs (Morimura et al. 2006). In contrast, no homologues to SALMs have been found in invertebrates, although they, like members of the LRIG family, share a high sequence identity with the *Drosophila* Kekkon protein family (Gur et al. 2004; Wang and Wenthold 2009).

Structurally, all five members of the family are composed of an ectodomain containing six LLRs with NF and CF domains, followed by an Ig module and a membrane-proximal Fn3 module. The ectodomain is followed by a single transmembrane domain and a ~68–234 amino acids long cytoplasmic domain, which in three members of the family, SALM1-3, includes a PDZ-domain-binding motif at the distal C-terminus (Wang et al. 2006).

SALMs have pairwise amino acid sequence identities of 50–60 %. The identities are predominantly restricted to the extracellular regions, whereas the sequences in the C-terminals are highly variable, demonstrating virtually no amino acid sequence identities except for the PDZ-domain-binding motifs, suggesting that the individual SALMs have distinct functions (Nam et al. 2011; Seabold et al. 2012; Mah et al. 2010).

Table 14.10 The human SALM/LRFN family

Gene symbol	Chromosomal location	Synonyms and previous names	Recommended protein name
LRFN1	19q13.2	KIAA1484, synaptic adhesion-like molecule 2, SALM2	Leucine-rich repeat and fibronectin type III domain-containing protein 1
LRFN2	6p21.2-p21.1	KIAA1246, fibronectin type III, immunoglobulin, and leucine-rich repeat domains 2 FIGLER2, Synaptic adhesion-like molecule 1, SALM1	Leucine-rich repeat and fibronectin type-III domain-containing protein 2
LRFN3	19q13.13	Fibronectin type III, immunoglobulin, and leucine-rich repeat domains 1, FIGLER1, MGC2656, synaptic adhesion-like molecule 4 SALM4	Leucine-rich repeat and fibronectin type-III domain-containing protein 3
LRFN4	11q13.1	Fibronectin type III, immunoglobulin and leucine-rich repeat domains 6″, FIGLER6, MGC3103, synaptic adhesion-like molecule 3, SALM3	Leucine-rich repeat and fibronectin type III domain-containing protein 4
LRFN5	14q21.1	C14orf146, fibronectin type III, immunoglobulin, and leucine-rich repeat domains 8, FIGLER8, synaptic adhesion-like molecule 5, SALM5	Leucine-rich repeat and fibronectin type-III domain-containing protein 5

Moreover, the cytoplasmic domains are highly variable in lengths, ranging in human SALMs from ~68 amino acids (SALM4) to ~234 amino acids (SALM1) (UniProtKB accession numbers: Q9ULH4, Q9P244, Q9BTN0, Q6PJG9, and Q96NI6).

A phylogenetic analysis utilizing the N-terminal parts of the SALM LRR domains has shown that SALMs are closely related to NgR, and to members of the SLIT, and AMIGO families. Thus, the amino acid identities between SALM1 and SLIT1-3, AMIGO1-3, and NgR, respectively, are ~24–28 % (Wang et al. 2006).

SALMs contain several potential sites for *N*-glycosylation, and treatment of cells with tunicamycin (an *N*-glycosylation inhibitor) or treatment of rat brain homogenates with N-glycosidase F (cleaving *N*-linked glycosylations) results in a reduction of apparent molecular weight of SALMs, demonstrating that at least some of the potential sites for *N*-glycosylation are glycosylated (Morimura et al. 2006; Mah et al. 2010).

Expression

In embryonic mice a weak expression of *SALM3* and *-4* is detected before E10.5. Around E10.5 an increased expression of SALM2-4 is seen, and a weak expression of *SALM1* and *-5* is observed around E11.5-12.5 (Morimura et al. 2006).

From a whole-mount in situ hybridization study of E10 embryos, *SALM1* has been found to be expressed in the forebrain, midbrain, spinal cord, and DRG. *SALM2-4* are expressed in the same regions as *SALM1* as well as in the hindbrain, nasal placode, optic cup (only *SALM3*), otocyt (only *SALM2* and *-3*), cranial ganglia, branchial arches, limb buds, and somites. Furthermore, *SALM3* is expressed in the heart, and *SALM3* and *-4* are expressed in the liver, gut, and mesonephros. Finally, *SALM5* is expressed in the hindbrain, spinal cord, cranial ganglia, and DRG (Homma et al. 2009).

In adult mouse, northern blot analysis has demonstrated the family to be strongly expressed in the brain, and *SALM1*, *-2*, and *-5* are exclusively detected in the brain. *SALM3* and *-4* are also found in the testis, and *SALM4* is additionally expressed in the gastrointestinal tract and the kidneys. In situ hybridization investigations of adult mouse forebrain have shown that *SALM1* and *-5* are strongly expressed in the pyramidal layer and dentate gyrus of the hippocampus. Furthermore, the family members all exhibit a broad but weak level of expression in the cerebral cortex and diencephalic nuclei (Homma et al. 2009; Morimura et al. 2006).

Immunoblotting analyses have shown that the expression levels of all SALMs gradually increase during the first three weeks of embryonic rat brain development, and SALM2, -3, and -5 are predominantly expressed in the brain (Mah et al. 2010; Ko et al. 2006).

Subcellularly, SALMs are located both intracellularly and at the cell surface, where they generate a punctate staining as a result of clustering. They are expressed on both axons, dendrites, growth cones, and the soma of neurons (Wang et al. 2006, 2008; Ko et al. 2006; Seabold et al. 2008).

Interactions

Members of the SALM family form both homophilic and heterophilic extracellular interactions, and they generate complexes that are probably dimers or larger. Results from studies with SALM-overexpressing cells suggest that all members of the family can be co-precipitated with any other member of the family as a result of *cis-interactions* between their extracellular domains. However, only SALM4 and -5 form *trans*-homophilic interactions. These interactions are Ca^{2+} independent and highly specific. Thus, SALM4 does not interact with SALM5 and vice versa (Seabold et al. 2012).

Reticulon 3 (RTN3) was in a yeast two-hybrid screening identified as a binding partner for SALM1, and in subsequent immunoprecipitation experiments the N-terminal part of RTN3 was found to interact with the LRR domains of SALM1-4 expressed in the brain (Chang et al. 2010). RTN3 is a transmembrane protein abundant in the ER that is involved in the regulation of ER structure and intracellular trafficking (Oertle and Schwab 2003; Dodd et al. 2005; Yang and Strittmatter 2007). Not surprisingly, the interactions have therefore been found not to mediate cell adhesion, and indeed they most likely do not exist extracellularly (Chang et al. 2010).

The extracellular part of SALM1 has also been shown to interact with the NMDA receptor subunit NR1/GluN1, and overexpression of SALM1 in cultured neurons promotes surface expression and clustering of NMDA receptors on dendrites and stimulation of neurite outgrowth. However, overexpression of SALM1, where the PDZ-domain-binding motif is deleted, does not affect NMDA receptor clustering or neurite outgrowth, suggesting that the PDZ-domain-binding motif is also required for the SALM1-mediated clustering of NMDA receptors (Wang et al. 2006).

The cytoplasmic domains of SALM1-3, which all contain PDZ-domain-binding motifs, have by yeast two-hybrid screenings been found to bind PSD-95 (Wang et al. 2006). In the brain, co-immunoprecipitation experiments have demonstrated that SALM1 not only can interact with PSD-95, and thereby promote a subcellular redistribution of PSD-95 (Morimura et al. 2006), but can also interact with the related synaptic proteins SAP102 and SAP97 (Wang et al. 2006). SALM2 also interacts with PSD-95 family proteins, including PSD-95, Chapsyn-110/PSD93, and SAP97, and SALM3 binds PSD-95. These observations have led to the suggestion that the expression of SALM1-3 can affect the function of PSD-95 family proteins (Mah et al. 2010; Ko et al. 2006).

SALM2 has been found to induce clustering of guanylate kinase-associated protein (GKAP)/Disks large-associated protein 1 (DAP-1) (a postsynaptic protein located in excitatory synapses, which also interacts with PSD-95 family proteins (Kim et al. 1997)) and to a lesser extent the AMPA receptor subunit GluR1 and the NMDA receptor subunit NR1. The clustering requires the presence of the PDZ-domain-binding motifs in the C-terminal of SALM2, suggesting that the interactions are indirect and mediated by PSD-95 family proteins (Ko et al. 2006).

Function

The expression of SALMs affects both neurite outgrowth and the formation and maintenance of synapses (Wang et al. 2008).

When overexpressed, SALMs all stimulate neuritogenesis, outgrowth of dendrites and axons, and of neurite branching, but each SALM produces distinct cellular phenotypes. For instance, SALM4 promotes the formation of many primary neurites and induces more neurite branching than other SALMs, whereas the mean process length of SALM4-overexpressing neurons is lower than for neurons overexpressing other SALMs (Wang et al. 2008). Knockdown of the protein Flotilin-1 specifically abrogates SALM4-mediated neurite branching, and it has been suggested that SALM4 works upstream of Flotilin-1, activating signaling pathways that facilitate the localization of Flotilin-1 in lipid raft, from where it can contribute to the branching process (Swanwick et al. 2010).

Overexpression of SALM2 leads to an increase in the formation of dendrite spines and excitatory synapses (Ko et al. 2006), and the SALM1-mediated clustering of NMDA subunits (Wang et al. 2006) also suggests a role for SALM1 in the formation as well as the maintenance of synapses. In contrast, SALM3 and -5 induce presynaptic differentiation, but do not induce the clustering of postsynaptic proteins (like NMDA receptor subunits) in contacting dendrites. Moreover, an overexpression of SALM3 or -5 increases the number of both excitatory and inhibitory presynaptic contacts, suggesting that they have a more general synaptogenic potential than SALM2 (Mah et al. 2010).

Diseases

In one study, a patient with developmental delay, microcephaly, receding forehead, learning disabilities, seizures, and narcolepsy was found to have an apparent *de novo* ~890 kb deletion in chromosome 14 encompassing several exons of *LRFN5* (Mikhail et al. 2011). In another study, a patient with severe autism and mental retardation was found in part to have a balanced t(14;21)(q21.1;p11.2) translocation. This translocation only affected the expression level of *LRFN5*, which in fibroblasts was reduced by 10-fold in the patient when compared to control fibroblasts (de Bruijn et al. 2010). Moreover, SNPs have in genome-wide association studies been identified in *LRFN5* of patients with ASD (Wang et al. 2009b), and investigations of copy number variants have led to the suggestion that *LRFN5* expression might be related to schizophrenia (Xu et al. 2009). Together these studies suggest that a modulated expression or function of SALM5 can have serious effects on the development and function of the nervous system.

RTN3, the protein believed to interact with the extracellular domain of SALM1-4 when the proteins are located in intracellular compartments, belongs to a family of four proteins. All four members of the RTN family have been found to be co-immunoprecipitated with BACE1, and when an RTN3 is overexpressed, the levels of Aβ are decreased, whereas a downregulation of RTN3 has been observed in the temporal lobes of AD patients. Moreover, RTN proteins also affect apoptosis through an interaction with Bcl-2 [see Yang and Strittmatter (2007) for review]. However, whether the interaction between SALM1 and RTN3 affects any of these RTN3-regulated processes has not been investigated.

In summary, the SALM family consists of five members, which all are believed to be able to interact with each other through *cis*-interactions. Moreover, SALM4 and -5 can form *trans*-homophilic interactions. Intracellularly, SALM1-3 bind PSD-95 family proteins and thereby facilitate clustering of, e.g., glutamate receptors. The expression of SALMs affects neuritogenesis and the formation and maintenance of synapses, and alterations in the expression levels can potentially contribute to mental retardation, ASD, or schizophrenia.

14.4.7 *The LRRN/NLRR Family*

The LRR neuronal (LRRN) or neuronal LRR (NLRR) protein family consists of four members: LRRN1-4/NLRR1-4 (Table 14.11). *LRRN1* and *-2* were identified already in 1996 (Taguchi et al. 1996), and later *LRRN1*, *-2*, and *-3* have been cloned

Table 14.11 The human LRRN/NLRR family

Gene symbol	Chromosomal location	Synonyms and previous names	Recommended protein name
LRRN1	3p26.2	Fibronectin type III, immunoglobulin, and leucine-rich repeat domains 3, FIGLER3, KIAA1497, neuronal leucine-rich repeat protein 1	Leucine-rich repeat neuronal protein 1
LRRN2	1q32.1	Fibronectin type III, immunoglobulin, and leucine-rich repeat domain 7, FIGLER7, glioma amplified on chromosome 1 protein, GAC1, leucine-rich and ankyrin repeats 1, LRANK1, leucine-rich repeat neuronal 5, LRRN5	Leucine-rich repeat neuronal protein 2
LRRN3	7q31.1	Fibronectin type III, immunoglobulin, and leucine-rich repeat domains 5, FIGLER5, FLJ11129, neuronal leucine-rich repeat protein 3, NLRR3	Leucine-rich repeat neuronal protein 3
LRRN4	20p12.3	C20orf75, neuronal leucine-rich repeat protein 4, NLRR4	Leucine-rich repeat neuronal protein 4

by different groups (Munfus et al. 2007; Ohira et al. 2003; Almeida et al. 1998; Haines et al. 2005) and identified in several large-scale studies (Clark et al. 2003; Ota et al. 2004; Gerhard et al. 2004; Bechtel et al. 2007; Gregory et al. 2006; Nagase et al. 2000a; Deloukas et al. 2001).

LRRN1, -2, and -3 belong to the Fn3, Ig, and LRR domain (FIGLER) proteins (named FIGLER3, -7, and -5, respectively) (Munfus et al. 2007). However, the last member of the family, *LRRN4*, which was not cloned until 2005 (Bando et al. 2005), has a slightly different structural organization. Consequently, it does not belong to the FIGLER group and should maybe not even be classified as a member of the LRRN family (Bando et al. 2012). The LRRN family has only been the topic of few studies, but the proteins have been related to neuronal development, memory formation, carcinogenesis, and, potentially, ASD.

Genes and Proteins

In contrast to most other genes encoding FIGLER proteins, human *LRRN1*, *-2*, and *-3* are encoded by a single exon, suggesting that they have been incorporated in the genome following reverse transcription (Munfus et al. 2007).

LRRN1 transcription can be activated by the transcription factor N-Myc (Hossain et al. 2008) and the homeobox transcription factor Nkx2.5 (Barth et al. 2010). N-Myc is essential during neurogenesis and is known to be hyperactive in, e.g., neuroblastomas, medulloblastomas, and retinoblastomas (Pession and Tonelli 2005). Nkx2.5 is expressed during heart development (Scott 2012).

LRRN2 has in chicken been shown to be regulated by the transcription factor homeobox B1 (Hoxb1) (Andreae et al. 2009).

Structurally, LRRN proteins are composed of an ectodomain followed by a single transmembrane domain and a cytoplasmic domain. LRRN1, -2, and -3 have an ectodomain composed of 11-12 LRRs within NF and CF domains, one Ig module, and one Fn3 module (Almeida et al. 1998; Haines et al. 2005; Fukamachi et al. 2001). The ectodomain of LRRN4 contains only 10 LRRs and supposedly lacks an Ig module (Bando et al. 2005) (although the number of amino acids between the Fn3 module and the transmembrane domain is approximately the same as for LRRN1-3).

The cytoplasmic domain of LRRN1 and -3, but not LRRN2, contains motifs for regulated Clathrin-mediated endocytosis (Haines et al. 2005; Fukamachi et al. 2001) and has been found to be essential for endocytosis of LRRN3 (Fukamachi et al. 2002).

Expression

In mouse, the expression of *LRRN1* and *-2* has been detected from E11.5 throughout development to adulthood. During development *LRRN1* is expressed in the CNS (forebrain, hindbrain, but not midbrain), DRG, and cartilage, whereas *LRRN2* is restricted to the CNS (Taguchi et al. 1996; Haines et al. 2005).

During mouse development, *LRRN3* is around E10.5 expressed in the trigeminal and facio-acoustic ganglia, in the migrating germ cells in the tail, the DRG, the developing motor horn in the neural tube, the hindbrain, eye, and the olfactory system (Haines et al. 2005). Moreover, *LNRR1* and *-3*, but not *LNRR2*, are during mouse development expressed in bone marrow cells (Munfus et al. 2007).

During adulthood *LRRN1* and *-2* are expressed throughout the brain. *LRRN1* is predominantly expressed in the granular cell layer of the dentate gyrus of the hippocampus, in granule neurons and Purkinje cells in the cerebellum, and in regions of the entorhinal cortex, whereas *LRRN2* predominant is expressed in the pyramidal cell layer of the hippocampus and Purkinje cells in the cerebellum (Taguchi et al. 1996; Haines et al. 2005).

Interestingly, in whole blood samples an increase in *LRRN3* expression is the most significant genetic change associated with smoking, and the gene retains a high level of expression even in nonsmoking previous smokers (Beineke et al. 2012).

Mouse *LRRN4* is expressed mainly in the lungs, heart, and ovaries. In the CNS it is expressed in the CA1 and CA3 regions, the dentate gyrus of the hippocampus, layers V and VI of the neocortex, the piriform cortex, cerebellum, hypothalamus, spinal cord, and DRG (Bando et al. 2005).

Interactions

In *Danio rerio* the ectodomain of Lrrn1 has been found to interact with the Netrin-G ligands Lrrc4a and Lrrc4c (Soellner and Wright 2009). No extracellular interactions have been demonstrated for mammalian versions of LRRN1, -2, or -3, but in a large-scale proteomic screening LNRR4 was reported to form homophilic interactions and to bind dickkopf1 (DKK1), a secreted inhibitor of WNT signaling, and tubulointerstitial nephritis antigen-like 1 (TINAGL1), another secreted protein (Rual et al. 2005).

Intracellularly, LRRN3, and probably also LRRN1, can interact with the Clathrin-adaptor protein β-Adaptin via the endocytosis-regulating sequences mentioned above (Fukamachi et al. 2002).

Functions and Diseases

The expression of LRRN1, -2, and -3 is related to carcinogenesis. Since *LRRN1* expression in part is regulated by N-Myc, it is not surprising that *NLRR1* is highly expressed in neuroblastomas and that the expression levels of *NLRR1* correlate with a poor clinical prognosis. However, in addition increased levels of NLRR1 seem to promote serum-independent cell proliferation, whereas decreased levels of NLRR1 reduce proliferation (Hossain et al. 2008). NLRR1 is therefore probably not just a marker of malignant neuroblastomas, but contributes directly to the cancer pathogenesis. *LRRN1* expression can together with the expression levels from 18 other genes also be used for the detection of, and discrimination between, different types of primary lung tumors. However, surprisingly *LRRN1* expression is generally downregulated in these tumors (Dmitriev et al. 2012).

LRRN2 was originally called Glioblastoma amplification on chromosome 1 (GAC1) and is, as the name implies, sometimes overexpressed in gliomas (Almeida et al. 1998; Arjona et al. 2005).

LRRN3 is differentially expressed in pheochromocytomas, and the expression of *LRRN3* together with the expression of four other genes has been suggested as a diagnostic marker enabling the identification of benign versus malignant pheochromocytomas (Suh et al. 2009). In COS-7 cells overexpression of LRRN3 increases the sensitivity of the cells to EGF, leading to EGF-dependent sustained activation of the Ras-MAPK pathway in a manner independent of EGF receptor phosphorylation. The process is dependent on the cytoplasmic domain of LRRN3 and has been suggested to be a result of LRRN3-mediated internalization of EGF (Fukamachi et al. 2002).

Consistent with their expression in the CNS, LRRN1 and -2 affect neuronal development. In rat hippocampal neurons LRRN1 has been found to act as a negative regulator of neuritogenesis (Buchser et al. 2010), and in chicken abnormal expression of LRRN2 can result in axonal rerouting, suggesting that the protein is involved in axon guidance (Andreae et al. 2007).

LRRN1 is a candidate gene for ASD. Thus, an analysis of copy number variations in children from the Autism Genetic Resource Exchange (AGRE) revealed a duplication of a part of chromosome 3 including *LRRN1*. However, the breakpoint was inside *LRRN1*, suggesting that the duplication may in fact lead to an LRRN1 loss of function (Davis et al. 2009).

LRRN4-knockout mice develop normally, are fertile, and demonstrate no increased mortality. However, the animals suffer several cognitive defects, including defects in contextual fear conditioning (electrical shock) and spatial memory (hidden-platform Morris water maze test), tasks known to involve the hippocampus. In contrast, LRRN4 deficiency has no effect on auditory cued fear conditioning (a tone combined with electrical shock), a task known to be hippocampus independent. Consistent with the fact that LRRN4 is expressed in the hippocampus, these observations suggest that LRRN4 is important for hippocampus-dependent memory formation (Bando et al. 2005). Moreover, in DRG neurons of wild-type mice, approximately 42 % and 58 % of the LRRN4-expressing cells are TrkA-positive, peptidergic nociceptors, and Ret-positive nonpeptidergic nociceptors, respectively. After unilateral sciatic nerve axotomy, the expression of LRRN4 is significantly reduced in the injured side, leading to the suggestion that LRRN4 is involved in the maintenance of nociceptive circuits (Bando et al. 2012).

In summary, the LRRN family consists of four members. The extracellular and intracellular interactions mediated by the proteins are not well characterized, but the expression of the individual members has been related to carcinogenesis, neuritogenesis, axon guidance, and memory formation, and at least one member of the family might be related to the autism.

14.5 Conclusion

The ten protein families described above all contain members that are CAMs or potential CAMs with extracellular LRR domains, which in part or exclusively are expressed in the nervous system. Another common characteristic for many of the described proteins is that they were identified recently; most members of the AMIGO, FLRT, LINGO, LRIT, LRRC15, LRRTM, NGL, SALM, and SLITRK were cloned between 1999 and 2003 (Lauren et al. 2003; Aruga and Mikoshiba 2003; Satoh et al. 2002; Kuja-Panula et al. 2003; Ono et al. 2003; Lin et al. 2003; Carim-Todd et al. 2003; Lacy et al. 1999; Gomi et al. 2000; Kikuno et al. 1999), and whereas some members of the LRRN family were identified already in 1996 (Taguchi et al. 1996), some members of the LRIT family were not cloned until 2009 (Homma et al. 2009). Consequently, many of the proteins are not well characterized, and the information regarding their functions and characteristics will in the following years most likely increase considerably. However, already now it is clear that many LRR-containing CAMs play pivotal roles in the development and maintenance of the nervous system and that they are implicated in numerous functions in the nervous system as well as in the pathogenesis of several diseases ranging from neurodegenerative diseases like AD, PD, and MS to ASD, schizophrenia, and obsessive–compulsive disorders. The fact that the LRR-containing CAMs within ~10 years of their cloning have been demonstrated to be involved in so many biological processes and diseases suggests that the list of functions is far from complete and that further research in the field will reveal additional important functions of this interesting group of proteins.

Compliance with Ethics Requirements The authors declare that they have no conflicts of interest.

References

Abelson JF, Kwan KY, O'Roak BJ, Baek DY, Stillman AA, Morgan TM, Mathews CA, Pauls DL, Rasin MR, Gunel M, Davis NR, Ercan-Sencicek AG, Guez DH, Spertus JA, Leckman JF, Dure LS IV, Kurlan R, Singer HS, Gilbert DL, Farhi A, Louvi A, Lifton RP, Sestan N, State MW (2005) Sequence variants in SLITRK1 are associated with Tourette's syndrome. Science 310(5746):317–320. doi:10.1126/science.1116502

Adriano F, Caltagirone C, Spalletta G (2012) Hippocampal volume reduction in first-episode and chronic schizophrenia: a review and meta-analysis. Neuroscientist 18(2):180–200. doi:10.1177/1073858410395147

Ahmad W, Ijaz B, Hassan S (2012) Gene expression profiling of HCV genotype 3a initial liver fibrosis and cirrhosis patients using microarray. J Transl Med 10:41. doi:10.1186/1479-5876-10-41

Almeida A, Zhu XX, Vogt N, Tyagi R, Muleris M, Dutrillaux AM, Dutrillaux B, Ross D, Malfoy B, Hanash S (1998) GAC1, a new member of the leucine-rich repeat superfamily on chromosome band 1q32.1, is amplified and overexpressed in malignant gliomas. Oncogene 16(23): 2997–3002. doi:10.1038/sj.onc.1201828

Ameye L, Young MF (2002) Mice deficient in small leucine-rich proteoglycans: novel in vivo models for osteoporosis, osteoarthritis, Ehlers-Danlos syndrome, muscular dystrophy, and corneal diseases. Glycobiology 12(9):107R–116R

Andreae LC, Peukert D, Lumsden A, Gilthorpe JD (2007) Analysis of Lrrn1 expression and its relationship to neuromeric boundaries during chick neural development. Neural Dev 2:22. doi:10.1186/1749-8104-2-22

Andreae LC, Lumsden A, Gilthorpe JD (2009) Chick Lrrn2, a novel downstream effector of Hoxb1 and Shh, functions in the selective targeting of rhombomere 4 motor neurons. Neural Dev 4:27. doi:10.1186/1749-8104-4-27

Anney R, Klei L, Pinto D, Regan R, Conroy J, Magalhaes TR, Correia C, Abrahams BS, Sykes N, Pagnamenta AT, Almeida J, Bacchelli E, Bailey AJ, Baird G, Battaglia A, Berney T, Bolshakova N, Bolte S, Bolton PF, Bourgeron T, Brennan S, Brian J, Carson AR, Casallo G, Casey J, Chu SH, Cochrane L, Corsello C, Crawford EL, Crossett A, Dawson G, de Jonge M, Delorme R, Drmic I, Duketis E, Duque F, Estes A, Farrar P, Fernandez BA, Folstein SE, Fombonne E, Freitag CM, Gilbert J, Gillberg C, Glessner JT, Goldberg J, Green J, Guter SJ, Hakonarson H, Heron EA, Hill M, Holt R, Howe JL, Hughes G, Hus V, Igliozzi R, Kim C, Klauck SM, Kolevzon A, Korvatska O, Kustanovich V, Lajonchere CM, Lamb JA, Laskawiec M, Leboyer M, Le Couteur A, Leventhal BL, Lionel AC, Liu XQ, Lord C, Lotspeich L, Lund SC, Maestrini E, Mahoney W, Mantoulan C, Marshall CR, McConachie H, McDougle CJ, McGrath J, McMahon WM, Melhem NM, Merikangas A, Migita O, Minshew NJ, Mirza GK, Munson J, Nelson SF, Noakes C, Noor A, Nygren G, Oliveira G, Papanikolaou K, Parr JR, Parrini B, Paton T, Pickles A, Piven J, Posey DJ, Poustka A, Poustka F, Prasad A, Ragoussis J, Renshaw K, Rickaby J, Roberts W, Roeder K, Roge B, Rutter ML, Bierut LJ, Rice JP, Salt J, Sansom K, Sato D, Segurado R, Senman L, Shah N, Sheffield VC, Soorya L, Sousa I, Stoppioni V, Strawbridge C, Tancredi R, Tansey K, Thiruvahindrapduram B, Thompson AP, Thomson S, Tryfon A, Tsiantis J, Van Engeland H, Vincent JB, Volkmar F, Wallace S, Wang K, Wang Z, Wassink TH, Wing K, Wittemeyer K, Wood S, Yaspan BL, Zurawiecki D, Zwaigenbaum L, Betancur C, Buxbaum JD, Cantor RM, Cook EH, Coon H, Cuccaro ML, Gallagher L, Geschwind DH, Gill M, Haines JL, Miller J, Monaco AP, Nurnberger JI Jr, Paterson AD, Pericak-Vance MA, Schellenberg GD, Scherer SW, Sutcliffe JS, Szatmari P, Vicente AM, Vieland VJ, Wijsman EM, Devlin B, Ennis S, Hallmayer J (2010) A genome-wide scan for common alleles affecting risk for autism. Hum Mol Genet 19(20):4072–4082. doi:10.1093/hmg/ddq307

Arjona D, Bello MJ, Alonso ME, Isla A, De Campos JM, Vaquero J, Sarasa JL, Gutierrez M, Rey JA (2005) Real-time quantitative PCR analysis of regions involved in gene amplification reveals gene overdose in low-grade astrocytic gliomas. Diagn Mol Pathol 14(4):224–229

Aruga J (2003) Slitrk6 expression profile in the mouse embryo and its relationship to that of Nlrr3. Gene Expr Patterns 3(6):727–733

Aruga J, Mikoshiba K (2003) Identification and characterization of Slitrk, a novel neuronal transmembrane protein family controlling neurite outgrowth. Mol Cell Neurosci 24(1):117–129

Aruga J, Yokota N, Mikoshiba K (2003) Human SLITRK family genes: genomic organization and expression profiling in normal brain and brain tumor tissue. Gene 315:87–94

Bai Y, Markham K, Chen F, Weerasekera R, Watts J, Horne P, Wakutani Y, Bagshaw R, Mathews PM, Fraser PE, Westaway D, St George-Hyslop P, Schmitt-Ulms G (2008) The in vivo brain interactome of the amyloid precursor protein. Mol Cell Proteomics 7(1):15–34. doi:10.1074/mcp.M700077-MCP200

Bando T, Sekine K, Kobayashi S, Watabe AM, Rump A, Tanaka M, Suda Y, Kato S, Morikawa Y, Manabe T, Miyajima A (2005) Neuronal leucine-rich repeat protein 4 functions in hippocampus-dependent long-lasting memory. Mol Cell Biol 25(10):4166–4175. doi:10.1128/mcb.25.10.4166-4175.2005

Bando T, Morikawa Y, Hisaoka T, Komori T, Miyajima A, Senba E (2012) Expression pattern of leucine-rich repeat neuronal protein 4 in adult mouse dorsal root ganglia. Neurosci Lett 531(1):24–29. doi:10.1016/j.neulet.2012.10.009

Bandtlow C, Dechant G (2004) From cell death to neuronal regeneration, effects of the p75 neurotrophin receptor depend on interactions with partner subunits. Sci STKE 2004(235):pe24. doi:10.1126/stke.2352004pe24

Barker N, Clevers H (2010) Leucine-rich repeat-containing G-protein-coupled receptors as markers of adult stem cells. Gastroenterology 138(5):1681–1696. doi:10.1053/j.gastro.2010.03.002

Barth JL, Clark CD, Fresco VM, Knoll EP, Lee B, Argraves WS, Lee KH (2010) Jarid2 is among a set of genes differentially regulated by Nkx2.5 during outflow tract morphogenesis. Dev Dyn 239(7):2024–2033. doi:10.1002/dvdy.22341

Beaubien F, Cloutier JF (2009) Differential expression of Slitrk family members in the mouse nervous system. Dev Dyn 238(12):3285–3296. doi:10.1002/dvdy.22160

Bech-Hansen NT, Naylor MJ, Maybaum TA, Sparkes RL, Koop B, Birch DG, Bergen AA, Prinsen CF, Polomeno RC, Gal A, Drack AV, Musarella MA, Jacobson SG, Young RS, Weleber RG (2000) Mutations in NYX, encoding the leucine-rich proteoglycan nyctalopin, cause X-linked complete congenital stationary night blindness. Nat Genet 26(3):319–323. doi:10.1038/81619

Bechtel S, Rosenfelder H, Duda A, Schmidt CP, Ernst U, Wellenreuther R, Mehrle A, Schuster C, Bahr A, Blocker H, Heubner D, Hoerlein A, Michel G, Wedler H, Kohrer K, Ottenwalder B, Poustka A, Wiemann S, Schupp I (2007) The full-ORF clone resource of the German cDNA Consortium. BMC Genomics 8:399. doi:10.1186/1471-2164-8-399

Beineke P, Fitch K, Tao H, Elashoff MR, Rosenberg S, Kraus WE, Wingrove JA (2012) A whole blood gene expression-based signature for smoking status. BMC Med Genomics 5:58. doi:10.1186/1755-8794-5-58

Bella J, Hindle K, McEwan P, Lovell S (2008) The leucine-rich repeat structure. Cell Mol Life Sci 65(15):2307–2333, doi:10.1007/s00018-008-8019-0

Berg A, Zelano J, Cullheim S (2010) Netrin G-2 ligand mRNA is downregulated in spinal motoneurons after sciatic nerve lesion. Neuroreport 21(11):782–785. doi:10.1097/WNR.0b013e32833cadd8

Bessero AC, Clarke PG (2010) Neuroprotection for optic nerve disorders. Curr Opin Neurol 23(1):10–15. doi:10.1097/WCO.0b013e3283344461

Bierkens M, Krijgsman O, Wilting SM, Bosch L, Jaspers A, Meijer GA, Meijer CJLM, Snijders PJF, Ylstra B, Steenbergen RDM (2013) Focal aberrations indicate EYA2 and hsa-miR-375 as oncogene and tumor suppressor in cervical carcinogenesis. Genes Chromosomes Cancer 52(1):56–68. doi:10.1002/gcc.22006

Bierne H, Sabet C, Personnic N, Cossart P (2007) Internalins: a complex family of leucine-rich repeat-containing proteins in Listeria monocytogenes. Microbes Infect 9(10):1156–1166. doi:10.1016/j.micinf.2007.05.003

Blumcke I, Vinters HV, Armstrong D, Aronica E, Thom M, Spreafico R (2009) Malformations of cortical development and epilepsies: neuropathological findings with emphasis on focal cortical dysplasia. Epileptic Disord 11(3):181–193. doi:10.1684/epd.2009.0261

Bonardi V, Cherkis K, Nishimura MT, Dangl JL (2012) A new eye on NLR proteins: focused on clarity or diffused by complexity? Curr Opin Immunol 24(1):41–50. doi:10.1016/j.coi.2011.12.006

Botos I, Segal David M, Davies David R (2011) The structural biology of toll-like receptors. Structure 19(4):447–459. doi:10.1016/j.str.2011.02.004

Bottcher RT, Pollet N, Delius H, Niehrs C (2004) The transmembrane protein XFLRT3 forms a complex with FGF receptors and promotes FGF signalling. Nat Cell Biol 6(1):38–44

Bottos A, Rissone A, Bussolino F, Arese M (2011) Neurexins and neuroligins: synapses look out of the nervous system. Cell Mol Life Sci 68(16):2655–2666, doi:10.1007/s00018-011-0664-z

Brasch J, Harrison OJ, Ahlsen G, Liu Q, Shapiro L (2011) Crystal structure of the ligand binding domain of netrin G2. J Mol Biol 414(5):723–734. doi:10.1016/j.jmb.2011.10.030

Brognard J, Newton AC (2008) PHLiPPing the switch on Akt and protein kinase C signaling. Trends Endocrinol Metab 19(6):223–230. doi:10.1016/j.tem.2008.04.001

Buchser WJ, Slepak TI, Gutierrez-Arenas O, Bixby JL, Lemmon VP (2010) Kinase/phosphatase overexpression reveals pathways regulating hippocampal neuron morphology. Mol Syst Biol 6:391. doi:10.1038/msb.2010.52

Bustos D, Bailey M, Sugden D, Carter D, Rath M, Møller M, Coon S, Weller J, Klein D (2011) Global daily dynamics of the pineal transcriptome. Cell Tissue Res 344(1):1–11. doi:10.1007/s00441-010-1094-1

Cai M, Han L, Chen R, Ye F, Wang B, Han F, Lei T, Guo D (2009a) Inhibition of LRIG3 gene expression via RNA interference modulates the proliferation, cell cycle, cell apoptosis, adhesion and invasion of glioblastoma cell (GL15). Cancer Lett 278(1):104–112. doi:10.1016/j.canlet.2009.01.001

Cai MJ, Xie RF, Han L, Chen RD, Wang BF, Ye F, Guo DS, Lei T (2009b) Effect of RNAi-mediated LRIG3 gene silencing on proliferation of glioma GL15 cells and expression of PCNA and Ki-67. Ai Zheng 28(1):1–4

Carim-Todd L, Escarceller M, Estivill X, Sumoy L (2003) LRRN6A/LERN1 (leucine-rich repeat neuronal protein 1), a novel gene with enriched expression in limbic system and neocortex. Eur J Neurosci 18(12):3167–3182. doi:10.1111/j.1460-9568.2003.03003.x

Chang K, Seabold GK, Wang CY, Wenthold RJ (2010) Reticulon 3 is an interacting partner of the SALM family of adhesion molecules. J Neurosci Res 88(2):266–274. doi:10.1002/jnr.22209

Chen Y, Aulia S, Li L, Tang BL (2006) AMIGO and friends: an emerging family of brain-enriched, neuronal growth modulating, type I transmembrane proteins with leucine-rich repeats (LRR) and cell adhesion molecule motifs. Brain Res Rev 51(2):265–274

Chen R, Jiang X, Sun D, Han G, Wang F, Ye M, Wang L, Zou H (2009) Glycoproteomics analysis of human liver tissue by combination of multiple enzyme digestion and hydrazide chemistry. J Proteome Res 8(2):651–661. doi:10.1021/pr8008012

Chen Y, Hor HH, Tang BL (2012) AMIGO is expressed in multiple brain cell types and may regulate dendritic growth and neuronal survival. J Cell Physiol 227(5):2217–2229. doi:10.1002/jcp.22958

Choi Y-J (2009) Efficacy of treatments for patients with obsessive-compulsive disorder: A systematic review. J Am Acad Nurse Pract 21(4):207–213. doi:10.1111/j.1745-7599.2009.00408.x

Clark HF, Gurney AL, Abaya E, Baker K, Baldwin D, Brush J, Chen J, Chow B, Chui C, Crowley C, Currell B, Deuel B, Dowd P, Eaton D, Foster J, Grimaldi C, Gu Q, Hass PE, Heldens S, Huang A, Kim HS, Klimowski L, Jin Y, Johnson S, Lee J, Lewis L, Liao D, Mark M, Robbie E, Sanchez C, Schoenfeld J, Seshagiri S, Simmons L, Singh J, Smith V, Stinson J, Vagts A, Vandlen R, Watanabe C, Wieand D, Woods K, Xie MH, Yansura D, Yi S, Yu G, Yuan J, Zhang M, Zhang Z, Goddard A, Wood WI, Godowski P, Gray A (2003) The secreted protein discovery initiative (SPDI), a large-scale effort to identify novel human secreted and transmembrane proteins: a bioinformatics assessment. Genome Res 13(10):2265–2270. doi:10.1101/gr.1293003

Colafrancesco V, Parisi V, Sposato V, Rossi S, Russo MA, Coassin M, Lambiase A, Aloe L (2011) Ocular application of nerve growth factor protects degenerating retinal ganglion cells in a rat model of glaucoma. J Glaucoma 20(2):100–108. doi:10.1097/IJG.0b013e3181d787e5
Connor RM, Allen CL, Devine CA, Claxton C, Key B (2005) BOC, brother of CDO, is a dorsoventral axon-guidance molecule in the embryonic vertebrate brain. J Comp Neurol 485(1):32–42. doi:10.1002/cne.20503
Cooper PR, McLachlan JL, Simon S, Graham LW, Smith AJ (2011) Mediators of inflammation and regeneration. Adv Dent Res 23(3):290–295. doi:10.1177/0022034511405389
Costantini C, Della-Bianca V, Formaggio E, Chiamulera C, Montresor A, Rossi F (2005) The expression of p75 neurotrophin receptor protects against the neurotoxicity of soluble oligomers of β-amyloid. Exp Cell Res 311(1):126–134. doi:10.1016/j.yexcr.2005.09.004
Craig AM, Kang Y (2007) Neurexin–neuroligin signaling in synapse development. Curr Opin Neurobiol 17(1):43–52. doi:10.1016/j.conb.2007.01.011
Crow TJ, Close JP, Dagnall AM, Priddle TH (2009) Where and what is the right shift factor or cerebral dominance gene? A critique of Francks et al. (2007). Laterality 14(1):3–10. doi:10.1080/13576500802574984
Croxtall J (2011) Clonidine extended-release. Pediatr Drugs 13(5):329–336. doi:10.2165/11208100-000000000-00000
Davidson I, Kobi D, Fadloun A, Mengus G (2005) New insights into TAFs as regulators of cell cycle and signalling pathways. Cell Cycle 4(11):1486–1490
Davies MN, Volta M, Pidsley R, Lunnon K, Dixit A, Lovestone S, Coarfa C, Harris RA, Milosavljevic A, Troakes C, Al-Sarraj S, Dobson R, Schalkwyk LC, Mill J (2012) Functional annotation of the human brain methylome identifies tissue-specific epigenetic variation across brain and blood. Genome Biol 13(6):R43. doi:10.1186/gb-2012-13-6-r43
Davis LK, Meyer KJ, Rudd DS, Librant AL, Epping EA, Sheffield VC, Wassink TH (2009) Novel copy number variants in children with autism and additional developmental anomalies. J Neurodev Disord 1(4):292–301. doi:10.1007/s11689-009-9013-z
de Bruijn DR, van Dijk AH, Pfundt R, Hoischen A, Merkx GF, Gradek GA, Lybaek H, Stray-Pedersen A, Brunner HG, Houge G (2010) Severe progressive autism associated with two de novo changes: a 2.6-Mb 2q31.1 deletion and a balanced t(14;21)(q21.1;p11.2) translocation with long-range epigenetic silencing of LRFN5 expression. Mol Syndromol 1(1):46–57. doi:10.1159/000280290
de Wit J, Sylwestrak E, O'Sullivan ML, Otto S, Tiglio K, Savas JN, Yates JR III, Comoletti D, Taylor P, Ghosh A (2009) LRRTM2 interacts with Neurexin1 and regulates excitatory synapse formation. Neuron 64(6):799–806. doi:10.1016/j.neuron.2009.12.019
Dellett M, Hu W, Papadaki V, S-i O (2012) Small leucine rich proteoglycan family regulates multiple signalling pathways in neural development and maintenance. Dev Growth Differ 54(3):327–340. doi:10.1111/j.1440-169X.2012.01339.x
Deloukas P, Matthews LH, Ashurst J, Burton J, Gilbert JG, Jones M, Stavrides G, Almeida JP, Babbage AK, Bagguley CL, Bailey J, Barlow KF, Bates KN, Beard LM, Beare DM, Beasley OP, Bird CP, Blakey SE, Bridgeman AM, Brown AJ, Buck D, Burrill W, Butler AP, Carder C, Carter NP, Chapman JC, Clamp M, Clark G, Clark LN, Clark SY, Clee CM, Clegg S, Cobley VE, Collier RE, Connor R, Corby NR, Coulson A, Coville GJ, Deadman R, Dhami P, Dunn M, Ellington AG, Frankland JA, Fraser A, French L, Garner P, Grafham DV, Griffiths C, Griffiths MN, Gwilliam R, Hall RE, Hammond S, Harley JL, Heath PD, Ho S, Holden JL, Howden PJ, Huckle E, Hunt AR, Hunt SE, Jekosch K, Johnson CM, Johnson D, Kay MP, Kimberley AM, King A, Knights A, Laird GK, Lawlor S, Lehvaslaiho MH, Leversha M, Lloyd C, Lloyd DM, Lovell JD, Marsh VL, Martin SL, McConnachie LJ, McLay K, McMurray AA, Milne S, Mistry D, Moore MJ, Mullikin JC, Nickerson T, Oliver K, Parker A, Patel R, Pearce TA, Peck AI, Phillimore BJ, Prathalingam SR, Plumb RW, Ramsay H, Rice CM, Ross MT, Scott CE, Sehra HK, Shownkeen R, Sims S, Skuce CD, Smith ML, Soderlund C, Steward CA, Sulston JE, Swann M, Sycamore N, Taylor R, Tee L, Thomas DW, Thorpe A, Tracey A, Tromans AC, Vaudin M, Wall M, Wallis JM, Whitehead SL, Whittaker P, Willey DL, Williams L, Williams

SA, Wilming L, Wray PW, Hubbard T, Durbin RM, Bentley DR, Beck S, Rogers J (2001) The DNA sequence and comparative analysis of human chromosome 20. Nature 414(6866): 865–871, http://www.nature.com/nature/journal/v414/n6866/suppinfo/414865a_S1.html

Deloukas P, Earthrowl ME, Grafham DV, Rubenfield M, French L, Steward CA, Sims SK, Jones MC, Searle S, Scott C, Howe K, Hunt SE, Andrews TD, Gilbert JG, Swarbreck D, Ashurst JL, Taylor A, Battles J, Bird CP, Ainscough R, Almeida JP, Ashwell RI, Ambrose KD, Babbage AK, Bagguley CL, Bailey J, Banerjee R, Bates K, Beasley H, Bray-Allen S, Brown AJ, Brown JY, Burford DC, Burrill W, Burton J, Cahill P, Camire D, Carter NP, Chapman JC, Clark SY, Clarke G, Clee CM, Clegg S, Corby N, Coulson A, Dhami P, Dutta I, Dunn M, Faulkner L, Frankish A, Frankland JA, Garner P, Garnett J, Gribble S, Griffiths C, Grocock R, Gustafson E, Hammond S, Harley JL, Hart E, Heath PD, Ho TP, Hopkins B, Horne J, Howden PJ, Huckle E, Hynds C, Johnson C, Johnson D, Kana A, Kay M, Kimberley AM, Kershaw JK, Kokkinaki M, Laird GK, Lawlor S, Lee HM, Leongamornlert DA, Laird G, Lloyd C, Lloyd DM, Loveland J, Lovell J, McLaren S, McLay KE, McMurray A, Mashreghi-Mohammadi M, Matthews L, Milne S, Nickerson T, Nguyen M, Overton-Larty E, Palmer SA, Pearce AV, Peck AI, Pelan S, Phillimore B, Porter K, Rice CM, Rogosin A, Ross MT, Sarafidou T, Sehra HK, Shownkeen R, Skuce CD, Smith M, Standring L, Sycamore N, Tester J, Thorpe A, Torcasso W, Tracey A, Tromans A, Tsolas J, Wall M, Walsh J, Wang H, Weinstock K, West AP, Willey DL, Whitehead SL, Wilming L, Wray PW, Young L, Chen Y, Lovering RC, Moschonas NK, Siebert R, Fechtel K, Bentley D, Durbin R, Hubbard T, Doucette-Stamm L, Beck S, Smith DR, Rogers J (2004) The DNA sequence and comparative analysis of human chromosome 10. Nature 429(6990):375–381. doi:10.1038/nature02462

Delprat B, Michel V, Goodyear R, Yamasaki Y, Michalski N, El-Amraoui A, Perfettini I, Legrain P, Richardson G, Hardelin JP, Petit C (2005) Myosin XVa and whirlin, two deafness gene products required for hair bundle growth, are located at the stereocilia tips and interact directly. Hum Mol Genet 14(3):401–410. doi:10.1093/hmg/ddi036

DeNardo LA, de Wit J, Otto-Hitt S, Ghosh A (2012) NGL-2 regulates input-specific synapse development in CA1 pyramidal neurons. Neuron 76(4):762–775. doi:10.1016/j.neuron.2012.10.013

Deng H, Le WD, Xie WJ, Jankovic J (2006) Examination of the SLITRK1 gene in Caucasian patients with Tourette syndrome. Acta Neurol Scand 114(6):400–402. doi:10.1111/j.1600-0404.2006.00706.x

Deng H, Gu S, Jankovic J (2012) LINGO1 variants in essential tremor and Parkinson's disease. Acta Neurol Scand 125(1):1–7. doi:10.1111/j.1600-0404.2011.01516.x

Dievart A, Gilbert N, Droc G, Attard A, Gourgues M, Guiderdoni E, Perin C (2011) Leucine-Rich repeat receptor kinases are sporadically distributed in eukaryotic genomes. BMC Evol Biol 11(1):367

Dmitriev AA, Kashuba VI, Haraldson K, Senchenko VN, Pavlova TV, Kudryavtseva AV, Anedchenko EA, Krasnov GS, Pronina IV, Loginov VI, Kondratieva TT, Kazubskaya TP, Braga EA, Yenamandra SP, Ignatjev I, Ernberg I, Klein G, Lerman MI, Zabarovsky ER (2012) Genetic and epigenetic analysis of non-small cell lung cancer with NotI-microarrays. Epigenetics 7(5):502–513. doi:10.4161/epi.19801

Dodd DA, Niederoest B, Bloechlinger S, Dupuis L, Loeffler JP, Schwab ME (2005) Nogo-A, -B, and -C are found on the cell surface and interact together in many different cell types. J Biol Chem 280(13):12494–12502. doi:10.1074/jbc.M411827200

Dolan J, Walshe K, Alsbury S, Hokamp K, O'Keeffe S, Okafuji T, Miller SF, Tear G, Mitchell KJ (2007) The extracellular leucine-rich repeat superfamily; a comparative survey and analysis of evolutionary relationships and expression patterns. BMC Genomics 8:320

Domeniconi M, Cao Z, Spencer T, Sivasankaran R, Wang K, Nikulina E, Kimura N, Cai H, Deng K, Gao Y, He Z, Filbin M (2002) Myelin-associated glycoprotein interacts with the Nogo66 receptor to inhibit neurite outgrowth. Neuron 35(2):283–290

Dosemeci A, Makusky AJ, Jankowska-Stephens E, Yang X, Slotta DJ, Markey SP (2007) Composition of the synaptic PSD-95 complex. Mol Cell Proteomics 6(10):1749–1760. doi:10.1074/mcp.M700040-MCP200

Dragovic M, Hammond G (2005) Handedness in schizophrenia: a quantitative review of evidence. Acta Psychiatr Scand 111(6):410–419. doi:10.1111/j.1600-0447.2005.00519.x

Edwards TL, Pericak-Vance M, Gilbert JR, Haines JL, Martin ER, Ritchie MD (2009) An association analysis of Alzheimer disease candidate genes detects an ancestral risk haplotype clade in ACE and putative multilocus association between ACE, A2M, and LRRTM3. Am J Med Genet B Neuropsychiatr Genet 150B(5):721–735. doi:10.1002/ajmg.b.30899

Egea J, Erlacher C, Montanez E, Burtscher I, Yamagishi S, Hess M, Hampel F, Sanchez R, Rodriguez-Manzaneque MT, Bosl MR, Fassler R, Lickert H, Klein R (2008) Genetic ablation of FLRT3 reveals a novel morphogenetic function for the anterior visceral endoderm in suppressing mesoderm differentiation. Genes Dev 22(23):3349–3362

Fabbrini G, Pasquini M, Aurilia C, Berardelli I, Breedveld G, Oostra BA, Bonifati V, Berardelli A (2007) A large Italian family with Gilles de la Tourette syndrome: clinical study and analysis of the SLITRK1 gene. Mov Disord 22(15):2229–2234. doi:10.1002/mds.21697

Fadloun A, Kobi D, Pointud J-C, Indra AK, Teletin M, Bole-Feysot C, Testoni B, Mantovani R, Metzger D, Mengus G, Davidson I (2007) The TFIID subunit TAF4 regulates keratinocyte proliferation and has cell-autonomous and non-cell-autonomous tumour suppressor activity in mouse epidermis. Development 134(16):2947–2958. doi:10.1242/dev.005041

Fancy SP, Baranzini SE, Zhao C, Yuk DI, Irvine KA, Kaing S, Sanai N, Franklin RJ, Rowitch DH (2009) Dysregulation of the Wnt pathway inhibits timely myelination and remyelination in the mammalian CNS. Genes Dev 23(13):1571–1585. doi:10.1101/gad.1806309

Fournier AE, GrandPre T, Strittmatter SM (2001) Identification of a receptor mediating Nogo-66 inhibition of axonal regeneration. Nature 409(6818):341–346. doi:10.1038/35053072

Francks C (2009) Understanding the genetics of behavioural and psychiatric traits will only be achieved through a realistic assessment of their complexity. Laterality 14(1):11–16. doi:10.1080/13576500802536439

Francks C, Maegawa S, Lauren J, Abrahams BS, Velayos-Baeza A, Medland SE, Colella S, Groszer M, McAuley EZ, Caffrey TM, Timmusk T, Pruunsild P, Koppel I, Lind PA, Matsumoto-Itaba N, Nicod J, Xiong L, Joober R, Enard W, Krinsky B, Nanba E, Richardson AJ, Riley BP, Martin NG, Strittmatter SM, Moller HJ, Rujescu D, St Clair D, Muglia P, Roos JL, Fisher SE, Wade-Martins R, Rouleau GA, Stein JF, Karayiorgou M, Geschwind DH, Ragoussis J, Kendler KS, Airaksinen MS, Oshimura M, DeLisi LE, Monaco AP (2007) LRRTM1 on chromosome 2p12 is a maternally suppressed gene that is associated paternally with handedness and schizophrenia. Mol Psychiatry 12(12):1129–1139, 1057. doi:10.1038/sj.mp.4002053

Fu QL, Hu B, Wu W, Pepinsky RB, Mi S, So KF (2008) Blocking LINGO-1 function promotes retinal ganglion cell survival following ocular hypertension and optic nerve transection. Invest Ophthalmol Vis Sci 49(3):975–985. doi:10.1167/iovs.07-1199

Fu QL, Li X, Yip HK, Shao Z, Wu W, Mi S, So KF (2009) Combined effect of brain-derived neurotrophic factor and LINGO-1 fusion protein on long-term survival of retinal ganglion cells in chronic glaucoma. Neuroscience 162(2):375–382. doi:10.1016/j.neuroscience.2009.04.075

Fu QL, Hu B, Li X, Shao Z, Shi JB, Wu W, So KF, Mi S (2010) LINGO-1 negatively regulates TrkB phosphorylation after ocular hypertension. Eur J Neurosci 31(6):1091–1097. doi:10.1111/j.1460-9568.2010.07127.x

Fukamachi K, Matsuoka Y, Kitanaka C, Kuchino Y, Tsuda H (2001) Rat neuronal leucine-rich repeat protein-3: cloning and regulation of the gene expression. Biochem Biophys Res Commun 287(1):257–263. doi:10.1006/bbrc.2001.5579

Fukamachi K, Matsuoka Y, Ohno H, Hamaguchi T, Tsuda H (2002) Neuronal leucine-rich repeat protein-3 amplifies MAPK activation by epidermal growth factor through a carboxyl-terminal region containing endocytosis motifs. J Biol Chem 277(46):43549–43552. doi:10.1074/jbc.C200502200

Fukata Y, Adesnik H, Iwanaga T, Bredt DS, Nicoll RA, Fukata M (2006) Epilepsy-related ligand/receptor complex LGI1 and ADAM22 regulate synaptic transmission. Science 313(5794):1792–1795. doi:10.1126/science.1129947

Gallwitz M, Enoksson M, Thorpe M, Hellman L (2012) The extended cleavage specificity of human thrombin. PLoS One 7(2):e31756. doi:10.1371/journal.pone.0031756

Gerhard DS, Wagner L, Feingold EA, Shenmen CM, Grouse LH, Schuler G, Klein SL, Old S, Rasooly R, Good P, Guyer M, Peck AM, Derge JG, Lipman D, Collins FS, Jang W, Sherry S, Feolo M, Misquitta L, Lee E, Rotmistrovsky K, Greenhut SF, Schaefer CF, Buetow K, Bonner TI, Haussler D, Kent J, Kiekhaus M, Furey T, Brent M, Prange C, Schreiber K, Shapiro N, Bhat NK, Hopkins RF, Hsie F, Driscoll T, Soares MB, Casavant TL, Scheetz TE, Brown-stein MJ, Usdin TB, Toshiyuki S, Carninci P, Piao Y, Dudekula DB, Ko MS, Kawakami K, Suzuki Y, Sugano S, Gruber CE, Smith MR, Simmons B, Moore T, Waterman R, Johnson SL, Ruan Y, Wei CL, Mathavan S, Gunaratne PH, Wu J, Garcia AM, Hulyk SW, Fuh E, Yuan Y, Sneed A, Kowis C, Hodgson A, Muzny DM, McPherson J, Gibbs RA, Fahey J, Helton E, Ketteman M, Madan A, Rodrigues S, Sanchez A, Whiting M, Madari A, Young AC, Wetherby KD, Granite SJ, Kwong PN, Brinkley CP, Pearson RL, Bouffard GG, Blakesly RW, Green ED, Dickson MC, Rodriguez AC, Grimwood J, Schmutz J, Myers RM, Butterfield YS, Griffith M, Griffith OL, Krzywinski MI, Liao N, Morin R, Palmquist D, Petrescu AS, Skalska U, Smailus DE, Stott JM, Schnerch A, Schein JE, Jones SJ, Holt RA, Baross A, Marra MA, Clifton S, Makowski KA, Bosak S, Malek J (2004) The status, quality, and expansion of the NIH full-length cDNA project: the Mammalian Gene Collection (MGC). Genome Res 14(10B):2121–2127. doi:10.1101/gr.2596504

Gerin CG, Madueke IC, Perkins T, Hill S, Smith K, Haley B, Allen SA, Garcia RP, Paunesku T, Woloschak G (2011) Combination strategies for repair, plasticity, and regeneration using regulation of gene expression during the chronic phase after spinal cord injury. Synapse 65(12):1255–1281. doi:10.1002/syn.20903

Ghasimi S, Haapasalo H, Eray M, Korhonen K, Brannstrom T, Hedman H, Andersson U (2012) Immunohistochemical analysis of LRIG proteins in meningiomas: correlation between estrogen receptor status and LRIG expression. J Neurooncol 108(3):435–441. doi:10.1007/s11060-012-0856-x

Gimelli S, Makrythanasis P, Stouder C, Antonarakis SE, Bottani A, Bena F (2011) A de novo 12q13.11 microdeletion in a patient with severe mental retardation, cleft palate, and high myopia. Eur J Med Genet 54(1):94–96. doi:10.1016/j.ejmg.2010.09.008

Gomi F, Imaizumi K, Yoneda T, Taniguchi M, Mori Y, Miyoshi K, Hitomi J, Fujikado T, Tano Y, Tohyama M (2000) Molecular cloning of a novel membrane glycoprotein, pal, specifically expressed in photoreceptor cells of the retina and containing leucine-rich repeat. J Neurosci 20(9):3206–3213

Gong SG, Mai S, Chung K, Wei K (2009) Flrt2 and Flrt3 have overlapping and non-overlapping expression during craniofacial development. Gene Expr Patterns 9(7):497–502

GrandPre T, Nakamura F, Vartanian T, Strittmatter SM (2000) Identification of the Nogo inhibitor of axon regeneration as a Reticulon protein. Nature 403(6768):439–444

Gregory SG, Barlow KF, McLay KE, Kaul R, Swarbreck D, Dunham A, Scott CE, Howe KL, Woodfine K, Spencer CCA, Jones MC, Gillson C, Searle S, Zhou Y, Kokocinski F, McDonald L, Evans R, Phillips K, Atkinson A, Cooper R, Jones C, Hall RE, Andrews TD, Lloyd C, Ainscough R, Almeida JP, Ambrose KD, Anderson F, Andrew RW, Ashwell RIS, Aubin K, Babbage AK, Bagguley CL, Bailey J, Beasley H, Bethel G, Bird CP, Bray-Allen S, Brown JY, Brown AJ, Buckley D, Burton J, Bye J, Carder C, Chapman JC, Clark SY, Clarke G, Clee C, Cobley V, Collier RE, Corby N, Coville GJ, Davies J, Deadman R, Dunn M, Earthrowl M, Ellington AG, Errington H, Frankish A, Frankland J, French L, Garner P, Garnett J, Gay L, Ghori MRJ, Gibson R, Gilby LM, Gillett W, Glithero RJ, Grafham DV, Griffiths C, Griffiths-Jones S, Grocock R, Hammond S, Harrison ESI, Hart E, Haugen E, Heath PD, Holmes S, Holt K, Howden PJ, Hunt AR, Hunt SE, Hunter G, Isherwood J, James R, Johnson C, Johnson D, Joy A, Kay M, Kershaw JK, Kibukawa M, Kimberley AM, King A, Knights AJ, Lad H, Laird G, Lawlor S, Leongamornlert DA, Lloyd DM, Loveland J, Lovell J, Lush MJ, Lyne R, Martin S, Mashreghi-Mohammadi M, Matthews L, Matthews NSW, McLaren S, Milne S, Mistry S, MJF M o, Nickerson T, O'Dell CN, Oliver K, Palmeiri A, Palmer SA, Parker A, Patel D, Pearce AV, Peck AI, Pelan S, Phelps K, Phillimore BJ, Plumb R, Rajan J, Raymond C, Rouse G, Saenphimmachak C, Sehra HK, Sheridan E, Shownkeen R, Sims S, Skuce CD, Smith M, Steward C, Subramanian S, Sycamore N, Tracey A, Tromans A, Van Helmond Z, Wall M,

Wallis JM, White S, Whitehead SL, Wilkinson JE, Willey DL, Williams H, Wilming L, Wray PW, Wu Z, Coulson A, Vaudin M, Sulston JE, Durbin R, Hubbard T, Wooster R, Dunham I, Carter NP, McVean G, Ross MT, Harrow J, Olson MV, Beck S, Rogers J, Bentley DR (2006) The DNA sequence and biological annotation of human chromosome 1. Nature 441(7091): 315–321, http://www.nature.com/nature/journal/v441/n7091/suppinfo/nature04727_S1.html

Grimwood J, Gordon LA, Olsen A, Terry A, Schmutz J, Lamerdin J, Hellsten U, Goodstein D, Couronne O, Tran-Gyamfi M, Aerts A, Altherr M, Ashworth L, Bajorek E, Black S, Branscomb E, Caenepeel S, Carrano A, Caoile C, Chan YM, Christensen M, Cleland CA, Copeland A, Dalin E, Dehal P, Denys M, Detter JC, Escobar J, Flowers D, Fotopulos D, Garcia C, Georgescu AM, Glavina T, Gomez M, Gonzales E, Groza M, Hammon N, Hawkins T, Haydu L, Ho I, Huang W, Israni S, Jett J, Kadner K, Kimball H, Kobayashi A, Larionov V, Leem SH, Lopez F, Lou Y, Lowry S, Malfatti S, Martinez D, McCready P, Medina C, Morgan J, Nelson K, Nolan M, Ovcharenko I, Pitluck S, Pollard M, Popkie AP, Predki P, Quan G, Ramirez L, Rash S, Retterer J, Rodriguez A, Rogers S, Salamov A, Salazar A, She X, Smith D, Slezak T, Solovyev V, Thayer N, Tice H, Tsai M, Ustaszewska A, Vo N, Wagner M, Wheeler J, Wu K, Xie G, Yang J, Dubchak I, Furey TS, DeJong P, Dickson M, Gordon D, Eichler EE, Pennacchio LA, Richardson P, Stubbs L, Rokhsar DS, Myers RM, Rubin EM, Lucas SM (2004) The DNA sequence and biology of human chromosome 19. Nature 428(6982):529–535. doi:10.1038/nature02399

Guo D, Holmlund C, Henriksson R, Hedman H (2004) The LRIG gene family has three vertebrate paralogs widely expressed in human and mouse tissues and a homolog in Ascidiacea. Genomics 84(1):157–165. doi:10.1016/j.ygeno.2004.01.013

Gur G, Rubin C, Katz M, Amit I, Citri A, Nilsson J, Amariglio N, Henriksson R, Rechavi G, Hedman H, Wides R, Yarden Y (2004) LRIG1 restricts growth factor signaling by enhancing receptor ubiquitylation and degradation. EMBO J 23(16):3270–3281. doi:10.1038/sj.emboj.7600342

Haines BP, Rigby PW (2007) Developmentally regulated expression of the LRRTM gene family during mid-gestation mouse embryogenesis. Gene Expr Patterns 7(1–2):23–29. doi:10.1016/j.modgep.2006.05.004

Haines BP, Rigby PW (2008) Expression of the Lingo/LERN gene family during mouse embryogenesis. Gene Expr Patterns 8(2):79–86

Haines BP, Gupta R, Jones CM, Summerbell D, Rigby PW (2005) The NLRR gene family and mouse development: Modified differential display PCR identifies NLRR-1 as a gene expressed in early somitic myoblasts. Dev Biol 281(2):145–159

Haines BP, Wheldon LM, Summerbell D, Heath JK, Rigby PW (2006) Regulated expression of FLRT genes implies a functional role in the regulation of FGF signalling during mouse development. Dev Biol 297(1):14–25

Hanamsagar R, Hanke ML, Kielian T (2012) Toll-like receptor (TLR) and inflammasome actions in the central nervous system. Trends Immunol 33(7):333–342. doi:10.1016/j.it.2012.03.001

Harrison PJ, Law AJ, Eastwood SL (2003) Glutamate receptors and transporters in the hippocampus in schizophrenia. Ann N Y Acad Sci 1003:94–101

Hayes-Jordan A, Anderson PM (2011) The diagnosis and management of desmoplastic small round cell tumor: a review. Curr Opin Oncol 23(4):385–389. doi:10.1097/CCO.0b013e3283477aab

Hedman H, Lindstrom AK, Tot T, Stendahl U, Henriksson R, Hellberg D (2010) LRIG2 in contrast to LRIG1 predicts poor survival in early-stage squamous cell carcinoma of the uterine cervix. Acta Oncol 49(6):812–815. doi:10.3109/0284186x.2010.492789

Hillier LW, Graves TA, Fulton RS, Fulton LA, Pepin KH, Minx P, Wagner-McPherson C, Layman D, Wylie K, Sekhon M, Becker MC, Fewell GA, Delehaunty KD, Miner TL, Nash WE, Kremitzki C, Oddy L, Du H, Sun H, Bradshaw-Cordum H, Ali J, Carter J, Cordes M, Harris A, Isak A, van Brunt A, Nguyen C, Du F, Courtney L, Kalicki J, Ozersky P, Abbott S, Armstrong J, Belter EA, Caruso L, Cedroni M, Cotton M, Davidson T, Desai A, Elliott G, Erb T, Fronick C, Gaige T, Haakenson W, Haglund K, Holmes A, Harkins R, Kim K, Kruchowski SS, Strong CM, Grewal N, Goyea E, Hou S, Levy A, Martinka S, Mead K, McLellan MD, Meyer R, Randall-Maher J, Tomlinson C, Dauphin-Kohlberg S, Kozlowicz-Reilly A, Shah N, Swearengen-Shahid

S, Snider J, Strong JT, Thompson J, Yoakum M, Leonard S, Pearman C, Trani L, Radionenko M, Waligorski JE, Wang C, Rock SM, Tin-Wollam AM, Maupin R, Latreille P, Wendl MC, Yang SP, Pohl C, Wallis JW, Spieth J, Bieri TA, Berkowicz N, Nelson JO, Osborne J, Ding L, Sabo A, Shotland Y, Sinha P, Wohldmann PE, Cook LL, Hickenbotham MT, Eldred J, Williams D, Jones TA, She X, Ciccarelli FD, Izaurralde E, Taylor J, Schmutz J, Myers RM, Cox DR, Huang X, McPherson JD, Mardis ER, Clifton SW, Warren WC, Chinwalla AT, Eddy SR, Marra MA, Ovcharenko I, Furey TS, Miller W, Eichler EE, Bork P, Suyama M, Torrents D, Waterston RH, Wilson RK (2005) Generation and annotation of the DNA sequences of human chromosomes 2 and 4. Nature 434(7034):724–731. doi:10.1038/nature03466

Hirosawa M, Nagase T, K-i I, Kikuno R, Nomura N, Ohara O (1999) Characterization of cDNA clones selected by the genemark analysis from size-fractionated cDNA libraries from human brain. DNA Res 6(5):329–336. doi:10.1093/dnares/6.5.329

Hohenester E (2008) Structural insight into Slit-Robo signalling. Biochem Soc Trans 36(Pt 2): 251–256. doi:10.1042/bst0360251

Holmlund C, Haapasalo H, Yi W, Raheem O, Brannstrom T, Bragge H, Henriksson R, Hedman H (2009) Cytoplasmic LRIG2 expression is associated with poor oligodendroglioma patient survival. Neuropathology 29(3):242–247. doi:10.1111/j.1440-1789.2008.00970.x

Homma S, Shimada T, Hikake T, Yaginuma H (2009) Expression pattern of LRR and Ig domain-containing protein (LRRIG protein) in the early mouse embryo. Gene Expr Patterns 9(1):1–26. doi:10.1016/j.gep.2008.09.004

Hossain MS, Ozaki T, Wang H, Nakagawa A, Takenobu H, Ohira M, Kamijo T, Nakagawara A (2008) N-MYC promotes cell proliferation through a direct transactivation of neuronal leucine-rich repeat protein-1 (NLRR1) gene in neuroblastoma. Oncogene 27(46):6075–6082. doi:10.1038/onc.2008.200

Hossain S, Ahmed MU, Alam S, Watanabe A, Harashima A, Yonekura H, Yamamoto H (2012) Expressions and roles of AMIGO gene family in vascular endothelial cells. Int J Biosci Biochem Bioinform 2(1):1–5

Huang JK, Franklin RJ (2011) Regenerative medicine in multiple sclerosis: identifying pharmacological targets of adult neural stem cell differentiation. Neurochem Int 59(3):329–332. doi:10.1016/j.neuint.2011.01.017

Huc-Chabrolle M, Charon C, Guilmatre A, Vourc'h P, Tripi G, Barthez MA, Sizaret E, Thepault RA, Le Gallic S, Hager J, Toutain A, Raynaud M, Andres C, Campion D, Laumonnier F, Bonnet-Brilhault F (2013) Xq27 FRAXA locus is a strong candidate for dyslexia: evidence from a genome-wide scan in French families. Behav Genet. doi:10.1007/s10519-012-9575-5

Inohara N, Nunez G (2003) NODs: intracellular proteins involved in inflammation and apoptosis. Nat Rev Immunol 3(5):371–382. doi:10.1038/nri1086

Inoue H, Lin L, Lee X, Shao Z, Mendes S, Snodgrass-Belt P, Sweigard H, Engber T, Pepinsky B, Yang L, Beal MF, Mi S, Isacson O (2007) Inhibition of the leucine-rich repeat protein LINGO-1 enhances survival, structure, and function of dopaminergic neurons in Parkinson's disease models. Proc Natl Acad Sci 104(36):14430–14435. doi:10.1073/pnas.0700901104

Ishikawa K, Nagase T, Nakajima D, Seki N, Ohira M, Miyajima N, Tanaka A, Kotani H, Nomura N, Ohara O (1997) Prediction of the coding sequences of unidentified human genes. VIII. 78 new cDNA clones from brain which code for large proteins in vitro. DNA Res 4(5):307–313

Jepson S, Vought B, Gross CH, Gan L, Austen D, Frantz JD, Zwahlen J, Lowe D, Markland W, Krauss R (2012) LINGO-1, a transmembrane signaling protein, inhibits oligodendrocyte differentiation and myelination through intercellular self-interactions. J Biol Chem 287(26): 22184–22195. doi:10.1074/jbc.M112.366179

Ji B, Li M, Wu WT, Yick LW, Lee X, Shao Z, Wang J, So KF, McCoy JM, Pepinsky RB, Mi S, Relton JK (2006) LINGO-1 antagonist promotes functional recovery and axonal sprouting after spinal cord injury. Mol Cell Neurosci 33(3):311–320. doi:10.1016/j.mcn.2006.08.003

Johnson RJ, McCoy JG, Bingman CA, Phillips GN Jr, Raines RT (2007) Inhibition of human pancreatic ribonuclease by the human ribonuclease inhibitor protein. J Mol Biol 368(2): 434–449. doi:10.1016/j.jmb.2007.02.005

Johnson EC, Guo Y, Cepurna WO, Morrison JC (2009) Neurotrophin roles in retinal ganglion cell survival: lessons from rat glaucoma models. Exp Eye Res 88(4):808–815. doi:10.1016/j.exer.2009.02.004

Kajander T, Kuja-Panula J, Rauvala H, Goldman A (2011) Crystal structure and role of glycans and dimerization in folding of neuronal leucine-rich repeat protein AMIGO-1. J Mol Biol 413(5):1001–1015. doi:10.1016/j.jmb.2011.09.032

Kajava AV (1998) Structural diversity of leucine-rich repeat proteins. J Mol Biol 277(3):519–527. doi:10.1006/jmbi.1998.1643

Kajiwara Y, Buxbaum JD, Grice DE (2009) SLITRK1 binds 14-3-3 and regulates neurite outgrowth in a phosphorylation-dependent manner. Biol Psychiatry 66(10):918–925. doi:10.1016/j.biopsych.2009.05.033

Karagiannidis I, Rizzo R, Tarnok Z, Wolanczyk T, Hebebrand J, Nothen MM, Lehmkuhl G, Farkas L, Nagy P, Barta C, Szymanska U, Panteloglou G, Miranda DM, Feng Y, Sandor P, Barr C, Paschou P (2012) Replication of association between a SLITRK1 haplotype and Tourette Syndrome in a large sample of families. Mol Psychiatry 17(7):665–668. doi:10.1038/mp.2011.151

Karaulanov EE, Bottcher RT, Niehrs C (2006) A role for fibronectin-leucine-rich transmembrane cell-surface proteins in homotypic cell adhesion. EMBO Rep 7(3):283–290

Karaulanov E, Bottcher RT, Stannek P, Wu W, Rau M, Ogata S, Cho KW, Niehrs C (2009) Unc5B interacts with FLRT3 and Rnd1 to modulate cell adhesion in Xenopus embryos. PLoS One 4(5):e5742

Kask M, Pruunsild P, Timmusk T (2011) Bidirectional transcription from human LRRTM2/CTNNA1 and LRRTM1/CTNNA2 gene loci leads to expression of N-terminally truncated CTNNA1 and CTNNA2 isoforms. Biochem Biophys Res Commun 411(1):56–61. doi:10.1016/j.bbrc.2011.06.085

Katayama K, Zine A, Ota M, Matsumoto Y, Inoue T, Fritzsch B, Aruga J (2009) Disorganized innervation and neuronal loss in the inner ear of Slitrk6-deficient mice. PLoS One 4(11):e7786. doi:10.1371/journal.pone.0007786

Katayama K, Yamada K, Ornthanalai VG, Inoue T, Ota M, Murphy NP, Aruga J (2010) Slitrk1-deficient mice display elevated anxiety-like behavior and noradrenergic abnormalities. Mol Psychiatry 15(2):177–184. doi:10.1038/mp.2008.97

Keen-Kim D, Mathews CA, Reus VI, Lowe TL, Herrera LD, Budman CL, Gross-Tsur V, Pulver AE, Bruun RD, Erenberg G, Naarden A, Sabatti C, Freimer NB (2006) Overrepresentation of rare variants in a specific ethnic group may confuse interpretation of association analyses. Hum Mol Genet 15(22):3324–3328. doi:10.1093/hmg/ddl408

Kikuno R, Nagase T, Ishikawa K, Hirosawa M, Miyajima N, Tanaka A, Kotani H, Nomura N, Ohara O (1999) Prediction of the coding sequences of unidentified human genes. XIV. The complete sequences of 100 new cDNA clones from brain which code for large proteins in vitro. DNA Res 6(3):197–205

Kim E, Naisbitt S, Hsueh Y-P, Rao A, Rothschild A, Craig AM, Sheng M (1997) GKAP, a novel synaptic protein that interacts with the guanylate kinase-like domain of the PSD-95/SAP90 family of channel clustering molecules. J Cell Biol 136(3):669–678. doi:10.1083/jcb.136.3.669

Kim S, Burette A, Chung HS, Kwon SK, Woo J, Lee HW, Kim K, Kim H, Weinberg RJ, Kim E (2006) NGL family PSD-95-interacting adhesion molecules regulate excitatory synapse formation. Nat Neurosci 9(10):1294–1301. doi:10.1038/nn1763

Kim S-D, Liu JL, Roscioli T, Buckley MF, Yagnik G, Boyadjiev SA, Kim J (2012) Leucine-rich repeat, immunoglobulin-like and transmembrane domain 3 (LRIT3) is a modulator of FGFR1. FEBS Lett 586(10):1516–1521. doi:10.1016/j.febslet.2012.04.010

Klein A, Olendrowitz C, Schmutzler R, Hampl J, Schlag PM, Maass N, Arnold N, Wessel R, Ramser J, Meindl A, Scherneck S, Seitz S (2009) Identification of brain- and bone-specific breast cancer metastasis genes. Cancer Lett 276(2):212–220. doi:10.1016/j.canlet.2008.11.017

Knight D, Xie W, Boulianne G (2011) Neurexins and neuroligins: recent insights from invertebrates. Mol Neurobiol 44(3):426–440. doi:10.1007/s12035-011-8213-1

Ko J (2012) The leucine-rich repeat superfamily of synaptic adhesion molecules: LRRTMs and Slitrks. Mol Cells 34(4):335–340. doi:10.1007/s10059-012-0113-3

Ko J, Kim S, Chung HS, Kim K, Han K, Kim H, Jun H, Kaang BK, Kim E (2006) SALM synaptic cell adhesion-like molecules regulate the differentiation of excitatory synapses. Neuron 50(2):233–245. doi:10.1016/j.neuron.2006.04.005

Ko J, Fuccillo MV, Malenka RC, Sudhof TC (2009) LRRTM2 functions as a neurexin ligand in promoting excitatory synapse formation. Neuron 64(6):791–798. doi:10.1016/j.neuron.2009.12.012

Ko J, Soler-Llavina GJ, Fuccillo MV, Malenka RC, Sudhof TC (2011) Neuroligins/LRRTMs prevent activity- and Ca2+/calmodulin-dependent synapse elimination in cultured neurons. J Cell Biol 194(2):323–334. doi:10.1083/jcb.201101072

Kobe B, Deisenhofer J (1996) Mechanism of ribonuclease inhibition by ribonuclease inhibitor protein based on the crystal structure of its complex with Ribonuclease A. J Mol Biol 264(5):1028–1043. doi:10.1006/jmbi.1996.0694

Kobe B, Kajava AV (2001) The leucine-rich repeat as a protein recognition motif. Curr Opin Struct Biol 11(6):725–732. doi:10.1016/s0959-440x(01)00266-4

Kong RCK, Shilling PJ, Lobb DK, Gooley PR, Bathgate RAD (2010) Membrane receptors: structure and function of the relaxin family peptide receptors. Mol Cell Endocrinol 320(1–2):1–15. doi:10.1016/j.mce.2010.02.003

Kopecki Z, Cowin AJ (2008) Flightless I: an actin-remodelling protein and an important negative regulator of wound repair. Int J Biochem Cell Biol 40(8):1415–1419. doi:10.1016/j.biocel.2007.04.011

Kostyukova A (2008) Tropomodulins and tropomodulin/tropomyosin interactions. Cell Mol Life Sci 65(4):563–569. doi:10.1007/s00018-007-7347-9

Krig SR, Frietze S, Simion C, Miller JK, Fry WH, Rafidi H, Kotelawala L, Qi L, Griffith OL, Gray JW, Carraway KL III, Sweeney C (2011) Lrig1 is an estrogen-regulated growth suppressor and correlates with longer relapse-free survival in ERalpha-positive breast cancer. Mol Cancer Res 9(10):1406–1417. doi:10.1158/1541-7786.mcr-11-0227

Kuja-Panula J, Kiiltomaki M, Yamashiro T, Rouhiainen A, Rauvala H (2003) AMIGO, a transmembrane protein implicated in axon tract development, defines a novel protein family with leucine-rich repeats. J Cell Biol 160(6):963–973. doi:10.1083/jcb.200209074

Laakso T, Muggalla P, Kysenius K, Lauren J, Paatero A, Huttunen HJ, Airaksinen MS (2012) LRRTM3 is dispensable for amyloid-beta production in mice. J Alzheimers Dis 31(4):759–764. doi:10.3233/jad-2012-120193

Lacy SE, Bonnemann CG, Buzney EA, Kunkel LM (1999) Identification of FLRT1, FLRT2, and FLRT3: a novel family of transmembrane leucine-rich repeat proteins. Genomics 62(3):417–426

Laederich MB, Funes-Duran M, Yen L, Ingalla E, Wu X, Carraway KL III, Sweeney C (2004) The leucine-rich repeat protein LRIG1 is a negative regulator of ErbB family receptor tyrosine kinases. J Biol Chem 279(45):47050–47056. doi:10.1074/jbc.M409703200

Laeremans A, Nys J, Luyten W, D'Hooge R, Paulussen M, Arckens L (2013) AMIGO2 mRNA expression in hippocampal CA2 and CA3a. Brain Struct Funct 218(1):123–130. doi:10.1007/s00429-012-0387-4

Larrieu-Lahargue F, Thomas KR, Li DY (2012) Netrin ligands and receptors: lessons from neurons to the endothelium. Trends Cardiovasc Med 22(2):44–47. doi:10.1016/j.tcm.2012.06.010

Lauren J, Airaksinen MS, Saarma M, Timmusk T (2003) A novel gene family encoding leucine-rich repeat transmembrane proteins differentially expressed in the nervous system. Genomics 81(4):411–421

Ledda F, Bieraugel O, Fard SS, Vilar M, Paratcha G (2008) Lrig1 is an endogenous inhibitor of Ret receptor tyrosine kinase activation, downstream signaling, and biological responses to GDNF. J Neurosci 28(1):39–49. doi:10.1523/jneurosci.2196-07.2008

Lee JY, Petratos S (2013) Multiple sclerosis: does nogo play a role? Neuroscientist. doi:10.1177/1073858413477207

Lee X, Yang Z, Shao Z, Rosenberg SS, Levesque M, Pepinsky RB, Qiu M, Miller RH, Chan JR, Mi S (2007) NGF regulates the expression of axonal LINGO-1 to inhibit oligodendrocyte differentiation and myelination. J Neurosci 27(1):220–225. doi:10.1523/jneurosci.4175-06.2007
Lesch KP, Merker S, Reif A, Novak M (2012) Dances with black widow spiders: dysregulation of glutamate signalling enters centre stage in ADHD. Eur Neuropsychopharmacol. doi:10.1016/j.euroneuro.2012.07.013
Liang X, Draghi NA, Resh MD (2004) Signaling from integrins to Fyn to Rho family GTPases regulates morphologic differentiation of oligodendrocytes. J Neurosci 24(32):7140–7149. doi:10.1523/jneurosci.5319-03.2004
Liang X, Martin ER, Schnetz-Boutaud N, Bartlett J, Anderson B, Zuchner S, Gwirtsman H, Schmechel D, Carney R, Gilbert JR, Pericak-Vance MA, Haines JL (2007) Effect of heterogeneity on the chromosome 10 risk in late-onset alzheimer disease. Hum Mutat 28(11): 1065–1073. doi:10.1002/humu.20567
Lin JC, Ho WH, Gurney A, Rosenthal A (2003) The netrin-G1 ligand NGL-1 promotes the outgrowth of thalamocortical axons. Nat Neurosci 6(12):1270–1276. doi:10.1038/nn1148
Linhoff MW, Laurén J, Cassidy RM, Dobie FA, Takahashi H, Nygaard HB, Airaksinen MS, Strittmatter SM, Craig AM (2009) An unbiased expression screen for synaptogenic proteins identifies the LRRTM protein family as synaptic organizers. Neuron 61(5):734–749. doi:10.1016/j.neuron.2009.01.017
Llorens F, Gil V, Iraola S, Carim-Todd L, Marti E, Estivill X, Soriano E, del Rio JA, Sumoy L (2008) Developmental analysis of Lingo-1/Lern1 protein expression in the mouse brain: interaction of its intracellular domain with Myt1l. Dev Neurobiol 68(4):521–541. doi:10.1002/dneu.20607
Llorens F, Gil V, del Río JA (2011) Emerging functions of myelin-associated proteins during development, neuronal plasticity, and neurodegeneration. FASEB J 25(2):463–475. doi:10.1096/fj.10-162792
Loov C, Fernqvist M, Walmsley A, Marklund N, Erlandsson A (2012) Neutralization of LINGO-1 during in vitro differentiation of neural stem cells results in proliferation of immature neurons. PLoS One 7(1):e29771. doi:10.1371/journal.pone.0029771
Loset M, Mundal SB, Johnson MP, Fenstad MH, Freed KA, Lian IA, Eide IP, Bjorge L, Blangero J, Moses EK, Austgulen R (2011) A transcriptional profile of the decidua in preeclampsia. Am J Obstet Gynecol 204(1):84. doi:10.1016/j.ajog.2010.08.043, e1–27
Ludwig KU, Mattheisen M, Muhleisen TW, Roeske D, Schmal C, Breuer R, Schulte-Korne G, Muller-Myhsok B, Nothen MM, Hoffmann P, Rietschel M, Cichon S (2009) Supporting evidence for LRRTM1 imprinting effects in schizophrenia. Mol Psychiatry 14(8):743–745. doi:10.1038/mp.2009.28
Mah AK (2010) SLITRK5, a protein that links striatal deficits to OCD-like behaviours in mice. Clin Genet 78(4):350–352. doi:10.1111/j.1399-0004.2010.01507.x
Mah W, Ko J, Nam J, Han K, Chung WS, Kim E (2010) Selected SALM (synaptic adhesion-like molecule) family proteins regulate synapse formation. J Neurosci 30(16):5559–5568. doi:10.1523/jneurosci.4839-09.2010
Majercak J, Ray WJ, Espeseth A, Simon A, Shi XP, Wolffe C, Getty K, Marine S, Stec E, Ferrer M, Strulovici B, Bartz S, Gates A, Xu M, Huang Q, Ma L, Shughrue P, Burchard J, Colussi D, Pietrak B, Kahana J, Beher D, Rosahl T, Shearman M, Hazuda D, Sachs AB, Koblan KS, Seabrook GR, Stone DJ (2006) LRRTM3 promotes processing of amyloid-precursor protein by BACE1 and is a positional candidate gene for late-onset alzheimer's disease. Proc Natl Acad Sci U S A 103(47):17967–17972. doi:10.1073/pnas.0605461103
Mandrioli R, Mercolini L, Saracino MA, Raggi MA (2012) Selective serotonin reuptake inhibitors (SSRIs): therapeutic drug monitoring and pharmacological interactions. Curr Med Chem 19(12): 1846–1863
Maretto S, Muller PS, Aricescu AR, Cho KW, Bikoff EK, Robertson EJ (2008) Ventral closure, headfold fusion and definitive endoderm migration defects in mouse embryos lacking the fibronectin leucine-rich transmembrane protein FLRT3. Dev Biol 318(1):184–193

Marteyn A, Maury Y, Gauthier MM, Lecuyer C, Vernet R, Denis JA, Pietu G, Peschanski M, Martinat C (2011) Mutant human embryonic stem cells reveal neurite and synapse formation defects in type 1 myotonic dystrophy. Cell Stem Cell 8(4):434–444. doi:10.1016/j.stem.2011.02.004

Mathis C, Schroter A, Thallmair M, Schwab ME (2010) Nogo-a regulates neural precursor migration in the embryonic mouse cortex. Cereb Cortex 20(10):2380–2390. doi:10.1093/cercor/bhp307

Matilla A, Radrizzani M (2005) The Anp32 family of proteins containing leucine-rich repeats. Cerebellum 4(1):7–18. doi:10.1080/14734220410019020

Matsumoto Y, Katayama K, Okamoto T, Yamada K, Takashima N, Nagao S, Aruga J (2011) Impaired auditory-vestibular functions and behavioral abnormalities of Slitrk6-deficient mice. PLoS One 6(1):e16497. doi:10.1371/journal.pone.0016497

Matsushima N, Tachi N, Kuroki Y, Enkhbayar P, Osaki M, Kamiya M, Kretsinger R (2005) Structural analysis of leucine-rich-repeat variants in proteins associated with human diseases. Cell Mol Life Sci 62(23):2771–2791. doi:10.1007/s00018-005-5187-z

Mburu P, Mustapha M, Varela A, Weil D, El-Amraoui A, Holme RH, Rump A, Hardisty RE, Blanchard S, Coimbra RS, Perfettini I, Parkinson N, Mallon AM, Glenister P, Rogers MJ, Paige AJ, Moir L, Clay J, Rosenthal A, Liu XZ, Blanco G, Steel KP, Petit C, Brown SD (2003) Defects in whirlin, a PDZ domain molecule involved in stereocilia elongation, cause deafness in the whirler mouse and families with DFNB31. Nat Genet 34(4):421–428. doi:10.1038/ng1208

McDonald CL, Bandtlow C, Reindl M (2011) Targeting the Nogo receptor complex in diseases of the central nervous system. Curr Med Chem 18(2):234–244

McEwan PA, Scott PG, Bishop PN, Bella J (2006) Structural correlations in the family of small leucine-rich repeat proteins and proteoglycans. J Struct Biol 155(2):294–305. doi:10.1016/j.jsb.2006.01.016

McManus C, Nicholls M, Vallortigara G (2009) Editorial commentary: is LRRTM1 the gene for handedness? Laterality 14(1):1–2. doi:10.1080/13576500802542577

Metlapally R, Michaelides M, Bulusu A, Li YJ, Schwartz M, Rosenberg T, Hunt DM, Moore AT, Zuchner S, Rickman CB, Young TL (2009) Evaluation of the X-linked high-grade myopia locus (MYP1) with cone dysfunction and color vision deficiencies. Invest Ophthalmol Vis Sci 50(4):1552–1558. doi:10.1167/iovs.08-2455

Mi S, Lee X, Shao Z, Thill G, Ji B, Relton J, Levesque M, Allaire N, Perrin S, Sands B, Crowell T, Cate RL, McCoy JM, Pepinsky RB (2004) LINGO-1 is a component of the Nogo-66 receptor/p75 signaling complex. Nat Neurosci 7(3):221–228. doi:10.1038/nn1188

Mi S, Miller RH, Lee X, Scott ML, Shulag-Morskaya S, Shao Z, Chang J, Thill G, Levesque M, Zhang M, Hession C, Sah D, Trapp B, He Z, Jung V, McCoy JM, Pepinsky RB (2005) LINGO-1 negatively regulates myelination by oligodendrocytes. Nat Neurosci 8(6):745–751. doi:10.1038/nn1460

Mi S, Hu B, Hahm K, Luo Y, Kam Hui ES, Yuan Q, Wong WM, Wang L, Su H, Chu TH, Guo J, Zhang W, So KF, Pepinsky B, Shao Z, Graff C, Garber E, Jung V, Wu EX, Wu W (2007) LINGO-1 antagonist promotes spinal cord remyelination and axonal integrity in MOG-induced experimental autoimmune encephalomyelitis. Nat Med 13(10):1228–1233. doi:10.1038/nm1664

Mi S, Sandrock A, Miller RH (2008) LINGO-1 and its role in CNS repair. The Int J Biochem Cell Biol 40(10):1971–1978. doi:10.1016/j.biocel.2008.03.018

Mi S, Miller RH, Tang W, Lee X, Hu B, Wu W, Zhang Y, Shields CB, Miklasz S, Shea D, Mason J, Franklin RJ, Ji B, Shao Z, Chedotal A, Bernard F, Roulois A, Xu J, Jung V, Pepinsky B (2009) Promotion of central nervous system remyelination by induced differentiation of oligodendrocyte precursor cells. Ann Neurol 65(3):304–315. doi:10.1002/ana.21581

Mikhail FM, Lose EJ, Robin NH, Descartes MD, Rutledge KD, Rutledge SL, Korf BR, Carroll AJ (2011) Clinically relevant single gene or intragenic deletions encompassing critical neurodevelopmental genes in patients with developmental delay, mental retardation, and/or autism spectrum disorders. Am J Med Genet A 155A(10):2386–2396. doi:10.1002/ajmg.a.34177

Milde T, Shmelkov SV, Jensen KK, Zlotchenko G, Petit I, Rafii S (2007) A novel family of slitrk genes is expressed on hematopoietic stem cells and leukemias. Leukemia 21(4):824–827. doi:10.1038/sj.leu.2404525

Miller JK, Shattuck DL, Ingalla EQ, Yen L, Borowsky AD, Young LJ, Cardiff RD, Carraway KL III, Sweeney C (2008) Suppression of the negative regulator LRIG1 contributes to ErbB2 overexpression in breast cancer. Cancer Res 68(20):8286–8294. doi:10.1158/0008-5472.can-07-6316

Miranda DM, Wigg K, Kabia EM, Feng Y, Sandor P, Barr CL (2009) Association of SLITRK1 to Gilles de la Tourette Syndrome. Am J Med Genet B Neuropsychiatr Genet 150B(4):483–486. doi:10.1002/ajmg.b.30840

Miraoui H, Marie PJ (2010) Fibroblast growth factor receptor signaling crosstalk in skeletogenesis. Sci Signal 3(146):re9. doi:10.1126/scisignal.3146re9

Moore SW, Tessier-Lavigne M, Kennedy TE (2007) Netrins and their receptors. Adv Exp Med Biol 621:17–31, doi:10.1007/978-0-387-76715-4_2

Morimura N, Inoue T, Katayama K, Aruga J (2006) Comparative analysis of structure, expression and PSD95-binding capacity of Lrfn, a novel family of neuronal transmembrane proteins. Gene 380(2):72–83. doi:10.1016/j.gene.2006.05.014

Mosyak L, Wood A, Dwyer B, Buddha M, Johnson M, Aulabaugh A, Zhong X, Presman E, Benard S, Kelleher K, Wilhelm J, Stahl ML, Kriz R, Gao Y, Cao Z, Ling HP, Pangalos MN, Walsh FS, Somers WS (2006) The structure of the Lingo-1 ectodomain, a module implicated in central nervous system repair inhibition. J Biol Chem 281(47):36378–36390. doi:10.1074/jbc.M607314200

Muller PS, Schulz R, Maretto S, Costello I, Srinivas S, Bikoff E, Robertson E (2011) The fibronectin leucine-rich repeat transmembrane protein Flrt2 is required in the epicardium to promote heart morphogenesis. Development. doi:10.1242/dev.059386

Muller S, Lindquist D, Kanter L, Flores-Staino C, Henriksson R, Hedman H, Andersson S (2013) Expression of LRIG1 and LRIG3 correlates with human papillomavirus status and patient survival in cervical adenocarcinoma. Int J Oncol 42(1):247–252. doi:10.3892/ijo.2012.1702

Munfus DL, Haga CL, Burrows PD, Cooper MD (2007) A conserved gene family encodes transmembrane proteins with fibronectin, immunoglobulin and leucine-rich repeat domains (FIGLER). BMC Biol 5:36. doi:10.1186/1741-7007-5-36

Mungall AJ, Palmer SA, Sims SK, Edwards CA, Ashurst JL, Wilming L, Jones MC, Horton R, Hunt SE, Scott CE, Gilbert JG, Clamp ME, Bethel G, Milne S, Ainscough R, Almeida JP, Ambrose KD, Andrews TD, Ashwell RI, Babbage AK, Bagguley CL, Bailey J, Banerjee R, Barker DJ, Barlow KF, Bates K, Beare DM, Beasley H, Beasley O, Bird CP, Blakey S, Bray-Allen S, Brook J, Brown AJ, Brown JY, Burford DC, Burrill W, Burton J, Carder C, Carter NP, Chapman JC, Clark SY, Clark G, Clee CM, Clegg S, Cobley V, Collier RE, Collins JE, Colman LK, Corby NR, Coville GJ, Culley KM, Dhami P, Davies J, Dunn M, Earthrowl ME, Ellington AE, Evans KA, Faulkner L, Francis MD, Frankish A, Frankland J, French L, Garner P, Garnett J, Ghori MJ, Gilby LM, Gillson CJ, Glithero RJ, Grafham DV, Grant M, Gribble S, Griffiths C, Griffiths M, Hall R, Halls KS, Hammond S, Harley JL, Hart EA, Heath PD, Heathcott R, Holmes SJ, Howden PJ, Howe KL, Howell GR, Huckle E, Humphray SJ, Humphries MD, Hunt AR, Johnson CM, Joy AA, Kay M, Keenan SJ, Kimberley AM, King A, Laird GK, Langford C, Lawlor S, Leongamornlert DA, Leversha M, Lloyd CR, Lloyd DM, Loveland JE, Lovell J, Martin S, Mashreghi-Mohammadi M, Maslen GL, Matthews L, McCann OT, McLaren SJ, McLay K, McMurray A, Moore MJ, Mullikin JC, Niblett D, Nickerson T, Novik KL, Oliver K, Overton-Larty EK, Parker A, Patel R, Pearce AV, Peck AI, Phillimore B, Phillips S, Plumb RW, Porter KM, Ramsey Y, Ranby SA, Rice CM, Ross MT, Searle SM, Sehra HK, Sheridan E, Skuce CD, Smith S, Smith M, Spraggon L, Squares SL, Steward CA, Sycamore N, Tamlyn-Hall G, Tester J, Theaker AJ, Thomas DW, Thorpe A, Tracey A, Tromans A, Tubby B, Wall M, Wallis JM, West AP, White SS, Whitehead SL, Whittaker H, Wild A, Willey DJ, Wilmer TE, Wood JM, Wray PW, Wyatt JC, Young L, Younger RM, Bentley DR, Coulson A, Durbin R, Hubbard T, Sulston JE, Dunham I, Rogers J, Beck S (2003) The DNA sequence and analysis of human chromosome 6. Nature 425(6960):805–811. doi:10.1038/nature02055

Muzny DM, Scherer SE, Kaul R, Wang J, Yu J, Sudbrak R, Buhay CJ, Chen R, Cree A, Ding Y, Dugan-Rocha S, Gill R, Gunaratne P, Harris RA, Hawes AC, Hernandez J, Hodgson AV, Hume J, Jackson A, Khan ZM, Kovar-Smith C, Lewis LR, Lozado RJ, Metzker ML, Milosavljevic A, Miner GR, Morgan MB, Nazareth LV, Scott G, Sodergren E, Song X-Z, Steffen D, Wei S,

Wheeler DA, Wright MW, Worley KC, Yuan Y, Zhang Z, Adams CQ, Ansari-Lari MA, Ayele M, Brown MJ, Chen G, Chen Z, Clendenning J, Clerc-Blankenburg KP, Chen R, Chen Z, Davis C, Delgado O, Dinh HH, Dong W, Draper H, Ernst S, Fu G, Gonzalez-Garay ML, Garcia DK, Gillett W, Gu J, Hao B, Haugen E, Havlak P, He X, Hennig S, Hu S, Huang W, Jackson LR, Jacob LS, Kelly SH, Kube M, Levy R, Li Z, Liu B, Liu J, Liu W, Lu J, Maheshwari M, Nguyen B-V, Okwuonu GO, Palmeiri A, Pasternak S, Perez LM, Phelps KA, Plopper FJH, Qiang B, Raymond C, Rodriguez R, Saenphimmachak C, Santibanez J, Shen H, Shen Y, Subramanian S, Tabor PE, Verduzco D, Waldron L, Wang J, Wang J, Wang Q, Williams GA, Wong GKS, Yao Z, Zhang J, Zhang X, Zhao G, Zhou J, Zhou Y, Nelson D, Lehrach H, Reinhardt R, Naylor SL, Yang H, Olson M, Weinstock G, Gibbs RA (2006) The DNA sequence, annotation and analysis of human chromosome 3. Nature 440:1194–1198

Nagase T, Ishikawa K, Suyama M, Kikuno R, Hirosawa M, Miyajima N, Tanaka A, Kotani H, Nomura N, Ohara O (1998) Prediction of the coding sequences of unidentified human genes. XII. The complete sequences of 100 new cDNA clones from brain which code for large proteins in vitro. DNA Res 5(6):355–364

Nagase T, Ishikawa K, Kikuno R, Hirosawa M, Nomura N, Ohara O (1999) Prediction of the coding sequences of unidentified human genes. XV. The complete sequences of 100 new cDNA clones from brain which code for large proteins in vitro. DNA Res 6(5):337–345

Nagase T, Kikuno R, Ishikawa K, Hirosawa M, Ohara O (2000a) Prediction of the coding sequences of unidentified human genes. XVII. The complete sequences of 100 new cDNA clones from brain which code for large proteins in vitro. DNA Res 7(2):143–150

Nagase T, Kikuno R, Nakayama M, Hirosawa M, Ohara O (2000b) Prediction of the coding sequences of unidentified human genes. XVIII. The complete sequences of 100 new cDNA clones from brain which code for large proteins in vitro. DNA Res 7(4):273–281

Nagase T, Kikuno R, Ohara O (2001a) Prediction of the coding sequences of unidentified human genes. XXI. The complete sequences of 60 new cDNA clones from brain which code for large proteins. DNA Res 8(4):179–187

Nagase T, Nakayama M, Nakajima D, Kikuno R, Ohara O (2001b) Prediction of the coding sequences of unidentified human genes. XX. The complete sequences of 100 new cDNA clones from brain which code for large proteins in vitro. DNA Res 8(2):85–95

Nakaya N, Sultana A, Lee HS, Tomarev SI (2012) Olfactomedin 1 interacts with the Nogo A receptor complex to regulate axon growth. J Biol Chem 287(44):37171–37184. doi:10.1074/jbc.M112.389916

Nam J, Mah W, Kim E (2011) The SALM/Lrfn family of leucine-rich repeat-containing cell adhesion molecules. Semin Cell Dev Biol 22(5):492–498. doi:10.1016/j.semcdb.2011.06.005

Ng ACY, Eisenberg JM, Heath RJW, Huett A, Robinson CM, Nau GJ, Xavier RJ (2011) Human leucine-rich repeat proteins: a genome-wide bioinformatic categorization and functional analysis in innate immunity. Proc Natl Acad Sci 108(1):4631–4638. doi:10.1073/pnas.1000093107

Niemann HH, Schubert W-D, Heinz DW (2004) Adhesins and invasins of pathogenic bacteria: a structural view. Microbes Infect 6(1):101–112. doi:10.1016/j.micinf.2003.11.001

Nishimura-Akiyoshi S, Niimi K, Nakashiba T, Itohara S (2007) Axonal netrin-Gs transneuronally determine lamina-specific subdendritic segments. Proc Natl Acad Sci U S A 104(37):14801–14806. doi:10.1073/pnas.0706919104

Oertle T, Schwab ME (2003) Nogo and its paRTNers. Trends Cell Biol 13(4):187–194. doi:10.1016/S0962-8924(03)00035-7

Oertle T, van der Haar ME, Bandtlow CE, Robeva A, Burfeind P, Buss A, Huber AB, Simonen M, Schnell L, Brosamle C, Kaupmann K, Vallon R, Schwab ME (2003) Nogo-A inhibits neurite outgrowth and cell spreading with three discrete regions. J Neurosci 23(13):5393–5406

Ogata S, Morokuma J, Hayata T, Kolle G, Niehrs C, Ueno N, Cho KW (2007) TGF-beta signaling-mediated morphogenesis: modulation of cell adhesion via cadherin endocytosis. Genes Dev 21(14):1817–1831

Ohira M, Morohashi A, Nakamura Y, Isogai E, Furuya K, Hamano S, Machida T, Aoyama M, Fukumura M, Miyazaki K, Suzuki Y, Sugano S, Hirato J, Nakagawara A (2003) Neuroblastoma oligo-capping cDNA project: toward the understanding of the genesis and biology of neuroblastoma. Cancer Lett 197(1–2):63–68

Okafuji T, Tanaka H (2005) Expression pattern of LINGO-1 in the developing nervous system of the chick embryo. Gene Expr Patterns 6(1):57–62. doi:10.1016/j.modgep.2005.04.016

Ono T, Sekino-Suzuki N, Kikkawa Y, Yonekawa H, Kawashima S (2003) Alivin 1, a novel neuronal activity-dependent gene, inhibits apoptosis and promotes survival of cerebellar granule neurons. J Neurosci 23(13):5887–5896

O'Prey J, Wilkinson S, Ryan KM (2008) Tumor antigen LRRC15 impedes adenoviral infection: implications for virus-based cancer therapy. J Virol 82(12):5933–5939. doi:10.1128/jvi.02273-07

O'Roak BJ, Morgan TM, Fishman DO, Saus E, Alonso P, Gratacos M, Estivill X, Teltsh O, Kohn Y, Kidd KK, Cho J, Lifton RP, State MW (2010) Additional support for the association of SLITRK1 var321 and Tourette syndrome. Mol Psychiatry 15(5):447–450. doi:10.1038/mp.2009.105

Orth M, Djarmati A, Baumer T, Winkler S, Grunewald A, Lohmann-Hedrich K, Kabakci K, Hagenah J, Klein C, Munchau A (2007) Autosomal dominant myoclonus-dystonia and Tourette syndrome in a family without linkage to the SGCE gene. Mov Disord 22(14):2090–2096. doi:10.1002/mds.21674

O'Sullivan ML, de Wit J, Savas JN, Comoletti D, Otto-Hitt S, Yates JR III, Ghosh A (2012) FLRT proteins are endogenous latrophilin ligands and regulate excitatory synapse development. Neuron 73(5):903–910. doi:10.1016/j.neuron.2012.01.018

Ota T, Suzuki Y, Nishikawa T, Otsuki T, Sugiyama T, Irie R, Wakamatsu A, Hayashi K, Sato H, Nagai K, Kimura K, Makita H, Sekine M, Obayashi M, Nishi T, Shibahara T, Tanaka T, Ishii S, Yamamoto J, Saito K, Kawai Y, Isono Y, Nakamura Y, Nagahari K, Murakami K, Yasuda T, Iwayanagi T, Wagatsuma M, Shiratori A, Sudo H, Hosoiri T, Kaku Y, Kodaira H, Kondo H, Sugawara M, Takahashi M, Kanda K, Yokoi T, Furuya T, Kikkawa E, Omura Y, Abe K, Kamihara K, Katsuta N, Sato K, Tanikawa M, Yamazaki M, Ninomiya K, Ishibashi T, Yamashita H, Murakawa K, Fujimori K, Tanai H, Kimata M, Watanabe M, Hiraoka S, Chiba Y, Ishida S, Ono Y, Takiguchi S, Watanabe S, Yosida M, Hotuta T, Kusano J, Kanehori K, Takahashi-Fujii A, Hara H, Tanase TO, Nomura Y, Togiya S, Komai F, Hara R, Takeuchi K, Arita M, Imose N, Musashino K, Yuuki H, Oshima A, Sasaki N, Aotsuka S, Yoshikawa Y, Matsunawa H, Ichihara T, Shiohata N, Sano S, Moriya S, Momiyama H, Satoh N, Takami S, Terashima Y, Suzuki O, Nakagawa S, Senoh A, Mizoguchi H, Goto Y, Shimizu F, Wakebe H, Hishigaki H, Watanabe T, Sugiyama A, Takemoto M, Kawakami B, Watanabe K, Kumagai A, Itakura S, Fukuzumi Y, Fujimori Y, Komiyama M, Tashiro H, Tanigami A, Fujiwara T, Ono T, Yamada K, Fujii Y, Ozaki K, Hirao M, Ohmori Y, Kawabata A, Hikiji T, Kobatake N, Inagaki H, Ikema Y, Okamoto S, Okitani R, Kawakami T, Noguchi S, Itoh T, Shigeta K, Senba T, Matsumura K, Nakajima Y, Mizuno T, Morinaga M, Sasaki M, Togashi T, Oyama M, Hata H, Komatsu T, Mizushima-Sugano J, Satoh T, Shirai Y, Takahashi Y, Nakagawa K, Okumura K, Nagase T, Nomura N, Kikuchi H, Masuho Y, Yamashita R, Nakai K, Yada T, Ohara O, Isogai T, Sugano S, Ota T, Suzuki Y, Nishikawa T, Otsuki T, Sugiyama T, Irie R, Wakamatsu A, Hayashi K, Sato H, Nagai K, Kimura K, Makita H, Sekine M, Obayashi M, Nishi T, Shibahara T, Tanaka T, Ishii S, Yamamoto J, Saito K, Kawai Y, Isono Y, Nakamura Y, Nagahari K, Murakami K, Yasuda T, Iwayanagi T, Wagatsuma M, Shiratori A, Sudo H, Hosoiri T, Kaku Y, Kodaira H, Kondo H, Sugawara M, Takahashi M, Kanda K, Yokoi T, Furuya T, Kikkawa E, Omura Y, Abe K, Kamihara K, Katsuta N, Sato K, Tanikawa M, Yamazaki M, Ninomiya K, Ishibashi T, Yamashita H, Murakawa K, Fujimori K, Tanai H, Kimata M, Watanabe M, Hiraoka S, Chiba Y, Ishida S, Ono Y, Takiguchi S, Watanabe S, Yosida M, Hotuta T, Kusano J, Kanehori K, Takahashi-Fujii A, Hara H, Tanase TO, Nomura Y, Togiya S, Komai F, Hara R, Takeuchi K, Arita M, Imose N, Musashino K, Yuuki H, Oshima A, Sasaki N, Aotsuka S, Yoshikawa Y, Matsunawa H, Ichihara T, Shiohata N, Sano S, Moriya S, Momiyama H, Satoh N, Takami S, Terashima Y, Suzuki O, Nakagawa S, Senoh A, Mizoguchi H, Goto Y, Shimizu F, Wakebe H, Hishigaki H, Watanabe T, Sugiyama A, Takemoto M, Kawakami B, Watanabe K, Kumagai A, Itakura S, Fukuzumi Y, Fujimori Y, Komiyama M, Tashiro H, Tanigami A, Fujiwara T, Ono

T, Yamada K, Fujii Y, Ozaki K, Hirao M, Ohmori Y, Kawabata A, Hikiji T, Kobatake N, Inagaki H, Ikema Y, Okamoto S, Okitani R, Kawakami T, Noguchi S, Itoh T, Shigeta K, Senba T, Matsumura K, Nakajima Y, Mizuno T, Morinaga M, Sasaki M, Togashi T, Oyama M, Hata H, Komatsu T, Mizushima-Sugano J, Satoh T, Shirai Y, Takahashi Y, Nakagawa K, Okumura K, Nagase T, Nomura N, Kikuchi H, Masuho Y, Yamashita R, Nakai K, Yada T, Ohara O, Isogai T, Sugano S, Ota T, Suzuki Y, Nishikawa T, Otsuki T, Sugiyama T, Irie R, Wakamatsu A, Hayashi K, Sato H, Nagai K, Kimura K, Makita H, Sekine M, Obayashi M, Nishi T, Shibahara T, Tanaka T, Ishii S, Yamamoto J, Saito K, Kawai Y, Isono Y, Nakamura Y, Nagahari K, Murakami K, Yasuda T, Iwayanagi T, Wagatsuma M, Shiratori A, Sudo H, Hosoiri T, Kaku Y, Kodaira H, Kondo H, Sugawara M, Takahashi M, Kanda K, Yokoi T, Furuya T, Kikkawa E, Omura Y, Abe K, Kamihara K, Katsuta N, Sato K, Tanikawa M, Yamazaki M, Ninomiya K, Ishibashi T, Yamashita H, Murakawa K, Fujimori K, Tanai H, Kimata M, Watanabe M, Hiraoka S, Chiba Y, Ishida S, Ono Y, Takiguchi S, Watanabe S, Yosida M, Hotuta T, Kusano J, Kanehori K, Takahashi-Fujii A, Hara H, Tanase TO, Nomura Y, Togiya S, Komai F, Hara R, Takeuchi K, Arita M, Imose N, Musashino K, Yuuki H, Oshima A, Sasaki N, Aotsuka S, Yoshikawa Y, Matsunawa H, Ichihara T, Shiohata N, Sano S, Moriya S, Momiyama H, Satoh N, Takami S, Terashima Y, Suzuki O, Nakagawa S, Senoh A, Mizoguchi H, Goto Y, Shimizu F, Wakebe H, Hishigaki H, Watanabe T, Sugiyama A, Takemoto M, Kawakami B, Watanabe K, Kumagai A, Itakura S, Fukuzumi Y, Fujimori Y, Komiyama M, Tashiro H, Tanigami A, Fujiwara T, Ono T, Yamada K, Fujii Y, Ozaki K, Hirao M, Ohmori Y, Kawabata A, Hikiji T, Kobatake N, Inagaki H, Ikema Y, Okamoto S, Okitani R, Kawakami T, Noguchi S, Itoh T, Shigeta K, Senba T, Matsumura K, Nakajima Y, Mizuno T, Morinaga M, Sasaki M, Togashi T, Oyama M, Hata H, Komatsu T, Mizushima-Sugano J, Satoh T, Shirai Y, Takahashi Y, Nakagawa K, Okumura K, Nagase T, Nomura N, Kikuchi H, Masuho Y, Yamashita R, Nakai K, Yada T, Ohara O, Isogai T, Sugano S (2004) Complete sequencing and characterization of 21,243 full-length human cDNAs. Nat Genet 36(1):40–45. doi:10.1038/ng1285

Otsuki T, Ota T, Nishikawa T, Hayashi K, Suzuki Y, Yamamoto J, Wakamatsu A, Kimura K, Sakamoto K, Hatano N, Kawai Y, Ishii S, Saito K, Kojima S, Sugiyama T, Ono T, Okano K, Yoshikawa Y, Aotsuka S, Sasaki N, Hattori A, Okumura K, Nagai K, Sugano S, Isogai T (2005) Signal sequence and keyword trap in silico for selection of full-length human cDNAs encoding secretion or membrane proteins from oligo-capped cDNA libraries. DNA Res 12(2): 117–126. doi:10.1093/dnares/12.2.117

Pan L, Zhang M (2012) Structures of Usher Syndrome 1 proteins and their complexes. Physiology 27(1):25–42. doi:10.1152/physiol.00037.2011

Park JB, Yiu G, Kaneko S, Wang J, Chang J, He XL, Garcia KC, He Z (2005) A TNF receptor family member, TROY, is a coreceptor with Nogo receptor in mediating the inhibitory activity of myelin inhibitors. Neuron 45(3):345–351. doi:10.1016/j.neuron.2004.12.040

Park JH, Widi GA, Gimbel DA, Harel NY, Lee DH, Strittmatter SM (2006) Subcutaneous Nogo receptor removes brain amyloid-beta and improves spatial memory in Alzheimer's transgenic mice. J Neurosci 26(51):13279–13286. doi:10.1523/jneurosci.4504-06.2006

Park H, Huxley-Jones J, Boot-Handford RP, Bishop PN, Attwood TK, Bella J (2008) LRRCE: a leucine-rich repeat cysteine capping motif unique to the chordate lineage. BMC Genomics 9:599. doi:10.1186/1471-2164-9-599

Parrilla-Reverter G, Agudo M, Sobrado-Calvo P, Salinas-Navarro M, Villegas-Perez MP, Vidal-Sanz M (2009) Effects of different neurotrophic factors on the survival of retinal ganglion cells after a complete intraorbital nerve crush injury: a quantitative in vivo study. Exp Eye Res 89(1): 32–41. doi:10.1016/j.exer.2009.02.015

Paschou P (2013) The genetic basis of Gilles de la Tourette Syndrome. Neurosci Biobehav Rev. doi:10.1016/j.neubiorev.2013.01.016

Pasquini M, Fabbrini G, Berardelli I, Bonifati V, Biondi M, Berardelli A (2008) Psychopathological features of obsessive-compulsive disorder in an Italian family with Gilles de la Tourette syndrome not linked to the SLITRK1 gene. Psychiatry Res 161(1):109–111. doi:10.1016/j.psychres.2008.02.012

Peltola MA, Kuja-Panula J, Lauri SE, Taira T, Rauvala H (2011) AMIGO is an auxiliary subunit of the Kv2.1 potassium channel. EMBO Rep 12(12):1293–1299. doi:10.1038/embor.2011.204

Pepinsky RB, Silvian L, Berkowitz SA, Farrington G, Lugovskoy A, Walus L, Eldredge J, Capili A, Mi S, Graff C, Garber E (2010) Improving the solubility of anti-LINGO-1 monoclonal antibody Li33 by isotype switching and targeted mutagenesis. Protein Sci 19(5):954–966. doi:10.1002/pro.372

Pepinsky RB, Shao Z, Ji B, Wang Q, Meng G, Walus L, Lee X, Hu Y, Graff C, Garber E, Meier W, Mi S (2011a) Exposure levels of anti-LINGO-1 Li81 antibody in the central nervous system and dose-efficacy relationships in rat spinal cord remyelination models after systemic administration. J Pharmacol Exp Ther 339(2):519–529. doi:10.1124/jpet.111.183483

Pepinsky RB, Walus L, Shao Z, Ji B, Gu S, Sun Y, Wen D, Lee X, Wang Q, Garber E, Mi S (2011b) Production of a PEGylated Fab' of the anti-LINGO-1 Li33 antibody and assessment of its biochemical and functional properties in vitro and in a rat model of remyelination. Bioconjug Chem 22(2):200–210. doi:10.1021/bc1002746

Pericak-Vance MA, Bass MP, Yamaoka LH, Gaskell PC, Scott WK, Terwedow HA, Menold MM, Conneally PM, Small GW, Vance JM, Saunders AM, Roses AD, Haines JL (1997) Complete genomic screen in late-onset familial Alzheimer disease. Evidence for a new locus on chromosome 12. JAMA 278(15):1237–1241

Pession A, Tonelli R (2005) The MYCN oncogene as a specific and selective drug target for peripheral and central nervous system tumors. Curr Cancer Drug Targets 5(4):273–283

Piton A, Gauthier J, Hamdan FF, Lafreniere RG, Yang Y, Henrion E, Laurent S, Noreau A, Thibodeau P, Karemera L, Spiegelman D, Kuku F, Duguay J, Destroismaisons L, Jolivet P, Cote M, Lachapelle K, Diallo O, Raymond A, Marineau C, Champagne N, Xiong L, Gaspar C, Riviere JB, Tarabeux J, Cossette P, Krebs MO, Rapoport JL, Addington A, Delisi LE, Mottron L, Joober R, Fombonne E, Drapeau P, Rouleau GA (2011) Systematic resequencing of X-chromosome synaptic genes in autism spectrum disorder and schizophrenia. Mol Psychiatry 16(8):867–880. doi:10.1038/mp.2010.54

Pokutta S, Drees F, Yamada S, Nelson WJ, Weis WI (2008) Biochemical and structural analysis of alpha-catenin in cell-cell contacts. Biochem Soc Trans 36(Pt 2):141–147. doi:10.1042/bst0360141

Powell AE, Wang Y, Li Y, Poulin EJ, Means AL, Washington MK, Higginbotham JN, Juchheim A, Prasad N, Levy SE, Guo Y, Shyr Y, Aronow BJ, Haigis KM, Franklin JL, Coffey RJ (2012) The pan-ErbB negative regulator Lrig1 is an intestinal stem cell marker that functions as a tumor suppressor. Cell 149(1):146–158. doi:10.1016/j.cell.2012.02.042

Proenca CC, Gao KP, Shmelkov SV, Rafii S, Lee FS (2011) Slitrks as emerging candidate genes involved in neuropsychiatric disorders. Trends Neurosci 34(3):143–153. doi:10.1016/j.tins.2011.01.001

Rabenau KE, O'Toole JM, Bassi R, Kotanides H, Witte L, Ludwig DL, Pereira DS (2004) DEGA/AMIGO-2, a leucine-rich repeat family member, differentially expressed in human gastric adenocarcinoma: effects on ploidy, chromosomal stability, cell adhesion/migration and tumorigenicity. Oncogene 23(29):5056–5067. doi:10.1038/sj.onc.1207681

Rajasekharan S, Kennedy TE (2009) The netrin protein family. Genome Biol 10(9):239. doi:10.1186/gb-2009-10-9-239

Reddy GK, Balk S (2006) Clinical utility of microarray-derived genetic signatures in predicting outcomes in prostate cancer. Clin Genitourin Cancer 5(3):187–189. doi:10.3816/CGC.2006.n.035

Reitz C, Conrad C, Roszkowski K, Rogers RS, Mayeux R (2012) Effect of genetic variation in LRRTM3 on risk of Alzheimer disease. Arch Neurol 69(7):894–900. doi:10.1001/archneurol.2011.2463

Renaudineau Y, Dugue C, Dueymes M, Youinou P (2002) Antiendothelial cell antibodies in systemic lupus erythematosus. Autoimmun Rev 1(6):365–372

Reynolds PA, Smolen GA, Palmer RE, Sgroi D, Yajnik V, Gerald WL, Haber DA (2003) Identification of a DNA-binding site and transcriptional target for the EWS-WT1(+KTS) oncoprotein. Genes Dev 17(17):2094–2107. doi:10.1101/gad.1110703

Ricciardi S, Ungaro F, Hambrock M, Rademacher N, Stefanelli G, Brambilla D, Sessa A, Magagnotti C, Bachi A, Giarda E, Verpelli C, Kilstrup-Nielsen C, Sala C, Kalscheuer VM, Broccoli V (2012) CDKL5 ensures excitatory synapse stability by reinforcing NGL-1-PSD95 interaction in the postsynaptic compartment and is impaired in patient iPSC-derived neurons. Nat Cell Biol 14(9):911–923. doi:10.1038/ncb2566

Rivers LE, Young KM, Rizzi M, Jamen F, Psachoulia K, Wade A, Kessaris N, Richardson WD (2008) PDGFRA/NG2 glia generate myelinating oligodendrocytes and piriform projection neurons in adult mice. Nat Neurosci 11(12):1392–1401. doi:10.1038/nn.2220

Robinson M, Parsons Perez MC, Tebar L, Palmer J, Patel A, Marks D, Sheasby A, De Felipe C, Coffin R, Livesey FJ, Hunt SP (2004) FLRT3 is expressed in sensory neurons after peripheral nerve injury and regulates neurite outgrowth. Mol Cell Neurosci 27(2):202–214

Romeo V (2012) Myotonic dystrophy Type 1 or Steinert's disease. In: Ahmad S (ed) Neurodegenerative diseases, vol 726, Advances in experimental medicine and biology. Springer, US, pp 239–257, doi:10.1007/978-1-4614-0653-2_18

Rual JF, Venkatesan K, Hao T, Hirozane-Kishikawa T, Dricot A, Li N, Berriz GF, Gibbons FD, Dreze M, Ayivi-Guedehoussou N, Klitgord N, Simon C, Boxem M, Milstein S, Rosenberg J, Goldberg DS, Zhang LV, Wong SL, Franklin G, Li S, Albala JS, Lim J, Fraughton C, Llamosas E, Cevik S, Bex C, Lamesch P, Sikorski RS, Vandenhaute J, Zoghbi HY, Smolyar A, Bosak S, Sequerra R, Doucette-Stamm L, Cusick ME, Hill DE, Roth FP, Vidal M (2005) Towards a proteome-scale map of the human protein-protein interaction network. Nature 437(7062): 1173–1178. doi:10.1038/nature04209

Rubio-Perez JM, Morillas-Ruiz JM (2012) A review: inflammatory process in alzheimer's disease, role of cytokines. Scientific World J 2012:15. doi:10.1100/2012/756357

Rudick RA, Mi S, Sandrock AW (2008) LINGO-1 antagonists as therapy for multiple sclerosis: in vitro and in vivo evidence. Expert Opin Biol Ther 8(10):1561–1570. doi:10.1517/14712598.8.10.1561

Sajan SA, Rubenstein JL, Warchol ME, Lovett M (2011) Identification of direct downstream targets of Dlx5 during early inner ear development. Hum Mol Genet 20(7):1262–1273. doi:10.1093/hmg/ddq567

Sandor P (2003) Pharmacological management of tics in patients with TS. J Psychosom Res 55(1):41–48. doi:10.1016/S0022-3999(03)00060-6

Santa-Coloma T (2003) Anp32e (Cpd1) and related protein phosphatase 2 inhibitors. Cerebellum 2(4):310–320. doi:10.1080/14734220310017212

Satoh K, Hata M, Yokota H (2002) A novel member of the leucine-rich repeat superfamily induced in rat astrocytes by beta-amyloid. Biochem Biophys Res Commun 290(2):756–762. doi:10.1006/bbrc.2001.6272

Satoh K, Hata M, Yokota H (2004) High lib mRNA expression in breast carcinomas. DNA Res 11(3):199–203

Satoh K, Hata M, Shimizu T, Yokota H, Akatsu H, Yamamoto T, Kosaka K, Yamada T (2005) Lib, transcriptionally induced in senile plaque-associated astrocytes, promotes glial migration through extracellular matrix. Biochem Biophys Res Commun 335(2):631–636. doi:10.1016/j.bbrc.2005.07.117

Scharf JM, Moorjani P, Fagerness J, Platko JV, Illmann C, Galloway B, Jenike E, Stewart SE, Pauls DL (2008) Lack of association between SLITRK1var321 and Tourette syndrome in a large family-based sample. Neurology 70(16 Pt 2):1495–1496. doi:10.1212/01.wnl.0000296833.25484.bb

Scherer SW, Cheung J, MacDonald JR, Osborne LR, Nakabayashi K, Herbrick JA, Carson AR, Parker-Katiraee L, Skaug J, Khaja R, Zhang J, Hudek AK, Li M, Haddad M, Duggan GE, Fernandez BA, Kanematsu E, Gentles S, Christopoulos CC, Choufani S, Kwasnicka D, Zheng XH, Lai Z, Nusskern D, Zhang Q, Gu Z, Lu F, Zeesman S, Nowaczyk MJ, Teshima I, Chitayat D, Shuman C, Weksberg R, Zackai EH, Grebe TA, Cox SR, Kirkpatrick SJ, Rahman N, Friedman JM, Heng HH, Pelicci PG, Lo-Coco F, Belloni E, Shaffer LG, Pober B, Morton CC, Gusella JF, Bruns GA, Korf BR, Quade BJ, Ligon AH, Ferguson H, Higgins AW, Leach NT, Herrick SR, Lemyre E, Farra CG, Kim HG, Summers AM, Gripp KW, Roberts W, Szatmari

P, Winsor EJ, Grzeschik KH, Teebi A, Minassian BA, Kere J, Armengol L, Pujana MA, Estivill X, Wilson MD, Koop BF, Tosi S, Moore GE, Boright AP, Zlotorynski E, Kerem B, Kroisel PM, Petek E, Oscier DG, Mould SJ, Dohner H, Dohner K, Rommens JM, Vincent JB, Venter JC, Li PW, Mural RJ, Adams MD, Tsui LC (2003) Human chromosome 7: DNA sequence and biology. Science 300(5620):767–772. doi:10.1126/science.1083423

Schnaar RL, Lopez PHH (2009) Myelin-associated glycoprotein and its axonal receptors. J Neurosci Res 87(15):3267–3276. doi:10.1002/jnr.21992

Schuetz CS, Bonin M, Clare SE, Nieselt K, Sotlar K, Walter M, Fehm T, Solomayer E, Riess O, Wallwiener D, Kurek R, Neubauer HJ (2006) Progression-specific genes identified by expression profiling of matched ductal carcinomas in situ and invasive breast tumors, combining laser capture microdissection and oligonucleotide microarray analysis. Cancer Res 66(10): 5278–5286. doi:10.1158/0008-5472.can-05-4610

Scott IC (2012) Chapter one – Life before Nkx2.5: cardiovascular progenitor cells: embryonic origins and development. In: Benoit GB (ed) Current topics in developmental biology, vol 100. Academic, pp 1–31. doi:10.1016/B978-0-12-387786-4.00001-4

Seabold GK, Wang PY, Chang K, Wang CY, Wang YX, Petralia RS, Wenthold RJ (2008) The SALM family of adhesion-like molecules forms heteromeric and homomeric complexes. J Biol Chem 283(13):8395–8405. doi:10.1074/jbc.M709456200

Seabold GK, Wang PY, Petralia RS, Chang K, Zhou A, McDermott MI, Wang YX, Milgram SL, Wenthold RJ (2012) Dileucine and PDZ-binding motifs mediate synaptic adhesion-like molecule 1 (SALM1) trafficking in hippocampal neurons. J Biol Chem 287(7):4470–4484. doi:10.1074/jbc.M111.279661

Seiradake E, von Philipsborn AC, Henry M, Fritz M, Lortat-Jacob H, Jamin M, Hemrika W, Bastmeyer M, Cusack S, McCarthy AA (2009) Structure and functional relevance of the Slit2 homodimerization domain. EMBO Rep 10:736–741

Seiradake E, Coles CH, Perestenko PV, Harlos K, McIlhinney RA, Aricescu AR, Jones EY (2011) Structural basis for cell surface patterning through NetrinG-NGL interactions. EMBO J 30(21): 4479–4488. doi:10.1038/emboj.2011.346

Sghendo L, Mifsud J (2012) Understanding the molecular pharmacology of the serotonergic system: using fluoxetine as a model. J Pharm Pharmacol 64(3):317–325. doi:10.1111/j.2042-7158.2011.01384.x

Shao Z, Browning JL, Lee X, Scott ML, Shulga-Morskaya S, Allaire N, Thill G, Levesque M, Sah D, McCoy JM, Murray B, Jung V, Pepinsky RB, Mi S (2005) TAJ/TROY, an orphan TNF receptor family member, binds Nogo-66 receptor 1 and regulates axonal regeneration. Neuron 45(3):353–359. doi:10.1016/j.neuron.2004.12.050

Shattuck DL, Miller JK, Laederich M, Funes M, Petersen H, Carraway KL III, Sweeney C (2007) LRIG1 is a novel negative regulator of the Met receptor and opposes Met and Her2 synergy. Mol Cell Biol 27(5):1934–1946. doi:10.1128/mcb.00757-06

Sherwood M, Thornton AE, Honer WG (2012) A quantitative review of the profile and time course of symptom change in schizophrenia treated with clozapine. J Psychopharmacol 26(9): 1175–1184. doi:10.1177/0269881112440513

Shirai T, Fujii H, Ono M, Nakamura K, Watanabe R, Tajima Y, Takasawa N, Ishii T, Harigae H (2012) A novel autoantibody against fibronectin leucine-rich transmembrane protein 2 expressed on the endothelial cell surface identified by retroviral vector system in systemic lupus erythematosus. Arthritis Res Ther 14(4):R157. doi:10.1186/ar3897

Shirai T, Fujii H, Ono M, Watanabe R, Ishii T, Harigae H (2013) An Innovative Method to Identify Autoantigens Expressed on the Endothelial Cell Surface: Serological Identification System for Autoantigens Using a Retroviral Vector and Flow Cytometry (SARF). Clin Dev Immunol 2013:453058. doi:10.1155/2013/453058

Shmelkov SV, Hormigo A, Jing D, Proenca CC, Bath KG, Milde T, Shmelkov E, Kushner JS, Baljevic M, Dincheva I, Murphy AJ, Valenzuela DM, Gale NW, Yancopoulos GD, Ninan I, Lee FS, Rafii S (2010) Slitrk5 deficiency impairs corticostriatal circuitry and leads to obsessive-compulsive-like behaviors in mice. Nat Med 16(5):598–602, 591p following 602. doi:10.1038/nm.2125

Siddiqui TJ, Pancaroglu R, Kang Y, Rooyakkers A, Craig AM (2010) LRRTMs and neuroligins bind neurexins with a differential code to cooperate in glutamate synapse development. J Neurosci 30(22):7495–7506. doi:10.1523/jneurosci.0470-10.2010

Silva JP, Ushkaryov YA (2010) The latrophilins, "split-personality" receptors. Adv Exp Med Biol 706:59–75

Sirerol-Piquer MS, Ayerdi-Izquierdo A, Morante-Redolat JM, Herranz-Perez V, Favell K, Barker PA, Perez-Tur J (2006) The epilepsy gene LGI1 encodes a secreted glycoprotein that binds to the cell surface. Hum Mol Genet 15(23):3436–3445. doi:10.1093/hmg/ddl421

Skaper S (2012a) The neurotrophin family of neurotrophic factors: an overview. In: Skaper SD (ed) Neurotrophic factors, vol 846. Methods in molecular biology. Humana Press, pp 1–12. doi:10.1007/978-1-61779-536-7_1

Skaper SD (2012b) The neurotrophin family of neurotrophic factors: an overview. Methods Mol Biol 846:1–12, doi:10.1007/978-1-61779-536-7_1

Smith TG, Tickle C (2006) The expression of Flrt3 during chick limb development. Int J Dev Biol 50(8):701–704

Smith EN, Bloss CS, Badner JA, Barrett T, Belmonte PL, Berrettini W, Byerley W, Coryell W, Craig D, Edenberg HJ, Eskin E, Foroud T, Gershon E, Greenwood TA, Hipolito M, Koller DL, Lawson WB, Liu C, Lohoff F, McInnis MG, McMahon FJ, Mirel DB, Murray SS, Nievergelt C, Nurnberger J, Nwulia EA, Paschall J, Potash JB, Rice J, Schulze TG, Scheftner W, Panganiban C, Zaitlen N, Zandi PP, Zollner S, Schork NJ, Kelsoe JR (2009) Genome-wide association study of bipolar disorder in European American and African American individuals. Mol Psychiatry 14(8):755–763. doi:10.1038/mp.2009.43

Smith AJ, Daut J, Schwappach B (2011a) Membrane Proteins as 14-3-3 Clients in Functional Regulation and Intracellular Transport. Physiology 26(3):181–191. doi:10.1152/physiol.00042.2010

Smith JD, Meehan MH, Crean J, McCann A (2011b) Alpha T-catenin (CTNNA3): a gene in the hand is worth two in the nest. Cell Mol Life Sci 68(15):2493–2498. doi:10.1007/s00018-011-0728-0

Soellner C, Wright G (2009) A cell surface interaction network of neural leucine-rich repeat receptors. Genome Biol 10(9):R99

Soler-Llavina GJ, Fuccillo MV, Ko J, Sudhof TC, Malenka RC (2011) The neurexin ligands, neuroligins and leucine-rich repeat transmembrane proteins, perform convergent and divergent synaptic functions in vivo. Proc Natl Acad Sci U S A 108(40):16502–16509. doi:10.1073/pnas.1114028108

Sousa I, Clark TG, Holt R, Pagnamenta AT, Mulder EJ, Minderaa RB, Bailey AJ, Battaglia A, Klauck SM, Poustka F, Monaco AP (2010) Polymorphisms in leucine-rich repeat genes are associated with autism spectrum disorder susceptibility in populations of European ancestry. Mol Autism 1(1):7. doi:10.1186/2040-2392-1-7

Stanbrough M, Bubley GJ, Ross K, Golub TR, Rubin MA, Penning TM, Febbo PG, Balk SP (2006) Increased expression of genes converting adrenal androgens to testosterone in androgen-independent prostate cancer. Cancer Res 66(5):2815–2825. doi:10.1158/0008-5472.can-05-4000

Staub E, Pérez-Tur J, Siebert R, Nobile C, Moschonas NK, Deloukas P, Hinzmann B (2002) The novel EPTP repeat defines a superfamily of proteins implicated in epileptic disorders. Trends Biochem Sci 27(9):441–444. doi:10.1016/s0968-0004(02)02163-1

Stefansson H, Steinberg S, Petursson H, Gustafsson O, Gudjonsdottir IH, Jonsdottir GA, Palsson ST, Jonsson T, Saemundsdottir J, Bjornsdottir G, Bottcher Y, Thorlacius T, Haubenberger D, Zimprich A, Auff E, Hotzy C, Testa CM, Miyatake LA, Rosen AR, Kristleifsson K, Rye D, Asmus F, Schols L, Dichgans M, Jakobsson F, Benedikz J, Thorsteinsdottir U, Gulcher J, Kong A, Stefansson K (2009) Variant in the sequence of the LINGO1 gene confers risk of essential tremor. Nat Genet 41(3):277–279. doi:10.1038/ng.299

Stillman AA, Krsnik Z, Sun J, Rasin MR, State MW, Sestan N, Louvi A (2009) Developmentally regulated and evolutionarily conserved expression of SLITRK1 in brain circuits implicated in Tourette syndrome. J Comp Neurol 513(1):21–37. doi:10.1002/cne.21919

Strausberg RL, Feingold EA, Grouse LH, Derge JG, Klausner RD, Collins FS, Wagner L, Shenmen CM, Schuler GD, Altschul SF, Zeeberg B, Buetow KH, Schaefer CF, Bhat NK, Hopkins RF,

Jordan H, Moore T, Max SI, Wang J, Hsieh F, Diatchenko L, Marusina K, Farmer AA, Rubin GM, Hong L, Stapleton M, Soares MB, Bonaldo MF, Casavant TL, Scheetz TE, Brownstein MJ, Usdin TB, Toshiyuki S, Carninci P, Prange C, Raha SS, Loquellano NA, Peters GJ, Abramson RD, Mullahy SJ, Bosak SA, McEwan PJ, McKernan KJ, Malek JA, Gunaratne PH, Richards S, Worley KC, Hale S, Garcia AM, Gay LJ, Hulyk SW, Villalon DK, Muzny DM, Sodergren EJ, Lu X, Gibbs RA, Fahey J, Helton E, Ketteman M, Madan A, Rodrigues S, Sanchez A, Whiting M, Young AC, Shevchenko Y, Bouffard GG, Blakesley RW, Touchman JW, Green ED, Dickson MC, Rodriguez AC, Grimwood J, Schmutz J, Myers RM, Butterfield YS, Krzywinski MI, Skalska U, Smailus DE, Schnerch A, Schein JE, Jones SJ, Marra MA (2002) Generation and initial analysis of more than 15,000 full-length human and mouse cDNA sequences. Proc Natl Acad Sci U S A 99(26):16899–16903. doi:10.1073/pnas.242603899

Suh I, Shibru D, Eisenhofer G, Pacak K, Duh QY, Clark OH, Kebebew E (2009) Candidate genes associated with malignant pheochromocytomas by genome-wide expression profiling. Ann Surg 250(6):983–990. doi:10.1097/SLA.0b013e3181b248bb

Sun T, Walsh CA (2006) Molecular approaches to brain asymmetry and handedness. Nat Rev Neurosci 7(655–562)

Swanwick CC, Shapiro ME, Vicini S, Wenthold RJ (2010) Flotillin-1 mediates neurite branching induced by synaptic adhesion-like molecule 4 in hippocampal neurons. Mol Cell Neurosci 45(3):213–225. doi:10.1016/j.mcn.2010.06.012

Taguchi A, Wanaka A, Mori T, Matsumoto K, Imai Y, Tagaki T, Tohyama M (1996) Molecular cloning of novel leucine-rich repeat proteins and their expression in the developing mouse nervous system. Brain Res Mol Brain Res 35(1–2):31–40

Takahashi N, Takahashi Y, Putnam FW (1985) Periodicity of leucine and tandem repetition of a 24-amino acid segment in the primary structure of leucine-rich alpha 2-glycoprotein of human serum. Proc Natl Acad Sci 82(7):1906–1910

Takahashi H, Katayama K, Sohya K, Miyamoto H, Prasad T, Matsumoto Y, Ota M, Yasuda H, Tsumoto T, Aruga J, Craig AM (2012) Selective control of inhibitory synapse development by Slitrk3-PTPdelta trans-synaptic interaction. Nat Neurosci 15(3):389–398. doi:10.1038/nn.3040, S381-382

Takashima N, Odaka YS, Sakoori K, Akagi T, Hashikawa T, Morimura N, Yamada K, Aruga J (2011) Impaired cognitive function and altered hippocampal synapse morphology in mice lacking Lrrtm1, a gene associated with schizophrenia. PLoS One 6(7):e22716. doi:10.1371/journal.pone.0022716

Tan J, Evin G (2012) β-Site APP-cleaving enzyme 1 trafficking and Alzheimer's disease pathogenesis. J Neurochem 120(6):869–880. doi:10.1111/j.1471-4159.2011.07623.x

Tanabe K, Bonilla I, Winkles JA, Strittmatter SM (2003) Fibroblast growth factor-inducible-14 is induced in axotomized neurons and promotes neurite outgrowth. J Neurosci 23(29):9675–9686

Tanemura A, Nagasawa T, Inui S, Itami S (2005) LRIG-1 provides a novel prognostic predictor in squamous cell carcinoma of the skin: immunohistochemical analysis for 38 cases. Dermatol Surg 31(4):423–430

Tang H, Liu X, Wang Z, She X, Zeng X, Deng M, Liao Q, Guo X, Wang R, Li X, Zeng F, Wu M, Li G (2011) Interaction of hsa-miR-381 and glioma suppressor LRRC4 is involved in glioma growth. Brain Res 1390:21–32. doi:10.1016/j.brainres.2011.03.034

Tang H, Wang Z, Liu X, Liu Q, Xu G, Li G, Wu M (2012) LRRC4 inhibits glioma cell growth and invasion through a miR-185-dependent pathway. Curr Cancer Drug Targets 12(8):1032–1042

Taylor TD, Noguchi H, Totoki Y, Toyoda A, Kuroki Y, Dewar K, Lloyd C, Itoh T, Takeda T, Kim DW, She X, Barlow KF, Bloom T, Bruford E, Chang JL, Cuomo CA, Eichler E, FitzGerald MG, Jaffe DB, LaButti K, Nicol R, Park HS, Seaman C, Sougnez C, Yang X, Zimmer AR, Zody MC, Birren BW, Nusbaum C, Fujiyama A, Hattori M, Rogers J, Lander ES, Sakaki Y (2006) Human chromosome 11 DNA sequence and analysis including novel gene identification. Nature 440(7083):497–500. doi:10.1038/nature04632

Thalhammer A, Trinidad JC, Burlingame AL, Schoepfer R (2009) Densin-180: revised membrane topology, domain structure and phosphorylation status. J Neurochem 109(2):297–302. doi:10.1111/j.1471-4159.2009.05951.x

Thomasson M, Wang B, Hammarsten P, Dahlman A, Persson JL, Josefsson A, Stattin P, Granfors T, Egevad L, Henriksson R, Bergh A, Hedman H (2011) LRIG1 and the liar paradox in prostate cancer: a study of the expression and clinical significance of LRIG1 in prostate cancer. Int J Cancer 128(12):2843–2852. doi:10.1002/ijc.25820

Thornton-Wells TA, Moore JH, Martin ER, Pericak-Vance MA, Haines JL (2008) Confronting complexity in late-onset alzheimer disease: application of two-stage analysis approach addressing heterogeneity and epistasis. Genet Epidemiol 32(3):187–203. doi:10.1002/gepi.20294

Ting JP, Kastner DL, Hoffman HM (2006) CATERPILLERs, pyrin and hereditary immunological disorders. Nat Rev Immunol 6(3):183–195. doi:10.1038/nri1788

Tomas AR, Certal AC, Rodriguez-Leon J (2011) FLRT3 as a key player on chick limb development. Dev Biol 355(2):324–333. doi:10.1016/j.ydbio.2011.04.031

Tsika E, Moore D (2012) Mechanisms of LRRK2-Mediated Neurodegeneration. Curr Neurol Neurosci Rep 12(3):251–260. doi:10.1007/s11910-012-0265-8

Tsuji L, Yamashita T, Kubo T, Madura T, Tanaka H, Hosokawa K, Tohyama M (2004) FLRT3, a cell surface molecule containing LRR repeats and a FNIII domain, promotes neurite outgrowth. Biochem Biophys Res Commun 313(4):1086–1091

van der Hoorn A, Burger H, Leenders KL, de Jong BM (2012) Handedness correlates with the dominant Parkinson side: A systematic review and meta-analysis. Mov Disord 27(2):206–210. doi:10.1002/mds.24007

van Dijk M, van Bezu J, Poutsma A, Veerhuis R, Rozemuller AJ, Scheper W, Blankenstein MA, Oudejans CB (2010) The pre-eclampsia gene STOX1 controls a conserved pathway in placenta and brain upregulated in late-onset alzheimer's disease. J Alzheimers Dis 19(2):673–679. doi:10.3233/jad-2010-1265

van Rooijen E, Giles RH, Voest EE, van Rooijen C, Schulte-Merker S, van Eeden FJ (2008) LRRC50, a conserved ciliary protein implicated in polycystic kidney disease. J Am Soc Nephrol 19(6):1128–1138. doi:10.1681/asn.2007080917

Vauhkonen H, Vauhkonen M, Sipponen P, Knuutila S (2007) Oligonucleotide array comparative genomic hybridization refines the structure of 8p23.1, 17q12 and 20q13.2 amplifications in gastric carcinomas. Cytogenet Genome Res 119(1–2):39–45. doi:10.1159/000109617

Vilarino-Guell C, Wider C, Ross OA, Jasinska-Myga B, Kachergus J, Cobb SA, Soto-Ortolaza AI, Behrouz B, Heckman MG, Diehl NN, Testa CM, Wszolek ZK, Uitti RJ, Jankovic J, Louis ED, Clark LN, Rajput A, Farrer MJ (2010) LINGO1 and LINGO2 variants are associated with essential tremor and Parkinson disease. Neurogenetics 11(4):401–408. doi:10.1007/s10048-010-0241-x

Voikar V, Kulesskaya N, Laakso T, Lauren J, Strittmatter SM, Airaksinen MS (2013) LRRTM1-deficient mice show a rare phenotype of avoiding small enclosures-A tentative mouse model for claustrophobia-like behaviour. Behav Brain Res 238:69–78. doi:10.1016/j.bbr.2012.10.013

Wadelin F, Fulton J, McEwan P, Spriggs K, Emsley J, Heery D (2010) Leucine-rich repeat protein PRAME: expression, potential functions and clinical implications for leukaemia. Mol Cancer 9(1):226

Wang P, Wenthold R (2009) Synaptic Adhesion-Like Molecules (SALMs). In: Umemori H, Hortsch M (eds) The Sticky Synapse. Springer, New York, pp 367–383

Wang KC, Kim JA, Sivasankaran R, Segal R, He Z (2002) P75 interacts with the Nogo receptor as a co-receptor for Nogo, MAG and OMgp. Nature 420(6911):74–78. doi:10.1038/nature01176

Wang CY, Chang K, Petralia RS, Wang YX, Seabold GK, Wenthold RJ (2006) A novel family of adhesion-like molecules that interacts with the NMDA receptor. J Neurosci 26(8):2174–2183. doi:10.1523/jneurosci.3799-05.2006

Wang PY, Seabold GK, Wenthold RJ (2008) Synaptic adhesion-like molecules (SALMs) promote neurite outgrowth. Mol Cell Neurosci 39(1):83–94. doi:10.1016/j.mcn.2008.05.019

Wang B, Han L, Chen R, Cai M, Han F, Lei T, Guo D (2009a) Downregulation of LRIG2 expression by RNA interference inhibits glioblastoma cell (GL15) growth, causes cell cycle redistribution, increases cell apoptosis and enhances cell adhesion and invasion in vitro. Cancer Biol Ther 8(11):1018–1023

Wang K, Zhang H, Ma D, Bucan M, Glessner JT, Abrahams BS, Salyakina D, Imielinski M, Bradfield JP, Sleiman PMA, Kim CE, Hou C, Frackelton E, Chiavacci R, Takahashi N, Sakurai T, Rappaport E, Lajonchere CM, Munson J, Estes A, Korvatska O, Piven J, Sonnenblick LI, Alvarez Retuerto AI, Herman EI, Dong H, Hutman T, Sigman M, Ozonoff S, Klin A, Owley T, Sweeney JA, Brune CW, Cantor RM, Bernier R, Gilbert JR, Cuccaro ML, McMahon WM, Miller J, State MW, Wassink TH, Coon H, Levy SE, Schultz RT, Nurnberger JI, Haines JL, Sutcliffe JS, Cook EH, Minshew NJ, Buxbaum JD, Dawson G, Grant SFA, Geschwind DH, Pericak-Vance MA, Schellenberg GD, Hakonarson H (2009b) Common genetic variants on 5p14.1 associate with autism spectrum disorders. Nature 459(7246):528–533, http://www.nature.com/nature/journal/v459/n7246/suppinfo/nature07999_S1.html

Wang R, Wang Z, Yang J, Liu X, Wang L, Guo X, Zeng F, Wu M, Li G (2011) LRRC4 inhibits the proliferation of human glioma cells by modulating the expression of STMN1 and microtubule polymerization. J Cell Biochem 112(12):3621–3629. doi:10.1002/jcb.23293

Weber AJ, Harman CD, Viswanathan S (2008) Effects of optic nerve injury, glaucoma, and neuroprotection on the survival, structure, and function of ganglion cells in the mammalian retina. J Physiol 586(Pt 18):4393–4400. doi:10.1113/jphysiol.2008.156729

Wei K, Xu Y, Tse H, Manolson MF, Gong SG (2011) Mouse FLRT2 interacts with the extracellular and intracellular regions of FGFR2. J Dent Res 90(10):1234–1239. doi:10.1177/0022034511415272

Wekerle H (2008) Lessons from multiple sclerosis: models, concepts, observations. Ann Rheum Dis 67(Suppl 3):iii56–60. doi:10.1136/ard.2008.098020

Wendland JR, Kruse MR, Murphy DL (2006) Functional SLITRK1 var321, varCDfs and SLC6A4 G56A variants and susceptibility to obsessive-compulsive disorder. Mol Psychiatry 11(9): 802–804. doi:10.1038/sj.mp.4001848

Wheldon LM, Haines BP, Rajappa R, Mason I, Rigby PW, Heath JK (2010) Critical role of FLRT1 phosphorylation in the interdependent regulation of FLRT1 function and FGF receptor signalling. PLoS One 5(4):e10264. doi:10.1371/journal.pone.0010264

Wickenden AD (2002) Potassium channels as anti-epileptic drug targets. Neuropharmacology 43(7):1055–1060

Wijeyewickrema LC, Berndt MC, Andrews RK (2005) Snake venom probes of platelet adhesion receptors and their ligands. Toxicon 45(8):1051–1061. doi:10.1016/j.toxicon.2005.02.025

Williams C, Edvardsson K, Lewandowski SA, Strom A, Gustafsson JA (2008) A genome-wide study of the repressive effects of estrogen receptor beta on estrogen receptor alpha signaling in breast cancer cells. Oncogene 27:1019–1032

Woo J, Kwon S-K, Kim E (2009a) The NGL family of leucine-rich repeat-containing synaptic adhesion molecules. Mol Cell Neurosci 42(1):1–10. doi:10.1016/j.mcn.2009.05.008

Woo J, Kwon SK, Choi S, Kim S, Lee JR, Dunah AW, Sheng M, Kim E (2009b) Trans-synaptic adhesion between NGL-3 and LAR regulates the formation of excitatory synapses. Nat Neurosci 12(4):428–437. doi:10.1038/nn.2279

Wright GJ, Washbourne P (2011) Neurexins, Neuroligins and LRRTMs: synaptic adhesion getting fishy. J Neurochem 117(5):765–778. doi:10.1111/j.1471-4159.2010.07141.x

Wu M, Huang C, Gan K, Huang H, Chen Q, Ouyang J, Tang Y, Li X, Yang Y, Zhou H, Zhou Y, Zeng Z, Xiao L, Li D, Tang K, Shen S, Li G (2006) LRRC4, a putative tumor suppressor gene, requires a functional leucine-rich repeat cassette domain to inhibit proliferation of glioma cells in vitro by modulating the extracellular signal-regulated kinase/protein kinase B/nuclear factor-kappaB pathway. Mol Biol Cell 17(8):3534–3542. doi:10.1091/mbc.E05-11-1082

Wu M, Huang H, Chen Q, Li D, Zheng Z, Xiong W, Zhou Y, Li X, Zhou M, Lu J, Shen S, Li G (2007) Leucine-rich repeat C4 protein is involved in nervous tissue development and neurite outgrowth, and induction of glioma cell differentiation. Acta Biochim Biophys Sin (Shanghai) 39(10):731–738

Wu M, Huang C, Li X, Gan K, Chen Q, Tang Y, Tang K, Shen S, Li G (2008) LRRC4 inhibits glioblastoma cell proliferation, migration, and angiogenesis by downregulating pleiotropic cytokine expression and responses. J Cell Physiol 214(1):65–74. doi:10.1002/jcp.21163

Wu X, Hedman H, Bergqvist M, Bergstrom S, Henriksson R, Gullbo J, Lennartsson J, Hesselius P, Ekman S (2012) Expression of EGFR and LRIG proteins in oesophageal carcinoma with emphasis on patient survival and cellular chemosensitivity. Acta Oncol 51(1):69–76. doi:10.3109/0284186x.2011.562239

Xie R, Yang H, Xiao Q, Mao F, Zhang S, Ye F, Wan F, Wang B, Lei T, Guo D (2013) Downregulation of LRIG1 expression by RNA interference promotes the aggressive properties of glioma cells via EGFR/Akt/c-Myc activation. Oncol Rep 29(1):177–184. doi:10.3892/or.2012.2102

Xu B, Woodroffe A, Rodriguez-Murillo L, Roos JL, van Rensburg EJ, Abecasis GR, Gogos JA, Karayiorgou M (2009) Elucidating the genetic architecture of familial schizophrenia using rare copy number variant and linkage scans. Proc Natl Acad Sci 106(39):16746–16751. doi:10.1073/pnas.0908584106

Yamagishi S, Hampel F, Hata K, Del Toro D, Schwark M, Kvachnina E, Bastmeyer M, Yamashita T, Tarabykin V, Klein R, Egea J (2011) FLRT2 and FLRT3 act as repulsive guidance cues for Unc5-positive neurons. EMBO J. doi:10.1038/emboj.2011.189

Yamashiro S, Gokhin DS, Kimura S, Nowak RB, Fowler VM (2012) Tropomodulins: pointed-end capping proteins that regulate actin filament architecture in diverse cell types. Cytoskeleton (Hoboken) 69(6):337–370. doi:10.1002/cm.21031

Yamashita T, Higuchi H, Tohyama M (2002) The p75 receptor transduces the signal from myelin-associated glycoprotein to Rho. J Cell Biol 157(4):565–570. doi:10.1083/jcb.200202010

Yang XW, Lu XH (2011) Molecular and cellular basis of obsessive-compulsive disorder-like behaviors: emerging view from mouse models. Curr Opin Neurol 24(2):114–118. doi:10.1097/WCO.0b013e32834451fb

Yang YS, Strittmatter SM (2007) The reticulons: a family of proteins with diverse functions. Genome Biol 8(12):234. doi:10.1186/gb-2007-8-12-234

Ypsilanti AR, Chedotal A (2013) ROUNDABOUT receptors. In: Berezin V, Walmod PS (eds) Cell adhesion molecules: implications in neurological diseases. Advances in neurobiology. Springer, New York, NY

Yu SX, Li S, Shu HF, Zhang CQ, Liu SY, Yang H (2012) Expression of the Nogo-A system in cortical lesions of pediatric patients with tuberous sclerosis complex and focal cortical dysplasia type IIb. J Neuropathol Exp Neurol 71(7):665–677. doi:10.1097/NEN.0b013e31825d6585

Zariwala MA, Knowles MR, Leigh MW. Primary Ciliary Dyskinesia. 2007 Jan 24 [updated 2013 Feb 28]. In: Pagon RA, Adam MP, Bird TD, Dolan CR, Fong CT, Stephens K, editors. GeneReviews™ [Internet]. Seattle (WA): University of Washington, Seattle; 1993–2013. Available from http://www.ncbi.nlm.nih.gov/books/NBK1122/ PubMed PMID: 20301301

Zawadzka M, Rivers LE, Fancy SP, Zhao C, Tripathi R, Jamen F, Young K, Goncharevich A, Pohl H, Rizzi M, Rowitch DH, Kessaris N, Suter U, Richardson WD, Franklin RJ (2010) CNS-resident glial progenitor/stem cells produce Schwann cells as well as oligodendrocytes during repair of CNS demyelination. Cell Stem Cell 6(6):578–590. doi:10.1016/j.stem.2010.04.002

Zeitz C, Jacobson SG, Hamel CP, Bujakowska K, Neuille M, Orhan E, Zanlonghi X, Lancelot ME, Michiels C, Schwartz SB, Bocquet B, Antonio A, Audier C, Letexier M, Saraiva JP, Luu TD, Sennlaub F, Nguyen H, Poch O, Dollfus H, Lecompte O, Kohl S, Sahel JA, Bhattacharya SS, Audo I (2013) Whole-Exome Sequencing Identifies LRIT3 Mutations as a Cause of Autosomal-Recessive Complete Congenital Stationary Night Blindness. Am J Hum Genet. doi:10.1016/j.ajhg.2012.10.023

Zeuner KE, Deuschl G (2012) An update on tremors. Curr Opin Neurol 25(4):475–482. doi:10.1097/WCO.0b013e3283550c7e

Zhang Q, Wang J, Fan S, Wang L, Cao L, Tang K, Peng C, Li Z, Li W, Gan K, Liu Z, Li X, Shen S, Li G (2005a) Expression and functional characterization of LRRC4, a novel brain-specific member of the LRR superfamily. FEBS Lett 579(17):3674–3682. doi:10.1016/j.febslet.2005.05.058

Zhang QH, Wang LL, Cao L, Peng C, Li XL, Tang K, Li WF, Liao P, Wang JR, Li GY (2005b) Study of a novel brain relatively specific gene LRRC4 involved in glioma tumorigenesis suppression using the Tet-on system. Acta Biochim Biophys Sin (Shanghai) 37(8):532–540

Zhang QH, Wu MH, Wang LL, Cao L, Tang K, Peng C, Gan K, Li XL, Li GY (2005c) Profiling of differentially expressed genes in LRRC4 overexpressed glioblastoma cells by cDNA array. Acta Biochim Biophys Sin (Shanghai) 37(10):680–687

Zhang Z, Xu X, Zhang Y, Zhou J, Yu Z, He C (2009) LINGO-1 Interacts with WNK1 to Regulate Nogo-induced Inhibition of Neurite Extension. J Biol Chem 284(23):15717–15728. doi:10.1074/jbc.M808751200

Zhang K, Zhang L, Weinreb RN (2012) Ophthalmic drug discovery: novel targets and mechanisms for retinal diseases and glaucoma. Nat Rev Drug Discov 11(7):541–559. doi:10.1038/nrd3745

Zhao XH, Jin WL, Wu J, Mi S, Ju G (2008) Inactivation of glycogen synthase kinase-3beta and up-regulation of LINGO-1 are involved in LINGO-1 antagonist regulated survival of cerebellar granular neurons. Cell Mol Neurobiol 28(5):727–735. doi:10.1007/s10571-007-9258-6

Zhao X, Kuja-Panula J, Rouhiainen A, Chen YC, Panula P, Rauvala H (2011) High mobility group box-1 (HMGB1; amphoterin) is required for zebrafish brain development. J Biol Chem 286(26): 23200–23213. doi:10.1074/jbc.M111.223834

Zhou H, Welcher AA, Shooter EM (1997) BDNF/NT4-5 receptor TrkB and cadherin participate in cell-cell adhesion. J Neurosci Res 49(3):281–291. doi:10.1002/(sici)1097-4547(19970801)49:3<281::aid-jnr3>3.0.co;2-8

Zimprich A, Hatala K, Riederer F, Stogmann E, Aschauer HN, Stamenkovic M (2008) Sequence analysis of the complete SLITRK1 gene in Austrian patients with Tourette's disorder. Psychiatr Genet 18(6):308–309. doi:10.1097/YPG.0b013e3283060f6f

Zuchner S, Cuccaro ML, Tran-Viet KN, Cope H, Krishnan RR, Pericak-Vance MA, Wright HH, Ashley-Koch A (2006) SLITRK1 mutations in trichotillomania. Mol Psychiatry 11(10):887–889. doi:10.1038/sj.mp.4001865

Index

V. Berezin and P.S. Walmod (eds.), *Cell Adhesion Molecules: Implications in Neurological Diseases*, Advances in Neurobiology 8, DOI 10.1007/978-1-4614-8090-7,

MIX
Papier aus verantwortungsvollen Quellen
Paper from responsible sources
FSC® C105338

If you have any concerns about our products,
you can contact us on
ProductSafety@springernature.com

In case Publisher is established outside the EU,
the EU authorized representative is:
Springer Nature Customer Service Center GmbH
Europaplatz 3, 69115 Heidelberg, Germany

Printed by Libri Plureos GmbH
in Hamburg, Germany